A WILEY PUBLICATION IN APPLIED STATISTICS

Probability and Mathematical Statistics (Continued)

SCHEFFE · The Analysis of Variance
SEBER · Linear Regression Analysis
WILKS · Mathematical Statistics
WILLIAMS · Diffusions, Markov Processes, and Martingales, Volume I: Foundations
ZACKS · The Theory of Statistical Inference

Applied Probability and Statistics

BAILEY · The Elements of Stochastic Processes with Applications to the Natural Sciences
BAILEY · Mathematics, Statistics and Systems for Health
BARNETT and LEWIS · Outliers in Statistical Data
BARTHOLOMEW · Stochastic Models for Social Processes, *Second Edition*
BARTHOLOMEW and FORBES · Statistical Techniques in Manpower Planning
BECK and ARNOLD · Parameter Estimation in Engineering and Science
BELSLEY, KUH, and WELSCH · Regression Diagnostics: Identifying Influential Data and Sources of Collinearity
BENNETT and FRANKLIN · Statistical Analysis in Chemistry and the Chemical Industry
BHAT · Elements of Applied Stochastic Processes
BLOOMFIELD · Fourier Analysis of Time Series: An Introduction
BOX · R. A. Fisher, The Life of a Scientist
BOX and DRAPER · Evolutionary Operation: A Statistical Method for Process Improvement
BOX, HUNTER, and HUNTER · Statistics for Experimenters: An Introduction to Design, Data Analysis, and Model Building
BROWN and HOLLANDER · Statistics: A Biomedical Introduction
BROWNLEE · Statistical Theory and Methodology in Science and Engineering, *Second Edition*
BURY · Statistical Models in Applied Science
CHAMBERS · Computational Methods for Data Analysis
CHATTERJEE and PRICE · Regression Analysis by Example
CHERNOFF and MOSES · Elementary Decision Theory
CHOW · Analysis and Control of Dynamic Economic Systems
CLELLAND, deCANI, BROWN, BURSK, and MURRAY · Basic Statistics with Business Applications, *Second Edition*
COCHRAN · Sampling Techniques, *Third Edition*
COCHRAN and COX · Experimental Designs, *Second Edition*
COX · Planning of Experiments
COX and MILLER · The Theory of Stochastic Processes, *Second Edition*
DANIEL · Biostatistics: A Foundation for Analysis in the Health Sciences, *Second Edition*
DANIEL · Applications of Statistics to Industrial Experimentation
DANIEL and Wood · Fitting Equations to Data: Computer Analysis of Multifactor Data, *Second Edition*
DAVID · Order Statistics
DEMING · Sample Design in Business Research
DODGE and ROMIG · Sampling Inspection Tables, *Second Edition*
DRAPER and SMITH · Applied Regression Analysis

continued on back

Multiple Decision Procedures:

Theory and Methodology of Selecting and Ranking Populations

Multiple Decision Procedures: Theory and Methodology of Selecting and Ranking Populations

SHANTI S. GUPTA
Purdue University
S. PANCHAPAKESAN
Southern Illinois University

John Wiley & Sons, New York · Chichester · Brisbane · Toronto

Library of Congress Cataloging in Publication Data:

Gupta, Shanti Swarup, 1925-
Multiple decision procedures.

(Wiley series in probability and mathematical statistics) (A Wiley publication in applied statistics)
Bibliography: p.
Includes index.
1. Statistical inference. 2. Order statistics.
I. Panchapakesan, S., joint author. II. Title.
QA279.7.G86 519.5′4 79-13119
ISBN 0-471-05177-2

Printed in the United States of America

10 9 8 7 6 5 4 3 2 1

To our families

Preface

The ranking and selection approach in multiple decision theory gained recognition as an active area of research in the fifties. The basic contributions of the first author of this monograph were made during that period. With the zeal and devotion of the early researchers in this area the literature grew steadily, and by the mid-sixties a second generation of researchers emerged, mostly from Cornell University, Minnesota, and, of course, Purdue University.

As is usually the case in a developing field of research, there is always a time lag between the published research literature and availability of the same in a compact book form. This lack of a readily accessible source for recent results and developments motivated the authors to provide a survey monograph giving significant results, along with a comprehensive bibliography not only as a reference for users, but also as a useful aid in delineating directions for future research. The present monograph found its origin in the initial effort to realize this objective starting in 1968.

Because of the increased pace of research in the area, quite significant strides were made in the advancement of the field in the last ten years. This has necessitated a constant updating of the material, without which the monograph would have been of very limited use while its publication would have been a little premature. We now feel that the progress in the field has reached a point where it is essential to have this material available to the researchers and the practitioners.

Since the very start of this venture about ten years ago, two books have been published dealing mainly with the field of ranking and selection. The first of these is a monograph by Bechhofer, Kiefer, and Sobel on sequential identification and ranking procedures. Its emphasis is on sequential procedures for parametric families of distributions and the approach is primarily the indifference-zóne formulation. The second book, by Gibbons, Olkin, and Sobel, more elementary in nature, was published in 1977. It also places emphasis on indifference-zone formulation. The present monograph differs from these two works in its objective and coverage. It covers the entire field, placing equal emphasis on the subset selection and the indifference-zone formulations. It also deals with related problems, such as estimation of the unknown ordered parameters. We have attempted to summarize and survey in a unifying manner a great majority of some six hundred main references in the bibliography and have also provided some one hundred and forty related references in the body of the monograph.

The authors feel that this monograph will provide a suitable text for a two-semester graduate course on ranking and selection, as it has done at Purdue University for the last several years. It is our sincere hope that the publication of this book will also lead to further research in this area and, more importantly, encourage the applications of these techniques in agriculture, industry, engineering, physical, social, and behavioral sciences and other fields where statistics is so widely used.

The authors gratefully acknowledge the financial support received for this project over the years under the Office of Naval Research Contracts NONR–1100(26), N00014–67–A–00014, and N00014–75–C–0455, all at Purdue, and in particular wish to acknowledge the assistance and cooperation of Drs. Robert J. Lundegard, Edward J. Wegman, and Bruce J. McDonald of the Office of Naval Research, Arlington, Virginia, and Mr. Robert L. Bachman, ONR Resident Representative at Purdue University.

For various useful comments and discussions that have helped to improve the presentation of this book, the authors thank Professors R. E. Bechhofer, R. J. Carroll, J. C. Hsu, D. Y. Huang, W. C. Kim, G. Kulldorf, G. C. McDonald, K. J. Miescke, I. Olkin, T. J. Santner, and W. J. Studden.

Finally, the authors wish to place on record their deep debt of gratitude to Norma Lucas for her excellent faultless typing, and unfaltering cooperation and patience in going through several revisions of the manuscript.

SHANTI S. GUPTA
S. PANCHAPAKESAN

Purdue University
Southern Illinois University
March 1979

Contents

List of Abbreviations and Symbols

a.e.	almost everywhere
a.s.	almost surely
cdf	cumultative distribution function
r.v.	random variable
w.p. 1	with probability 1
$\forall$	for all
$\underset{p}{\rightarrow}$	convergence in probability
$\overset{\mathfrak{L}}{\rightarrow}$	convergence in distribution
$a_n \approx b_n$	as $n \rightarrow \infty$ means $\lim\limits_{n \rightarrow \infty} a_n / b_n = 1$

Multiple Decision Procedures:

Theory and Methodology of
Selecting and Ranking Populations

CHAPTER 1

Introduction

1.1 ORIGIN OF THE PROBLEM

In the practical world of everyday life, full of complexities, we face the baffling problem of making myriad of decisions. Each situation requires formulation of the problem under a suitable theoretical model followed by sifting and a careful analysis of the evidence leading to a decision whose validity is based on reason. In the advancement of scientific reasoning, statistical inference provides the methodology developed to meet these requirements. Significant original contributions, which have lasting impact on the subsequent phenomenal growth of this branch of science in the last 50 years, were undoubtedly made first and foremost by the British statistician, Sir Ronald Fisher.

Among others whose fundamental contributions have provided breakthroughs and inspired new avenues of research are Jerzy Neyman and Abraham Wald. The signal contributions of Neyman are the Neyman–Pearson theory of testing hypotheses and the theory of confidence intervals which form the backbone of any statistical training today. Abraham Wald influenced the course of modern theory of statistical decisions.

Historically, then, the problems of statistical inference were basically being formulated as those of estimation or testing of hypotheses. This formulation, however, does not exactly suit the objectives of an experimenter in many situations when he is faced with the problem of comparing several populations. These are generally the populations of the responses to certain "treatments." The treatments may be, for example, different varieties of wheat in an agricultural experiment, or different competing designs of engines to be used in an automobile plant, or different drugs for a specific disease. Some other examples are comparison of chicken stocks [Becker (1961)†], common stocks for investment [Dalal and Srinivasan (1977)] and traffic fatality rates of several states [Gupta and Hsu (1977b) and McDonald (1977b)]. In all these problems, we have k ($\geqslant 2$) populations and each population is characterized by the value of a parameter θ, which may denote, for example, in the agricultural experiment, the population average yield for a variety of wheat. In the example of drugs, θ may denote an appropriate measure of the effectiveness of a drug. For comparing chicken stocks Becker (1961) considers the henhouse egg production as the measure of comparison. For the comparison of common stocks, an appropriate measure is the average ratio of price to earnings.

The classical approach in all the preceding situations is to test the hypothesis H_0: $\theta_1 = \cdots = \theta_k$, where $\theta_1, \ldots, \theta_k$ are the values of the parameter θ for these populations. The hypothesis H_0 is usually referred to as the null hypothesis or the homogeneity hypothesis. If the populations are assumed to be normal with means $\theta_1, \ldots, \theta_k$, and a common variance σ^2, then the test can be carried out by using the analysis of variance technique developed by R. A. Fisher. In cases of other distributions for which θ may denote a different measure, one can develop a test of the null hypothesis H_0 using the Neyman–Pearson theory. Such tests have been developed for various situations and many of these are available in the statistical literature.

It should, however, be recognized that a satisfactory solution to any statistical inference problem depends on the goal of the experiment. In this sense, the classical tests of homogeneity cannot provide a panacea for the problems of the experimenter. In many situations, the goal of the experimenter is not just to accept or refute the homogeneity hypothesis. For example, in the agricultural experiment, one would expect the varieties of grain to be essentially different and so would expect to reject the null hypothesis at any preassigned level of significance by taking a sufficiently large sample from each variety. This makes the test of homogeneity unrealistic [see Chew (1977a)]. Further, the experimenter's real goal often

†For an explanation of the listing of Bibliography and other related references, see § 1.8 (p. 15).

is to identify the variety with the largest average, the most effective drug, and so on. In view of this, if a test of homogeneity is carried out and the test result is significant, the experimenter faces some real problems. In the past, the method of least significant differences based on t-tests has been used to detect differences between the average yields and thus to choose the "best" variety. This method is at best indirect and less efficient, because it lacks protection in terms of a guaranteed probability against picking out a wrong variety. Although the deficiencies of the classical approach of tests of homogeneity are evident, it should be borne in mind that these drawbacks are not in the design aspects of the procedure, but rather in the types of decisions that are made on the basis of the data.

The continued use of the classical tests in instances where the formulation was inadequate for the problem was largely due to the lack of procedures under more realistic formulations. The origin of selection and ranking procedures lies in seeking formulations based on more realistic goals.

1.2 SLIPPAGE TESTS

The first step in the direction of a somewhat more realistic approach came in the form of k-sample slippage tests where the alternative hypothesis states that one of the populations has "slipped" to the right or to the left. In other words, we test the hypothesis H_0: $\theta_1 = \cdots = \theta_k$ against one or more of the alternatives H_{1i}: $\theta_1 = \cdots = \theta_{i-1} = \theta_{i+1} = \cdots = \theta_k = \theta_i - \Delta$ where $\Delta > 0$ implies slippage to the right and $\Delta < 0$ implies slippage to the left. Such a formulation also arises from the necessity to decide whether one or more apparently outlying observations come from a different population than that generating the others.

Even though the problem of outliers has been of interest for a long time, Mosteller [1948] may be mentioned as the first author who made more formal investigations in this area. He considered k continuous populations with densities $f(x-\theta_i)$, $i = 1, \ldots, k$, where the θ_i are unknown. The interest is in devising a test to decide on the identification of a θ_i such that $\theta_i > \theta_r$, $1 \leqslant r \leqslant k$, $r \neq i$. Later, Paulson [1952] studied a similar problem in the case of normal means θ_i with the requirement that the null hypothesis H_0: $\theta_1 = \cdots = \theta_k$ be chosen with probability $1-\alpha$ $(0 < \alpha < 1)$ when H_0 is true. Since then, many authors have investigated such procedures in several cases. Some early investigators in this area are Truax [1953], Kudô [1956b], and Doornbos and Prins [1956]. A decision-theoretic formulation of Karlin and Truax [1960] and some other related results are briefly discussed in Chapter 22. A survey of some of the important results is given by Doornbos [1966].

A related development is that of multiple comparison tests and simultaneous confidence intervals. Among the early authors here, significant contributions were made by Duncan [1947, 1951, 1952, 1955], Tukey [1949], and Scheffé [1953]. While Duncan developed the use of multiple range tests, Tukey was largely responsible for the use of multiple comparison techniques in practice. Scheffé gave a method of judging all contrasts in the analysis of variance by considering simultaneous confidence intervals. The interest in multiple comparison techniques arises from the desire to draw inference about the populations when the homogeneity hypothesis is rejected. For details regarding several multiple comparison techniques, the reader is referred to the book of Miller (1966) and a later bibliography by him [1977].

1.3 MULTIPLE DECISION (SELECTION AND RANKING) PROCEDURES

The formulation of a k-sample problem as a multiple decision problem enables the experimenter to answer his natural questions regarding the best populations. Among the early investigators of multiple decision procedures are Paulson (1949), Bahadur (1950), and Bahadur and Robbins (1950). Suppose we have k independent normal populations with unknown means, μ_i, $i=1,\ldots,k$, and a common variance σ^2 (known or unknown). Paulson's goal is to classify the k populations into two groups: "superior" and "inferior." In doing so, he is interested in defining a procedure that will select a nonempty superior group with a preassigned (small) probability of not including all the populations in the superior group when $\mu_1=\cdots=\mu_k$. As a measure of the performance of his rule, he considers the probability of not having a superior group consisting only of the population associated with $\mu_{[k]}$, the largest mean, when $\mu_{[k]}=\mu+\Delta$ $(\Delta>0)$ and all other means are equal to μ. (Given a set of quantities $x_1,\ldots,x_k$, their ordered values will always be denoted, unless otherwise stated, by $x_{[1]}\leqslant\cdots\leqslant x_{[k]}$.)

Bahadur (1950) considered a class of "impartial" decision rules for the k-sample problem. Let $\pi_1,\ldots,\pi_k$ be a given set of populations where π_i is characterized by the absolutely continuous distribution function (cdf) $F(x,\theta_i)=h((x-b_i)/c_i)$, $\theta_i=(b_i,c_i)$, $c_i>0$, $i=1,\ldots,k$. Based on independent sample observations $\{X_{ij}\}, j=1,\ldots,n$, from π_i, let T_i be a sufficient statistic for θ_i. The problem is to construct a suitable decision rule $d=d(\{T_i\})$ for selecting one or more populations subject to minimizing the expected value of the random distribution function

$$G(\mathbf{t}|d)=\frac{\sum_{i=1}^{k} Z_i(d)G(t_i,\theta_i)}{\sum_{i=1}^{k} Z_i(d)}, \tag{1.1}$$

where $\mathbf{t}=(t_1,\ldots,t_k)$, $G(t_i,\theta_i)$ is the distribution function of T_i, and where $Z_i(d)=1$ if π_i is selected by d, and $=0$, otherwise. The decision $\mathbf{s}(d)$ according to d is given by $\mathbf{s}(d)=(p_1(d),\ldots,p_k(d))$, which is, in general, a random vector. Thus, for any fixed $\mathbf{t}$, $G(\mathbf{t}|\mathbf{s}(d))$ is a random variable. Let

$$H(\mathbf{t}|d)\equiv E[G(\mathbf{t}|\mathbf{s}(d))]\equiv \sum_{i=1}^{k} G(t_i,\theta_i)E[p_i(d)] \tag{1.2}$$

which is the average overall effect of using the decision rule d. In terms of mixtures of distributions, $H(\mathbf{t}|d)$ is the mixture of $G(\mathbf{t}|\mathbf{s})$ with respect to δ, where δ is the probability measure induced by the decision rule d on the class of Borel sets in the space of all possible decisions $\mathbf{s}$.

In general, one may wish to find a rule d^* in a class D such that $\alpha(d^*)=\sup_{d\in D}\alpha(d)$, where

$$\alpha(d)=\int u(\mathbf{t})dH(\mathbf{t}|d)$$

and $u(\mathbf{t})$ is some appropriate function. Setting

$$u_i=\int u(t_i)dG(t_i,\theta_i),$$

it can be shown that our object is equivalent to constructing d in such a way that the expected value of the "weight function"

$$W(\omega,\mathbf{s})=\max_i u_i-\sum_{i=1}^{k} p_i u_i \tag{1.3}$$

is minimized for every parameter point $\omega=(\theta_1,\ldots,\theta_k)$.

Bahadur (1950) considered decision rules under the following restrictions.

1. It is assumed that $T_i=\phi(X_{i1},\ldots,X_{in})$, $i=1,\ldots,k$, and $Y=\psi(\{X_{ij}\})$ form a set of independent random variables having densities. The choice of ϕ and ψ will, of course, depend upon particular cases.
2. Given the statistics $\{T_i\}$, and Y, the class $D(\phi;\psi)$ is the class of all impartial decision rules based on them.

A decision rule $d=d(\{T_i\},Y)$ is said to be *impartial* if it has the following structure. Let $T_{[1]}<T_{[2]}<\cdots<T_{[k]}$ be the ordered T_i. Then d defines nonnegative random variables $\lambda_j(T_{[1]},\ldots,T_{[k]};Y)$, $j=1,\ldots,k$, such

that $\Sigma_{j=1}^{k}\lambda_j=1$. Here λ_j is the proportion $p_j(d)$ which is assigned by d to the population corresponding to $T_{[j]}$. The impartiality of such a rule lies in the fact that it determines the proportions $(\lambda_1,\ldots,\lambda_k)$ without regard to which T_i belongs to which population, and then assigns these proportions in strict order of the T_i.

Suppose the populations $\pi_1,\ldots,\pi_k$ are normal with means μ_i and a common unknown variance σ^2. Then, for the problem of the largest mean, D is the class of all impartial decision rules that are based on the statistics

$$T_i=\bar{X}_i=\sum_{j=1}^{n}\frac{X_{ij}}{n},\qquad i=1,\ldots,k,$$

$$Y=s^2=\sum_{i=1}^{k}\sum_{j=1}^{n}\frac{\left(X_{ij}-\bar{X}_i\right)^2}{k(n-1)}.$$

The main theorem of Bahadur (1950) demonstrates that in the class D, the rule d_k which selects always the population corresponding to $\bar{X}_{[k]}$ cannot be improved upon, no matter what the observations or the parameter values may be. It also follows that d_k is the uniformly best decision rule in the class of impartial decision rules for all weight functions of the type $W=\mu_{[k]}-(\Sigma_{i=1}^{k}z_i\mu_i/\Sigma_{i=1}^{k}z_i)$. This theorem of Bahadur is given in Chapter 3 where we discuss some optimal properties of fixed subset size selection rules.

To see an application of the preceding results, let us consider the following example of Bahadur (1950). Grain is to be raised on a given area A of land and k varieties, $\pi_1,\ldots,\pi_k$, are available. It is assumed that the yields per unit area are normally distributed with unknown means μ_i and a common variance σ^2, also unknown. A preliminary field experiment has been carried out with n plots of unit area assigned to each variety, and $\{X_{ij}\}, j=1,\ldots,n;\ i=1,\ldots,k$, is the set of independent plot-yields obtained. The problem is to divide the land between the k varieties with the object of making the total expected yield as large as possible. In other words, we want to choose $\mathbf{s}=(p_1,\ldots,p_k)$ with $0\leqslant p_i\leqslant 1$ and $\Sigma_{i=1}^{k}p_i=1$ to minimize the loss

$$W(\omega,\mathbf{s})=\max_i A\mu_i-\sum_{i=1}^{p}A\mu_ip_i.$$

In the class of all impartial rules based on $\{\bar{X}_i\}$ and s^2, the uniformly best procedure assigns the whole area to the variety with the greatest observed yield.

Bahadur and Robbins (1950) considered a minimax rule for the problem of the larger of two normal means, which is a special case in Wald's general theory of decision functions. However, the results of Bahadur and Robbins are derived by very simple direct methods without recourse to Wald's general theory.

The formulation of multiple decision procedures in the framework of what has now come to be known as selection and ranking procedures has been generally accomplished under one of two basic approaches, namely, the *indifference zone* formulation and the *subset selection* formulation. Most of the literature surveyed in this monograph comes under one of these two approaches and related modifications and problems. Recently, Gibbons, Olkin, and Sobel (1977) have used the terminology of 'ordering of populations' instead of the traditional 'ranking of populations' employed by us.

1.4 THE INDIFFERENCE ZONE FORMULATION

The indifference zone approach was introduced by Bechhofer in his now classical paper of 1954 in which he considered the problem of ranking k normal means. To explain the basic formulation, let us consider the simple problem of selecting the population with the largest mean from k normal populations with unknown means μ_i, $i=1,\ldots,k$, and a common known variance σ^2. Let $\bar{X}_i$, $i=1,\ldots,k$, denote the means of independent samples of size n from these populations. The "natural" procedure will be to select the population that yields the largest $\bar{X}_i$. Naturally, the experimenter would need a guarantee that this procedure will pick with a high probability P^* the population associated with $\mu_{[k]}$, which will be called the best population. Let $\bar{X}_{(i)}$ denote the mean of the sample observations from the population with mean $\mu_{[i]}$. It is easy to see that

$$\Pr\{\text{the best population is selected}\} \tag{1.4}$$

$$=\Pr\{\bar{X}_{(k)} \geqslant \bar{X}_{(i)}, \qquad i=1,\ldots,k-1\}$$

$$=\int_{-\infty}^{\infty} \prod_{i=1}^{k-1} \Phi\left\{z+\frac{\sqrt{n}}{\sigma}(\mu_{[k]}-\mu_{[i]})\right\} d\Phi(z),$$

where $\Phi(\cdot)$ denotes the cdf of a standard normal variate. Now for the problem to be meaningful P^* lies between $1/k$ and 1. Since the true values of the μ_i are not known, we need the probability of selecting the best population to be at least P^* whatever be the values of the μ_i. Thus we are interested in the configuration of the μ_i for which the probability in (1.4) is a minimum. Such a configuration will be called a *least favorable configuration* (LFC). It is obvious that the LFC is given by $\mu_{[1]}=\cdots=\mu_{[k]}$. But unfortunately the minimum value of the probability is $1/k$. So we cannot meet the probability requirement whatever be the sample size unless some modification is made in the probability requirement.

A natural modification is to insist on the minimum probability P^* of selecting the best population whenever the best is sufficiently far apart from the next best. In other words, the experimenter specifies a positive constant Δ^* and requires that the probability of selecting the best population is at least P^* whenever $\mu_{[k]}-\mu_{[k-1]}\geqslant\delta^*$. The specification of δ^* provides a partition of the parameter space Ω into two parts, namely, Ω_{δ^*} in which $\mu_{[k]}-\mu_{[k-1]}\geqslant\delta^*$ and the complement $\Omega_{\delta^*}^c$ of it. The minimization of the probability of selecting the best population is over Ω_{δ^*}. For the obvious reason, $\Omega_{\delta^*}^c$ is called the *indifference zone* by Bechhofer (1954). Subsequent authors have termed Ω_{δ^*} the *preference zone.*

It is now easy to see that the least favorable configuration in Ω_{Δ^*} is given by $\mu_{[1]}=\cdots=\mu_{[k-1]}=\mu_{[k]}-\delta^*$ and the minimum sample size required is the smallest integer n for which

$$\int_{-\infty}^{\infty}\Phi^{k-1}\left(z+\frac{\sqrt{n}}{\sigma}\delta^*\right)d\Phi(z)\geqslant P^*. \tag{1.5}$$

This procedure will be discussed in greater detail in the next chapter.

REMARK 1.1. In the preceding discussion, we mean by the least favorable configuration in Ω_{δ^*} a set of vector points $\boldsymbol{\mu}=(\mu_1,\ldots,\mu_k)$ in Ω_{δ^*} satisfying the condition $\mu_{[1]}=\cdots=\mu_{[k-1]}=\mu_{[k]}-\delta^*$. For any $\boldsymbol{\mu}$ in this set, the probability of selecting the best population does not depend upon the value of $\mu_{[1]}$. However, this need not be the case in general. Suppose we have parameters $\theta_1,\ldots,\theta_k$ and the infimum of the probability of selecting the best population over $\Omega_{\Delta^*}=\{\boldsymbol{\theta}\colon \theta_{[k]}-\theta_{[k-1]}\geqslant\delta^*\}$ occurs at a point $\boldsymbol{\theta}$ satisfying

$$\theta_{[1]}=\cdots=\theta_{[k-1]}=\theta_{[k]}-\delta^*. \tag{1.6}$$

However, the probability may depend upon $\theta_{[1]}$. In this case, we must carry on the process of finding the vector-point(s) at which the minimum occurs. The LFC may contain only one point, as in the case of binomial populations discussed in Chapter 4. The configuration (1.6) will then be called the *generalized least favorable configuration* (GLFC). Of course, in the case of normal means the LFC coincides with the GLFC.

It should also be noted that the LFC in the case of normal means is the same for all k and n. This need not be true in general. □

Let us now describe the general ranking problem under the indifference zone approach. Suppose there are k populations $\pi_1,\ldots,\pi_k$ where π_i is characterized by the distribution function F_{θ_i}, $i=1,\ldots,k$, where θ_i is a real-valued parameter belonging to the set Θ. It is assumed that the θ_i are unknown. Let the ordered θ_i be denoted by $\theta_{[1]} \leqslant \theta_{[2]} \leqslant \cdots \leqslant \theta_{[k]}$ and the population associated with $\theta_{[i]}$ be denoted by $\pi_{(i)}$, $i=1,\ldots,k$. The populations are ranked in terms of their θ-values. To be specific, we define $\pi_{(j)}$ to be better than $\pi_{(i)}$ $(\pi_{(i)} \prec \pi_{(j)})$ if $i<j$. The experimenter has no prior knowledge relevant to the true pairing of the populations and the ordered θ_i. Usually, two goals are considered: Goal I is to select an unordered set of t populations associated with the set $\{\theta_{[k-t+1]},\ldots,\theta_{[k]}\}$, and Goal II is to select t populations and associate them with the set $\{\theta_{[k-t+1]},\ldots,\theta_{[k]}\}$ by a one-to-one correspondence. The statement of a ranking problem is not complete without specifying certain constants and an associated probability requirement. To spell these out, let $\delta_{j,i}$ denote a suitably defined measure of distance between the populations $\pi_{(i)}$ and $\pi_{(j)}$, $i<j$. (It should be pointed out that here and in the sequel the word "distance" is used as a measure of separation between populations. It may, in special cases, denote a distance function.) The natural choice of $\delta_{j,i}$ in the cases of location and scale parameters will be $\theta_{[j]}-\theta_{[i]}$ and $\theta_{[j]}/\theta_{[i]}$, respectively. In the case of Goal I, it is required that the probability of a correct selection (PCS), that is, a selection of t populations satisfying the goal, is at least a preassigned number P^*, whenever $\delta_{k-t+1,k-t} \geqslant \delta^*_{k-t-1,k-t}$, where $\delta^*_{k-t+1,k-t}$ is a positive constant specified in advance. Obviously, for a meaningful problem P^* must be restricted to the interval $\left(1/\binom{k}{t},1\right)$. For Goal II, the probability of a correct selection should be at least P^* whenever $\delta_{k-i+1,k-i} \geqslant \delta^*_{k-i+1,k-i}$, $i=1,\ldots,t$, where all the starred quantities are positive constants specified in advance. Of course, for this goal $P^* \in ((k-t)!/k!, 1)$.

One can consider a more general goal which includes Goals I and II as special cases, namely, to partition the set of k populations into s nonempty subsets denoted by $I_1,\ldots,I_s$, consisting of $k_1,\ldots,k_s(k_1+\cdots+k_s=k)$ populations, respectively. The probability of a correct selection is required to be at least $P^* \in (k_1!\ldots k_s!/k!, 1)$, whenever $\delta_{\hat{k}_i+1,\hat{k}_i} \geqslant \delta^*_{\hat{k}_i+1,\hat{k}_i}$, $i=1,\ldots,s-1$, where $\hat{k}_i=\Sigma^i_{\alpha=1}k_\alpha$ and the starred quantities are positive constants specified in advance. These goals have been discussed by Bechhofer (1954).

Let Ω denote the parameter space $\{\boldsymbol{\theta}: \boldsymbol{\theta}=(\theta_1,\ldots,\theta_k),\ \theta_i \in \Theta\}$. Denoting the probability of a correct selection using a procedure R by $P(CS|R)$, we can write the probability requirement described previously as

$$P(CS|R) \geqslant P^* \text{ whenever } \boldsymbol{\theta} \in \Omega^*, \tag{1.7}$$

where Ω^* denotes the subset of Ω defined by the bounds (denoted by starred quantities) on appropriate distances. The usual approach is to propose a rule R on heuristic grounds and investigate its properties. The rule R is generally based on statistics T_i, $i=1,\ldots,k$, where T_i is usually sufficient for θ_i. The main problem, as we have seen in the special case considered earlier, is the evaluation of the infimum of $P(CS|R)$ over Ω^*. If we assume that the samples from these populations are of the same size n, then we need to determine the smallest n for which

$$\inf_{\Omega^*} P(CS|R) \geqslant P^*. \tag{1.8}$$

Thus the experimenter has the choice of the sample size to control the level of the PCS.

A dual problem that arises is the following. Suppose the common sample size n and P^* are given. For a rule R designed to select the t best populations, we are interested in determining the minimum t for which (1.8) can be met.

1.5 SUBSET SELECTION FORMULATION

In the subset selection formulation one of the authors has made a large number of contributions starting with the work in Gupta (1956), where investigation was carried out for the case of normal means. To see the motivation behind this approach, consider k normal populations with unknown means $\mu_1,\ldots,\mu_k$ and common known variance σ^2. Let $\overline{X}_i$, $i=1,\ldots,k$, be the means of samples of size n from these populations. If we want to select a subset of fixed size 3 (assuming $k>3$), we will select the populations that yield the three largest sample means. Suppose the three largest sample means in an experiment are 35, 33, and 4. Now if $\sigma/\sqrt{n}=2$, say, one may be content with selecting the first two alone if there is no compelling need to take three. Thus the main feature of selecting a subset of random size is to allow the size to be determined by the observations themselves.

To describe the problem formally, let us assume as in the previous section that π_i, $i=1,\ldots,k$, are k populations where π_i is characterized by F_{θ_i}, $\theta_i \in \Theta$. The goal is to select a nonempty subset of these populations so that the selected subset includes with a high probability the t best populations, that is, the populations associated with the set $\{\theta_{[k-t+1]},\ldots,\theta_{[k]}\}$. If there are more than t contenders because of ties, it is assumed that t of these are appropriately tagged. The size of the selected subset is an integer-valued random variable taking on the values $1,2,\ldots,k$. It is required that

$$P(CS|R) \geqslant P^* \quad \text{for } \boldsymbol{\theta} \in \Omega, \tag{1.9}$$

where $P^* \in \left(1/\binom{k}{t}, 1\right)$ is preassigned.

As in the indifference zone approach, the rule R for selecting a subset containing the t best will depend on sufficient statistics T_i when they exist. Usually, the distribution G_{θ_i} of T_i belongs to a family that is stochastically ordered in θ. The distribution G_{θ_1} is *stochastically smaller* than G_{θ_2} (or equivalently, G_{θ_2} is *stochastically larger* than G_{θ_1}) if the distributions G_{θ_1} and G_{θ_2} are distinct, and $G_{\theta_2}(x) \leqslant G_{\theta_1}(x)$ for all x.

In the case of $t=1$ (i.e., selecting a subset containing the best), the rules proposed in many specific cases are one of the following two types, R_1 and R_2.

$\boldsymbol{R_1}$: Select π_i if and only if

$$T_i \geqslant \max_{1 \leqslant r \leqslant k} T_r - d, \tag{1.10}$$

$\boldsymbol{R_2}$: Select π_i if and only if

$$T_i \geqslant c \max_{1 \leqslant r \leqslant k} T_r. \tag{1.11}$$

In the preceding rules, d is a positive constant and c is a constant between 0 and 1. For given n and P^*, the constants c and d are to be determined to meet the requirement (1.9). Thus we are interested in evaluating the infimum of $P(CS|R)$ over Ω. Though we do not explicitly state it in defining the rules R_1 and R_2, it should be borne in mind that we are really defining the classes of rules $\{R_1(d)\}$ and $\{R_2(c)\}$. Thus, in the case of the former, we will choose the rule with the smallest d for which

$$\inf_{\Omega} P(CS|R_1) \geqslant P^*. \tag{1.12}$$

In the other case, we will choose the largest c for which the infimum of the PCS is greater than or equal to P^*.

When the θ_i are location parameters, we use an R_1-type procedure. The other type is used when the θ_i are scale parameters. Of course, when the θ_i are neither location nor scale parameters one could still decide on R_1 or R_2 depending on the nature of the supports of the distributions F_θ.

One could consider a general class of procedures defined through a class of real-valued functions $\{h_{a,b}(x)\}$, $a \geqslant 1$, $b \geqslant 0$. Of course, one needs to impose certain conditions on this class. Letting $h(x) \equiv h_{a,b}(x)$, we can define the following class of procedures.

$\boldsymbol{R_{a,b}}$: Select π_i if and only if

$$h(T_i) \geqslant \max_{1 \leqslant r \leqslant k} T_r. \tag{1.13}$$

It is easily seen that the class of rules $R_{a,b}$ includes the classes $\{R_1(d)\}$ and $\{R_2(c)\}$. These generalizations and the relevant theorems have been obtained by Gupta and Panchapakesan (1972a). These are dealt with in detail in Chapter 11.

1.6 SOME GENERAL REMARKS

As we have seen, the probability of a correct selection is required to satisfy (1.7) or (1.9) depending upon the formulation. This requirement is commonly referred to as the *basic probability requirement* or the *P^*-condition*. The infimum of the PCS in Bechhofer's formulation is evaluated over the preference zone, whereas in the subset selection formulation it is over the entire parameter space. Under both formulations we are interested in investigating the properties of a procedure. Suppose we are interested in selecting the best population. For a good rule R under either formulation one would like to have

$$\Pr(\pi_{(k)} \text{ is selected}) \geqslant \Pr(\pi_{(i)} \text{ is selected}), \qquad i = 1, \ldots, k-1. \tag{1.14}$$

This property is known as the *unbiasedness* of the rule R. A stronger property called the *monotonicity* is defined by

$$\Pr(\pi_{(i)} \text{ is selected}) \geqslant \Pr(\pi_{(j)} \text{ is selected}) \tag{1.15}$$

for any pair such that $\pi_{(j)} \prec \pi_{(i)}$.

To evaluate the performance of a subset selection rule R, one considers the supremum of the expected subset size. These and other properties are discussed in several places in this book.

In any given situation, the preference for one over the other of the two formulations is mainly dictated by the objectives of the experimenter. Selecting a fixed number of populations is generally identified with a terminal decision. On the other hand, selecting a random subset can be looked upon as a screening process as well as a terminal decision.

Closely related to the problem of selecting the best among k populations is the problem of comparing k populations with a control or standard population. The goal of the experimenter in these problems is to choose the best among the k populations only if it is better than the control or the standard population. More generally, one can define good and bad populations with respect to a standard or control with a possibility of a population not being substantially different from the standard in either direction. Then the goal of the experimenter may be to select a subset with a high proportion of good populations. All these and related aspects of comparing several populations with a standard or control are discussed in Chapter 20.

There are also some modifications and generalizations of the basic indifference zone and subset selection formulations investigated by several authors among whom are Lehmann (1961), Mahamunulu (1967), Gupta and Deverman (1968), Carroll, Gupta, and Huang (1975), Santner (1975), Huang and Panchapakesan (1976, 1978), and Panchapakesan and Santner (1977). Attempts have also been made to combine the features of the two formulations by Sobel (1969), Gupta and Santner (1973), and Santner (1975). These are discussed in several appropriate places in this monograph.

1.7 ON THE CHAPTERS TO FOLLOW

Chapters 2 through 10 comprising Part I are devoted to the indifference zone formulation. Chapter 2 deals with ranking of normal populations in terms of their (i) means, (ii) variances, and (iii) absolute means. Some problems in r-way classification models are also discussed. This chapter also discusses a concept of Δ-correct ranking, and selection of the largest interaction in a two-factor experiment. Some optimum properties of fixed subset selection rules are given in Chapter 3. These include most economic character of rules, uniformly best impartial rule, Bayes, minimax, and admissibility properties, and consistency of ranking procedures.

Chapter 4 describes ranking and selection problems for discrete distributions. The discussion on procedures for binomial populations includes not only fixed sample size procedures but also those using sequential procedures with different sampling schemes such as inverse sampling and play-the-winner sampling. This chapter also includes procedures for selecting the best multinomial cell.

Two-stage subsampling procedures for finite populations are discussed in Chapter 5 which also includes some problems of optimum sampling and estimation of the probability of a correct selection.

The remaining chapters of Part I contain sequential selection procedures (Chapter 6), procedures for selection from multivariate populations (Chapter 7), nonparametric selection procedures (Chapter 8), Bayesian procedures (Chapter 10), and a discussion of a generalized goal for fixed-size subset selection and other modifications (Chapter 9).

Part II consists of nine chapters (11 through 19) devoted to subset selection procedures. Chapter 11 describes the general theory of subset selection which includes the formulation discussed earlier as well as the modified goals of restricted subset selection and elimination of nonbest populations.

Selection from univariate continuous populations and selection from discrete populations are discussed in Chapters 12 and 13, respectively. Normal, exponential, and gamma distributions are the most important continuous distributions considered. The discrete distributions discussed are binomial, Poisson, negative binomial, hypergeometric, and multinomial.

The next five chapters are devoted to selection from multivariate normal populations (Chapter 14), nonparametric procedures (Chapter 15), selection from restricted families of probability distributions (Chapter 16), sequential procedures (Chapter 17), and Bayes, empirical Bayes and Γ-minimax procedures (Chapter 18). The last chapter of Part II is concerned with some modified formulations and related problems.

Part III comprises five chapters (10 through 24) dealing with several problems related to the procedures described in the preceding chapters. Chapter 20 discusses different aspects of the problem of comparing several populations with a control or standard. The next chapter deals with the problems of estimating ordered parameters and procedures with combined goals of selection and estimation.

Chapter 22 contains brief discussions on some general theory of multiple decision problems, selection of variables in linear regression, and slippage tests.

The purpose of the next two chapters is to assemble information mostly available in the earlier chapters regarding the tables needed for the applications of several of the procedures described therein (Chapter 23) and also to illustrate through examples the applications of some important procedures (Chapter 24).

In arranging the material in several chapters as indicated previously, we are naturally faced with the problem of determining the most appropriate place for results relating to a particular procedure, as for example, a sequential nonparametric procedure for multivariate populations, or a procedure for the best multinomial cell. In resolving the issue, we have been strongly guided by the general context of the problem and the main emphasis of the results as we perceived them.

In describing the results of several authors, the notations have been changed in many places for the sake of uniformity. The procedures have been relabeled. A procedure denoted by R_6 in a chapter might well have been called D or R or R_2 in the original paper. For referring to other sections of this book, we have used the notation such as §2.7.3. This refers to Chapter 2, Section 7, Subsection 3.

1.8 BIBLIOGRAPHY AND OTHER REFERENCES

The sources for the results described in the book are referenced by the author(s) and the year. The year is enclosed in either ( ) or [ ]. The first indicates that the item is listed under Bibliography whereas the second refers to an item under Other References. The first list includes all the items that are directly and mainly related to the theme of the monograph. The list of Other References includes publications that relate to some background materials and also those which are undeniably important to ranking and selection problems but are written in a wider context. The items under either list are specified by the name(s) of the author(s), the year, the title of the paper or book, the name of the publisher in the case of a book, and the page numbers. In case of multiple authors, an item is listed only once at an alphabetical location based on the first author's name. The Bibliography includes only a few abstracts. Technical reports that are known to have been published are omitted. Every effort has been made to bring the bibliography to date.

PART I

INDIFFERENCE ZONE FORMULATION

CHAPTER 2

Ranking of Normal Populations

Normal distribution is used as an appropriate model in many situations and as such one of the most studied parametric models in ranking and selection. For comparing normal populations several different measures can be used. In this chapter, we discuss ranking of normal populations according to measures such as the mean, the variance and the coefficient of variation. We also discuss two-factor experiments and selection of the levels of the factors that have the largest interaction.

Notation. We have k independent populations denoted by $\pi_1,\ldots,\pi_k$, where π_i is a normal population with mean μ_i and variance σ_i^2, $i=1,\ldots,k$. In this chapter as well as the rest of the book, unless otherwise stated, we use $\Phi(\cdot)$ and $\varphi(\cdot)$ to denote the cdf and the density of a standard normal random variable.

2.1 RANKING IN TERMS OF THE MEANS

Bechhofer (1954), introducing the indifference zone formulation, considered the problem of ranking means of normal populations with known variances. The problem of selecting the t $(1 \leqslant t \leqslant k)$ best populations becomes a special case of the general ranking problem. In this section, we first discuss the single-stage procedure of Bechhofer (1954) when the variances are known, and then proceed to some two-stage procedures in the case of unknown variances for which there does not exist a single-stage procedure that will satisfy the probability requirement. We also discuss some two-stage procedures when the variances are known; in these procedures, the first stage is used for screening the populations. In ranking the populations, π_j is considered better than π_i if $\mu_j > \mu_i$.

2.1.1 A Single-Stage Ranking Procedure: Known Variances

The goal of a general ranking problem is to partition the set of k normal populations into s nonempty subsets denoted by $I_1, \ldots, I_s$, and consisting of $k_1, \ldots, k_s$ $(k_1 + \cdots + k_s = k)$ populations, respectively, so that any population in the set I_j is better than any population in the set I_i for $i<j$. Let $\overline{X}_i$ be the mean of a sample of n_i independent observations from π_i, $i=1,\ldots,k$. Bechhofer (1954) proposed the following rule R_1.

***R*₁**: Include in I_r, $r=1,\ldots,s$, the populations associated with

$$\overline{X}_{[\hat{k}_{r-1}+1]}, \overline{X}_{[\hat{k}_{r-1}+2]}, \ldots, \overline{X}_{[\hat{k}_r]}, \tag{2.1}$$

where $\hat{k}_r = \Sigma_{i=1}^{r} k_i$ and $\hat{k}_0 = 0$.

In particular, for selecting the best population $(s=2,\ k_1=k-1,\ k_2=1)$, the rule R_1 selects the population that yields the largest $\overline{X}_i$ and the PCS is given by

$$P(CS|R_1) = \Pr(Y_1 > 0, \ldots, Y_{k-1} > 0), \tag{2.2}$$

where the joint distribution of $Y_1, \ldots, Y_{k-1}$ is the $(k-1)$-variate normal distribution with mean vector $(\delta_{k,1}, \ldots, \delta_{k,k-1})$ and covariance matrix $\Sigma = (\sigma_{ij})$, and

$$\begin{aligned} \delta_{k,j} &= \mu_{[k]} - \mu_{[j]}, \qquad j=1,\ldots,k-1, \\ \sigma_{ii} &= \frac{\sigma_{(k)}^2}{n_{(k)}} + \frac{\sigma_{(i)}^2}{n_{(i)}}, \qquad i=1,\ldots,k-1, \\ \sigma_{ij} &= \frac{\sigma_{(k)}^2}{n_{(k)}}, \qquad i,j=1,\ldots,k-1;\ i \neq j. \end{aligned} \tag{2.3}$$

In (2.3), $\sigma_{(i)}^2$ and $n_{(i)}$ denote the variance of the population associated with $\mu_{[i]}$ and the associated sample size, respectively. If $\sigma_{(i)}^2/n_{(i)}$ is same for all $i=1,\dots,k$, then $Y_1,\dots,Y_{k-1}$ are equicorrelated with correlation equal to $1/2$. The PCS is to be minimized over $\Omega_{\delta*}$, where

$$\Omega_{\delta*}=\{\boldsymbol{\mu}:\ \mu_{[k]}-\mu_{[k-1]}\geqslant\delta^*\} \tag{2.4}$$

and $\delta^*>0$ is specified in advance. The LFC is given by

$$\mu_{[1]}=\cdots=\mu_{[k-1]}=\mu_{[k]}-\delta^*. \tag{2.5}$$

For selecting the t best populations without ordering them ($s=2$, $k_1=k-t$, $k_2=t$), the rule R_1 chooses the populations corresponding to the t largest values among the $\overline{X}_i$. The PCS in this case is minimized over $\Omega_{\delta*}=\{\boldsymbol{\mu}:\ \mu_{[k-t+1]}-\mu_{[k-t]}\geqslant\delta^*\}$ and the LFC is given by

$$\mu_{[1]}=\cdots=\mu_{[k-t]}=\mu; \qquad \mu_{[k-t+1]}=\cdots=\mu_{[k]}=\mu+\delta^*.$$

Bechhofer (1954) has tabulated the values of $d=\sqrt{n}\ \delta^*/\sigma$ for several selected values of k, t, and P^*, where $n=n_1=\cdots=n_k$, and $\sigma=\sigma_1=\cdots=\sigma_k$. In the special case of $t=1$, it can be easily seen that the minimum sample size required by R_1 is the smallest n such that

$$\Pr(Z_1\leqslant H,\dots,Z_{k-1}\leqslant H)\geqslant P^*, \tag{2.6}$$

where $H=(n/2)^{1/2}(\delta^*/\sigma)$ and $Z_1,\dots,Z_{k-1}$ are standard normal variates with equal correlation $\rho=\frac{1}{2}$. Evaluation of the probability on the left-hand side of (2.6) has been discussed by Gupta [1963a], who has tabulated the H-values for $\rho=\frac{1}{2}$ for selected values of k and P^*. The H-values for $\rho=\frac{1}{2}$ have also been tabulated by Milton [1963]. Tables of Gupta [1963a] also give the probability for selected values of k, ρ, and H. On the other hand, Gupta, Nagel, and Panchapakesan [1973] give the values of H for which the probability in (2.6) is equal to P^*, for selected values of k, ρ, and P^*. These tables are discussed in detail in §23.2.

Approximations for Minimum Sample Size and H. Several approximations are available for the H-value satisfying (2.6). In terms of n (the smallest sample size), the following presents five different approximations denoted by n_1 through n_5:

$$n_1=-\left(\frac{2\sigma}{\delta^*}\right)^2\log(1-P^*), \text{ Dudewicz (1969a);}$$

$$n_2=\left(\frac{2\sigma^2}{\delta^{*2}}\right)\left[\Phi^{-1}\left(\frac{1-P^*}{k-1}\right)\right]^2, \text{ Ramberg (1972);}$$

$$n_3=\left(\frac{2\sigma^2}{\delta^{*2}}\right)\left[\Phi^{-1}(P^*)^{1/k-1}\right]^2, \text{ Ramberg (1972);} \tag{2.7}$$

$$n_4=\frac{\sigma^2}{\delta^{*2}}\left[4\log\left\{\frac{k-1}{1-P^*}\right\}-2\log\log\left\{\frac{k-1}{1-P^*}\right\}-2\log 4\pi\right],$$

Bechhofer, Kiefer, and Sobel (1968); ·

$$n_5=\left(\frac{2\sigma^2}{\delta^{*2}}\right)\log(k-1), \text{ Robbins and Siegmund (1968).}$$

It has been shown (by respective authors) that $\lim_{k\to\infty} n_5/n=1$ and $\lim_{P^*\to 1} n_i/n=1$, $i=1,\ldots,4$. The approximation n_2 is based on a Bonferroni inequality whereas n_3 uses the Slepian inequality (see Slepian [1962] or Gupta [1963a]). These approximations can be used for any k and P^*, and n_3 provides a better approximation [Dudewicz and Zaino (1971), and Ramberg (1972)]. Asymptotically ($P^*\to 1$), n_1, n_3, n_4, and n are equal. As $k\to\infty$, n_3/n and n_4/n tend to 2. For some numerical comparisons, see Dudewicz and Zaino (1971). It should be noted that the approximation n_2 was obtained in terms of the corresponding H-value by Paulson (1952) who observed that the approximation seemed better for small values of k.

2.1.2 Two-Stage Ranking Procedures: Unknown Variances

It is unrealistic to assume that we always have a common known variance. Generally, the variances are unknown and are not necessarily all equal as, for example, in computer simulation experiments dealing with models for business and economic systems.

When the variances σ_i^2 are unknown, it is not possible to predetermine the sample sizes for a single-stage procedure since the standard error of the sample means are unknown [see, for example, Dudewicz (1971a)]. Bechhofer, Dunnett, and Sobel (1954) have considered a two-stage procedure when $\sigma_i^2=a_i\sigma^2$, where the a_i are assumed to be known positive rational numbers, which without loss of generality, can be taken to be integers. For ranking the populations according to their means, their procedure R_2 is described as follows.

1. Take a sample of $a_i n_0$ observations from π_i, $i=1,\ldots,k$, where n_0 is any integer such that $n=n_0\Sigma_{i=1}^k a_i-k>0$.

2. Calculate

$$s_0^2 = n^{-1} \sum_{i=1}^{k} a_i^{-1} \sum_{j=1}^{a_i n_0} \left[X_j - (a_i n_0)^{-1} \sum_{j=1}^{a_i n_0} X_{ij} \right]^2,$$

where $X_{ij}, j=1,\dots,a_i n_0$, are observations from π_i, $i=1,\dots,k$.

3. Take a second sample of $(N-n_0)a_i$ observations from π_i, $i=1,\dots,k$, where $N=\max\{n_0, \langle 2s_0^2h^2/\delta^{*2}\rangle\}$, where $\langle y\rangle$ denotes the smallest integer $\geqslant y$.

4. Calculate $\overline{X}_i = \sum_{j=1}^{a_i N} X_{ij}/a_i N$, $i=1,\dots,k$.

5. Rank the populations according to the rankings of the $\overline{X}_i$.

The constant $h=h(n,k,P^*)$ is chosen so as to satisfy the P^*-condition. For $k=2$, the evaluation of h reduces to the determination of the appropriate percentage point of the univariate Student's t. For $k=3$, we have a bivariate generalization of t with correlation $\rho=\frac{1}{2}$ for the goal of selecting the population with the largest mean and $\rho=-\frac{1}{2}$ for the goal of completely ranking the populations with $\delta^*_{3,2}=\delta^*_{2,1}=\delta^*$ in the notations of Chapter 1. In these cases, the values of h are tabulated to three decimal places by Dunnett and Sobel [1954] for $P^*=0.50$, 0.75, 0.90, 0.95, 0.99 and $n=1(1)30(3)60(15)120$, 150, 300, 600, and ∞. For the case of $a_i=1$, Freeman, Kuzmack, and Maurice (1967) have tabulated the values of h in the case of complete ranking goal for $P^*=0.95$, $k=3(1)6$, and $n_0=10(10)100$, 200, 500. For the evaluation of multivariate t integrals, see also Dunnett (1955), Dunnett and Sobel [1955], Gupta [1963a], Gupta and Sobel (1957), John [1964], [1966], and Krishnaiah and Armitage [1966].

For selecting the normal population having the largest mean when the variances σ_i^2 are unknown, Dudewicz and Dalal (1975) have studied a generalized Stein-type procedure. Of course, such a possibility had earlier been mentioned by Bechhofer, Dunnett, and Sobel (1953).

Let $h=h_m(k,P^*)$ be the unique solution of

$$\int_{-\infty}^{\infty} [F_m(z+h)]^{k-1} f_m(z)\,dz = P^* \tag{2.8}$$

where $F_m(\cdot)$ and $f_m(\cdot)$ are the cdf and the density, respectively, of a Student-t random variable with $m=n_0-1$ degrees of freedom. Let $\overline{X}_i(n_0)$ and s_i^2 denote the sample mean and the sample variance based on n_0 observations from π_i, $i=1,\dots,k$. Define

$$n_i = \max\left\{ n_0+1, \left\langle \left(\frac{s_i h}{\delta^*} \right)^2 \right\rangle \right\}, \tag{2.9}$$

where $\langle y\rangle$ denotes the smallest integer $\geqslant y$.

The procedure R_3 of Dudewicz and Dalal (1975) is based on a first stage sample of n_0 observations from each of the populations and an additional sample of $n_i - n_0$ observations from π_i, $i=1,\ldots,k$. The procedure chooses the population associated with the largest $\tilde{\bar{X}}_i$, where $\tilde{\bar{X}}_i$ is a weighted average of the observations X_{ij}, $j=1,\ldots,n_i$, from π_i given by $\tilde{\bar{X}}_i = \sum_{j=1}^{n_i} a_{ij} X_{ij}$, the a_{ij} being subject to the conditions:

$$\sum_{j=1}^{n_i} a_{ij} = 1, \; a_{i1} = \cdots = a_{in_0} \tag{2.10}$$

and

$$s_i^2 \sum_{j=1}^{n_i} a_{ij} = \left(\frac{\delta^*}{h}\right)^2. \tag{2.11}$$

A possible choice of the a_{ij} is obtained by letting $a_{i1} = \cdots = a_{i,n_i-1} = c_i$, and then choosing c_i subject to (2.11). For this procedure, the PCS has been shown to be independent of $\sigma_1^2,\ldots,\sigma_k^2$. Its infimum over Ω_{δ^*} in (2.4) occurs when $\mu_{[1]} = \cdots = \mu_{[k-1]} = \mu_{[k]} - \delta^*$ and is equal to P^*. Dudewicz and Dalal (1975) have tabulated the values of P^* in (2.8) for $k=2(1)25$, $n_0 = 2(1)15(5)30$, and $h = 0.0(0.1)5.1$.

An alternative procedure R_4 considered by Dudewicz and Dalal (1975) is same as R_3 except that the n_i are defined by

$$n_i = \max\left\{ n_0, \left\langle \left(\frac{s_i h}{\delta^*}\right)^2 \right\rangle \right\} \tag{2.12}$$

and $\bar{X}_i(n_i)$ is used in the place of $\tilde{\bar{X}}_i$, $i=1,\ldots,k$. For $k=2$, it has been shown that

$$P(CS \mid R_4, s_1^2, s_2^2) \geqslant P(CS \mid R_3, s_1^2, s_2^2) \tag{2.13}$$

uniformly in s_1^2 and s_2^2; however, a similar assertion does not hold for $k>2$ [see Rinott (1978)].

Let us define two more rules R_4' and R_4'' which are same as R_4 except that h is replaced by h' and h'', respectively, where

$$\int_0^\infty \left[\int_0^\infty \Phi\left(\frac{h'}{\{(n_0-1)(1/x+1/y)\}^{1/2}} \right) g(x)\,dx \right]^{k-1} g(y)\,dy = P^* \tag{2.14}$$

and

$$\int_{-\infty}^{\infty} F_m(t+h'')f_m(t)\,dt=(P^*)^{1/k-1}. \tag{2.15}$$

Here g is the density function of a chi-square random variable with n_0-1 degrees of freedom. Again, F_m and f_m are the cdf and the density, respectively, of a Student t variable with n_0-1 degrees of freedom.

Both R_4' and R_4'' satisfy the basic probability requirement. Also $h \leqslant h' \leqslant h''$ with equalities holding in the case of $k=2$ and strict inequalities in the case of $k>2$. If n_i and n_i' are the sizes of the samples from π_i for the procedures R_4 and R_4', respectively, then Rinott (1978) has shown that $E(n_i)+1 \geqslant E(n_i')$.

A few other two-stage procedures in the case of unequal and unknown variances have been investigated by Maurice (1958a), Ofosu (1973), and Chiu (1974d). Errors in the proofs regarding the evaluations of the infimum of the PCS invalidate these precedures. For further information regarding these the reader is referred to Bechhofer (1974), Dudewicz and Dalal (1975), Lam and Chiu (1976), Tong and Wetzell (1979), and the correction notes of Ofosu (1975), and Chiu (1975).

2.1.3 Two-Stage Selection Procedures: Common Known Variance

For selecting the population having the largest mean from normal populations with equal known variance σ^2, Cohen (1959), Alam (1970b), and Tamhane and Bechhofer (1977) have all studied a two-stage procedure R_5 which uses a subset selection procedure of the type proposed by Gupta (1956, 1965) for screening the populations at the first stage. The procedure R_5 is described as follows:

$\boldsymbol{R_5}$**:** Take a first-stage sample of n_1 independent observations from each population. Define $I=\{i\colon\ \bar{X}_i \geqslant \bar{X}_{[k]}-h\sigma\}$, where the $\bar{X}_i$ are the sample means and $h>0$ is to be specified prior to the experimentation. Let $\Pi_i=\{\pi_i\colon\ i\in I\}$. If Π_i consists of only one population, stop sampling and select that population. If Π_i consists of more than one population, take an additional sample of n_2 observations from each of the populations in the set Π_i and choose the population associated with the maximum of $\bar{Y}_i$, where the $\bar{Y}_i$ are the means based on cumulative samples of n_1+n_2 observations for those populations in Π_i.

The fixed-sample procedure R_1 of Bechhofer (1954) described earlier becomes a special case of R_5 corresponding to $h=0$ or $h=\infty$. Analytical and computational difficulties restricted most of the results of Cohen (1959) and Alam (1970b) to the special case of $k=2$.

Let S' denote the cardinality of the set I and define

$$S=\begin{cases}0 & \text{if } S'=1,\\ S' & \text{if } S'>1.\end{cases}$$

Then $T=kn_1+Sn_2$ is the total sample size required by the procedure R_5. For any given $\{k,\delta^*,P^*\}$, we have an infinite number of combinations of (n_1,n_2,h) which will guarantee the minimum PCS when $\boldsymbol{\mu}$ is a vector in $\Omega(\delta^*)$ defined in (2.4). The design criterion proposed by Alam (1970b) is to

$$\textit{minimize} \quad \sup_{\boldsymbol{\mu}\in\Omega(\delta^*)} E(T|R_5)$$

$$\textit{subject to} \quad \inf_{\boldsymbol{\mu}\in\Omega(\delta^*)} P(CS|R_5)\geqslant P^*$$

where n_1 and n_2 are nonnegative integers and $h\geqslant 0$. This insures that $E(T|R_5)\leqslant kn$ for $\boldsymbol{\mu}\in\Omega(\delta^*)$ where n is the minimum sample size from each population required by the fixed sample size procedure R_1 of Bechhofer (1954). However, the true $\boldsymbol{\mu}$ may not be in $\Omega(\delta^*)$. Also, it is possible to have $\mu_{[1]}=\mu_{[k]}$ and $E(T|R_5)$ very large compared to kn for P^* sufficiently close to unity. To avoid this difficulty, Tamhane and Bechhofer (1977) have used an unrestricted minimax criterion which requires minimization of $\sup_{\boldsymbol{\mu}\in\Omega}E(T|R_5)$ rather than $\sup_{\boldsymbol{\mu}\in\Omega(\delta^*)}E(T|R_5)$.

Let $\mu(\delta^*)=\{\boldsymbol{\mu}\colon \mu_{[1]}=\cdots=\mu_{[k-1]}=\mu_{[k]}-\delta^*\}$. In the case of $k=2$, any $\boldsymbol{\mu}\in\mu(\delta^*)$ gives an LFC. Such a result is only conjectured to hold for $k\geqslant 3$ by Alam (1970b) and Tamhane and Bechhofer (1977). However, the latter authors have obtained a lower bound for $P(CS|R_5)$ that attains its infimum over $\Omega(\delta^*)$ at $\boldsymbol{\mu}\in\mu(\delta^*)$. Thus their result provides a conservative procedure.

Tamhane and Bechhofer (1977) have further obtained an approximate solution by removing the restriction that n_1 and n_2 must be integers and solving a continuous version of the optimization problem.

Define

$$c_1=\frac{\delta^*\sqrt{n_1}}{\sigma}, \qquad c_2=\frac{\delta^*\sqrt{n_2}}{\sigma}, \qquad d=\frac{h\sqrt{n_1}}{\sigma}. \tag{2.16}$$

Let $R_{5\mathrm{EU}}$, $R_{5\mathrm{ER}}$, and $R_{5\mathrm{CU}}$ denote the various cases of the procedure R_5 that arise. The subscripts E and C refer to exact and conversative procedures; R and U refer to the restricted and unrestricted minimax criteria, respectively. Tamhane and Bechhofer (1977) have tabulated the values of c_1, c_2, and d to be used in $R_{5\mathrm{EU}}$ and $R_{5\mathrm{ER}}$ for $k=2$ and $P^*=0.55(0.05)0.95$, 0.99, 0.999, 0.9995, 0.9999. They have also given the constants to implement R_{5CU} for $k=3(1)5(5)15, 25$ and the following values of P^*: 0.75, 0.90, 0.95, 0.99 for all the values of k given previously, 0.60 and 0.45 for $k=10$, 15, 25, and 0.60 for $k=5$.

Their numerical evaluations for different parametric configurations indicate that R_{5EU} and R_{5CU} are highly effective as screening procedures. The effectiveness of R_{5CU} seems to be increasing with k. Also, it is recognized by the authors that a sharper bound for $P(CS|R_5)$ can be obtained.†

Zinger and St-Pierre (1958) have discussed the problem of choosing the best among three normal populations with known variances with a slight modification in the formulation. Suppose π_1, π_2, π_3 are three independent normal populations, $N(\mu_i, \sigma_i^2)$, $i=1,2,3$, where the variances are known. The objective is (i) to select the population with the largest mean subject to an upper bound α for the probability of taking a wrong decision, and (ii) to determine the smallest sample sizes required to detect the population with the largest mean in the case of some given values for the noncentrality parameters. Let $\bar{X}_{[i]}$, $i=1,2,3$, be the sample means based on n_i observations from π_i determined by $\sigma_i^2/n_i=\sigma^2$, $i=1,2,3$. Let

$$u=\frac{(\bar{X}_{[3]}-\bar{X}_{[2]})}{\sigma},$$

$$\gamma=\frac{(\mu_{[3]}-\mu_{[2]})}{\sigma}, \tag{2.17}$$

$$\delta=\frac{(\mu_{[2]}-\mu_{[1]})}{\sigma}.$$

The procedure R_5 of Zinger and St-Pierre is as follows:

R_5: Select the population associated with $\bar{X}_{[3]}$ as the best if $u \geqslant c$; otherwise, do not make a decision. (2.18)

The constant c is chosen such that $W(c,\gamma,\delta) \equiv \Pr(\text{taking a wrong decision}) \leqslant \alpha$, $0<\alpha<1$. Based on numerical results, $W(c,\gamma,\delta)$ is maximized at (γ_0,δ_0) given by $\gamma_0=0=\delta_0$ or by $\gamma_0=0$, $\delta_0=\infty$ depending on whether α is greater than 0.2726 or not. Tables are available for the values of c corresponding to selected values of α and vice versa. The least sample sizes required are determined such that the probability of making a correct decision is $\geqslant 1-\beta$ $(0<\alpha\leqslant\beta<1)$ whenever $\mu_{[3]}-\mu_{[2]}\geqslant\Delta$, besides the requirement on $W(c,\gamma,\delta)$ being satisfied. If $\alpha=\beta$, then $c=0$ and a decision is always taken and the procedure is comparable to that of Bechhofer (1954).

†Recently, Tamhane and Bechhofer (1979) have given tables based on this sharper bound.

2.2 RANKING IN TERMS OF THE VARIANCES

The problem of ranking the normal populations according to their variances was first investigated by Bechhofer and Sobel (1954a). Here we consider π_i to be better than π_j if $\sigma_i^2 \leqslant \sigma_j^2$. Two goals are considered by Bechhofer and Sobel, namely, (I) to divide the set of k populations into two groups, the t best and the $k-t$ worst, both groups unordered ($1 \leqslant t \leqslant k-1$), and (II) to divide the set of k populations into $t+1$ groups, the t-best ordered and the $k-t$ worst unordered. The two goals coincide for $t=1$. Let $\theta_{ij} = \sigma_{[i]}^2 / \sigma_{[j]}^2$. It is required that the PCS should be at least P^* whenever $\theta_{t+1,t} \geqslant \theta_{t+1,t}^*$ in the case of Goal I, and whenever $\theta_{i+1,i} \geqslant \theta_{i+1,i}^*$, $i=1,\ldots,t$, in the case of Goal II, where all the starred quantities are $\geqslant 1$ and specified in advance by the experimenter.

Let s_i^2 be the best unbiased estimator of σ_i^2 based on a sample of size n_i from π_i, $i=1,\ldots,k$. Let ν_i be the degrees of freedom associated with s_i^2. The procedures of Bechhofer and Sobel denoted by R_6 and R_7, for Goals I and II, respectively, are as follows.

R_6: Include the populations associated with $s_{[1]}^2, \ldots, s_{[t]}^2$ in the best group and the remaining in the worst group. (2.19)

R_7: Select the populations associated with $s_{[1]}^2, \ldots, s_{[t]}^2$ as the best, second best, $\ldots$, the tth best, and group the remaining as the worst. (2.20)

In the case of $\nu_i = \nu$ for all i, it has been shown that the LFC is given by

$$\sigma_{[1]}^2 = \cdots = \sigma_{[t]}^2 = \sigma^2; \qquad \sigma_{[t+1]}^2 = \cdots = \sigma_{[k]}^2 = \sigma^2 \theta_{t+1,t}^* \tag{2.21}$$

in the case of Goal I, and by

$$\sigma_{[i+1]}^2 = \theta_{i+1,i}^* \sigma_{[i]}^2, \qquad i=1,\ldots,t; \; \sigma_{[t+1]}^2 = \cdots = \sigma_{[k]}^2 = \sigma^2 \tag{2.22}$$

in the case of Goal II.

For Goal I, assuming that $n_i = n$ for all i, the PCS corresponding to the LFC is given by

$$P_0(CS|R_6) = t\int_{1/\theta^*}^{\infty} \cdots \int_{1/\theta^*}^{\infty} \int_0^1 \cdots \int_0^1 g_n(u_1,\ldots,u_{k-1})\, du_1 \ldots du_{t-1} du_t \ldots du_{k-1}, \tag{2.23}$$

where $\theta^* = \theta^*_{t+1,t}$ and

$$g_n(u_1,\dots,u_{k-1}) = \frac{\Gamma(kn/2)}{[\Gamma(n/2)]^k} \frac{\prod_{i=1}^{k-1} u_i^{(n-2)/2}}{\left(1+\sum_{i=1}^{k-1} u_i\right)^{nk/2}}. \tag{2.24}$$

In the case of Goal II, for $n_i = n$, the corresponding expression for the PCS is given by

$$P_0(CS|R_7) = \int \cdots \int \frac{\Gamma(kn/2)}{[\Gamma(n/2)]^k} \frac{\prod_{i=t}^{k-1} u_i^{(n-2)/2} \prod_{j=1}^{t-1} u_j^{[(k-j)n-2]}}{\left[1+\sum_{i=1}^{k-1}\left(\prod_{j=1}^{i} u_j \Big/ \prod_{\alpha=t}^{i-1} u_\alpha\right)\right]^{nk/2}} du_1,\dots,du_{k-1}, \tag{2.25}$$

where $\prod_{\alpha=m}^{n} u_\alpha = 1$, if $m>n$, and the range of integration for u_i is $(\theta^{*-1}_{i+1,i}, \infty)$, $i=1,\dots,t-1$ and for each of $u_t,\dots,u_{k-1}$ it is $(\theta^{*-1}_{t+1,t}, \infty)$.

Thus, the determination of the smallest n for which the P^*-condition is met involves the evaluation of the integral in (2.23) or (2.25) depending on the goal. It should be noted that for $t=1$ the two goals coincide. For Goal I, Bechhofer and Sobel (1954a) have tabulated the values of $P_0(CS|R_6)$ for $\theta^* = 1.0(0.1)2.2$ and $(k,t) = (2,1)$, $(3,1)$, $(4,1)$, and $(3,2)$. In the cases of $k=2$ and 3, both exact and approximate values are given for $\nu = 1(1)20$. In the case of $k=4$, exact values are given for $\nu = 2(2)12$ and approximate values for $\nu = 1(1)20$. For Goal II, the values of $P_0(CS|R_7)$ (both exact and approximate) are tabulated for $k=t=3$, $\nu=1(1)20$, and the same values of $\theta^*(=\theta^*_{3,2} = \theta^*_{2,1})$ as previously. The reader is also referred to the discussion of these tables in §23.5.

2.3 RANKING IN TERMS OF THE ABSOLUTE VALUES OF THE MEANS

Let π_i, $i=1,\dots,k$, be k independent normal populations with means $\mu_1,\dots,\mu_k$ and a common known variance σ^2. Let $\theta_i = |\mu_i|$; $i=1,\dots,k$. Rizvi (1971) has considered the problem of selecting the t best populations, that is, the populations associated with the t largest θ-values. It should be noted that because of the common variance the problem of ranking the populations according to the values of the θ_i is equivalent to that of ranking them according to the values of the $\lambda_i = |\mu_i|/\sigma$. The λ_i are the absolute values of the reciprocals of the coefficients of variation and can be looked upon as the (univariate) Mahalanobis distances of the populations from a normal population with mean zero and variance σ^2. Selection procedures for multivariate normal populations in terms of the Mahalanobis distances are discussed in §7.2.1. In the univariate case, λ_i denotes the signal-to-noise ratio, and thus ranking in terms of λ_i is meaningful for comparison of electronic devices.

For selecting the populations with t largest θ-values the P^* condition is to be met whenever $\boldsymbol{\theta} \in \Omega_{\delta^*} = \{\boldsymbol{\theta}: \boldsymbol{\theta} = (\theta_1, \ldots, \theta_k); \theta_{[k-t+1]} - \theta_{[k-t]} \geq \delta^* > 0\}$, where δ^* is specified in advance. Assuming without loss of generality that $\sigma^2 = 1$, let $W_i = |\bar{X}_i|$, where $\bar{X}_i$ is the sample mean based on n independent observations from π_i, $i = 1, \ldots, k$. The procedure proposed by Rizvi is

$\boldsymbol{R_8}$: Select the t populations associated with

$$W_{[k-t+1]}, \ldots, W_{[k]}. \tag{2.26}$$

He has shown that

$$\inf_{\Omega_{\delta^*}} P(CS|R_8)$$

$$= 2(k-t)\int_0^\infty [2\Phi(u)-1]^{k-t-1}[\Phi(-u+\lambda)+\Phi(-u-\lambda)]^t d\Phi(u), \tag{2.27}$$

where $\Phi(\cdot)$ is the standard normal cdf and $\lambda = \sqrt{n}\,\delta^*$. Thus, the sample size required is the smallest integer n for which the right-hand side of (2.27) is $\geq P^*$. Tables are provided for the two special cases: (1) $t = 1$ and (2) $t = k-1$. The second case corresponds to selecting the population with the smallest θ. The tables give the PCS for $k = 2(1)10$ and selected values of λ and the value of λ for selected values of k and P^* such that the infimum of the PCS is P^*.

If the variances $\sigma_i^2 = a_i \sigma^2$, $i = 1, \ldots, k$, where σ^2 and a_i (rational) are known constants, then one could choose the sample sizes n_i, $i = 1, \ldots, k$, to make the variances of the sample means equal. This is not necessarily an optimal way but it makes the tables in the preceding case applicable.

2.4 RANKING OF NORMAL MEANS UNDER r-WAY CLASSIFICATION WITHOUT INTERACTION

Bechhofer (1954) has discussed two single-stage procedures for ranking normal populations with a common known variance, for problems involving an r-way ($r \geq 2$) classification without interaction. For selecting the best level of each of the r-factors, one of the procedures involves conducting one r-factor experiment and using the observations obtained therefrom to select the r best levels, while the other procedure consists of conducting r one-factor experiments and using the observations from each experiment to select the best level of each of the associated r-factors.

Bechhofer considered the two-way classification in particular. Let X_{ijm} be normally and independently distributed random variables with

$$E(X_{ijm}) = \mu + \alpha_i + \beta_j,$$

$$V(X_{ijm}) = \sigma_{ij}^2, \tag{2.28}$$

for $i=1,\ldots,a$; $j=1,\ldots,b$; $m=1,\ldots,n_{ij}$, where the parameters satisfy the restriction $\Sigma_{i=1}^{r}\alpha_i = \Sigma_{j=1}^{r}\beta_j = 0$. It is assumed that μ, the α_i, and the β_j are all unknown; the σ_{ij}^2 are known and may be equal or unequal. Let $\alpha_{[1]} \leqslant \cdots \leqslant \alpha_{[a]}$ and $\beta_{[1]} \leqslant \cdots \leqslant \beta_{[b]}$ be the ranked α_i and β_j, respectively. Define

$$\begin{aligned}
\bar{X}_{ij} &= \sum_{m=1}^{n_{ij}} \frac{X_{ijm}}{n_{ij}}, \qquad && i=1,\ldots,a;\, j=1,\ldots,b;\\
\bar{X}_{i.} &= \sum_{j=1}^{b} \frac{\bar{X}_{ij}}{b}, && i=1,\ldots,a;\\
\bar{X}_{.j} &= \sum_{i=1}^{a} \frac{\bar{X}_{ij}}{a}, && j=1,\ldots,b.
\end{aligned} \tag{2.29}$$

The sample mean, the population variance, and the sample size associated with the population having the mean $\mu + \alpha_{[i]} + \beta_{[j]}$ are denoted by $\bar{X}_{(i)(j)}$, $\sigma_{(i)(j)}^2$, and $n_{(i)(j)}$, respectively. We also define $\bar{X}_{(i).}$ and $\bar{X}_{.(j)}$ similar to $\bar{X}_{i.}$ and $\bar{X}_{.j}$ with $\bar{X}_{(i)(j)}$ instead of $\bar{X}_{ij}$. The ranked $\bar{X}_{i.}$ and $\bar{X}_{.j}$ are denoted by $\bar{X}_{[1].} \leqslant \bar{X}_{[2].} \leqslant \cdots \leqslant \bar{X}_{[a].}$ and $\bar{X}_{.[1]} \leqslant \bar{X}_{.[2]} \leqslant \cdots \leqslant \bar{X}_{.[b]}$, respectively.

Any goal for the two-way classification is of the same type as for the one-way classification except that they consist of two parts. For example, the experimenter's goal may be to find the best set of populations according to the first classification, and the best set of populations according to the second classification. The procedure for this goal will choose the set associated with $\bar{X}_{[a].}$ as the best set of populations according to the first classification, and will choose the set associated with $\bar{X}_{.[b]}$ as the best set according to the second classification. It can be shown that

$$\Pr\left(\bar{X}_{(1).}<\bar{X}_{(2).}<\cdots<\bar{X}_{(a).} \quad \text{and} \quad \bar{X}_{.(1)}<\bar{X}_{.(2)}<\cdots<\bar{X}_{.(b)}\right)$$

$$=\Pr\left(\bar{X}_{(1).}<\bar{X}_{(2).}<\cdots<\bar{X}_{(a).}\right)\Pr\left(\bar{X}_{.(1)}<\bar{X}_{.(2)}<\cdots<\bar{X}_{.(b)}\right), \tag{2.30}$$

if the means in each cell have the same variance, that is, if

$$\frac{\sigma_{ij}^2}{n_{ij}}=\sigma_{\bar{X}}^2 \text{ (say)}, \qquad i=1,\ldots,a; j=1,\ldots,b. \tag{2.31}$$

The first of the probabilities on the right-hand side of (2.30) can be increased (for fixed b) by decreasing $\sigma_{\bar{X}}^2$, that is, by increasing n_{ij}. But increasing n_{ij} has the effect of increasing the second probability on the right-hand side of (2.30). Thus the factorial design of the experiment makes the data "work twice" and in this sense more efficient than two separate experiments.

In this connection Bawa (1972) studied the efficiency of the factorial experiment relative to several one-factor experiments for the ranking problem. The *relative efficiency* is defined here as the ratio of the minimum total number of observations needed by each procedure to guarantee a certain specified PCS, P^*. *Asymptotic relative efficiency* is then defined as the limit of the relative efficiency as $P^*\to 1$. In the symmetric case (same number of levels for each factor and the same δ^* for defining the indifference zones), the r-factor experiment is r times more efficient than r one-factor experiments.

In the next section we discuss the problem of ranking multiply classified variances of normal populations investigated by Bechhofer (1968a). The use of the log transformation of the sample variances makes it possible, when n is moderately large, to regard the problem of ranking normal variances as one of ranking means; this is the consequence of the fact that the distribution of the logarithms of the sample variances approaches the normal distribution very rapidly as n increases.

2.5 RANKING MULTIPLY CLASSIFIED VARIANCES OF NORMAL POPULATIONS

Ranking multiply classified variances is a generalization of the problem investigated by Bechhofer and Sobel (1954a) which was discussed in §2.2. Let X_{ijm} ($1\leqslant i\leqslant a$; $1\leqslant j\leqslant b$; $1\leqslant m\leqslant n_{ij}$) be distributed normally with mean

μ_{ij} and variance σ_{ij}^2. Let

$$\sigma^2 = \left(\prod_{i=1}^{a} \prod_{j=1}^{b} \sigma_{ij}^2 \right)^{1/ab}, \qquad \alpha_i = \sigma^{-2} \left(\prod_{j=1}^{b} \sigma_{ij}^2 \right)^{1/b},$$

$$\beta_j = \sigma^{-2} \left(\prod_{i=1}^{a} \sigma_{ij}^2 \right)^{1/a}, \qquad \gamma_{ij} = \frac{\sigma_{ij}^2}{\sigma^2 \alpha_i \beta_j}. \tag{2.32}$$

The parameters satisfy the equations

$$\prod_{i=1}^{a} \alpha_i = \prod_{j=1}^{b} \beta_j = \prod_{i=1}^{a} \gamma_{ij} = \prod_{j=1}^{b} \gamma_{ij} = 1,$$

$$\sigma_{ij}^2 = \sigma^2 \alpha_i \beta_j \gamma_{ij}. \tag{2.33}$$

We also define

$$\delta_{ij} = \alpha_i \beta_j \gamma_{ij}, \tag{2.34}$$

corresponding to $i = 1, \ldots, a$; $j = 1, \ldots, b$. In some problems, it is convenient to relabel the δ_{ij} as δ_r, $r = 1, \ldots, ab$, and regard them as the main effects of Factor D. The α_i and the β_j are the main effects of Factors A and B, respectively. Under the indifference zone formulation, Bechhofer (1968a) considers the following goals:

I. Select the level of Factor A associated with $\alpha_{[1]}$.
II. Select the level of Factor B associated with $\beta_{[1]}$.
III. Select simultaneously the levels of Factors A and B associated with $\alpha_{[1]}$ and $\beta_{[1]}$.
IV. Select the level of Factor D associated with $\delta_{[1]}$.

Goals I and II are meaningful, whether or not there is an interaction between the levels of A and B. Goal III can be used when there is no interaction. When there is an interaction, it is meaningful to consider Goal IV. The ranking procedures of Bechhofer in these cases are based on the observed values of the ab-independent estimators s_{ij}^2 of σ_{ij}^2, $1 \leqslant i \leqslant a$; $1 \leqslant j \leqslant b$. Here $n_{ij} s_{ij}^2 / \sigma_{ij}^2$ has a chi-square distribution with n_{ij} degrees of freedom and the value of n_{ij} depends on the particular $E(X_{ij})$ regression model which is appropriate in the (i,j)th cell.

Now, define

$$s_{i.}^2 = \left(\prod_{j=1}^{b} s_{ij}^2 \right)^{1/b}, \quad s_{.j}^2 = \left(\prod_{i=1}^{a} s_{ij}^2 \right)^{1/a},$$

$$s_{..}^2 = \left(\prod_{i=1}^{a} \prod_{j=1}^{b} s_{ij}^2 \right)^{1/ab}. \tag{2.35}$$

Let

$$\hat{\alpha}_i = \frac{s_{i.}^2}{s_{..}^2}, \quad \hat{\beta}_j = \frac{s_{.j}^2}{s_{..}^2},$$

$$\hat{\gamma}_{ij} = \frac{s_{ij}^2 s_{..}^2}{s_{i.}^2 s_{.j}^2}, \quad \hat{\delta}_{ij} = \frac{s_{ij}^2}{s_{..}^2}. \tag{2.36}$$

Bechhofer's rule for any goal is to rank according to the relevant estimates. Goals I and II are similar. For these goals, assuming $n_{ij} = n$ for all i and j, he has considered several special cases, namely,

(i) $a = 2$.
(ii) $a = b = 2$, $n = 1(1)6, 8$
(iii) $a = 2$, $b = 3$, $n = 2, 4$.

For Goal IV, the experimenter is essentially regarding the two-factor experiment in Factors A and B as a one-factor experiment involving the new Factor D. Thus the results of Bechhofer and Sobel (1954a) are applicable to this case. For Goal III, the special case discussed is that of $a = b = 2$, $n_{ij} = n$.

Tables are given by Bechhofer (1968a) for the exact PCS for Goal I corresponding to LFC configuration $\alpha_{[1]}\theta^* = \alpha_{[2]} = \cdots = \alpha_{[a]}$ for $\theta^* = 1.0(0.2)3.0(0.5)8.0(2.0)26.0, 30.0(5.0)50.0(10.0)100.0$ in the following cases: (i) $a = b = 2$, $n = 1(1)6, 8$ and (ii) $a = 2$, $b = 3$, $n = 2, 4$. When $b = 1$, the earlier tables of Bechhofer and Sobel (1954a) discussed in §2.2 are applicable with a and n playing the roles of k and ν, respectively.

When n is moderately large, one can use the asymptotic distribution of the log transform of the sample variance to obtain approximate value of the PCS. The log transformation reduces the problem to that of ranking means, when n is moderately large.

2.6 SELECTING THE LARGEST INTERACTION IN A TWO-FACTOR EXPERIMENT

Bechhofer, Santner, and Turnbull (1977) have considered a two-factor experiment, both factors qualitative, involving r levels of the first factor and c levels of the second. In the motivating example of the preceding authors, the first factor is sex having $r=2$ levels, the second factor is method of treatment with c levels, and the variable observed is a physical response. It is assumed that the effect of the treatment on the mean response not only varies from treatment to treatment, but is also different for men than for women. In other words, there is an interaction between sex and method of treatment. It is naturally of interest to identify the sex–treatment combination for which this interaction is largest because such information may provide some understanding of the effectiveness of the methods of treatments.

Assuming a fixed-effects linear model, we write

$$Y_{ijm} = \mu + \alpha_i + \beta_j + \gamma_{ij} + \epsilon_{ijm}, \qquad 1 \leqslant i \leqslant r;\ 1 \leqslant j \leqslant c;\ 1 \leqslant m \leqslant n, \quad (2.37)$$

where the ϵ_{ijm} are independent normal random variables with $E(\epsilon_{ijm})=0$ and $V(\epsilon_{ijm})=\sigma^2$. The parameters satisfy the conditions

$$\sum_{i=1}^{r} \alpha_i = \sum_{j=1}^{c} \beta_j = \sum_{i=1}^{r} \gamma_{ij} = \sum_{j=1}^{c} \gamma_{ij} = 0. \qquad (2.38)$$

It is also assumed that μ, the α_i, β_j, and γ_{ij} are unknown, and that σ^2 is known.

Bechhofer, Santner, and Turnbull (1977) considered the goal of selecting the factor-level combination associated with $\gamma_{[rc]}$ with a preference-zone for the probability requirement different from that of Bechhofer (1954). It is now required that

$$\text{PCS} \geqslant P^* \text{ whenever } \gamma_{[rc]} \geqslant \Delta^* \text{ and}$$
$$\gamma_{[rc]} - \gamma_{[rc-1]} \geqslant \delta^*. \qquad (2.39)$$

The three quantities $\{\Delta^*, \delta^*, P^*\}$ are to be specified by the experimenter such that

$$0 < \Delta^* < \infty, \quad 0 < \delta^* < \frac{(r-1)(c-1)-1}{(r-1)(c-1)}\Delta^*,$$

$1/rc < P^* < 1$. Here "correct selection" is the event of selecting the levels of the two factors associated with $\gamma_{[rc]}$ when $\gamma_{[rc]} - \gamma_{[rc-1]} > 0$. The preference-zone specified in (2.39) is appropriate for the present problem. Since we want to identify the largest positive interaction, we are interested in this interaction only when it is sufficiently larger than *zero*, the value corresponding to "no interaction." In a way, the situation is similar to the one to be discussed later in §20.2 since the zero value defines the standard.

Let Y_{ijm} $(1 \leqslant m \leqslant n)$ be n independent observations for (i,j)-combination, $i = 1,\ldots,r$ and $j = 1,\ldots,c$. Define

$$\hat{\gamma}_{ij} = \bar{Y}_{ij.} - \bar{Y}_{i..} - \bar{Y}_{.j.} + \bar{Y}_{...} \tag{2.40}$$

where

$$\bar{Y}_{ij.} = n^{-1} \sum_{m=1}^{n} Y_{ijm},$$

$$\bar{Y}_{i..} = (nc)^{-1} \sum_{j=1}^{c} \sum_{m=1}^{n} Y_{ijm},$$

$$\bar{Y}_{.j.} = (nr)^{-1} \sum_{i=1}^{r} \sum_{m=1}^{n} Y_{ijm},$$

$$\bar{Y}_{...} = (ncr)^{-1} \sum_{i=1}^{r} \sum_{j=1}^{c} \sum_{m=1}^{n} Y_{ijm}.$$

The rule proposed by Bechhofer, Santner, and Turnbull (1977) is the "natural" rule:

***R*:** Select the factor-level combination which produced $\hat{\gamma}_{[rc]}$. (2.41)

In view of the constraints on the values of δ^*/Δ^*, we have $c \geqslant 3$ and $c \geqslant r \geqslant 2$. The exact PCS can be expressed in terms of $rc - 1$ linear stochastic inequalities involving $(r-1)(c-1)$ correlated normal variates. The specific results of the preceding authors relate to the two special cases: (A) $r = 2$, $c \geqslant 3$, and (B) $r = c = 3$.

***Case A:* $r = 2$, $c \geqslant 3$.** In this case, we have

$$\mathrm{PCS} = \Pr\left\{ \sum_{j=2}^{c} (X_{2j} + \gamma_{2j}) - (X_{2b} + \gamma_{2b}) > 0,\ (2 \leqslant b \leqslant c); \right.$$

$$\left. \sum_{j=2}^{c} (X_{2j} + \gamma_{2j}) + (X_{2b} + \gamma_{2b}) > 0,\ (2 \leqslant b \leqslant c) \right\}, \tag{2.42}$$

where the $X_{2j}=\hat{\gamma}_{2j}-\gamma_{2j}$ have a $(c-1)$-variate normal distribution with $E(X_{2j})=0,\ 2\leqslant j\leqslant c,$

$$V(X_{2j})=\frac{c-1}{cn}\sigma^2,\quad 2\leqslant j\leqslant c,$$

$$\operatorname{cov}(X_{2j_1},X_{2j_2})=-\frac{1}{cn}\sigma^2, j_1\neq j_2;\quad 2\leqslant j_1,j_2\leqslant c. \tag{2.43}$$

We need to introduce the following notation before stating the result of Bechhofer, Santner, and Turnbull (1977) regarding the LFC. Let $\boldsymbol{\gamma}=(\gamma_{22},\gamma_{23},\ldots,\gamma_{2c})$. For each $\gamma_{11}\geqslant\Delta^*$, let

$$E(\gamma_{11})=\left\{\boldsymbol{\gamma}\,|\,\sum_{j=2}^{c}\gamma_{2j}=\gamma_{11};\ \delta^*-\gamma_{11}\leqslant\gamma_{2b}\leqslant\gamma_{11}-\delta^*,\quad(2\leqslant b\leqslant c)\right\}. \tag{2.44}$$

Then $E(\gamma_{11})$ is the set of configurations in the preference zone on the plane $\Sigma_{j=2}^{c}\gamma_{2j}=\gamma_{11}$ and $E=\cup_{\gamma_{11}>\Delta^*}E(\gamma_{11})$ is the entire preference zone. We further let $p=(c-3)/2$ or $(c-4)/2$ according as c is odd or even, and define the integer s as follows:

$$s=\begin{cases} p, & \text{if}\quad 0<\delta^*<\dfrac{c-2-2p}{c-1-2p}\gamma_{11} \\ t\ (0\leqslant t<p), & \text{if}\quad \dfrac{c-4-2t}{c-3-2t}\gamma_{11}\leqslant\delta^*<\dfrac{c-2-2t}{c-1-2t}\gamma_{11}. \end{cases} \tag{2.45}$$

The integer s always exists and is uniquely defined because of the fact that $0<\delta^*<(c-2)\Delta^*/(c-1)$. Furthermore, let

$$\boldsymbol{\gamma}(s,\gamma_{11})=(\gamma_{11}-\delta^*,\ldots,\gamma_{11}-\delta^*,d_s,\delta^*-\gamma_{11},\ldots,\delta^*-\gamma_{11}) \tag{2.46}$$

where $d_s=\delta^*(c-2-2s)-\gamma_{11}(c-3-2s)$. The vector $\boldsymbol{\gamma}(s,\gamma_{11})$ has the first $c-2-s$ elements equal to $\gamma_{11}-\delta^*$, and the last s elements equal to $\delta^*-\gamma_{11}$. Finally, define

$$\boldsymbol{\gamma}_e(t)=\boldsymbol{\gamma}\left(t,\frac{\delta^*(c-3-2t)}{(c-4-2t)}\right). \tag{2.47}$$

Now, we are ready to state the main theorem of Bechhofer, Santner, and Turnbull (1977).

Theorem 2.1. For the $2\times c$ case ($c \geqslant 3$) and any specified (Δ^*,δ^*) with $0<\Delta^*<\infty$, $0<\delta^*<(c-2)\Delta^*/(c-1)$, define s by (2.45) with γ_{11} set equal to Δ^*.

A. If $s=p$ (i.e., $\Delta^* \geqslant \delta^*(c-1-2p)/(c-2-2p)$) then γ^*, the LFC of the γ over E, is given by $\gamma^*=\gamma(p,\Delta^*)=(\Delta^*-\delta^*,\ldots,\Delta^*-\delta^*,d_p,\delta^*-\Delta^*,\ldots,\delta^*-\Delta^*)$.

B. If $s=t<p$, then γ^* occurs at one of the following points: $\gamma(t,\Delta^*),\gamma_e(t),\gamma_e(t+1),\ldots,\gamma_e(p-1)$.

Case B: r=c=3. In this case

$$\mathrm{PCS}=\Pr\left\{\sum_{i=2}^{3}\sum_{j=2}^{3}(X_{ij}+\gamma_{ij})-(X_{ab}+\gamma_{ab})>0,\ (2\leqslant a,b\leqslant 3)\right\}$$

where $\mathbf{X}=(X_{22},X_{23},X_{32},X_{33})$ has a 4-variate normal distribution with $E(X_{ij})=0$ $(2\leqslant i,j\leqslant 3)$ and the covariance matrix Σ given by

$$\Sigma=\frac{\sigma^2}{9n}\begin{bmatrix} 4 & -2 & -2 & 1\\ -2 & 4 & 1 & -2\\ -2 & 1 & 4 & -2\\ 1 & -2 & -2 & 4\end{bmatrix}.$$

In this case the LFC is $\gamma_{22}^*=\gamma_{23}^*=\gamma_{32}^*=\Delta^*-\delta^*$, $\gamma_{33}^*=3\delta^*-2\Delta^*$ for all δ^* $(0<\delta^*<3\Delta^*/4)$.

REMARK 2.1. In the 2×3, 2×4, and 3×3 cases, there is only *one* distinct γ^* independent of δ^*/Δ^*. It is a conjecture that these are the only ones of the general $r\times c$ case having only one distinct γ^*. To implement the rule R, one has to evaluate the exact PCS numerically which can be done with the existing tables in the case $(r,c)=(2,3)$, $(2,4)$, and $(3,3)$. □

Santner (1978) has discussed the general $r\times c$ experiment, treating the problem of minimizing the PCS as one of nonlinear programming. The LFC is shown to occur at an extreme point of a certain convex polytope. In the special case of a 3×4 experiment, Santner (1978) has made a detailed analysis of LFC for various values of Δ^* and δ^*. He has also considered the problem in the light of Fabian (1962), by defining a correct selection as the selection of any *preferred* factor-level combination. Any combination (i,j) is said to be *preferred* if *either* (a) $\gamma_{[rc]}<\Delta^*$, *or* (b) $\gamma_{[rc]}\geqslant\Delta^*$ and $\gamma_{ij}>\gamma_{[rc]}-\delta^*$. It has been shown by Santner (1978) that the sample size needed under the indifference zone formulation is also the sample size needed to guarantee the minimum specified PCS for selecting a preferred combination of levels.

2.7 Δ-CORRECT RANKING AND SOME REMARKS ON THE RANKING PROCEDURES FOR THE NORMAL MEANS

In this section, we discuss some results concerning the general ranking goal of Bechhofer (1954) and the associated probability requirement. These results are due to Fabian (1962).

Following our notations of §1.4, the general goal of Bechhofer (1954) is to partition a collection of k normal populations π_i: $N(\mu_i,\sigma^2)$, $i=1,\ldots,k$, into s nonempty sets $I_1,\ldots,I_s$ consisting of $k_1,\ldots,k_s$ $(k_1+\cdots+k_s=k)$ populations, respectively, such that for $\pi_i\in I_\alpha$ and $\pi_j\in I_\beta$, $1\leqslant\alpha<\beta\leqslant s$, we have $\mu_i<\mu_j$. It is required of any decision rule R that $P(CS|R)$ satisfies the condition

$$P(CS|R)\geqslant P^* \quad \text{for} \quad \boldsymbol{\mu}\in\Omega(\boldsymbol{\delta}^*), \tag{2.48}$$

where

$$\boldsymbol{\delta}^*=(\delta^*_{\hat{k}_1},\delta^*_{\hat{k}_2},\ldots,\delta^*_{\hat{k}_{s-1}}),\quad \delta^*_{\hat{k}_i}>0, \qquad i=1,\ldots,s-1,$$

$$\Omega(\boldsymbol{\delta}^*)=\{\boldsymbol{\mu}\colon \mu_{[\hat{k}_i+1]}-\mu_{[\hat{k}_i]}\geqslant\delta^*_{\hat{k}_i}, \qquad i=1,\ldots,s-1\},$$

and

$$\hat{k}_i=\sum_{\alpha=1}^{i}k_\alpha,\ (\hat{k}_0\equiv 0).$$

Let the set S of the sample means $\bar{X}_i$, $i=1,\ldots,k$, each based on n independent observations, be partitioned into $S=\{S_1,\ldots,S_s\}$, where $S_\alpha=\{\bar{X}_{[\hat{k}_{\alpha-1}+1]},\ldots,\bar{X}_{[\hat{k}_\alpha]}\}$. The rule R_1 proposed by Bechhofer includes in I_r, $r=1,\ldots,s$, all the populations associated with the means in S_r.

Now, in specifying $\Omega(\boldsymbol{\delta}^*)$, the numbers $\delta^*_{\hat{k}_i}$ are chosen by the experimenter as the smallest values of the differences $\mu_{[\hat{k}_i+1]}-\mu_{[\hat{k}_i]}$ which are "worth detecting." With this intuitive meaning of $\boldsymbol{\delta}^*$, we now define a Δ-correct ranking. It should first be noted that any ranking can be regarded as k classification statements of the form "π_i belongs to the group I_α." Let $\Delta=\{\Delta_1,\ldots,\Delta_{s-1}\}$, $\Delta_i>0$, $i=1,\ldots,s-1$.

Definition 2.1. We say that the classification of π_i and π_j in the groups I_α and $I_{\alpha+1}$, respectively, is Δ-correct if $\mu_j>\mu_i-\Delta_\alpha$.

Definition 2.2. The classification of π_i and π_j in the groups I_α and I_β, respectively, is Δ-correct, if for every $\gamma=\alpha,\alpha+1,\ldots,\beta-1$, the classification of π_i and π_j in the groups I_γ and $I_{\gamma+1}$, respectively, is Δ-correct, that is, if

$$\mu_j>\mu_i-\min\{\Delta_\gamma;\ \gamma=\alpha,\alpha+1,\ldots,\beta-1\}.$$

Any partitioning of the k populations into s groups is called a ranking $I=\{I_1,\ldots,I_s\}$. Fabian (1962) introduces the concept of Δ-correct ranking in terms of the preceding definitions.

Definition 2.3. A ranking I is Δ-correct if, for every $\pi_i \in I_\alpha$, $\pi_j \in I_\beta$, $1 \leqslant \alpha < \beta \leqslant s$, the classification of π_i and π_j in the groups I_α and I_β, respectively, is Δ-correct.

Let us denote by $P\{\Delta\text{-}CD;\ \mu\}$ the probability of a Δ-correct decision, that is, the probability that S is a Δ-correct ranking. According to this, S is Δ-correct if and only if the following implication holds:

$$1 \leqslant \alpha < \beta \leqslant s,\ \pi_i \in S_\alpha,\ \pi_j \in S_\beta \Rightarrow \mu_j > \mu_i - \min\{\Delta_\gamma\colon \gamma = \alpha, \alpha+1, \ldots, \beta-1\}. \tag{2.49}$$

Suppose one defines the Δ-correctness of the ranking I by the implication

$$\pi_i \in I_\alpha,\ \pi_j \in I_{\alpha+1}, \qquad 1 \leqslant \alpha < s \Rightarrow \mu_j > \mu_i - \Delta_\alpha. \tag{2.50}$$

It is shown by Fabian through an example that the ranking obtained by changing a Δ-correct ranking into a "better" one may not be Δ-correct if we adopt (2.50). A ranking better than a given ranking I is one that can be obtained by a finite number of successive changes consisting of transferring π_i and π_j—if they satisfy $\pi_i \in I_\alpha$, $\pi_j \in I_\beta$, $\mu_i > \mu_j$, $\alpha < \beta$—from I_α to I_β and from I_β to I_α, respectively.

The special case of $s=2$, $k_1 = k-1$ is of interest. Let $\mu_{(k)}$ be the mean of the population associated with $\bar{X}_{[k]}$. Then, for Bechhofer's procedure, we have

$$\Pr(\mu_{(k)} = \mu_{[k]}) \geqslant P^* \quad \text{for} \quad \boldsymbol{\mu} \in \Omega(\Delta), \tag{2.51}$$

where $\Omega(\Delta) = \{\boldsymbol{\mu}\colon \mu_{[k]} - \mu_{[k-1]} \geqslant \Delta\}$. It is easy to show that (2.51) is equivalent to

$$\Pr(\mu_{(k)} > \mu_{[k]} - \Delta) \geqslant P^* \quad \text{for} \quad \boldsymbol{\mu} \in \Omega. \tag{2.52}$$

Thus, for the special case of selecting the largest mean,

$$\inf_{\Omega(\Delta)} P(CS\,|\,R_1) = \inf_{\Omega} P(\Delta\text{-}CD\,|\,R_1). \tag{2.53}$$

The following theorem of Fabian shows that the assertion in (2.51) can be strengthened without decreasing the infimum of the PCS.

Theorem 2.2. Let $d=\max\{0,\Delta-(\bar{X}_{[k]}-\bar{X}_{[k-1]})\}$ and

$$I(d)=\begin{cases}(\mu_{[k]}-d,\mu_{[k]}), & d>0\\ (\mu_{[k]},\mu_{[k]}), & d=0.\end{cases}$$

Then (2.51), (2.52), and

$$\Pr(\mu_{(k)}\in I(d))\geqslant P^* \text{ for every } \boldsymbol{\mu}\in\Omega \tag{2.54}$$

are mutually equivalent.

The assertion $\mu_{(k)}\in I(d)$ is stronger than the assertion $\mu_{(k)}>\mu_{[k]}-\Delta$, since $d\leqslant\Delta$. For the general ranking problem, let $p(\Omega_\Delta)$ be the infimum of PCS in Bechhofer's formulation, where $\Delta=(\delta^*_{k_1},\ldots,\delta^{\leqslant})$, and let $p_\Delta=\inf_\Omega$ P(Δ-CD; $\boldsymbol{\mu}$). It has been shown by Fabian that, for every $\Delta=\{\Delta_1,\ldots,\Delta_{s-1}\}\in E_{s-1}$ (the Euclidean $(s-1)$-space), $\Delta_\alpha>0$, we have

$$p_\Delta\leqslant p(\Omega_\Delta) \tag{2.55}$$

with the equality holding in the special case of $s=2$, $k_1=k-1$. This means that, according to the definition of Δ-correctness, the LFC of the means μ_i does not necessarily lie in the region $\Omega(\Delta)$.

2.8 NOTES AND REMARKS

Sequential procedures for selection from normal populations are discussed in Chapter 6. An important reference for sequential procedures is the monograph of Bechhofer, Kiefer and Sobel (1968). In §2.1.3, we discussed two-stage selection procedures for the normal means problem in the common known variance case. Tamhane (1977) has discussed a class of multistage selection procedures for this problem. In another earlier paper (1976), he considered a three-stage elimination type procedure when the common variance is unknown.

In discussing two-stage procedures we saw that there arises the problem of determining the first-stage and the second-stage sample sizes for each population. There are other results regarding optimal choice of sample sizes. These are discussed in Chapter 5. Chiu (1974a, b, c) has discussed a generalized goal which is Δ-correct selection relative to the best population.

In this chapter we did not discuss procedures for selection from correlated normal populations. These procedures and several other procedures for selection from two or more multivariate normal populations form the content of Chapter 7. Selection procedures for normal populations under the subset selection approach are described in Chapter 12.

CHAPTER 3

Some Optimum Properties of Fixed Subset Size Selection Rules

In Chapter 2, we discussed ranking of normal populations in terms of several measures. Before discussing selection procedures for other specific distributions, we review some results concerning optimum properties of selection rules. These are also applicable in many specific cases.

3.1 MOST ECONOMICAL CHARACTER OF RULES

The procedures of Bechhofer (1954) and Bechhofer and Sobel (1954a) were proved by Hall (1959) to be *most economical* in the sense that no other single-stage rule can meet the basic probability requirement for a correct selection with a smaller sample size. The result of Hall is, in fact, applicable to any analogous problem of choosing the population with the largest parameter when, for each sample, there is a sufficient statistic with a monotone likelihood ratio (MLR) and the parameter is one of location or scale. To state the main result of Hall more precisely, we make the following assumptions under which it is true.

Let $X_{i1},\ldots,X_{in}$ be a random sample from π_i which has the associated density f_{θ_i}, $\theta_i \in \Omega \subset R_1$ (the real line), $i=1,\ldots,k$, with respect to a fixed measure. The region of positive density is assumed to be independent of the parameter θ. Let $t_i = t_i(X_{i1},\ldots,X_{in})$ be a statistic sufficient for θ_i. Assume that t_i has a monotone likelihood ratio and θ_i is a location parameter in the distribution of t_i. Let D_n denote any decision rule for choosing the population with the largest θ_i based on samples of size n from each population and let D_n^0 denote that D_n which chooses the population corresponding to the largest t_i (with ties broken by randomization) as the population with the largest θ_i. The following theorem on optimality of D_n is due to Hall (1959).

Theorem 3.1. Suppose N is the least n for which

$$\int G_n^{k-1}(t+\delta-0)\,dG_n(t)$$
$$+\sum_{r=2}^{k}\frac{1}{r}\binom{k-1}{r-1}\int\left[G_n(t+\delta)-G_n(t+\delta-0)\right]^{r-1}\left[G_n(t+\delta-0)\right]^{m-r}dG_n(t)$$
$$\geqslant\gamma,\quad \delta>0,\quad 0<\gamma<1, \tag{3.1}$$

where $G_{\theta,n}(t) = G_n(t-\theta)$ is the cdf of t with parameters θ and n, and $G_n(t+\delta-0) = \lim_{x\uparrow t+\delta} G_n(x)$. Then D_N^0 satisfies

(a) $P(CS|D_n)\geqslant\gamma$ for all $\boldsymbol{\theta}=(\theta_1,\ldots,\theta_k)\in\Omega(\delta)$, where

$$\Omega(\delta)=\{\boldsymbol{\theta}:\theta_{[k]}-\theta_{[k-1]}\geqslant\delta\},\text{ and}$$

(b) there does not exist a D_n satisfying (a) with $n<N$.

The preceding theorem is also valid for the scale parameter case with the following changes:

R		positive real line
location		scale
$\delta>0$		$\delta>1$
$G_n(t-\theta)$	changed into	$G_n(t/\theta)$
$t+\delta$		$t\delta$
$\theta_{[k]}-\theta_{[k-1]}$		$\theta_{[k]}/\theta_{[k-1]}$

Rizvi (1963) has proved the most economical character of his rule R_8 described earlier in §2.3 by demonstrating its minimax and admissible nature with respect to a simple loss function. In an earlier paper, Hall

(1958) has discussed multiple decision procedures in general for which the sample size is a minimum subject to lower bounds on the probabilities of making correct decisions. Recently, Miescke (1979a) has proved the most economical property in the case of distributions with monotone likelihood ratio property for the class of procedures that are permutation invariant and based on total orderings in the sample space.

3.2 UNIFORMLY BEST IMPARTIAL RULE

Consider the problem of selecting the population with the largest mean from k normal populations $N(\mu_i,\sigma^2)$, $i=1,\ldots,k$. If we take samples of equal size, the "natural" selection rule selects the population from which the sample with the largest mean came as the one associated with $\mu_{[k]}$. In §1.3 we discussed impartial decision rules and referred to the theorem of Bahadur (1950) which shows that the natural rule cannot be improved upon under the restriction of impartial decision rules. Before stating the theorem of Bahadur, we introduce a few more notations and definitions in addition to what we have in §1.3. Let a_{ij} denote the indicator function of the set $A_{ij}=\{T_i=T_{[j]}\}$.

Definition 3.1. Let $\boldsymbol{\beta}=(b_1,\ldots,b_k)$ be a vector of real numbers b_i, and $\mathbf{f}=(f_1,\ldots,f_k)$ a vector of real-valued functions $f_i(x)$ defined for every real x. We say that $\mathbf{f}\in M(\beta)$ if for any $r,s=1,2,\ldots,k$ for which $b_r\leqslant b_s$, the set $\{f_r(x)f_s(y)<f_r(y)f_s(x),\ x<y\}$ is of measure zero.

We now make a complete statement of Bahadur's theorem.

Theorem 3.2. Let $\omega=(\theta_1,\ldots,\theta_k)$. Suppose we have the following.

A. $\boldsymbol{\beta}(\omega)=(b_1,\ldots,b_k)$ and $\boldsymbol{\gamma}(\omega)=(u_1,\ldots,u_k)$ are defined for every ω such that

$$b_i\leqslant b_j\Rightarrow u_i\leqslant u_j,\qquad i,j=1,\ldots,k.$$

Given an $\mathbf{s}=(p_1,\ldots,p_k)$, with $p_i\geqslant 0$, $\sum_{i=1}^k p_i=1$, the loss $W(\omega,\mathbf{s})$ is defined by

$$W(\omega,\mathbf{s})=\max_i\{u_i\}-\sum_{i=1}^k p_iu_i.$$

B. $T_1,\ldots,T_k;\ Y$ are independent random variables each T_i having a density $f(t,\theta_i)=f_i(t)$, say, and $\mathbf{f}(\omega)=(f_1,\ldots,f_k)$.

C. D is the class of all decision rules $\mathbf{d}$ such that

$$\mathbf{d}=d(T_{[1]},\ldots,T_{[k]};\ Y)=[\lambda_1,\ldots,\lambda_k],$$

$$\lambda_j\geqslant 0,\ \sum_{j=1}^k\lambda_j\equiv 1,$$

and

$$\mathbf{s(d)} = (p_1(\mathbf{d}),\ldots,p_k(\mathbf{d})),$$

where $p_i(\mathbf{d}) = \sum_{j=1}^{k}\lambda_j a_{ij}$, $i=1,\ldots,k$.

Given $\mathbf{d}\in D$, the risk is given by $r(\mathbf{d}|\omega) = E[W(\omega,\mathbf{s(d)})|\omega]$.

D. For every ω, $\mathbf{f}\in M(\beta)$.

Then, for every ω,

$$r(\mathbf{d}_1|\omega) = \sup_{\mathbf{d}\in D} r(\mathbf{d}|\omega),$$

$$r(\mathbf{d}_k|\omega) = \inf_{\mathbf{d}\in D} r(\mathbf{d}|\omega), \tag{3.2}$$

where $\mathbf{d}_1 \equiv [1,0,\ldots,0]$ and $\mathbf{d}_k \equiv [0,0,\ldots,0,1]$.

Certain improvements of the preceding result are provided by Bahadur and Goodman (1952), who define impartiality in terms of permutations of the k samples rather than the k ordered values of an arbitrarily chosen real-valued statistic. We will slightly change the notations from the earlier paper of Bahadur (1950). Let $\mathbf{u}_i = (x_{i1},\ldots,x_{in})$ be the sample of n independent observations drawn from π_i, $i=1,\ldots,k$, and let the combined sample point be $\mathbf{v} = (\mathbf{u}_1,\ldots,\mathbf{u}_k)$. Let $\varphi(\mathbf{v}) = (p_1(\mathbf{v}),\ldots,p_k(\mathbf{v}))$ be an ordered set of functions p_i of the combined sample point $\mathbf{v}$ such that $0 \leqslant p_i(\mathbf{v}) \leqslant 1$ and $\sum_{i=1}^{k} p_i(\mathbf{v}) = 1$ for all $\mathbf{v}$. Then $\varphi(\mathbf{v})$ is a decision rule. If d_i is the decision to choose π_i, then $p_i(\mathbf{v})$ denotes the probability of taking the decision d_i when $\mathbf{v}$ is the observed sample point.

Definition 3.2. Two decision rules $\varphi(\mathbf{v}) = (p_1(\mathbf{v}),\ldots,p_k(\mathbf{v}))$ and $\varphi'(\mathbf{v}) = (p_1'(\mathbf{v}),\ldots,p_k'(\mathbf{v}))$ are *equivalent* if $E[p_i(\mathbf{v})|\omega] = E[p_i'(\mathbf{v})|\omega]$ for $i=1,\ldots,k$, and all $\omega\in\Omega$.

Definition 3.3. A decision rule $\varphi(\mathbf{v}) = (p_1(\mathbf{v}),\ldots,p_k(\mathbf{v}))$ is said to be *impartial* if, for any $\mathbf{v} = (\mathbf{u}_1,\ldots,\mathbf{u}_k)$ and any permutation $i_1,\ldots,i_k$ of $1,\ldots,k$, we have

$$p_j(\mathbf{u}_{i_1},\ldots,\mathbf{u}_{i_k}) = p_{i_j}(\mathbf{u}_1,\ldots,\mathbf{u}_k) \quad \text{for} \quad j=1,\ldots,k.$$

We assume that there exists a function $t(\mathbf{u})$, not necessarily real-valued, on n-dimensional sample space such that $t_i = t(\mathbf{u}_i)$ is a sufficient statistic for θ_i, $i=1,\ldots,k$, and that the conditional distribution of $\mathbf{u}_i$ given that $\mathbf{t}_i = c$ is the same for each i. (These assumptions are satisfied if, for example, the k populations are (i) normal, or (ii) rectangular, or (iii) exponential, or (iv) binomial, or (v) of Poisson type.) Then $t(\mathbf{v}) = (t_1,\ldots,t_k)$ is a sufficient statistic for ω. Let $\varphi^*(t(\mathbf{v}))$ be the conditional expectation of $\varphi(\mathbf{V})$ given that $t(\mathbf{V})\equiv t(\mathbf{v})$. It is easy to verify that $\varphi^*(t(\mathbf{v}))$ is a decision rule. It has been shown by Bahadur and Goodman (1952) that $\varphi(\mathbf{v})$ and $\varphi^*(t(\mathbf{v}))$ are equivalent. They further show that if $\varphi(\mathbf{v})$ is an impartial rule, so is the equivalent rule $\varphi^*(t(\mathbf{v}))$. Another of their results gives a characterization of impartial decision rules which relates the present definition to the one adopted by Bahadur (1950). To state the result, let $\psi(\mathbf{u})$ be a real-valued function on R^n and for any $\mathbf{v}=(\mathbf{u}_1,\ldots,\mathbf{u}_k)$, set $\psi_i=\psi(\mathbf{u}_i)$, $i=1,\ldots,k$. Let $\mathfrak{D}(\psi)$ be the class of all impartial decision rules that are based on $\psi_1,\ldots,\psi_k$ alone, that is, all impartial decision rules of the form $\varphi(\mathbf{v})=(p_1(\psi_1,\ldots,\psi_k),\ldots,p_k(\psi_1,\ldots,\psi_k))$. Since $\psi(\mathbf{u})$ is a given function, $\mathfrak{D}(\psi)$ will, in general, be a subclass of the class of all impartial decision rules, but may coincide with it. For any given $\mathbf{v}$, let the ordered ψ_i be denoted by $\psi_{[1]}\leqslant\cdots\leqslant\psi_{[k]}$ and let

$$a_{ij}=\begin{cases}1 & \text{if } \psi_i=\psi_{[j]},\\ 0 & \text{otherwise.}\end{cases}\qquad i,j=1,\ldots,k.$$

The following theorem of Bahadur and Goodman (1952) characterizes the class $\mathfrak{D}(\psi)$ of impartial decision rules.

Theorem 3.3. A decision rule $\varphi(\mathbf{v})=p_1(\mathbf{v}),\ldots,p_k(\mathbf{v}))$ is a member of $\mathfrak{D}(\psi)$ if and only if there exist functions $\lambda_j(z_1,\ldots,z_k)$, $j=1,\ldots,k$, such that $0\leqslant\lambda_j\leqslant 1$, $\sum_{j=1}^k\lambda_j=1$ and such that

$$p_i(\mathbf{v})=\sum_{j=1}^{k}\left[\frac{a_{ij}}{\sum_{i=1}^{k}a_{ij}}\right]\lambda_j(\psi_{[1]},\ldots,\psi_{[k]})$$

for each $i=1,\ldots,k$, and all $\mathbf{v}$.

These results and Theorem 3.2 of Bahadur yield the desired improvements. Thus, the results of Bahadur (1950) and Bahadur and Goodman (1952) show that for certain families of distributions the natural selection procedure uniformly minimizes the risk among all symmetric procedures for a large class of loss functions. Lehmann (1966) has restated the result of Bahadur and Goodman and given an alternative proof. His approach is based on invariant procedures.

Continuing the notations of the preceding discussion, let the distribution of the sufficient statistics $\mathbf{t}=(t_1,\ldots,t_k)$ have the density

$$h_{\boldsymbol{\theta}}(\mathbf{t})=c(\boldsymbol{\theta})f_{\theta_1}(t_1)\cdots f_{\theta_k}(t_k)\tag{3.3}$$

with respect to a σ-finite measure ν which is invariant under the group G of all permutations of $(t_1,\ldots,t_k)$. Let f_{θ_i} have monotone likelihood ratio in t_i, $i=1,\ldots,k$. For any $g \in G$, let $\bar{g}$ and g^* be defined as the same permutation of $(\theta_1,\ldots,\theta_k)$ and $(d_1,\ldots,d_k)$, respectively.

Definition 3.4. A procedure φ is said to be *invariant* if

$$g^*\varphi(\mathbf{t}) = \varphi(g\mathbf{t}) \quad \text{for all } \mathbf{t} \text{ and } g. \tag{3.4}$$

We now assume that the loss function L satisfies

$$L(\bar{g}\boldsymbol{\theta}, g^*d_j) = L(\boldsymbol{\theta}, d_j), \qquad j=1,\ldots,k,$$
$$\theta_i < \theta_j \Rightarrow L(\boldsymbol{\theta}, d_i) \geqslant L(\boldsymbol{\theta}, d_j). \tag{3.5}$$

Under the preceding setup, Lehmann (1966) has proved the following theorem.

Theorem 3.4. Let $\varphi^{(0)}$ be the procedure that takes decision d_i when t_i is the unique largest among $(t_1,\ldots,t_k)$ and decisions $d_{i_1},\ldots,d_{i_r}$ each with probability $1/r$ when $t_{i_1},\ldots,t_{i_r}$ are tied for the largest. Then $\varphi^{(0)}$ uniformly minimizes the risk among all procedures based on $\mathbf{t}$ that are invariant under G.

Theorem 3.4 is shown by Lehmann to imply many other optimum properties. For examples, the procedure $\varphi^{(0)}$ of the theorem maximizes the minimum probability of a correct selection whenever the best population is sufficiently much better than the second best, a property proved also by Hall (1959). It also follows that $\varphi^{(0)}$ minimizes $\sup[\theta_{[k]} - \Sigma_{j=1}^{k}\theta_j E_\theta \varphi_j(X)]$, a measure of regret considered by Bahadur (1950) and Bahadur and Robbins (1950).

A generalization of Theorem 3.4 is given by Eaton (1967a) whose treatment of the problem is also restricted to the symmetric case.

Consider a Borel subset $\mathfrak{X} \subseteq R^k$ (k-dimensional Euclidean space) and let μ be a σ-finite measure on $\mathfrak{X}$. Also, let Θ be a symmetric subset of R^k and A be an arbitrary set.

Definition 3.5. A family of real-valued density functions (with respect to μ) $\{f_\alpha(x,\theta) | \alpha \in A\}$ is said to have *Property M* if, for each $\alpha \in A$ and for each $i,j (i \neq j)$, $1 \leqslant i,j \leqslant k$, the following holds:

$$x_i \geqslant x_j \quad \text{and} \quad \theta_i \geqslant \theta_j \Rightarrow f_\alpha(x,\theta) \geqslant f_\alpha(x,(\theta^{(i,j)})) \tag{3.6}$$

where $(\theta^{(i,j)})$ is the vector θ with components θ_i and θ_j interchanged.

Lehmann (1966) considered densities of the form (3.3) where f is assumed to have a monotone likelihood ratio and $C(\theta)$ is constant when the coordinates of θ are permuted. Densities of this form have Property M. But the converse is not necessarily true. We will discuss presently a necessary and sufficient condition of Eaton for a class of densities to have Property M.

For the general ranking problem of Bechhofer (1954) for k populations whose densities belong to a class with Property M, Eaton has shown that the natural selection procedure uniformly minimizes the risk among symmetric decision rules. The other optimum properties of this procedure obtained by Eaton are discussed in the next section. We conclude this section with some additional remarks on Property M.

REMARK 3.1. Property M, as one can easily see, is a multivariate analog of the monotone likelihood ratio property. Suppose a family of density functions $\{f_\alpha(x,\theta)|\alpha\}\in A$ has the form

$$f_\alpha(x_1,\ldots,x_k;\theta_1,\ldots,\theta_k)=\prod_{i=1}^{k} g_\alpha(x_i,\theta_i), \tag{3.7}$$

where g_α is a density. Then the family $\{f_\alpha|\alpha\in A\}$ has Property M if and only if $\{g_\alpha|\alpha\in A\}$ has a monotone likelihood ratio. This result allows the construction of many densities $\{f_\alpha\}$ which have Property M. However, there are multivariate densities of interest that cannot be written in the form (3.7). The following theorem of Eaton (1967a) gives a necessary and sufficient condition that a certain class of densities has Property M. □

Theorem 3.5. Let f be a positive strictly decreasing function defined on $[0,\infty)$ and consider

$$h_\Lambda(x,\theta)\equiv c(\Lambda)f((x-\theta)\Lambda(x-\theta)') \tag{3.8}$$

where $\boldsymbol{\theta}$ and $\mathbf{x}$ are (row) vectors in R^k, $k\geqslant 2$, Λ is a $k\times k$ positive definite matrix and $c(\Lambda)$ is a positive constant. Assume that $h_\Lambda(\mathbf{x},\boldsymbol{\theta})$ is a density on R^k. The following are equivalent:

(i) $h_\Lambda(x,\theta)$ has Property M.
(ii) $\Lambda=c_1I-c_2\boldsymbol{e}'\boldsymbol{e}$ where $\boldsymbol{e}=(1,1,\ldots,1)$, $c_1>0$, $-\infty<c_2/c_1<1/k$, and I is the $k\times k$ identity matrix. (3.9)

Bechhofer, Kiefer, and Sobel (1968, p. 41) have defined a *rankability condition* (see §6.8) which implies Property M.

3.3 BAYES, MINIMAX, AND ADMISSIBILITY PROPERTIES

In this section, we discuss further results of Eaton (1967a) and their applications. We follow here the notation of Eaton for a fairly general treatment of the ranking problem.

Consider a random observable $Z=(X,Y)$ with values in a measurable space $(\mathcal{X}\times\mathcal{Y},\ \mathcal{B}(\mathcal{X})\times\mathcal{B}(\mathcal{Y}))$ so that X has values in $\mathcal{X}$ and Y has values in $\mathcal{Y}$. The space $\mathcal{X}$ is assumed to be a symmetric Borel subset of R^k and $\mathcal{B}(\mathcal{X})$ is the Borel field inherited from R^k while $\mathcal{Y}$ is arbitrary. It is assumed that Z has a density

$$p_{\boldsymbol{\alpha}}(\mathbf{x},\mathbf{y};\boldsymbol{\theta})\,d\mu(\mathbf{x},\mathbf{y}) \tag{3.10}$$

where μ is a σ-finite measure on $\mathcal{B}(\mathcal{X})\times\mathcal{B}(\mathcal{Y})$ and $(\boldsymbol{\theta},\boldsymbol{\alpha})$ is a parameter in a set $\Theta\times A$ ($\boldsymbol{\theta}\in\Theta$, $\boldsymbol{\alpha}\in A$). The set Θ is assumed to be a symmetric Borel subset of R^k. For the ranking problem, we consider $\boldsymbol{\theta}$ as a vector of parameters we want to rank and α as a nuisance parameter. In most problems, the observation $\mathbf{Z}$ represents a sufficient statistic for the parameter $(\boldsymbol{\theta},\boldsymbol{\alpha})$ based on a sample size of n from each of k populations.

The general ranking problem of Bechhofer (1954) is to partition the set of coordinate values of the parameter $\boldsymbol{\theta}=(\theta_1,\ldots,\theta_k)$ into s disjoint (nonempty) subsets $I_1,\ldots,I_s$ containing $k_1,\ldots,k_s(k_1+\cdots+k_s=k)$ components, respectively, such that, for $\theta_i\in I_i$ and $\theta_j\in I_j$, $1\leqslant i<j\leqslant k$, we have $\theta_i<\theta_j$. Equivalently, one can partition the set $\{1,\ldots,k\}$ into s disjoint subsets, say, $\gamma_1,\ldots,\gamma_s$, where γ_i has k_i elements $1\leqslant k_i<k$, $\Sigma_{i=1}^{s}k_i=k$, and then make the obvious association between γ_i and I_i. Thus the action space for the ranking problem is the set $\Gamma=\{\gamma\}$ of all partitions $\gamma=\{\gamma_1,\ldots,\gamma_s\}$ of $\{1,\ldots,k\}$ as described previously.

Let G be the group of permutations of the set $\{1,\ldots,k\}$ and g denote any element of G. The element of G which interchanges i and j, leaving all other members of $\{1,2,\ldots,k\}$ fixed, is denoted by (i,j). For $(\mathbf{x},\mathbf{y})\in\mathcal{X}\times\mathcal{Y}$ and $g\in G$, define $g(\mathbf{x},\mathbf{y})$ by $g(\mathbf{x},\mathbf{y})\equiv(g\mathbf{x},\mathbf{y})$ where $g\mathbf{x}$ is defined by $(g\mathbf{x})_i\equiv x_{g^{-1}i}$. Similarly for $(\boldsymbol{\theta},\boldsymbol{\alpha})\in\Theta\times A$ and $g\in G$, $g(\boldsymbol{\theta},\boldsymbol{\alpha})$ is defined by $g(\boldsymbol{\theta},\boldsymbol{\alpha})\equiv(g\boldsymbol{\theta},\boldsymbol{\alpha})$ where $(g\boldsymbol{\theta})_i\equiv\theta_{g^{-1}i}$. Also, for $\gamma=\{\gamma_1,\ldots,\gamma_s\}\in\Gamma$ and $g\in G$, define $g\gamma$ by $g\gamma\equiv\{g\gamma_1,\ldots,g\gamma_s\}$ where $g\gamma_1$ is the image of γ_1 under g. For the density p and the measure μ, the following invariance is assumed:

$$p_{\boldsymbol{\alpha}}(\mathbf{x},\mathbf{y};\boldsymbol{\theta})=p_{\boldsymbol{\alpha}}(g\mathbf{x},\mathbf{y};g\boldsymbol{\theta}), \tag{3.11}$$

$$d\mu(\mathbf{x},\mathbf{y})=d\mu(g\mathbf{x},\mathbf{y}). \tag{3.12}$$

Let $\mathcal{D}$ be the class of all decision functions, that is, functions φ which are measurable vector functions on $\mathcal{X}\times\mathcal{Y}$ such that $\varphi=\{\varphi_\gamma:\gamma\in\Gamma\}$ where

$0 \leq \varphi_\gamma \leq 1$ and $\Sigma_{\gamma \in \Gamma} \varphi_\gamma \equiv 1$. For $\varphi \in \mathfrak{D}$ and $g \in G$, define $g\varphi$ by $(g\varphi)_\gamma = \varphi_{g^{-1}\gamma}$. A decision function φ is invariant if $g\varphi(\mathbf{x},\mathbf{y}) = \varphi(g\mathbf{x},\mathbf{y})$, that is, $\varphi_{g^{-1}\gamma}(\mathbf{x},\mathbf{y}) = \varphi_\gamma(g\mathbf{x},\mathbf{y})$. Let $\mathfrak{D}_I$ be the set of invariant decision functions.

We need the following definition for stating the assumptions regarding the loss function $L_\gamma(\boldsymbol{\theta},\boldsymbol{\alpha})$, which is the loss for taking action $\gamma \in \Gamma$ at the parameter point $(\boldsymbol{\theta},\boldsymbol{\alpha})$.

Definition 3.6. If $\gamma = \{\gamma_1,\ldots,\gamma_s\}$ and $\gamma' = \{\gamma'_1,\ldots,\gamma'_s\}$ are elements of Γ, then γ *differs adjacently from* γ' at $[i,j]$ if there exists an integer $\beta(1 \leq \beta < s)$ such that: (i) $i \in \gamma_\beta$, $i \in \gamma'_{\beta+1}$, (ii) $j \in \gamma'_\beta, j \in \gamma_{\beta+1}$, and (iii) $\gamma'^{(i,j)} = \gamma$.

The loss functions $L_\gamma(\boldsymbol{\theta},\boldsymbol{\alpha})$ are assumed to satisfy the following:

$$0 \leq L_\gamma(\boldsymbol{\theta},\boldsymbol{\alpha}) = L_{g\gamma}(g\boldsymbol{\theta},\boldsymbol{\alpha}), \tag{3.13}$$

$$L_\gamma(\boldsymbol{\theta},\boldsymbol{\alpha}) \leq L_{\gamma'}(\boldsymbol{\theta},\boldsymbol{\alpha}) \text{ when } \gamma \text{ differs adjacently from } \gamma' \text{ at } [i,j] \text{ and } \theta_i \geq \theta_j. \tag{3.14}$$

The assumption (3.14) is natural for the ranking problem, whereas (3.13) is suggested by the natural invariance of the problem.

We define the risk function of $\varphi \in \mathfrak{D}$ by

$$\rho(\varphi,\boldsymbol{\theta},\boldsymbol{\alpha}) = \int \sum_\gamma \varphi_\gamma(\mathbf{x},\mathbf{y}) L_\gamma(\boldsymbol{\theta},\boldsymbol{\alpha}) p_{\boldsymbol{\alpha}}(\mathbf{x},\mathbf{y};\boldsymbol{\theta})\, d\mu(\mathbf{x},\mathbf{y}). \tag{3.15}$$

Also, if F is a probability measure on $\Theta \times A$ ($\Theta \times A$ is assumed to be a measurable space with $\mathfrak{B}(\Theta)$ being the Borel sets of Θ), then the Bayes risk of $\varphi \in \mathfrak{D}$ is

$$\rho(\varphi,F) = \int \rho(\varphi,\boldsymbol{\theta},\boldsymbol{\alpha})\, dF(\boldsymbol{\theta},\boldsymbol{\alpha}). \tag{3.16}$$

For each $\gamma = \{\gamma_1,\ldots,\gamma_s\} \in \Gamma$, let

$$B_\gamma = \left\{\mathbf{x} \mid \mathbf{x} \in \mathfrak{X},\ x_{i_1} \geq \cdots \geq x_{i_s} \quad \text{for all } i_j \in \gamma_j, j = 1,\ldots,s\right\}. \tag{3.17}$$

For each $x \in \mathfrak{X}$, let $H(\mathbf{x}) = \{\gamma \mid \gamma \in \Gamma,\ x \in B_\gamma\}$ and let $n(\mathbf{x})$ be the number of elements in the set $H(\mathbf{x})$ so that $n(\mathbf{x}) \geq 1$. The decision rule φ^* is defined by

$$\varphi^*_\gamma(\mathbf{x},\mathbf{y}) = \begin{cases} 1/n(\mathbf{x}) & \text{if } \gamma \in H(\mathbf{x}), \\ 0 & \text{if } \gamma \notin H(\mathbf{x}). \end{cases} \tag{3.18}$$

Thus $\varphi^* = \{\varphi^*_\gamma \mid \gamma \in \Gamma\}$ is not a function of $\mathbf{y}$. It is easy to see that $\varphi^* \in \mathfrak{D}_I$.

The decision rule φ^* is the natural rule which ranks the vector $\boldsymbol{\theta}$ according to the ranking of the observed vector $\mathbf{X}$.

Now, let $\mathcal{P}_0$ be the class of probability measures on $(\Theta \times A, \mathcal{B}(\Theta) \times \mathcal{B}(A))$ such that $F \in \mathcal{P}_0 \Leftrightarrow F = F_1 F_2$, where F_1 and F_2 are probability measures on Θ and A, respectively, and F_1 is invariant under the group G operating on Θ.

The following results regarding the optimum properties of the decision rule φ^* were obtained by Eaton (1967a).

Theorem 3.6. Suppose $p_{\alpha}(\mathbf{x}, \mathbf{y}; \boldsymbol{\theta})$ has Property M for all $\mathbf{y} \in \mathcal{Y}$. Then

a. φ^* is Bayes for $F \in \mathcal{P}_0$, that is,

$$\rho(\varphi^*, F) = \inf_{\varphi \in \mathcal{D}} \rho(\varphi, F),$$

b. For all $(\boldsymbol{\theta}, \boldsymbol{\alpha}) \in \Theta \times A$,

$$\rho(\varphi^*, \boldsymbol{\theta}, \boldsymbol{\alpha}) \leqslant \rho(\varphi, \boldsymbol{\theta}, \boldsymbol{\alpha}) \quad \text{for all } \boldsymbol{\varphi} \in \mathcal{D}_I,$$

c. φ^* is minimax and admissible in $\mathcal{D}$.

REMARK 3.2. Part (b) of Theorem 3.6 states that the natural selection rule uniformly minimizes the risk among invariant decision rules. This is a generalization of Theorem 3.4. If one assumes the result in (b), then Part (a) can be proved easily. From (b), one obtains the minimax nature and admissibility of φ^* within the class $\mathcal{D}_I$. Since the group G is finite, Part (c) follows. □

Eaton has also extended the results of Hall (1959) on most economical decision rules, to the class of densities with Property M and applied his results to the specific cases of (i) ranking main effects in analysis of variance, (ii) ranking variances in normal populations, and (iii) ranking correlation coefficients in bivariate normal populations. It is shown that, in the case of (i), no most economical sample size can exist if σ^2 (error variance) is unknown. However, if σ^2 and ρ (correlation between any two different error terms) are known and if the main effects are sufficiently separated, the decision rule φ^* is most economical. For (ii), if a most economical sample size exists, then φ^* based on sufficient statistics is most economical in the class of all decision rules. The rule φ^* based on sample correlation coefficients is minimax within the class of all decision rules for the problem (iii). Also, if the populations are sufficiently different (in terms of their correlation coefficients), then φ^* is most economical.

Alam (1973) has also obtained some optimal properties of the natural ranking procedure. His goal is a special case of Eaton's, namely, to partition the set of coordinate values of the parameter $\boldsymbol{\theta}=(\theta_1,\ldots,\theta_k)$ into two sets I_1 and I_2 consisting of $k-m$ and m components such that, for $\theta_i \in I_1$ and $\theta_j \in I_2$, we have $\theta_i < \theta_j$. The action space now is the set Γ of all partitions $\gamma=\{\gamma_1,\gamma_2\}$ of $\{1,\ldots,k\}$ as described earlier. Let $\mathbf{X}=(X_1,\ldots,X_k)$ be a random vector whose distribution depends on $\boldsymbol{\theta}$ and $\mathfrak{D}$ be the class of all decision functions $\varphi=\{\varphi_\gamma(\mathbf{x}):\gamma\in\Gamma\}$ where $\theta \leqslant \varphi_\gamma \leqslant 1$ and $\Sigma_{\gamma\in\Gamma}\varphi_\gamma \equiv 1$. The natural ranking procedure is $\varphi^*=\{\varphi^*_\gamma(\mathbf{x}):\gamma\in\Gamma\}$ which is defined as follows: Let

$$A_\gamma=\{\mathbf{x}\colon x_i \geqslant x_j \quad \text{for all } i\in\gamma_1 \text{ and } j\in\gamma_2\}, \tag{3.19}$$

$$C(\mathbf{x})=\{\gamma\colon \gamma\in\Gamma,\ \mathbf{x}\in A_\gamma\}$$

and let $n(\mathbf{x})$ be the number of elements in $C(\mathbf{x})$. Then

$$\varphi^*_\gamma(\mathbf{x})=\begin{cases} 1/n(\mathbf{x}) & \text{if } \gamma\in C(\mathbf{x}),\\ 0 & \text{otherwise.}\end{cases} \tag{3.20}$$

We now define the Stochastically Increasing Property (SIP) of a family of distributions introduced by Lehmann [1955]. Let a partial ordering $\prec$ be defined on $\mathfrak{X}$ by the relation: $\mathbf{x}\prec\mathbf{x}'$ if and only if $x_i \leqslant x_i'$, $i=1,\ldots,k$. Similarly, $\boldsymbol{\theta}\prec\boldsymbol{\theta}'$ if and only if $\theta_i \leqslant \theta_i'$, $i=1,\ldots,k$.

Definition 3.7. A measurable subset S of the sample space $\mathfrak{X}$ is said to be *monotonically increasing* (with respect to $\prec$) if $\mathbf{x}\in S$ and $\mathbf{x}\prec\mathbf{x}'$ implies $\mathbf{x}'\in S$.

Definition 3.8. The distribution of $\mathbf{X}$ is said to have SIP in $\boldsymbol{\theta}$ if $P_{\boldsymbol{\theta}}(S) \leqslant P_{\boldsymbol{\theta}'}(S)$ for every monotone nondecreasing set S and $\boldsymbol{\theta}\prec\boldsymbol{\theta}'$, where $P_{\boldsymbol{\theta}}(S)$ denotes the probability measure of S under the conditional distribution of $\mathbf{X}$, given $\boldsymbol{\theta}$.

As before, let G denote the group of permutations of the components of a k-vector. We assume that $\mathfrak{X}$ and Ω are symmetric Borel subsets of R^k. Let Π be a given symmetric prior distribution on Ω (that is, $\Pi(A)=\Pi(gA)$ for all measurable set A and $g\in G$, where $\Pi(A)$ is the probability measure of A under Π). Let $P_{\mathbf{x}}$ denote the family of posterior distributions of $\boldsymbol{\theta}$, corresponding to $\mathbf{x}$. We assume that the loss function $L_\gamma(\boldsymbol{\theta})$ satisfies the following property: For all $\gamma\in\Gamma$, $g\in G$, and $\boldsymbol{\theta}\in\Omega$,

$$\begin{aligned} &\text{(i)}\quad L_\gamma(\boldsymbol{\theta})=L_{g\gamma}(g\boldsymbol{\theta}),\\ &\text{(ii)}\quad L_\gamma(\boldsymbol{\theta}) \text{ is nonincreasing (nondecreasing) in } \theta_i \text{ for } i\in\gamma_1(\gamma_2).\end{aligned} \tag{3.21}$$

The following theorem has been proved by Alam (1973).

Theorem 3.7. If the loss function satisfies the property (3.21) and if the family of posterior distributions $P_{\mathbf{x}}$ with respect to a prior distribution Π, is invariant with respect to G, and stochastically increasing in $\mathbf{x}$, then φ^* is a Bayes procedure with respect to Π.

If follows from Theorem 3.7 that φ^* is admissible if (i) $\Omega = R^k$ and the support of Π is Ω, (ii) $\int \xi(\mathbf{x}) dP_{\boldsymbol{\theta}}(\mathbf{x})$ is a continuous function of $\boldsymbol{\theta}$ for all bounded measurable functions ξ, and (iii) $L_\gamma(\boldsymbol{\theta})$ is bounded and continuous in $\boldsymbol{\theta}$ for each $\gamma \in \Gamma$.

REMARK 3.3. If the family of distributions P_θ has Property M of Eaton, then Theorem 3.7 is not as strong as Eaton's. A family of distributions may have SIP but not Property M (as in the case of a location parameter θ in general) or may have both (as in the case of the multinomial distribution). □

3.4 CONSISTENCY OF SINGLE-STAGE RANKING PROCEDURES

To begin, let us restate the formulation of the problem. We have k populations with distributions $F(x;\theta_i)$, $\theta_i \in \Theta$, a subset of the real line. We assume that $\mathcal{F} = \{F(x,\theta),\ \theta \in \Theta\}$ is a stochastically increasing family. For every $\theta \in \Theta$, let $F(x,\theta)$ be absolutely continuous with respect to a fixed (Lebesgue or counting) measure and assume that $F(x,\theta)$ depends on θ only through its functional form. Further, let δ: $R^2 \to R^1$ (R^1 and R^2 are the one- and two-dimensional Euclidean spaces, respectively) be a distance function satisfying the conditions: (i) $\delta(a,b) \geq 0$, (ii) $\delta(a,b) = 0$ if and only if $a = b$, (iii) $\delta(a,b) = \delta(b,a)$, and (iv) $\delta(a,b)$ is strictly increasing (decreasing) in a for fixed b when $a \geq b (a \leq b)$. For any two distributions $F(x,\theta_1)$ and $F(x,\theta_2)$ in $\mathcal{F}$, the distance between them can reasonably be measured by $\delta(\theta_1,\theta_2)$. If we have a location parameter family, we can use $\delta(a,b) = |a-b|$. On the other hand, $\delta(a,b) = |\log a/b|$ can be used for the scale parameter family. In general δ is not a metric because the triangle inequality is not assumed. However, the triangle inequality is satisfied in most applications.

For an arbitrary but preassigned $\delta^* > 0$, we define

$$\Omega(\delta^*) = \{\omega = (\theta_1, \ldots, \theta_k):\ \delta(\theta_{[k]}, \theta_{[k-1]}) \geq \delta^*\}. \tag{3.22}$$

We are interested in a procedure R for selecting the population associated with $\theta_{[k]}$. Our probability requirement is

$$P_\omega(CS|R) \geq P^* \quad \text{for } \omega \in \Omega(\delta^*). \tag{3.23}$$

For fixed n, let $\{X_j\}$ and $\{X_{ij}\}$, $j=1,\ldots,n$, be independent random variables with distributions $F(x,\theta)$ and $F(x,\theta_i)$, $i=1,\ldots,k$, respectively. Let $T=T^{(n)}$ be a real-valued statistic and

$$t=t^{(n)}=T(X_1,\ldots,X_n),$$

$$t_i=t_i^{(n)}=T(X_{i1},\ldots,X_{in}), \qquad i=1,\ldots,k. \tag{3.24}$$

Let $G_n(y,\theta)$ and $g_n(y,\theta)$ denote the cdf and the density of t.

Definition 3.9. T is *consistent* with respect to $(\mathcal{F},\delta)$ if, for every $\delta^*>0$ and every $P^*\in(k^{-1},1)$, there exists an $N=N(\mathcal{F},\delta^*,P^*)$ such that

$$\inf_{\omega\in\Omega(\delta^*)} P_\omega(CS|R)\geqslant P^* \tag{3.25}$$

for every $n>N$.

Equivalently, we can state that

$$\lim_{n\to\infty}\ \inf_{\omega\in\Omega(\delta^*)} P_\omega(CS|R)=1. \tag{3.26}$$

Usually, the ranking statistic T is chosen to be a consistent estimator of θ. However, in general, a consistent estimator of θ is not always consistent for the ranking and selection problem. Further, in the case of certain families of distributions, the consistency of T depends on the distance function δ.

(i) If $\mathcal{F}$ is the Poisson family with parameter θ, then the means procedure is not consistent with respect to $\delta(\theta_{[k-1]},\theta_{[k]})=\theta_{[k]}-\theta_{[k-1]}$ or $\delta(\theta_{[k-1]},\theta_{[k]})=\log(\theta_{[k]}/\theta_{[k-1]})$ [see Sobel (1963)].

(ii) If $\mathcal{F}$ is the Cauchy family with location parameter θ, then the means procedure is not consistent but the procedure based on the medians is.

(iii) If $\mathcal{F}$ is the family of uniform distributions on $[0,\theta]$ for $\theta\in(0,\infty)$, then the means procedure is consistent with respect to $\delta(\theta_{[k-1]},\theta_{[k]})=\log(\theta_{[k]}/\theta_{[k-1]})$, but not with respect to $\delta(\theta_{[k-1]},\theta_{[k]})=\theta_{[k]}-\theta_{[k-1]}$.

These have been discussed by Tong (1972), who has investigated the conditions for the consistency of a ranking procedure. He has shown that the consistency of a ranking statistic T does not depend upon the number of populations involved; in other words, T is consistent with respect to $(\mathcal{F},\delta)$ for any k if and only if T is consistent with respect to $(\mathcal{F},\delta)$ for $k=2$. Thus we can restrict ourselves to the case $k=2$ with no loss of generality.

Tong also establishes the relationship between the consistency of ranking procedures and the uniform consistency of hypothesis-testing procedures. Consider a two-sample problem of testing H_1: $\theta_1 \leqslant \theta_2$ against H_2: $\theta_1 > \theta_2$ where the test ϕ depends on $\{X_{ij}\}, j=1,\ldots,n;\ i=1,2$, only through (t_1,t_2), and

$$\phi = \phi(t_1,t_2) = \begin{cases} 1 & \text{if } t_1 \leqslant t_2, \\ 0 & \text{otherwise,} \end{cases} \tag{3.27}$$

where H_1 is accepted if and only if $\phi=1$. Let

$$\Omega^{(1)} = \{(\theta_1,\theta_2)\colon \theta_1 < \theta_2, \delta(\theta_1,\theta_2) \geqslant \delta^*\},$$

$$\Omega^{(2)} = \{(\theta_1,\theta_2)\colon \theta_1 > \theta_2, \delta(\theta_1,\theta_2) \geqslant \delta^*\}. \tag{3.28}$$

Definition 3.10. The test ϕ is said to be *uniformly consistent* on $\Omega^{(1)} \cup \Omega^{(2)}$ if

$$\lim_{n\to\infty} \inf_{\omega\in\Omega^{(1)}} E_\omega \phi = 1 \quad \text{and} \quad \lim_{n\to\infty} \sup_{\omega\in\Omega^{(2)}} E_\omega \phi = 0.$$

Theorem 3.8. T is consistent with respect to $(\mathcal{F},\delta)$ if and only if the test ϕ defined in (3.27) is uniformly consistent on $\Omega^{(1)} \cup \Omega^{(2)}$ for every $\delta^* > 0$.

In many applications, it is not easy in general to justify whether a certain test is uniformly consistent even if it is known that a uniformly consistent test exists. Suppose there exists a group $\mathcal{G}$ of transformations on $\mathcal{X} = \{(t_1,t_2)\}$ and let $\bar{\mathcal{G}}$ be the induced group of transformations on the product parameter space. If the test ϕ defined in (3.28) is invariant and if the distance function δ is a maximal invariant with respect to $\bar{\mathcal{G}}$, then it can be shown that T is consistent with respect to $(\mathcal{F},\delta)$ if and only if (1) the test ϕ is constant, and (2) the power of the test ϕ is monotonically increasing (decreasing) in $\delta(\theta_1,\theta_2)$ for $(\theta_1,\theta_2) \in \Omega^{(1)}(\Omega^{(2)})$. It follows that ϕ is uniformly consistent on $\Omega^{(1)} \cup \Omega^{(2)}$.

The following theorems of Tong (1972) state sufficient conditions for the consistency of ranking procedures in terms of convergence in distributions uniformly in θ.

Theorem 3.9. Let T be a consistent estimator of θ. Assume that δ is a metric, that is, in addition to conditions (i)–(iv) on δ, the triangle inequality is also satisfied. If $\delta(t^{(n)},\theta)$ converges to 0 in probability uniformly in θ as $n\to\infty$, then T is consistent with respect to $(\mathcal{F},\delta)$.

Theorem 3.10. Let $\tau_n(\theta)=E_\theta t^{(n)}$, $\sigma_n^2(\theta)=E_\theta[t^{(n)}-\tau_n(\theta)]^2$, and

$$H_n(x,\theta)=\Pr\left[\frac{t^{(n)}-\tau_n(\theta)}{\sigma_n(\theta)}\leqslant x\right]\to\Phi(x)$$

uniformly in θ. Then T is consistent with respect to $(\mathcal{F},\delta)$ if and only if the absolute value of

$$c_n(\omega)=\frac{[\tau_n(\theta_2)-\tau_n(\theta_1)]}{[\sigma_n^2(\theta_1)+\sigma_n^2(\theta_2)]^{1/2}}$$

approaches infinity uniformly in ω for $\omega\in\Omega(\delta)$. In particular, if $t^{(n)}$ is an unbiased estimator of θ and $\delta(\theta_1,\theta_2)=|\theta_1-\theta_2|$, the preceding condition reduces to $\sigma_n^2(\theta)\to 0$ uniformly in θ.

The preceding results of Tong (1972) have been applied by him to demonstrate the consistency of some commonly used ranking procedures. The means procedure (under which the ranking statistic is the sample mean) has been used for ranking problems relating to normal (Bechhofer, 1954), binomial (Sobel and Huyett, 1957), Poisson (Sobel, 1963), and gamma (Gupta, 1963) populations. Lehmann (1963) has discussed means procedure with respect to location and scale parameter families. The ranking procedure based on the maximum likelihood estimator is consistent when $\delta(\theta_1,\theta_2)=|\theta_1-\theta_2|$, provided the conditions of Parzen [1954] are satisfied. These conditions ensure the uniform convergence of the estimator in probability to θ. The ranking procedure based on certain linear combination of order statistics is also shown to be consistent for a location parameter family. Some investigation has been done in the case of exponential family.

3.5 ASYMPTOTICALLY OPTIMAL PROCEDURES

Dalal and Hall (1977) have discussed asymptotically most economical procedures for location parameters. Let X_{ij}, $j=1,\ldots,n$, be n independent observations from π_i which has the associated cdf $F(x-\theta_i)$, $i=1,\ldots,k$. Let us consider the goal of selecting the population associated with the largest

θ_i. Let $\Omega(\Delta)=\{\boldsymbol{\theta}: \theta_{[k]}-\theta_{[k-1]} \geqslant \Delta\}$, where $\Delta>0$. The problem is formulated as follows: Find a sequence of selection procedures (one for each sample size) and a sample size formula $N=N(\Delta)$ such that

$$\varliminf_{\Delta\downarrow 0} \inf_{\substack{F\in\mathfrak{F}\\ \boldsymbol{\theta}\in\Omega(\Delta)}} PCS_{N(\Delta)}(F,\boldsymbol{\theta}) \geqslant P^*. \tag{3.29}$$

Definition 3.11. A procedure satisfying (3.29), for some suitable $\mathfrak{F}$ containing Φ (standard normal), is said to be *asymptotically robust* at $\Phi\in\mathfrak{F}$.

Definition 3.12. If every other procedure satisfying (3.29), but with $N=N'(\Delta)$, has the property $\varliminf_{\Delta\downarrow 0}[N'(\Delta)/N(\Delta)]\geqslant 1$, then the procedure is said to be *asymptotically most economical* for $\mathfrak{F}$.

The approach here is to choose a suitable family $\mathcal{F}$ of possible error distributions and determine N (minimally) so as to meet the PCS goal for every $F\in\mathcal{F}$ and in particular, for a least favorable $F^0\in\mathcal{F}$. Dalal and Hall (1977) considered $\mathcal{F}=\{F|F=(1-\gamma)\Phi+\gamma H\}$, where H is an arbitrary distribution function symmetric at the origin, and $0\leqslant\gamma<1$. In other words, we have a contaminated neighborhood of the ideal model Φ. Let $c=c(\gamma)$ be implicitly defined by $(1-\gamma)^{-1}=1-2\Phi(-c)+2\varphi(c)/c$. Dalal and Hall (1977) have shown that a procedure that selects the population that yields the largest Huber's M-estimate (Huber [1964]), with sample size chosen to meet the PCS requirement, meets the goal for every $F\in\mathcal{F}^0$, an equicontinuous-at-c subset of $\mathcal{F}$, and the appropriate asymptotic sample size N is $d^2\sigma_0^2/\Delta^2$, where $\sigma_0^2=1/[1-2(1-\gamma)\varphi(c)/c]$ and d is the solution of $\int_{-\infty}^{\infty}\Phi^{k-1}(x+d)\,d\Phi(x)=P^*$. Moreover, this N is minimal. This means that the procedure is (asymptotically) a most-economical robust selection procedure.

Asymptotically optimal (minimax) ranking and selection procedures have been investigated by Bawa (1970a) using the method proposed by Weiss and Wolfowitz [1969] in developing asymptotically minimax tests of composite hypotheses. The problem is solved for an arbitrary, but fixed, location of the ranking parameter. If the ranking parameter is a location or scale parameter admitting a sufficient statistic, then the results hold irrespective of the specified location of the parameters. If a least favorable location is not known, then the procedure gives an asymptotically optimal identification procedure, for any arbitrary, but fixed, location of ranking parameter. Bawa (1970a) has also considered the goal of selecting a fixed-size subset, a formulation discussed in Chapter 9.

3.6 NOTES AND REMARKS

Alam and Wallenius (1977) have considered the problem of optimal selection rule in a specified class for selecting the best population. Miescke (1979a) has shown that the optimum properties of selection procedures based on total orderings in the sample space are closely related to those of corresponding test procedures. In selection procedures, problems arise regarding optimum sampling. Such problems are discussed in Chapter 5. Optimal properties of subset selection procedures are discussed in Chapter 18.

CHAPTER 4

Ranking and Selection Problems for Discrete Distributions

Discrete distributions serve as appropriate models in many practical problems especially in biomedical and social sciences. Perhaps the most widely used discrete distributions are the binomial and Poisson distributions. Clinical trials with any drug give rise to dichotomous data, the response of each patient being a success or a failure. Thus the problem of comparing two or more drugs (treatments) is a problem of comparing two or more binomial populations in terms of their single-trial success probabilities. One can reasonably use a Poisson model to describe accident data. Some other well-known discrete distributions used as models are negative binomial, multinomial, and Fisher's logarithmic series distributions. Multinomial distribution is an appropriate model in pollen profile studies. Counts of the numbers of the various kinds of fossil pollen grains found at different depths in accumulated sediment form a fossil profile. The characteristics of this profile reflect the past changes in vegetation and climate that occurred in the area of sediment deposition. The sample counts of pollen at a given depth for each of k types of pollen are distributed according to a multinomial distribution. An interesting example of negative binomial distribution arises in market research studies. Ehrenberg [1959] observed that the negative binomial distribution gives a good fit to the data relating to the number of consumers in the sample who bought r units ($r=0,1,2,\ldots$) of a nondurable consumer product. The logarithmic series distribution is found to be a good model for rainfall distributions. Selection procedures have been investigated for all these distributions. From the preceding examples, it is easy to see how the problem of comparing different populations arise in each of these contexts.

In this chapter, we discuss the selection and ranking problems under the indifference zone approach for discrete populations. Although parametric and nonparametric sequential procedures are discussed in a later chapter, we have included here some of the sequential procedures for binomial and Poisson distributions because of the comparative studies made with regard to many competing procedures.

4.1 SEVERAL BINOMIAL DISTRIBUTIONS: FIXED SAMPLE SIZE PROCEDURE

Let $\pi_1,\ldots,\pi_k$ be k independent binomial populations with single-trial success probabilities p_i, $i=1,\ldots,k$, respectively. The goal is to select the population associated with $p_{[k]}$. In case of a tie, selection of any population with $p_i=p_{[k]}$ is a correct selection. The basic probability requirement is to be satisfied whenever $\mathbf{p}=(p_1,\ldots,p_k)\in\Omega_{\Delta^*}$, where $\Omega(\Delta^*)=\{\mathbf{p}\colon p_{[k]}-p_{[k-1]}\geqslant \Delta^*\}$ and $\Delta^*>0$ is preassigned. Let n be the number of independent observations from each population and let S_i be the number of successes in n observations from π_i, $i=1,\ldots,k$.

Sobel and Huyett (1957) proposed a rule (denoted here by R_1) based on the numbers of successes, namely,

R_1: Select π_i associated with the largest S_i, using randomization to break a tie. (4.1)

For this rule, we have

$$\begin{aligned} P(CS|R_1)=&\Pr(S_{(i)}<S_{(k)}, \quad i=1,2,\ldots,k-1)\\ &+\frac{1}{2}\sum_{\alpha=1}^{k-1}\Pr(S_{(\alpha)}=S_{(k)} \text{ and } S_{(i)}<S_{(k)}, \quad i=1,2,\ldots,k-1, i\neq\alpha)\\ &+\cdots+\frac{1}{k}\Pr(S_{(1)}=\cdots=S_{(k)}), \end{aligned} \tag{4.2}$$

where $S_{(i)}$ is the S_i associated with $p_{[i]}$, $i=1,\ldots,k$. The GLFC for this rule is shown to be

$$p_{[1]}=\cdots=p_{[k-1]}=p_{[k]}-\Delta^*. \tag{4.3}$$

The sample size required is the smallest n for which $P(CS|R_1)$ evaluated at the LFC is $\geqslant P^*$. The sample sizes (exact and approximate) are tabulated by Sobel and Huyett (1957) for $k=2,3,4,10$, $\Delta^*=0.05(0.05)0.50$, and $P^*=0.50,0.60,0.75,0.80,0.85,0.90,0.95,0.99$.

When n is large, the normal approximation to the PCS yields

$$n\approx\frac{c^2(1-\Delta^{*2})}{4\Delta^{*2}}, \tag{4.4}$$

where $c=c(k,P^*)$ is the constant satisfying

$$\int_{-\infty}^{\infty}\Phi^{k-1}(x+c)\,d\Phi(x)=P^*.$$

Thus the tables of Bechhofer (1954), Gupta [1963a], Milton [1963], and Gupta, Nagel, and Panchapakesan [1973] are relevant here. When k is large, the value of n required to meet any specification of Δ^* and P^* is shown to be approximately equal to a constant multiple of $\log k$.

Sobel and Huyett (1957) have suggested an alternative approximation to the PCS. Starr and Woodroofe (1974) have pointed out that the classical approach of letting the sample size tend to infinity and appealing to the central limit theorem yields the correct limiting PCS only if it is tacitly assumed that $\Delta^* \to 0$ at a suitable rate. However, the classical approximations used by Sobel and Huyett (1957) are found to be excellent for the choices of P^* which are of the greatest practical interest.

Sobel and Huyett (1957) also consider a modification of the specification of parametric region where the P^*-condition is to be met. The modified region is $\Omega(p^*_{[k-1]}, p^*_{[k]}) = \{\mathbf{p}: p_{[k-1]} \leqslant p^*_{[k-1]} \text{ and } p_{[k]} \geqslant p^*_{[k]}\}$. Such a specification is meaningful when we have some prior information of this type. The LFC in this case is given by

$$p_{[1]} = \cdots = p_{[k-1]} = p^*_{[k-1]}; \quad p_{[k]} = p^*_{[k]}. \tag{4.5}$$

In this case, the sample sizes are tabulated by Sobel and Huyett for $k=2,4$, $P^*=0.50, 0.60, 0.75, 0.80, 0.85, 0.90, 0.95, 0.99$, and $(p^*_{[k-1]}, p^*_{[k]}) = (0.60, 0.75), (0.80, 0.95), (0.80, 0.90), (0.80, 0.85), (0.90, 0.95)$.

REMARK 4.1. The sampling rule adopted in the procedure R_1 of §4.1 can be described as Vector-at-a-Time (VT) sampling which is terminated when the total number of observations is $N = kn$. This sampling rule and another one are explained in §4.3 which deals with several procedures for selecting the better of the two binomial populations. The importance of the problem lies in its applications to clinical trials where the experimenter is interested in deciding on the better drug, the efficiency of a drug being measured by p, the probability of a success when tried on a patient. Since all these procedures are strongly motivated by the practical problem, we often describe these procedures in terms of the clinical trials. □

4.2 SEVERAL BINOMIAL POPULATIONS: A MULTISTAGE PROCEDURE

Taylor and David (1962) have discussed a multistage procedure for the selection of the best of several binomial populations. The purpose of the multistage sampling is to allow more observations to be made on the populations with larger success probabilities. More descriptively, we can state the problem in terms of k treatments, $\pi_1, \ldots, \pi_k$, with probabilities of a success (in treating a patient) p_i, $i = 1, \ldots, k$, respectively. It is assumed that

1. Each patient has a response to the treatment that is dichotomuns: success or failure.
2. The time between application of the treatment and observation of response is small compared to the duration of the trial.

3. The probability that a given patient responds favorably to treatment π_i is constant throughout the trial.
4. All responses are independent.

The essential feature of the procedure of Taylor and David (1962) is to define a weighting function that will operate on the results accumulated up to any particular stage to determine the proportion of observations to be taken from each population for that stage. Let $n_{ij}(i=1,\ldots,k;\ j=1,\ldots,m)$ be the number of patients allocated to π_i for the jth period (stage) and let r_{ij} be the corresponding number of successes. Let N be the total number of patients to be allocated among the k treatments at each stage. Define

$$p_{ij}=\frac{\sum_{\alpha=1}^{j} r_{i\alpha}}{\sum_{\alpha=1}^{j} n_{i\alpha}}, \qquad i=1,\ldots,k;j=1,\ldots,m. \tag{4.6}$$

Now, several functions providing nonnegative weights for each treatment and period and summing to unity for each period, have been considered by Taylor and David.

a. The *rank weight* for π_i and jth period is defined by

$$p_{ij}=\frac{\left[k+1-R_{i,j-1}\right]}{\frac{1}{2}k(k+1)}, \qquad i=1,\ldots,k;j=1,\ldots,m, \tag{4.6a}$$

where R_{ij} is the rank of p_{ij} among p_{1j} to p_{kj} (the largest p_{ij} gets rank 1), and $R_{i,0}$ is defined to be $\frac{1}{2}(k+1)$ so that $p_{i1}=k^{-1}$. The allocation is determined by $n_{ij}=Np_{ij}$, $i=1,\ldots,k$. It should be noted that the upper and lower bounds on the number of patients that can be assigned to a particular treatment π_i are independent of p_i.

b. The *first power* weight for π_i and the jth period is defined by

$$w_{ij}=(1+d_{i,j-1})k^{-1}, \quad i=1,\ldots,k;j=1,\ldots,m; \tag{4.6b}$$

where $\bar{p}_{.j}=\sum_{i=1}^{k}p_{ij}/k$, $d_{ij}=p_{ij}-\bar{p}_{.j}$, and d_{i0} is defined to be zero. It is clear that the weights (4.6b) are not very different unless the treatment parameters are far apart.

c. *The rth power* weights are defined by

$$w_{ij}^{(r)}=w_{ij}^{r}\Big/\sum_{i=1}^{k}w_{ij}^{r}, \quad i=1,\ldots,k;j=1,\ldots,m;\ r\geqslant 0, \tag{4.6c}$$

where w_{ij} is defined in (4.6c). In particular, $w_{ij}^{(1)} = w_{ij}$. The case $r=0$ corresponds to equal allocation. For $r=\infty$, all patients are assigned to the leading treatment for each period following the first. If ties for the first place occur, the patients for that period are divided equally among the tied treatments.

The selection procedure R' of Taylor and David (1962) selects at the end of the experiment the treatment that yielded the highest proportion of successes. The procedure of Sobel and Huyett (1957) which we have discussed in §4.1, corresponds to the special case of one period ($m=1$) or equal allocation. For $m>1$, no result is available about the LFC. Monte Carlo results utilizing five somewhat arbitrary weighting functions (rank; first, second, fourth, and eighth powers) indicate a higher probability of a correct selection than in the case of equal sample experiments.

4.3 TWO BINOMIAL POPULATIONS: PLAY-THE-WINNER AND VECTOR-AT-A-TIME SAMPLING

The statement of a procedure for a selection problem has three parts, namely, (i) the sampling rule, (ii) the termination rule, and (iii) the decision rule. The procedures discussed in this section differ in this respect from one another. The main thrust of these investigations is to compare the performance of different sampling rules.

Play-the-Winner (PW) sampling rule was first suggested by Robbins [1956]. For comparing two treatments, the PW rule determines the treatment to be used for any trial on the basis of the previous trial and its result; a success generates another trial on the same treatment and a failure generates a switch to the other treatment. In particular, it is assumed that the result of each treatment is observed without delay.

Vector-at-a-Time (VT) sampling rule implies an equal number of observations on each treatment. We have a vector of observations as one stage.

Throughout this section we assume without any loss of generality that $p_1 \leqslant p_2$. Our goal is to select the treatment π_2 subject to the P^*-condition whenever $p_2 - p_1 \geqslant \Delta^*$, $\Delta^* > 0$. Let S_i denote the number of successes observed on treatment π_i, $i=1,2$, at any stage of the sampling.

The problem of finding a sequential rule for taking observations from two Bernoulli distributions to optimize a chosen criterion such as the expected number of or proportion of successes, is known as the two-armed bandit problem. Contributions to this problem have been made by several authors among whom are Berry [1972, 1978], Bradt, Johnson, and Karlin [1956], Fabius and van Zwet [1970], Feldman [1962], Robbins [1952, 1956], Samuels [1968], Smith and Pyke [1965], and Vogel [1960].

4.3.1 Rules with Termination Based on $|S_1 - S_2|$.

Sobel and Weiss (1970a) investigated two procedures, one using PW sampling and the other VT sampling. Both use the same rule to terminate sampling, namely, stop sampling when $|S_1 - S_2| = r$, a preassigned positive integer. Let us denote these rules by $R_{D,\mathrm{PW}}$ and $R_{D,\mathrm{VT}}$, respectively, where D indicates that the termination rule is based on the absolute difference $D = |S_1 - S_2|$. The decision rules for $R_{D,\mathrm{PW}}$ and $R_{D,\mathrm{VT}}$ are same, namely,

$$\text{Select } \pi_i \text{ if and only if } S_i = \max(S_1, S_2). \tag{4.7}$$

To derive the PCS for $R_{D,\mathrm{PW}}$, we define

$$\begin{aligned} P_n &= \Pr(\pi_2 \text{ is selected} \mid S_2 - S_1 = n,\ NT = 2), \\ Q_n &= \Pr(\pi_2 \text{ is selected} \mid S_2 - S_1 = n,\ NT = 1), \end{aligned} \tag{4.8}$$

where $NT = i$ denotes the fact that the next observation is taken on treatment π_i, $i = 1, 2$. The first treatment at the outset is chosen at random. The PCS at termination is given by

$$P(CS \mid R_{D,\mathrm{PW}}) = \frac{(P_0 + Q_0)}{2}. \tag{4.9}$$

Using the PW sampling rule and (4.8), and letting $q_i = 1 - p_i$, $i = 1, 2$, we have

$$\begin{aligned} P_n &= p_2 P_{n+1} + q_2 Q_n, \\ Q_n &= p_1 Q_{n-1} + q_1 P_n. \end{aligned} \tag{4.10}$$

The boundary conditions are

$$P_r = 1, \quad Q_{-r} = 0. \tag{4.11}$$

The integer r needed to satisfy the P^*-condition is the smallest integer $\geqslant$ the root in x of

$$\lambda^x = (4q_2P^*)^{-1}\left[(q_1+q_2) - \left\{(q_1+q_2)^2 - 16q_1q_2P^*(1-P^*)\right\}^{1/2}\right], \tag{4.12}$$

where $\lambda = p_1/p_2 < 1$. In the LFC, $p_2 - p_1 = \Delta^*$ and we minimize the PCS (or maximize the solution of (4.12) in x) as a function of p_2 for $\Delta^* \leqslant p_2 \leqslant 1$. This minimum occurs at a value close to 1. The corresponding maximum of r, denoted by r_m, is approximately given by

$$r_m \approx \left[\frac{\log\{2(1-P^*)\}}{\log(1-\Delta^*)} \right] + 1 \tag{4.13}$$

where $[t]$ denotes the largest integer $\leqslant t$.

For the procedure $R_{D,\mathrm{VT}}$, let

$$P_n' = \Pr(\pi_2 \text{ is selected} \mid S_2 - S_1 = n). \tag{4.14}$$

Then P_n' satisfies the recurrence relation

$$P_n' = p_2 q_1 P_{n+1}' + q_2 p_1 P_{n-1}' + (p_1 p_2 + q_1 q_2) P_n' \tag{4.15}$$

with the boundary conditions,

$$P_r' = 1,\ P_{-r}' = 0. \tag{4.16}$$

Then, corresponding to (4.12), the integer needed to satisfy the P^*-condition [see Sobel and Weiss (1972b)] is the smallest integer $\geqslant$ the root in x of

$$P^*(1+\delta^x) = 1, \tag{4.17}$$

where $\delta = p_1 q_2 / p_2 q_1 < 1$. Under the LFC, we set $p_1 = (1-\Delta^*)/2$ and $p_2 = (1+\Delta^*)/2$ so that x is given by

$$x = 2^{-1} \frac{\log[(1-P^*)/P^*]}{\log[(1-\Delta^*)/(1+\Delta^*)]}. \tag{4.18}$$

To compare the two sampling rules for the same pair (Δ^*, P^*) and the same termination rule, two different criteria are employed by Sobel and Weiss (1970a). Let N_1 and N_2 be the number of patients put under the poorer and the better treatment, respectively. The first criterion is the expected loss or risk $E(L)$ (which we denote here by $\bar{L}$) defined by

$$\bar{L} = (p_2 - p_1) E(N_1). \tag{4.19}$$

This criterion represents the difference in the expected number of successes between a conceptual set of trials in which the better treatment is always used and the actual set of trials; thus we have $\bar{L}_{D,\mathrm{PW}}$ and $\bar{L}_{D,\mathrm{VT}}$ corresponding to the two sampling schemes.

The second criterion is the expected total number of trials needed for termination which is denoted by $E(N)$ and is given by

$$E(N) = E(N_1) + E(N_2). \tag{4.20}$$

This is the standard criterion used by Bechhofer, Kiefer, and Sobel (1968) for comparing procedures that satisfy the same (Δ^*, P^*) requirement. For convenience, let r and s denote the r-values for the two procedures $R_{D,\mathrm{PW}}$ and $R_{D,\mathrm{VT}}$, respectively, so that $r = r_{\mathrm{PW}}$ and $s = r_{\mathrm{VT}}$. Then

$$\bar{L}_{D,\mathrm{PW}} = \frac{(p_2 + 2q_2 r)(1-\lambda^r)(q_1 - q_2\lambda^r)}{2(q_1 - q_2\lambda^{2r})},$$

$$E(N|R_{D,\mathrm{PW}}) = (1-\lambda^r)(q_1 - q_2\lambda^r)(\bar{p} + 2r\bar{q})\left[p_2(1-\lambda)(q_1 - q_2\lambda^{2r})\right]^{-1},$$

$$\bar{L}_{D,\mathrm{VT}} = s(1-\delta^s)(1+\delta^s)^{-1},$$

$$E(N|R_{D,\mathrm{VT}}) = 2s(1-\delta^s)(p_2 - p_1)^{-1}(1+\delta^s)^{-1}, \tag{4.21}$$

where $\bar{p} = (p_1 + p_2)/2$ and $\bar{q} = 1 - \bar{p} = (q_1 + q_2)/2$.

The asymptotic $(P^* \to 1)$ analysis of Sobel and Weiss (1970a) shows that $R_{D,\mathrm{PW}}$ is superior to $R_{D,\mathrm{VT}}$ if

$$p_2 > \frac{3}{4} - \frac{\Delta^*}{8} + O(\Delta^{*3}) \tag{4.22}$$

in terms of the criterion $\bar{L}$, and if

$$\frac{(p_1 + p_2)}{2} > \frac{3}{4} - \frac{\Delta^*}{8} + O(\Delta^{*3}) \tag{4.23}$$

in terms of the criterion $E(N)$. The procedure $R_{D,\mathrm{VT}}$ is superior to $R_{D,\mathrm{PW}}$ if the inequalities are reversed in (4.22) and (4.23). These results are corroborated by the tables of Sobel and Weiss (1972b) who have tabulated the exact values of $\bar{L}$ and $E(N)$ for the configuration in which $p_1 = p_2 - \Delta^*$ and $\Delta^* \leqslant p_2 \leqslant 1$ for some selected values of (Δ^*, P^*).

Hoel, Sobel, and Weiss (1972) proposed a two-stage procedure, in which the first stage is devoted to finding a rough estimate of p_2 and the second to carry out the remainder of the trial with the sampling scheme judged to be better. Let us call this procedure $R_{D,E}$ (D indicates that termination depends upon the value of D, and E denotes the fact that the sampling rule is either PW or VT depending on the first stage). The procedure $R_{D,E}$ can be described as follows.

First take M observations on each treatment. Let m and n denote the numbers of successes on the two treatments. The type of sampling used in the second stage depends on a critical parameter k_0 such that when

$$\max(m,n) \geqslant k_0 \tag{4.24}$$

the experimenter uses PW sampling for the second stage, whereas in the contrary case he uses VT sampling. We assume that termination can take place at any time during the second stage or at the end of the first stage. Let S_1 and S_2 denote the number of successes observed on treatments 1 and 2, respectively, at any stage of the experiment. Then the termination rules depending on the sampling method are:

$$\text{VT: terminate when } |S_1 - S_2| = s(m,n,k_0,M),$$

$$\text{PW: terminate when } |S_1 - S_2| = r(m,n,k_0,M).$$

The decision rule of $R_{D,E}$ is: Select the treatment associated with the larger S_i.

Detailed numerical results of Hoel, Sobel, and Weiss (1972) indicate that a relatively small amount of preliminary sampling leads to an efficient experiment as measured by the loss due to the necessity of conducting an experiment. Of course, an experimenter can design the experiment more efficiently if he has prior knowledge of bounds on p_2.

The procedure $R_{D,\text{PW}}$ yields very large expected total sample size when both p_1 and p_2 are small and never terminates when $p_1 = p_2 = 0$. To overcome this, Fushimi (1973a) proposed a truncated version of this rule, which is denoted here by $R'_{D,\text{PW}}$. For this procedure, sampling is continued until either $|S_1 - S_2| = r$ or $F_1 + F_2 = s$, where F_i is the number of failures observed on treatment π_i, $i = 1,2$, and r and s are preassigned positive integers. At termination of sampling, $R'_{D,\text{PW}}$ chooses the treatment with larger number of successes, while randomizing in case of a tie. Of course, one could modify the rule so as to choose the treatment with fewer failures in case of a tie in the number of successes. This would increase the PCS slightly when s is odd. However, the change has no influence on the asymptotic theory.

Similar to (4.8) in the case of $R_{D,\text{PW}}$, we now define

$$\begin{aligned} P_{mn} &= \Pr(\pi_2 \text{ is selected } | F_1 + F_2 = s - m,\ S_2 - S_1 = n,\ NT = 2), \\ Q_{mn} &= \Pr(\pi_2 \text{ is selected } | F_1 + F_2 = s - m,\ S_2 - S_1 = n,\ NT = 1), \end{aligned} \tag{4.25}$$

where, as before, $NT = i$ indicates that the next observation is taken on treatment π_i, $i=1,2$. These probabilities satisfy

$$\begin{aligned} P_{mn} &= p_2 P_{m,n+1} + q_2 Q_{m-1,n}, \\ Q_{mn} &= p_1 Q_{m,n-1} + q_1 P_{m-1,n}, \end{aligned} \tag{4.26}$$

with the boundary conditions

$$P_{mr} = Q_{mr} = 1(m=1,\ldots,s),\ P_{m,-r} = Q_{m,-r} = 0(m=1,\ldots,s),$$

$$P_{0n} = Q_{0n} = \begin{cases} 1, & n=1,\ldots,r-1, \\ \dfrac{1}{2}, & n=0 \\ 0, & n=-1,\ldots,-r+1. \end{cases}$$

Fushimi's computations suggest that one may be able to obtain a good approximation to the value of r which guarantees the probability requirement by letting $s\to\infty$ and, similarly, to the value of s by letting $r\to\infty$. If $s=\infty$, we note that this procedure reduces to $R_{D,\mathrm{PW}}$ of Sobel and Weiss (1970a). If $r=\infty$, then the procedure reduces to an inverse sampling procedure similar to $R_{I,\mathrm{PW}}$ of Sobel and Weiss (1971a), discussed in §4.3.2. Thus r_m given by (4.13) is an approximate lower bound for r of the procedure $R'_{D,\mathrm{PW}}$. If $\bar{r}=\bar{r}(\Delta^*,P^*)$ denotes the smallest integer r for which the P^*-condition is met, the table values of Fushimi show that r_m given by (4.13) is a good approximation to $\bar{r}$.

For the case of $r\to\infty$, let us assume for simplicity that s is even and $s=2c$. As pointed out earlier, in this case $R'_{D,\mathrm{PW}}$ reduces to a procedure similar to $R_{I,\mathrm{PW}}$ discussed in another section. For this procedure we have $c \geqslant c_0(P^*,\Delta^*) \approx \frac{8}{27}(\lambda/\Delta^*)^2$, where $\lambda=\lambda(P^*)$ denotes the 100λ percentile of the standard normal distribution.

4.3.2 Rules with Termination Based on Inverse Sampling

For the problem discussed previously, Sobel and Weiss (1971a) investigated two procedures, one using PW sampling and the other VT sampling, but both based on an inverse sampling termination rule. We shall denote these by $R_{I,\mathrm{PW}}$ and $R_{I,\mathrm{VT}}$, respectively. To be specific, the sampling is terminated when any one of the treatments yields r successes. The decision rule in either case is: Select π_i if and only if $S_i=r$.

For any pair (p_1,p_2) and any r, the PCS is the same for both $R_{I,\mathrm{PW}}$ and $R_{I,\mathrm{VT}}$, and is given by

$$\begin{aligned} P(CS|R_{I,\mathrm{PW}}) &= P(CS|R_{I,\mathrm{VT}}) \\ &= \tfrac{1}{2} E_r\{ I_{q_1}(X,r) + I_{q_1}(X+1,r) \}, \end{aligned} \tag{4.27}$$

where $I_p(x,y)$ is the usual incomplete beta function, $I_q(0,r)=1=1-I_p(r,0)$

for $r \geqslant 1$, and $E_r(X)$ denotes the expectation of the random variable X which has the discrete negative binomial probability law

$$f(x) = p_2^r \binom{x+r-1}{x} q_2^x, \; x = 0, 1, \ldots. \tag{4.28}$$

However, the two procedures do not have the same expression for $E(N)$ or $E(N_1)$. The exact expressions are given by Sobel and Weiss (1971a). Disregarding an error of order Δ^{*2}, the LFC is given by $p_1 = \frac{2}{3} - \Delta^*/2$ and $p_2 = \frac{2}{3} + \Delta^*/2$, and a lower bound to the PCS using either rule is given by $\Phi(\Delta^* \sqrt{27r/8}\,)$. Thus the appropriate r value is given by

$$r = 8\lambda^2 [27\Delta^{*2}]^{-1}, \tag{4.29}$$

where $\Phi(\lambda) = P^*$.

For large values of r,

$$E(N_1 | R_{I,\text{PW}}) < E(N_1 | R_{I,\text{VT}}),$$
$$E(N | R_{I,\text{PW}}) < E(N | R_{I,\text{VT}}); \tag{4.30}$$

that is, the procedure $R_{I,\text{PW}}$ is uniformly better than $R_{I,\text{VT}}$. As we noted earlier, such a result is not true for $R_{D,\text{PW}}$ and $R_{D,\text{VT}}$. It is also found that for small Δ^*, the procedures $R_{D,\text{PW}}$ and $R_{D,\text{VT}}$ are preferable to $R_{I,\text{PW}}$ and $R_{I,\text{VT}}$, whereas the reverse is true for $\Delta = p_2 - p_1$ small or zero, when Δ^* is fixed and P^* is sufficiently close to unity. Thus there is no result in terms of $E(N)$ and $\bar{L}$ that is uniform in both Δ^* and P^*.

It is obvious that the inverse sampling procedures $R_{I,\text{PW}}$ and $R_{I,\text{VT}}$ of Sobel and Weiss (1971a) have the problem of no termination when $p_1 = p_2 = 0$. Hoel (1972) proposed a modification in $R_{I,\text{PW}}$. His procedure (denoted by $R'_{I,\text{PW}}$) retains PW sampling rule and the inverse sampling nature of the termination rule. However, the termination now depends on $T_1 = S_1 + F_2$ and $T_2 = S_2 + F_1$, where $S_i(F_i)$ denotes the number of successes (failures) on treatment π_i, $i = 1, 2$. The sampling is continued until either T_1 or T_2 is equal to r. Consequently, the decision rule is: Select π_i if and only if $T_i = r$.

This procedure is obviously truncated; that is, at most $2r - 1$ observations are needed. The PCS is given by

$$P(CS | R'_{I,\text{PW}}) = \begin{cases} (2q_1)^{-1}\big[q_1 E_{r-1}\{ I_{q_1}(X+1, r-X)\} + E_r\{ I_{q_1}(X+1, r-X)\} \\ \quad - p_1 E_r\{ I_{q_1}(X+1, r-X-1)\}\big], & q_1 \neq 0, \\ p_2^r/2, & q_1 = 0, \end{cases} \tag{4.31}$$

where $I_p(b,c)$ is the incomplete beta function and $E_r(X)$ denotes the expectation of a binomial random variable X with parameters r and q_2. The LFC is given by $p_1=(1-\Delta^*)/2$ and $p_2=(1+\Delta^*)/2$. Under the LFC, (4.31) gives

$$P(CS|R'_{I,\text{PW}})=E_{r-1}\{I_{q_1}(X+1,r-X)\}. \tag{4.32}$$

Computations made for the purpose of comparing $R_{I,\text{PW}}$ and $R'_{I,\text{PW}}$ show that the expected sample sizes are larger for $R_{I,\text{PW}}$ when $\bar{p}=(p_1+p_2)/2\leqslant 0.5$. Neither procedure is uniformly better; however, $R'_{I,\text{PW}}$ is truncated and it makes substantial savings in sample sizes for low values of p_1 and p_2.

Another procedure (denoted by $R''_{I,\text{PW}}$) has been investigated by Berry and Sobel (1973). This procedure modifies $R_{I,\text{PW}}$ by truncating sampling whenever the PW scheme has gone through c cycles (i.e., whenever c failures have been obtained on both treatments). Under $R''_{I,\text{PW}}$, therefore, sampling terminates whenever either $\max(S_1,S_2)=r$ or $F_1=F_2=c$. In either case, the population with the larger number of successes is selected, using randomization to break a tie.

Many pairs (r,c) used in procedure $R''_{I,\text{PW}}$ satisfy the P^*-condition; one obvious such pair has the r used in $R_{I,\text{PW}}$ and $c=\infty$. The PCS is given by $P+P'$ where P is the probability of selecting π_2 before c cycles and P' is the probability of selecting π_2 based on the number of successes in exactly c cycles. For $i=1,2$, let $F_i(r)$ denote the number of failures on π_i before the rth success on π_i and let $S_i(c)$ denote the number of successes on π_i before the cth failure on π_i. Then

$$P(CS|R''_{I,\text{PW}})=P+P', \tag{4.33}$$

where

$$P=\Pr(F_2(r)<F_1(r),F_2(r)<c)+\tfrac{1}{2}\Pr(F_2(r)=F_1(r)<c)$$

and

$$P'=\Pr(S_1(c)<S_2(c)<r)+\tfrac{1}{2}\Pr(S_1(c)=S_2(c)<r).$$

Exact expressions for $E(N_1)$ and $E(N)$ have been obtained by Berry and Sobel (1973). It is shown that asymptotically $(r=c\to\infty)r=c=r_0$, where r_0 is given by (4.29), makes $R''_{I,\text{PW}}$ uniformly better than $R_{I,\text{PW}}$ in terms of $E(N)$ as well as $E(N_1)$. In the exact case, a slightly larger common value of r and c than r_0 is required under $R''_{I,\text{PW}}$ to make it satisfy the P^*-condition, thus making $E(N)$ and $E(N_1)$ slightly larger for some values of (p_1+p_2) under $R''_{I,\text{PW}}$. Also, $R''_{I,\text{PW}}$ is not uniformly better than $R_{I,\text{PW}}$ in the exact case.

4.3.3 Rules with Fixed Total Sample Size

We now discuss some procedures using PW and VT sampling with total sample size N fixed. For VT sampling, N is even. This case for any $k \geqslant 2$ (where N is a multiple of k) has been discussed in §4.1. Nebenzahl and Sobel (1972) have investigated an alternative procedure based on PW sampling where the sampling is terminated with N observations. We denote this procedure by $R_{F,\mathrm{PW}}$, where F denotes fixed total sample observations. In this section we relabel the procedure R_1 of §4.1 as $R_{F,\mathrm{VT}}$. For the procedure $R_{F,\mathrm{PW}}$, we start sampling from one of the two populations at random and continue sampling (switching the population with every failure) until N observations. The decision rule of $R_{F,\mathrm{PW}}$ is: Select the populations π_i associated with $\max(S_1, S_2)$, using randomization to break a tie. It is shown by Nebenzahl and Sobel (1972) that

$$P(CS|R_{F,\mathrm{PW}}) = P(CS|R_{F,\mathrm{VT}}) \tag{4.34}$$

and

$$E(N_1|R_{F,\mathrm{PW}}) \leqslant E(N_1|R_{F,\mathrm{VT}}). \tag{4.35}$$

They have also proposed a slight modification in the sampling rule of $R_{F,\mathrm{VT}}$ so that N can be odd or even. We select a population at random at the outset and then switch after every observation. The sampling rule of $R_{F,\mathrm{PW}}$ is already defined for odd N. The decision rules are now slightly modified to handle a tie in the number of successes when N is odd. A quantity θ is introduced representing the probability of selecting the population with fewer failures in case of a tie in the number of successes at termination. For an even N such a tie also implies a tie in the number of failures and there is no need to define θ. The modified procedures are denoted by $R_{F,\mathrm{PW}}(\theta)$ and $R_{F,\mathrm{VT}}(\theta)$. It should be noted that $R_{F,\mathrm{PW}}(\theta)$ is not, in general, the same as the original $R_{F,\mathrm{PW}}$ for odd N. Of course, for $\theta = \frac{1}{2}$, the procedure $R_{F,\mathrm{PW}}(\frac{1}{2})$ is same as $R_{F,\mathrm{PW}}$ for all N. It is shown that

$$P(CS|R_{F,\mathrm{PW}}(\theta)) = P(CS|R_{F,\mathrm{VT}}(\theta)) \tag{4.36}$$

if and only if $\theta = 1$. Also, for all N, and all θ,

$$P(CS|R_{F,\mathrm{PW}}(\theta)) \geqslant P(CS|R_{F,\mathrm{VT}}(\theta)). \tag{4.37}$$

The equality holds in (4.37) for all θ when N is even. Intuitively, one would want to take $\theta = 1$ for a slightly better procedure of either type, and in this case (4.36) holds for all N.

Pradhan and Sathe (1973) have derived exact expressions for the probabilities of various events connected with the selection of the better population. Their main result shows that it is possible to curtail the sampling in the case of $R_{F,\mathrm{PW}}$ and $R_{F,\mathrm{VT}}$. Let $N=2n$. For the PW rule, when the ith treatment is observed first, sampling can be curtailed after r trials if, for $i,j=1,2(i\neq j)$,

$$S_{i,r}-S_{j,r}=2n-r,\ r=n,\ldots,2n-1, \quad \text{or}$$

$$S_{j,r}-S_{i,r}=2n-r+1,\ r=n+1,\ldots,2n, \tag{4.38}$$

and decision $S_{i,2n}>S_{j,2n}$ or $S_{j,2n}>S_{i,2n}$, respectively, can be taken, where $S_{i,r}$ is the number of successes for the ith treatment in r trials.

For the VT rule the sampling can be curtailed after observing r pairs, if

$$S_{i,r}+F_{j,r}=n+1 \quad \text{or} \quad n+2(i,j=1,2,\ i\neq j) \tag{4.39}$$

and decision $S_i>S_j$ in n pairs can be taken; otherwise n pairs are obtained.

4.3.4 A Truncated Procedure with VT Sampling

The following procedure (denoted by $R_{DT,\mathrm{VT}}$) proposed by Kiefer and Weiss (1971) is based on VT sampling and can be considered as a truncated version of $R_{D,\mathrm{VT}}$ of §4.2.1. However, the type of decision made is slightly different. For this procedure $R_{DT,\mathrm{VT}}$, a decision is made at or before test N (where each test is assumed to contain both treatments). The sampling is terminated at test $j<N$ if $|S_1-S_2|=s$; otherwise, at test N. The decision rule of $R_{DT,\mathrm{VT}}$ is: If the sampling is terminated at $j\leqslant N$ with $|S_1-S_2|=s$, choose π_i associated with $\max(S_1,S_2)$. Otherwise, conclude that both treatments are equal. The PCS is given by

$$P(CS|R_{DT,\mathrm{VT}})=\theta^s(1+\theta^s)^{-1}-s^{-1}\theta^{(s+1)/2}\sum_{r=1}^{2s-1}(-1)^{r+1}\lambda_r^N\psi(r,\theta), \tag{4.40}$$

where

$$\theta=\frac{p_2q_1}{p_1q_2},$$

$$\lambda_r=p_1p_2+q_1q_2+2\sqrt{p_1p_2q_1q_2}\,\cos\left(\frac{\pi r}{2s}\right), \tag{4.41}$$

$$\psi(r,\theta)=\frac{\sin(\pi r/2)\sin(\pi r/2s)}{\left[1+\theta-2\sqrt{\theta}\,\cos(\pi r/2s)\right]}.$$

Let $W_s(N)$ denote the probability that no population is designated as better after N pairs of trials. Then

$$W_s(N) = s^{-1}\theta^{(1-s)/2} \sum_{r=1}^{2s-1} \left[1+(-1)^{r+1}\theta^s\right]\lambda_r^N \psi(r,\theta). \tag{4.42}$$

An upper bound for T = expected number of tests before termination, is given by

$$T \leqslant s^{-1}\theta^{(1-s)/2} \sum_{r=1}^{2s-1} (1-\lambda_r^N)(1-\lambda_r)^{-1}\left[1+(-1)^{r+1}\theta^s\right]\psi(r,\theta). \tag{4.43}$$

Now, three possibilities arise:

a. Fix N and determine s so that the PCS is a maximum.

b. Choose the smallest N (depending on s) consistent with the P^*-condition when $p_2 - p_1 \geqslant \Delta^*$.

c. Require that $W_s(N) \geqslant P_1^*$ when $p_1 = p_2$, in addition to meeting the P^*-condition when $p_2 - p_1 \geqslant \Delta^*$.

Kiefer and Weiss (1971) have discussed the first two possibilities. For Case (a), the PCS is minimized when $p_1 = (1-\Delta^*)/2$ and $p_2 = (1+\Delta^*)/2$ and the numerical calculations indicate that the minimum is unique. Also, for $p = 0.5$ and a given P^*, $W_s(N)$ decreases as Δ^* decreases. For Case (b), detailed numerical calculations show that the expected trial length is not appreciably shorter than that for unrestricted testing. The only advantage seems to be the option of declaring the two populations equal. As $N \to \infty$, the PCS for the truncated trials is always less than the PCS for the untruncated trials. This result is not proved when N is fixed.

Taheri and Young (1977) compared the PW and VT sampling for selecting the better of two binomial populations taking the ratio of the success probabilities to define the "distance" between the populations. The probability requirement is to be met whenever $p_{[2]}/p_{[1]} \geqslant \Delta^* > 1$. They considered termination rules based on the difference in the number of successes as well as on inverse sampling. The PW sampling is found to be better than the VT sampling under both termination rules. When Δ^* is close to 1, the rule based on the difference in the number of successes is better than the inverse sampling in the sense that the expected number of trials on the poorer treatments is less.

A survey of adaptive sampling for clinical trials is given by Hoel, Sobel, and Weiss (1975). Simon, Weiss, and Hoel (1975b) have discussed sequential likelihood ratio procedures for choosing the better of two competing dichotomous treatments, or to decide that they are equal. Their procedures are described in §4.5.

4.4 THREE OR MORE BINOMIAL DISTRIBUTIONS: INVERSE STOPPING RULES WITH PLAY-THE-WINNER AND VECTOR-AT-A-TIME SAMPLING

The procedures discussed in this section are generalizations of $R_{I,\text{PW}}$ and $R_{I,\text{VT}}$ of §4.2.2. These have been proposed and studied by Sobel and Weiss (1972a). The PW sampling in this case is a cyclic variation of the earlier one. This scheme (denoted by PWC) puts the populations in a random order at the outset, say, $\pi_1,\ldots,\pi_k$. Population π_j is sampled until a failure is observed and then π_{j+1} is sampled ($j=1,\ldots,k$): π_{k+1} is identified with π_1. Sampling terminates as soon as any one population has r successes and that population is selected as the best. We assume that p_k is the largest p_i. We determine r so that the P^*-condition is satisfied whenever $p_k - \max_{1\leqslant j<k-1} p_j \geqslant \Delta^* > 0$. The PCS is given by

$$P(CS|R_{I,\text{PWC}}) = \frac{1}{k} E_r\left\{ \prod_{j=1}^{k-1} I_{q_j}(X,r) \right\}$$

$$+ \sum_{\alpha=1}^{k-1} \left[\prod_{j=1}^{\alpha-1} I_{q_j}(X,r) \right]\left[\prod_{j=\alpha}^{k-1} I_{q_j}(X+1,r) \right], \tag{4.44}$$

where the random variable X has the negative binomial distributions with index $r>0$, success parameter p_k, and mean rq_k/p_k.

For $R_{I,\text{VT}}$, the decision rule selects the population yielding the largest number of successes, using randomization to break ties. The PCS is minimized by setting $p_i = p_{k-1}$, $i=1,\ldots,k-2$. This minimum denoted by $P_1(CS|R_{I,\text{VT}})$ is given by

$$P_1(CS|R_{I,\text{VT}}) = \frac{1}{k} E_r \left\{ \frac{I^k_{q_{k-1}}(X,r) - I^k_{q_{k-1}}(X+1,r)}{I_{q_{k-1}}(X,r) - I_{q_{k-1}}(X+1,r)} \right\}. \tag{4.45}$$

The right-hand side of (4.45) is also the minimum of $P(CS|R_{I,\text{PWC}})$ for $p_i \leqslant p_{k-1}$, $i=1,\ldots,k-2$. Thus the LFCs are the same and $r_{\text{PWC}} = r_{\text{VT}}$. This value of r needed by both is again approximately given by (4.29) for small Δ^* except that now $\lambda = \lambda(p^*,\rho,k)$ is the value of H that satisfies

$$\int_{-\infty}^{\infty} \Phi^{k-1}(x+\sqrt{2}\,H)\,d\Phi(x) = P^*. \tag{4.46}$$

The criteria for comparing $R_{I,\mathrm{VT}}$ and $R_{I,\mathrm{PWC}}$ will be $E(N)$ and $E(L)$, which is now given by

$$E(L)=\sum_{i=1}^{k-1}(p_k-p_i)E(N_i), \tag{4.47}$$

where N_i is the number of observations from π_i. It is shown by Sobel and Weiss (1972a) that asymptotically ($\Delta^*\to 0$, $\Delta^* r^{1/2}\to$constant>0) the PWC is uniformly better than VT with regard to both $E(N)$ and $E(L)$.

One can define a variation, $\hat{R}_{I,\mathrm{PWC}}$, of $R_{I,\mathrm{PWC}}$ by reordering the k populations according to the number of successes (using randomization for ties) after every complete cycle, consisting of k failures (one from each population). It is shown that this does not alter the PCS and thus the same value of r will be required for $\hat{R}_{I,\mathrm{PWC}}$. However, $E(N)$ and $E(L)$ are reduced, but for any k, large r and fixed $p_k>0$, the saving is small.

Another variation, $R^*_{I,\mathrm{PWC}}$, discussed by Sobel and Weiss (1972a) can be considered as a dual of $R_{I,\mathrm{PWC}}$. It is defined by waiting until every population has exactly r failures under PWC sampling. The population yielding the larger (or largest) number of successes is declared the best. It is shown asymptotically to be an improvement on $R_{I,\mathrm{PWC}}$ when $p_k<1/2$.

Eckardt (1976) considered a two-stage procedure modification of the rule $R_{I,\mathrm{PWC}}$ of Sobel and Weiss (1972a) referred to previously. The first stage is used for screening inferior populations. It consists of first taking a random order of the populations and use PW sampling which is terminated when one of the treatments yields r successes or m failures. Let s_1, s_2, and s_3 be the numbers of successes at the end of the first stage. Then the treatment i is eliminated if $s_i \leqslant c \max_{1\leqslant j\leqslant 3} s_j$, $0<c<1$. If there remain more than one population under consideration, we go on to the second stage by still using PW sampling which is terminated when any treatment yields r successes (including the first stage). At termination, the treatment that yielded r successes is selected. This modified procedure is neither uniformly superior nor uniformly inferior compared to other procedures.

4.5 SEQUENTIAL PROCEDURES BASED ON LIKELIHOOD RATIOS

Our main interest here is to consider the procedures for selecting from two or more binomial populations. We first describe some sequential likelihood ratio procedures employing PW and VT sampling schemes and results relating to their comparisons. We then discuss some sequential procedures again based on likelihood ratios for choosing the better of two competing dichotomous treatments, or deciding that they are equal.

4.5.1 Procedures with PW and VT Sampling

Sobel and Weiss (1972b) have discussed procedures based on likelihood using PWC sampling rule. These procedures, denoted here by $R_{L,\mathrm{PWC}}$ and $R_{\mathrm{LC,PWC}}$, have been discussed also by Hoel and Sobel (1972) while comparing several sequential procedures using PW and VT sampling rules. To write down these rules, let us consider the case $k=3$ and let $S_1 \leqslant S_2 \leqslant S_3$ denote the current (ordered) number of successes from the three populations. Also, let F_i denote the current number of failures from the population associated with S_i, $i=1,2,3$. If $S_3=S_2(>S_1)$, we associate S_3 with the smaller of the two corresponding F-values; similarly for $S_1=S_2=S_3$, we assign S_3 to the one with the smallest F-value. If it is still not determined, then we use randomization. It is shown by Hoel and Sobel (1972) that the rule never terminates when randomization is used for $P^* \geqslant 1/2$. The stopping rule for $R_{L,\mathrm{PWC}}$ is: Stop sampling as soon as $F_3 \leqslant \min(F_1, F_2)$ and

$$\sup_{\Delta^* < p < 1} \left\{ \left(1-\frac{\Delta^*}{p}\right)^{T_1} \left(\frac{1-p}{1-p+\Delta^*}\right)^{U_1} + \left(1-\frac{\Delta^*}{p}\right)^{T_2} \left(\frac{1-p}{1-p+\Delta^*}\right)^{U_2} \right\} \leqslant \frac{1-P^*}{P^*}, \tag{4.48}$$

where $T_i = S_3 - S_i$ and $U_i = F_i - F_3$, $i=1,2$.

A modification of $R_{L,\mathrm{PWC}}$ is $R_{\mathrm{LC,PWC}}$ (likelihood conservative) which replaces $(1-p)/(1-p+\Delta^*)$ in (4.48) by its upper bound 1. This procedure stops if $F_3 \leqslant \min(F_1, F_2)$ and

$$(1-\Delta^*)^{T_1} + (1-\Delta^*)^{T_2} \leqslant \frac{(1-P^*)}{P^*}. \tag{4.49}$$

Both procedures select the population associated with S_3.

Monte Carlo results for $k=3$ show that $R_{L,\mathrm{PWC}}$ and $R_{\mathrm{LC,PWC}}$ give similar results with $R_{L,\mathrm{PWC}}$ showing a reduction of about 20 percent in $E(N)$ over $R_{\mathrm{LC,PWC}}$.

Bechhofer, Kiefer, and Sobel (1968) have investigated a procedure based on likelihood using a VT sampling rule. In the case of this procedure (denoted here by $R_{L,\mathrm{VT}}$), the sampling is terminated as soon as

$$\left(\frac{1-\Delta^*}{1+\Delta^*}\right)^{2T_1} + \left(\frac{1-\Delta^*}{1+\Delta^*}\right)^{2T_2} \leqslant \frac{(1-P^*)}{P^*} \tag{4.50}$$

and the population associated with S_3 is selected as the best. For $P^* > \frac{1}{2}$, we will not stop when $S_3 = S_2$ and hence, randomization will not be required at termination. Numerical comparisons of $R_{L,\mathrm{VT}}$ and $R_{L,\mathrm{PWC}}$ show that PWC sampling rule is better for $p_{[3]} > 0.75$ and the VT sampling rule is better for $p_{[3]} < 0.75$ (approximately).

For $k > 2$, an elimination procedure has been proposed by Hoel and Sobel (1972). This procedure, denoted by $R_{E,\text{PWC}}$, is an extension of $R_{D,\text{PW}}$ of §4.2.1 defined for $k=2$. In this procedure, population π_j is eliminated if for some π_i (not yet eliminated) $S_i - S_j = r$. The P^*-condition is met by setting $r =$ the smallest integer greater than or equal to r_k, where

$$r_k = \frac{\log\left[2(1-P^*)(k-1)^{-1}\right]}{\log(1-\Delta^*)}. \tag{4.51}$$

Based on general results of another paper of his (1971a), Hoel (1971b) derived an elimination procedure using VT sampling. This procedure $R_{E,\text{VT}}$ eliminates population π_j if for some π_i (not yet eliminated)

$$S_i - S_j \geqslant c + dn^*, \tag{4.52}$$

where $c > 0$ and $d \leqslant 0$ are predetermined constants and n^* is the number of unlike pairs from π_i and π_j (that is, observations in the same vector of the form S,F or F,S).

Define τ_0 by

$$\tau_0 = \left[\frac{(1-\Delta^*)}{(1+\Delta^*)}\right]^2 < 1, \tag{4.53}$$

and let τ_1 denote any value such that

$$\tau_0 < \tau_1 \leqslant \tau_0^{-1} \tag{4.54}$$

It is shown by Hoel (1971b) that the P^*-condition is satisfied by setting

$$c = \frac{2\log\left(\dfrac{k-1}{1-P^*}\right)}{\log(\tau_1/\tau_0)}, \qquad d = \frac{2\log\left(\dfrac{1+\tau_1}{1+\tau_0}\right)}{\log(\tau_1/\tau_0)} - 1. \tag{4.55}$$

Asymptotically ($P^* \to 1$) for the configuration $p_{[1]} = \cdots = p_{[k-1]} = p_{[k]} - \Delta^*$, the risk $E(L)$ in (4.47) is minimized when $\tau_1 = \tau_0^{-1}$ (this implies $d = 0$).

Hoel and Sobel (1972) have compared the procedures $R_{I,\text{PW}}$, $R'_{I,\text{PW}}$, $R''_{I,\text{PW}}$, $\hat{R}_{I,\text{PWC}}$ and $R_{L,\text{PWC}}$ in the case of $k=2$ populations. Their table values show the following:

(i) For $\bar{p} = (p_1+p_2)/2 > \frac{1}{2}$, $R_{L,\text{PW}}$ is preferable using either of the two criteria, $E(L)$ and $E(N)$.

(ii) For $\bar{p}<\frac{1}{2}$, the value of $E(N)$ becomes infinite for $R_{L,\text{PWC}}$, $R_{I,\text{PW}}$, and $\hat{R}_{I,\text{PWC}}$ when $p_1=p_2$.

(iii) Procedures $R'_{I,\text{PW}}$ and $R''_{I,\text{PW}}$ have a bounded $E(N)$ function even for $p_1=p_2$ and the numerical improvement for small $\bar{p}$ (especially for $p_1=p_2$) is significant.

(iv) Procedure $\hat{R}_{I,\text{PWC}}$ shows only a small improvement over $R_{I,\text{PW}}$; however, it is uniform over the entire parameter space.

Hoel and Sobel (1972) have also made Monte Carlo comparisons of several procedures using PW as well as VT sampling in the case of $k=3$ populations. The PW sampling procedures included in the comparisons are $R_{I,\text{PWC}}$, $R_{L,\text{PWC}}$, $R_{\text{LC,PWC}}$, and $R_{E,\text{PWC}}$, whereas VT sampling procedures considered are $R_{L,\text{VT}}$, $R_{E,\text{VT}}$, and $R_{P,\text{VT}}$ (which is discussed in §4.6).

As a group, the PW sampling procedures are different from the group of VT sampling procedures. The risk and $E(N)$ change very little in value over varying values of $p_{[3]}$ in the case of the VT sampling procedures, whereas they appear to be monotonically decreasing with $p_{[3]}$ for the other group. The cross-over point is about 0.65 or 0.75 according as the criterion is the risk $E(L)$ or $E(N)$.

Among the PW sampling procedures, $R_{E,\text{PWC}}$ and $R_{L,\text{PWC}}$ are quite similar and uniformly better than both $R_{I,\text{PWC}}$ and $R_{\text{LC,PWC}}$. However, $R_{L,\text{PWC}}$ appears to be slightly better than $R_{E,\text{PWC}}$.

For the VT sampling procedures, $R_{E,\text{VT}}$ is preferable to both $R_{L,\text{VT}}$ and $R_{P,\text{VT}}$ using either $E(L)$ or $E(N)$ as the criterion. The latter two procedures perform about the same.

Sobel and Weiss (1972) have surveyed many of the procedures discussed in this and the preceding sections of this chapter. They also refer to a number of related problems under investigation. Some of them use other variations of the PW, VT, and allied sampling schemes.

4.5.2 Two Binomial Populations Case with Two Formulations

Simon, Weiss, and Hoel (1975) have discussed sequential likelihood ratio procedures for choosing the better of two competing dichotomous treatments, or deciding that they are equal. They have considered different assignments including an adaptive assignment. Their procedures are discrete analogs of designs proposed by Flehinger and Louis [1971, 1972] and Hoel and Weiss (1974).

Let p_i be the success probability for the treatment π_i, $i=1,2$. Simon, Weiss, and Hoel have considered two formulations; the first choosing between the two hypotheses H_1: $p_1-p_2\geqslant\Delta^*$ and H_2: $p_2-p_1\geqslant\Delta^*$, where Δ^* is the present discrimination level, and the second allowing a third hypothesis, H_0: $p_1=p_2$.

For the two-decision problem it is required that the PCS $\geqslant P^*$ whenever $|p_1 - p_2| \geqslant \Delta^*$. At any point in the trial, let S_i and F_i denote the numbers of successes and failures, respectively, for π_i. Define

$$L(a,b) = a^{S_1} b^{S_2} (1-a)^{F_1} (1-b)^{F_2},$$

$$\mathfrak{L}_1 = \max_{\Delta^* < p \leqslant 1} L(p, p-\Delta^*), \tag{4.56}$$

$$\mathfrak{L}_2 = \max_{\Delta^* < p \leqslant 1} L(p-\Delta^*, p).$$

Termination of the procedure is based on $\Lambda = \mathfrak{L}_1 / \mathfrak{L}_2$. Let N denote the number of tests made at any stage. The sampling continues as long as $N < N^*$, a predetermined number, and $c^{-1} \leqslant \Lambda \leqslant c$, where c is a stopping parameter determined by simulation. For any $N < N^*$, we choose π_1 if $\Lambda > c$. If $N = N^*$ without a decision, then the treatment with the larger success ratio is chosen using randomization to break a tie.

The stopping rule described previously can be used with any method for assigning treatments to successively arriving patients. Simon, Weiss, and Hoel (1975) have considered two different assignments. The first is an alternating assignment with a random start. The second is an adaptive assignment based on $\theta = \Lambda/(1+\Lambda)$. Treatment is allocated by assigning probabilities θ and $1-\theta$ to π_1 and π_2, respectively. The stopping rule constrains θ to satisfy $(k+1)^{-1} \leqslant \theta \leqslant k(k+1)^{-1}$.

For the three-decision problem, it is required that the PCS $\geqslant P^*$ whenever $|p_1 - p_2| \geqslant \Delta^*$ and that Pr(accepting H_0) $\geqslant P_1^*$ whenever $p_1 = p_2$. Termination is now based on

$$\Gamma = \frac{\max_{0 \leqslant p \leqslant 1} p^{S_1+S_2}(1-p)^{F_1+F_2}}{\max(\mathfrak{L}_1, \mathfrak{L}_2)}.$$

Sampling is continued as long as $N < N^*$ and $c_0 \leqslant \Lambda \leqslant c_1$, where c_0 and c_1 are the stopping parameters determined by simulation. If $N < N^*$ and $\Gamma \geqslant c_1$ then the two treatments are declared equal. If $\Gamma \leqslant c_0$, the trial is stopped and the treatment with the greater value of $\mathfrak{L}$ is selected as the better one. For this problem, Simon, Weiss, and Hoel (1975) consider three assignment methods: alternating assignment, random assignment in which a patient is assigned to either treatment with probability $\frac{1}{2}$, and adaptive assignment. The adaptive sampling is such that if $\mathfrak{L}_1 > \mathfrak{L}_2$ the next arriving patient is given π_1 with probability equal to $\min\{b, (2+\Gamma)/(2+2\Gamma)\}$, where b is a parameter allowing the degree of adaptiveness to be limited. If $\mathfrak{L}_1 < \mathfrak{L}_2$, this probability applies to π_2.

Simon, Weiss, and Hoel have compared their design with several other designs in the literature using the expected number of patients given the poorer treatment and the expected total number of patients used in the trial. Their results based on simulation show their design to be substantially better or equal to other designs except in the case of both treatments having high success probabilities.

4.6 ELIMINATING-TYPE SEQUENTIAL PROCEDURES

Paulson (1967) proposed and investigated an eliminating-type sequential procedure for selecting the best of k binomial populations under two different specifications of the preference zone considered by Sobel and Huyett (1957). The first one is to satisfy the P^*-condition whenever $p_{[k]} - p_{[k-1]} \geqslant \Delta^*$ and the other is to satisfy the condition whenever $p_{[k]} \geqslant p^*_{[k-1]} + \Delta^*$ and $p_{[k-1]} \leqslant p^*_{[k-1]}$. Paulson's sampling scheme is based on ad hoc independent Poisson random variables that have no intuitive relation to the problem.

Let $\{N_{ir}\}$, $i=1,\ldots,k$; $r=1,2,\ldots$, be a double sequence of independent random variables each having a Poisson distribution with mean M, where M is a positive integer fixed in advance of the experiment. Let S_{ir} and F_{ir} denote the numbers of successes and failures, respectively, when N_{ir} observations are taken from π_i. If $N_{ir}=0$, then $S_{ir}=F_{ir}=0$. For each i and r, it is easy to see that the (unconditional) joint probability distribution of S_{ir} and F_{ir} is that of two Poisson random variables with means Mp_i and $M(1-p_i)$, respectively. The pair (S_{ir}, F_{ir}) is independent of (S_{jt}, F_{jt}) unless $i=j$ and $r=t$.

Let us consider the first specification of the indifference zone, which is the usual specification. For any λ, $1<\lambda<(1+\Delta^*)(1-\Delta^*)^{-1}$, define

$$\alpha = (1-P^*)(k-1)^{-1},$$

$$A(\lambda) = M\left[\Delta^*(\lambda^2-1) - (\lambda-1)^2\right]\left[\lambda \log \lambda\right]^{-1}, \tag{4.57}$$

$$N(\lambda) = \text{largest integer} < \left[-\log \alpha\right]\left[A(\lambda)\log\lambda\right]^{-1}.$$

The restriction on the choice of λ makes $A(\lambda)>0$. We can assume that the parameters were so chosen that $N(\lambda)>0$. Paulson's sequential procedure denoted by R_P (called $R_{P,\mathrm{VT}}$ in the context of §4.5) can be described as follows.

We observe the values $N_{11},\ldots,N_{k1}$ of k independent Poisson random variables each with mean M. We take N_{i1} observations from π_i, $i=1,\ldots,k$.

After the first stage of the experiment, we reject any population π_i for which

$$S_{i1} - F_{i1} \leqslant \max_{1 \leqslant j \leqslant k} \left[S_{j1} - F_{j1} \right] + \frac{\log \alpha}{\log \lambda} + A(\lambda). \tag{4.58}$$

Proceeding by induction, suppose at the start of the rth stage there are $T(r)$ populations, $\pi_{r_1}, \ldots, \pi_{r_{T(r)}}$, left after the $(r-1)$st stage is completed. We observe the values of $T(r)$ additional Poisson random variables, namely, $N_{r_1 r}, N_{r_2 r}, \ldots, N_{r_{T(r)} r}$ and take $N_{r_i r}$ observations from π_{r_i}, $i = 1, \ldots, T(r)$. Reject any remaining population π_i for which

$$\sum_{\beta=1}^{r} (S_{i\beta} - F_{i\beta}) \leqslant \max \left[\sum_{\beta=1}^{r} (S_{j\beta} - F_{j\beta}) + \frac{\log \alpha}{\log \lambda} + rA(\lambda) \right], \tag{4.59}$$

where the maximum is taken over the $T(r)$ populations. Terminate as soon as only one population is left. Since $N(\lambda)$ is such that $[\log \alpha / \log \lambda] + rA(\lambda) < 0$ for $r \leqslant N(\lambda)$, the populations can never all be eliminated before the $[N(\lambda)+1]$st stage. If more than one population are left after $N(\lambda)$th stage, terminate at $[N(\lambda)+1]$st stage by selecting the population for which $\Sigma_{\beta=1}^{N(\lambda)+1}(S_{i\beta} - F_{i\beta})$ is a maximum. Randomization is used to break ties. An optimum choice of λ is not determined. Based on numerical calculations, Paulson suggested $\lambda = [1 + 0.75\Delta^*][1 - 0.75\Delta^*]^{-1}$.

The specification of the region of the parameter space in the preceding formulation seems satisfactory when nothing is known about the magnitudes of $p_1, \ldots, p_k$ or if there is some a priori information available indicating that $p_{[k]}$ and $p_{[k-1]}$ do not differ too much from 0.5. The alternative formulation seems reasonable when the a priori information indicates that $p_{[k]}$ and $p_{[k-1]}$ differ substantially from 0.5. This formulation considered first by Sobel and Huyett (1957) requires the P^*-condition to be satisfied whenever $p_{[k-1]} \leqslant p^*_{[k-1]}$ and $p_{[k]} \geqslant p^*_{[k-1]} + \Delta^*$. Paulson (1967) considered the formulation in a little more general form. The P^*-condition is to be met whenever $q_{[k]}/q_{[k-1]} \leqslant c_1 (0 < c_1 < 1)$ or $p_{[k]}/p_{[k-1]} \geqslant c_2 > 1$, depending on whether a priori information indicates that $p_{[k]}$ is large or small. Here $q_{[i]} = 1 - p_{[i]}$, $i = 1, \ldots, k$. By setting

$$c_1 = \frac{\left[1 - p^*_{[k-1]} - \Delta^*\right]}{\left[1 - p^*_{[k-1]}\right]} \quad \text{and} \quad c_2 = \frac{(p^*_{[k-1]} + \Delta^*)}{p^*_{[k-1]}}$$

it can be verified that either case includes the Sobel–Huyett formulation.

When $p_{[2]}$ is known to be large, restrict a_1 to the interval $c_1 < a_1 < 1$ and obtain a family of sequential procedures, one for each choice of a_1 in this interval, such that the basic probability requirement is satisfied for each procedure in the family. To obtain this family of procedures, the selection of M and the general sampling procedure is the same as for the main formulation. However, instead of (4.59), we now eliminate at the rth stage ($r=1,2,\ldots$) any population π_i remaining after the $(r-1)$st stage for which

$$\sum_{\beta=1}^{r} F_{i\beta} \geqslant \left\{ \min_j \left(\sum_{\beta=1}^{r} F_{j\beta} \right) \log a_1 + \log \alpha \right\} \left\{ \log \left[\frac{c_1}{(1+c_1-a_1)} \right] \right\}^{-1} \tag{4.60}$$

where the minimum is taken over all populations not eliminated in the first $(r-1)$ stages. The experiment is terminated as soon as only one population is left, in which case that population is selected as the best. However, if at any stage (4.60) is satisfied for all the remaining populations, we terminate the experiment and select one of these populations with minimum cumulative sum as the best.

For each choice of a_1 in the interval $c_1 < a_1 < 1$ the resulting sequential procedure is partially closed in the sense that the number of stages in which at least one failure occurs with at least one of the remaining populations is bounded.

As in the main formulation, the optimum choice of a_1 is unknown. The value suggested by Paulson (1967) is $a_1 = c_1 + 0.25(1-c_1)$.

The alternative formulation when $p_{[2]}$ is small can be handled in a similar way except that in the place of (4.60) we now have

$$\sum_{\beta=1}^{r} S_{i\beta} \leqslant \left\{ \max_j \left(\sum_{\beta=1}^{r} S_{j\beta} \right) \left[\log(1+c_2-a_2c_2) \right] + \log \alpha \right\} (-\log a_2)^{-1}, \tag{4.61}$$

where the maximum is taken over all the populations left after the first $(r-1)$ stages and a_2 is restricted to the interval $c_2^{-1} < a_2 < 1$. The value of a_2 recommended is $a_2 = c_2^{-1} + 0.25(1-c_2^{-1})$.

A more complicated procedure (R_p') was proposed by Paulson (1969). To write this procedure, we define

$$\bar{S}_{i,r} = \sum_{\nu=1}^{r} \frac{S_{i\nu}}{r}, \quad \bar{F}_{i,r} = \sum_{\nu=1}^{r} \frac{F_{i\nu}}{r}. \tag{4.62}$$

Let $x_0 = x_0(i,j,r)$ be the root of the equation

$$-x^{-1}\log(1-x^2) = \left(\tfrac{3}{2}\right)\Delta^* M\left[\bar{S}_{i,r-1} + \bar{S}_{j,r-1}\right]^{-1} \tag{4.63}$$

and $y_0 = y_0(i,j,r)$ be the root of the equation (4.63) with F in the place of S. Define

$$g_{ij}^{(r)} = \begin{cases} x_0, & \text{if } \bar{S}_{i,r-1} + \bar{S}_{j,r-1} > 0, \\ \frac{1}{2}, & \text{otherwise,} \end{cases} \tag{4.64}$$

and

$$h_{ij}^{(r)} = \begin{cases} y_0, & \text{if } \bar{F}_{i,r-1} + \bar{F}_{j,r-1} > 0, \\ \frac{1}{2}, & \text{otherwise.} \end{cases} \tag{4.65}$$

Also let

$$T_r(i,j,P^*) = \frac{A_r}{B_r} \tag{4.66}$$

where

$$A_r = \log\left[\frac{(k-1)}{(1-P^*)}\right] - \sum_{\nu=2}^{r} \{ S_{i\nu} \log(1 - g_{ij}^{(\nu)})$$

$$+ S_{j\nu} \log(1 + g_{ij}^{(\nu)}) + F_{i\nu} \log(1 + h_{ij}^{(\nu)}) + F_{j\nu} \log(1 - h_{ij}^{(\nu)}) \}$$

and

$$B_r = M \sum_{\nu=2}^{r} \left[g_{ij}^{(\nu)} + h_{ij}^{(\nu)} \right].$$

At the first stage we take $N_{11}, \ldots, N_{k1}$ observations from $\pi_1, \ldots, \pi_k$, respectively. At the rth stage ($r = 2, 3, \ldots$) take N_{jr} from each π_j not yet eliminated and then eliminate any population π_i for which $T_r(i,j,P^*) < \Delta^*$ for some $j \neq i$. When $(k-1)$ populations are eliminated the experiment terminates and the remaining population is selected.

4.7 SELECTING THE COIN WITH THE LARGEST BIAS

Let π_i be a Bernoulli population with probability of a success (head) equal to p_i, $i = 1, \ldots, k$. Let $\theta_i = |p_i - \frac{1}{2}|$. Then θ_i is the bias of π_i and the populations are ranked according to θ_i. We want to select the population associated with $\theta_{[k]}$, the largest bias. A large bias corresponds to a small variance. The parameter space is $\Omega = \{\boldsymbol{\theta}\colon \boldsymbol{\theta}(\theta_1, \ldots, \theta_k);\ 0 \leqslant \theta_i \leqslant \frac{1}{2}\}$. The

P^*-condition must be met whenever $\boldsymbol{\theta} \in \Omega_{\delta^*} = \{\boldsymbol{\theta}: \theta_{[k]} - \theta_{[k-1]} \geqslant \delta^*\}$. Based on n observations from each population, let S_i be the sample sum from π_i. Define $y_i = |S_i - n/2|$, $i = 1, \ldots, k$.

Let the ordered y_i be $0 \leqslant y_{[1]} \leqslant \cdots \leqslant y_{[k]} \leqslant n/2$. Sobel and Starr (1975) have proposed and investigated the following procedure.

R: Select π_i if and only if $y_i = y_{[k]}$, using randomization to break any ties.

For this procedure, the PCS is minimized for a configuration of the type $0 \leqslant \theta_{[1]} = \cdots = \theta_{[k-1]} = \theta_0 = \theta_{[k]} - \delta^*$. Thus one has to minimize the PCS over θ_0 where $0 \leqslant \theta_0 \leqslant \frac{1}{2} - \delta^*$. Sobel and Starr (1975) have shown that asymptotically ($\delta^* \to 0$ and therefore $n \to \infty$) this minimum takes place for $\theta_0 = 0$.

The entropy function associated with π_i is

$$\psi(p_i) = -p_i \log p_i - (1 - p_i)\log(1 - p_i).$$

It can be easily seen that selecting the coin with the largest bias is equivalent to selecting the population with the smallest associated entropy function. Subset selection procedures for selection in terms of entropy function have been studied by Gupta and Huang (1976b) and are described in §13.2.3.

4.8 POISSON POPULATIONS

Let us consider k independent Poisson processes π_i, $i = 1, \ldots, k$, with mean arrival times λ_i^{-1}, $i = 1, \ldots, k$, respectively. The best process is defined to be the one with the smallest mean arrival time, that is, the one associated with $\lambda_{[k]}$. The probability requirement is to be met whenever $\lambda_{[k]}/\lambda_{[k-1]} \geqslant \theta > 1$. Let N be a fixed positive integer. Observations are made at successive intervals of time $t (t = 1, 2, \ldots)$ and are continued until time t_0, say, when the number of arrivals from any one of the processes is equal to or greater than N. Let $X_{i,t}$ be the number of arrivals from π_i during time t. Define $I = \{i: X_{i,t_0} \geqslant N\}$. Alam (1971b) has studied the following two procedures, R_1 and R_2.

To describe R_1, let

$$\begin{aligned} n_i &= X_{i,t_0} - X_{i,t_0-1}, \\ m_i &= N - X_{i,t_0-1}, \end{aligned} \qquad i = 1, \ldots, k. \tag{4.67}$$

Let $U(m, n)$ denote the mth smallest observation in a sample of n independent observations from a uniform distribution on $(0, 1)$. Observe $U(m_i, n_i)$

and define $U_i = t_0 - 1 + U(m_i y n_i)$, $i = 1,\ldots,k$. The procedure R_1 selects π_i if and only if

$$U_i = \min_{j\in I} U_j. \tag{4.68}$$

For this procedure

$$P(CS|R_1) = \int_0^\infty \prod_{i=1}^{k-1}\left[1 - G_N\left(\frac{\lambda_{[i]}x}{\lambda_{[k]}}\right)\right] dG_N(x), \tag{4.69}$$

where $G_N(x)$ is the gamma cdf given by

$$G_N(x) = (\Gamma(N))^{-1}\int_0^x e^{-y}y^{N-1}dy. \tag{4.70}$$

The LFC is given by $\lambda_{[1]} = \cdots = \lambda_{[k-1]} = \theta^{-1}\lambda_{[k]}$, and N is the smallest positive integer satisfying

$$\int_0^\infty \left[1 - G_N\left(\frac{x}{\theta}\right)\right]^{k-1} dG_N(x) \geqslant P^*. \tag{4.71}$$

The value of N can be obtained approximately for selected values of θ and P^* from the tables of Gupta and Sobel (1962b) and Armitage and Krishnaiah [1964]. For large N, normal approximation gives

$$\int_{-\infty}^\infty \left[\Phi\left(-\frac{x}{\theta} + \left(1 - \frac{1}{\theta}\right)\sqrt{N}\right]^{k-1} \varphi(x)\,dx \geqslant P^*. \tag{4.72}$$

For procedure R_2, observations are continued as before until t_0 when the number of arrivals from any one of the process is $\geqslant N$. The rule R_2 is:
Select π_i if and only if

$$X_{i,t_0} = \max_{j\in I} X_{j,t_0} \tag{4.73}$$

using randomization to break ties. To write down the expression for the PCS, define

$$f_v(z) = \frac{e^{-v}v^z}{z!},$$

$$h_v(z) = f_v(z)I_q(z - N + 1),\ H_v(z) = \sum_{x=0}^{z} h_v(z),$$

$$q = (1 + t)^{-1},$$

$$I_q(m)=\begin{cases}1, & m\leqslant 0,\\ \dfrac{1}{B(m,n)}\displaystyle\int_0^q x^{m-1}(1-x)^{n-1}dx, & m>0,\end{cases} \tag{4.74}$$

$$v_i=(1+t)\lambda_{[i]}, \qquad i=1,\dots,k,$$

$$Q_t(v_1,\dots,v_k)=\sum \frac{1}{r}\prod_{i=1}^{k} h_{v_i}(z),$$

where the summation is over k-vectors $(z_1,\dots,z_k)$ for which $z_{[k]}\geqslant N$ and $(r-1)=$ the number of $z_{[i]}$'s $(i=1,\dots,k-1)$ equal to $z_{[k]}$. Then the PCS can be written as

$$P(CS|R_2)=\sum_{t=0}^{\infty} Q_t(v_1,\dots,v_k). \tag{4.75}$$

Note that $Q_t(v_1,\dots,v_k)$ is the probability that $t_0=t+1$ and that a correct selection is made. The PCS is minimized for the configuration

$$\lambda=\lambda_{[1]}=\cdots=\lambda_{[k-1]}=\theta^{-1}\lambda_{[k]}. \tag{4.76}$$

Thus we need the minimum of $\Sigma_{t=0}^{\infty}Q_t(\lambda)$, where

$$Q_t(\lambda)=k^{-1}e^{-(\theta-1)v_t}\sum_{z=N}^{\infty}\theta^z\left\{\left(H_{v_t}(z)\right)^k-\left(H_{v_t}(z-1)\right)^k\right\} \tag{4.77}$$

and $v_t=(1+t)\lambda$. The monotonicity of $P(CS|R_2)$ at LFC in terms of N and λ has not been established but surmised from the computations. If T_0' denotes the stopping time for the procedures R_1 and R_2, we have

$$E(T_0')=\sum_{t=0}^{\infty}\prod_{i=1}^{k}\left[1-G_N(t\lambda_i)\right]. \tag{4.78}$$

The selection problem discussed previously corresponds to the problem of selecting from k Poisson populations the one having the largest mean. For the corresponding procedures, observations are taken one at a time from each population until the sum of the observed values from any one of the populations is greater than or equal to N. Alam (1971b) has also studied a rule R_3 to be used when the processes are observed continuously until the number of arrivals from any process π_i, say, equals N. Then R_3 selects this process π_i. The PCS is same as that for R_1 given by (4.69). So the smallest N needed for the rule is given by (4.71).

The basic ranking procedure of Bechhofer, Kiefer, and Sobel (1968) for Koopman–Darmois populations is applicable to the Poisson problem and the application is discussed in their monograph.

4.9 SELECTION OF THE BEST MULTINOMIAL CELL

4.9.1 Most Probable Cell

Consider a multinomial distribution with k cells, where the cell π_i has probability p_i. For selecting the cell with the largest probability $p_{[k]}$, Bechhofer, Elmaghraby, and Morse (1959) proposed a fixed sample procedure which we call R_1 here. Let $Y_{i,N}$ denote the number of observations that arise in the cell π_i. The procedure R_1 is: Select the cell π_i for which

$$Y_{i,N} = \max_{1 \leqslant r \leqslant k} Y_{r,N} \tag{4.79}$$

with the provision that a tie will be broken by randomization. The PCS is given by

$$P(CS|R_1) = \sum \frac{1}{s} \frac{N!}{y_1! \cdots y_N!} p_{[1]}^{y_1} \cdots p_{[k]}^{y_k}, \tag{4.80}$$

where the summation is over all the vectors $\mathbf{y} = (y_1, \ldots, y_k)$ such that $\sum_{i=1}^{k} y_i = N$ and $y_k \geqslant y_i$, $i = 1, \ldots, k-1$, and s denotes the number of y_i's tied for the largest. The basic probability requirement is to be satisfied whenever $p_{[k]}/p_{[k-1]} \geqslant \theta^* > 1$. It is shown by Kesten and Morse (1959) that the LFC is given by

$$p_{[1]} = \cdots = p_{[k-1]} = (\theta^* + k - 1)^{-1}; \; p_{[k]} = \theta^*(\theta^* + k - 1)^{-1}. \tag{4.81}$$

Though tedious it is, one can compute in principle the PCS for the LFC and obtain the smallest N for which it is at least P^*. Bechhofer, Elmaghraby, and Morse (1959) have tabulated the values of PCS corresponding to the LFC for $k = 2(1)4$, $N = 1(1)30$, and $\theta^* = 1.02(0.02)1.10(0.10)2.00(0.20)3.00, 10.00$. The usual normal approximation yields

$$\min P(CS|R_1) \approx \Pr\left(U_i \geqslant -(\theta^* - 1)\left[\frac{N}{\{(k+2)\theta^* + (k-2)\}} \right]^{-1/2}, \quad i = 1, \ldots, k-1 \right), \tag{4.82}$$

where the U_i are standardized normal variates with equal correlation $\rho=[(k+1)\theta^*-1][(k+2)\theta^*+(k-2)]^{-1}$. Tables of Gupta [1963a] and Gupta, Nagel and Panchapakesan [1973] can be used to obtain approximate values of N for selected values of k,θ^*, and P^*. Bechhofer, Elmaghraby, and Morse (1959) have computed the approximate values in some cases by considering the arc sine transformation. In this case, the large sample approximation is obtained in terms of standardized normal variables with equal correlation$=0.5$.

An alternative procedure R_2 was proposed and investigated by Cacoullos and Sobel (1966) for the same problem. This procedure uses an inverse sampling rule. Observations are taken one at a time and the sampling is terminated when the count in any one of the cells reaches M (specified in advance). Let y_i be the count in π_i at termination. Then R_2 is defined as follows.

$$\boldsymbol{R_2}\text{:}\quad \text{Select the cell } \pi_i \text{ for which } y_i=M. \tag{4.83}$$

Of course, the probability requirement is the same as in the case of R_1. It is easy to see that

$$P(CS|R_2)=p_{[k]}^{M}\sum_{y_1=0}^{M-1}\cdots\sum_{y_{k-1}=0}^{M-1}\frac{\Gamma(M+y_0)}{\Gamma(M)}\prod_{i=1}^{k-1}\frac{p_{[i]}^{y_i}}{y_i!}, \tag{4.84}$$

where $y_0=y_1+\cdots+y_{k-1}$. The multiple sum in (4.84) can be written as a Dirichlet integral using the result of Olkin and Sobel [1965]. In other words,

$$P(CS|R_2)=\frac{\Gamma(kM)}{[\Gamma(M)]^k}\int_{\theta_1^{-1}}^{\infty}\cdots\int_{\theta_{k-1}^{-1}}^{\infty}\frac{\prod_{i=1}^{k-1}y_i^{M-1}}{(1+y_0)^{kM}}dy_1\cdots dy_{k-1}, \tag{4.85}$$

where $\theta_i=p_{[k]}/p_{[k-1]}$. It follows that the LFC is again given by (4.81). Thus the value of M required to satisfy the P^*-condition is the smallest positive integer $\geqslant n=n(k;\theta^*,P^*)$, the solution of the equation

$$\frac{\Gamma(kn)}{[\Gamma(n)]^k}\int_{\theta^{*-1}}^{\infty}\cdots\int_{\theta^{*-1}}^{\infty}\frac{\prod_{i=1}^{n-1}y_i^{n-1}}{(1+y_0)^{kn}}dy_1\cdots dy_{k-1}=P^*. \tag{4.86}$$

The values of θ^{*-1} satisfying (4.86) for selected values of k, P^*, and n can be obtained from the tables of Gupta and Sobel (1962b). Using these tables, one can obtain the M values for selected values of k, P^*, and θ^*. This has been done by Cacoullos and Sobel (1966) who have also tabulated approximate values of M in selected cases.

It is shown by Cacoullos and Sobel that, for $k=2$, R_2 is uniformly more efficient than R_1, in the sense that, for all configurations of the parameters,

$$N_0 > E(T), \tag{4.87}$$

where N_0 is the minimum sample size required for R_1 and T denotes the total number of observations in the case of R_2.

For $k>2$, only asymptotic comparisons are available and these are also only with respect to the LFC and the worst configuration (WC), $p_{[1]} = \cdots = p_{[k]}$. For θ^* sufficiently close to 1 (and hence M large),

$$\begin{aligned} E(T|\text{WC}) &< N_0 \quad \text{for } k \leqslant 3, \\ E(T|\text{WC}) &> N_0 \quad \text{for } k > 3. \end{aligned} \tag{4.88}$$

For the purpose of comparison under LFC, the criterion employed is the asymptotic ($\theta^* \to 1$) proportion saved (APS) defined as follows. For fixed P^* the *first-order asymptotic proportion saved* (APS_1) for any procedure R relative to R' is defined by

$$\text{APS}_1(R, R'; P^*) = \lim_{\theta^* \to 1} \left[\frac{E(T|R'; \text{LFC})}{E(T|R; \text{LFC})} - 1 \right]. \tag{4.89}$$

The procedure R is said to be *asymptotically first-order more efficient than* or *asymptotically first-order equivalent to* R' according as the limit in (4.89) is positive or zero. If the limit is zero, the *second-order asymptotic proportion saved* for R relative to R' is defined by

$$\text{APS}_2(R, R'; P^*) = \lim_{\theta^* \to 1} \frac{1}{(\theta^* - 1)} \left[\frac{E(T|R'; \text{LFC})}{E(T|R; \text{LFC})} - 1 \right]. \tag{4.90}$$

If $\text{APS}_1 = 0$ and $\text{APS}_2 > 0$, we say that R is *asymptotically second-order more efficient* than R'.

The following theorem has been proved by Cacoullos and Sobel (1966).

Theorem 4.1. For $k \geqslant 2$ and all P^*,

$$\text{APS}_1(R_2, R_1; P^*) = 0 \tag{4.91}$$

and

$$\mathrm{APS}_2(R_2, R_1; P^*) \geqslant \frac{k+2}{2[k-(k-1)P^*]} - 1, \tag{4.92}$$

which is positive for all $P^* \geqslant \frac{1}{2}$; in particular, for $k=2$ and 3, it is positive for all nontrivial P^* values, that is, for all $P^* > 1/k$.

Alam, Seo, and Thompson (1971) considered a procedure using a variation of the sampling scheme of Cacoullos and Sobel (1966). Let a positive number λ and a positive integer M be given. Let $N_1, N_2, \ldots$ denote a sequence of random observations taken from a Poisson distribution with mean λ. At stage $t (t=1,2,\ldots)$, we take N_t observations from the multinomial distribution. Let $Y_{i,t}$ denote the count in the cell π_i at stage t. The sampling is terminated when the total count for any one of the cells is equal to or greater than M. Let t_0 be the terminal stage. Then $X_{i,t_0-1} = \Sigma_{t=1}^{t_0-1} Y_{i,t} < M$ for $i=1,\ldots,k$, and $X_{i,t_0} \geqslant M$ for at least one value of i. If T is the total number of observations sampled from the multinomial distribution, then $E(T) = \lambda E(t_0)$. Let $I = \{i\colon X_{i,t_0} \geqslant M\}$ and $U_{m,n}$ be the mth smallest observed value in a sample of n independent observations from a uniform distribution on $(0,1)$. We observe $U(m_i, n_i)$ for each $i \in I$, where $m_i = M - X_{i,t_0-1}$ and $n_i = X_{i,t_0} - X_{i,t_0-1}$. Set $U_i = t_0 - 1 + U(m_i, n_i)$, $i=1,\ldots,k$. The rule R_3 of Alam, Seo, and Thompson (1971) is: Select π_i for which

$$U_i = \min_{j \in I} U_j. \tag{4.93}$$

Another procedure R_4 discussed by the same authors is: Select π_i for which

$$X_{i,t_0} = \max_{j \in I} X_{j,t_0} \tag{4.94}$$

using randomization to break a tie. The LFC for R_3 as well as R_4 is given by (4.81).

Yet another procedure R_5 has been proposed by Alam (1971a). This procedure consists of taking one observation at a time until the difference between the highest and the next highest cell counts is equal to a positive integer r. Let Y_i denote the count in the cell π_i at termination of sampling. The procedure R_5 is: Select π_i for which

$$Y_i = \max_{1 \leqslant j \leqslant k} Y_j. \tag{4.95}$$

It can be shown that the sampling for R_5 terminates with probability 1. For $k=2$, the LFC is again shown to be (4.81). However, it is only conjectured that the same is true for $k>2$. For $k=2$, the expected sample size for R_5 is less than or equal to that for R_2. Of course, R_2 is more efficient than R_1 for $k=2$ (see (4.87)).

It is known that if $X_1,\ldots,X_k$ are independent Poisson random variables with means $\lambda_1,\ldots,\lambda_k$, respectively, then the conditional distribution of $X_1,\ldots,X_k$ given $X_1+\cdots+X_k=N$ is multinomial with parameters p_i, $i=1,\ldots,k$, where $p_i=\lambda_i/(\lambda_1+\cdots+\lambda_k)$. Thus procedures for Poisson populations have applications to the multinomial problem. One can see this from the procedures R_1 and R_2 of §4.8 and, R_4 of this section.

4.9.2 Least Probable Cell

For selecting the multinomial cell with probability $p_{[1]}$, one can use a procedure analogous to R_1 defined in (4.79). This procedure (denoted by R_1') has the same sampling scheme as R_1 and selects the cell π_i for which

$$Y_{i,N}=\min_{1\leqslant r\leqslant k} Y_{r,N} \tag{4.96}$$

using randomization to break ties. It will now be required that R_1' satisfies the P^*-condition whenever $p_{[2]}/p_{[1]}\geqslant\theta>1$. In this case, it can be shown that the LFC is given by

$$\theta p_{[1]}=p_{[2]}=\cdots=p_{[k-1]};\ p_{[k]}=1-\left[1+(k-2)\theta\right]p_{[1]} \tag{4.97}$$

and that $P(CS|R_1')\to(k-1)^{-1}$ for all N as $p_{[1]}\to 0$. Thus, given $P^*>(k-1)^{-1}$, the probability requirement cannot be met.

To overcome the preceding difficulty Alam and Thompson (1972) have used a modified formulation. They require that $P(CS|R_1')\geqslant P^*$ whenever $p_{[2]}-p_{[1]}\geqslant c$, where $0<c<(k-1)^{-1}$. Under this formulation, they have shown that the LFC is given by

$$p_{[1]}=\left[1-(k-1)c\right]k^{-1},\quad p_{[2]}=\cdots=p_{[k]}=(1+c)k^{-1}. \tag{4.98}$$

For large N, using the normal approximation, we get under the LFC

$$P(CS|R',\text{LFC})\approx\int_{-\infty}^{\infty}\Phi^{k-1}\left(\frac{ax+c\sqrt{n}}{b}\right)d\Phi(x) \tag{4.99}$$

for $c\leqslant k^{-1}$, where $a^2=k^{-1}(1+c)(1-ck)$ and $b^2=k^{-1}(1+c)$.

Alam and Thompson (1972) have tabulated the minimum values of the sample size for $k=2(1)5$, $c=0.05(0.05)0.20$, and $P^*=0.70(0.05)0.95, 0.99$.

4.10 NOTES AND REMARKS

Sequential procedures for binomial, Poisson, and negative-binomial distributions are given by Bechhofer, Kiefer, and Sobel (1968) who have considered different distance functions for defining the indifference zone. Alam and Thompson (1973) have considered the problem of simultaneously selecting the Poisson process with the largest parameter λ_i and obtaining a confidence interval for the largest λ_i. This is discussed in §21.7. Contributions to subset selection procedures for discrete distributions are reviewed in Chapter 13. For some useful results on the monotonicity properties of the multinomial cell probabilities see Alam (1970a) and Olkin [1972]. Recently, Ramey and Alam (1979) have obtained some empirical results for selecting the most probable multinomial cell using a procedure that combines the features of R_2 and R_5 of §4.9.1 in its stopping rule.

CHAPTER 5

Selection from Univariate Populations, Optimum Sampling and Estimation of Probability of Correct Selection

In Chapter 2, we discussed selection from normal populations in terms of various ranking measures. Selection problems for discrete distributions were reviewed in the last chapter. Selection from multivariate normal populations, nonparametric procedures and sequential procedures are discussed in the next three chapters. There are some single-stage and two-stage procedures relating to other distributions such as exponential and uniform distributions. These and selection procedures for finite populations are discussed in this chapter. Besides these we also discuss some problems of optimum sampling and estimation of the probability of correct selection.

5.1 SELECTION FROM FINITE POPULATIONS AND UNIVARIATE CONTINUOUS POPULATIONS

Importance of sampling from finite populations cannot be overstated. It is a realistic problem to compare, for example, the average annual incomes of two or more industrial cities or states. One naturally resorts to an appropriate method of survey sampling, such as stratified sampling or cluster sampling. We discuss here a two-stage subsampling procedure for ranking the means of finite populations. We also discuss in this section some single-stage and two-stage procedures for exponential populations that are ranked according to their guaranteed lives.

5.1.1 A Two-Stage Subsampling Procedure for Ranking Means of Finite Populations

Let $\pi_1,\ldots,\pi_k$ be k finite populations. The situation we consider here is one in which each population is divided into U primary units and each primary unit in turn is divided into T secondary units. A simple random sample of $u \leqslant U$ primary units is drawn without replacement from each population. From each primary unit selected, a simple random sample of $t \leqslant T$ elements is drawn without replacement. This technique of sampling in two stages is known as subsampling in sample surveys. Subsampling has also applications in other contexts such as processes that involve physical, chemical, or biological tests that can be done on a fairly small amount of material.

Now, let X_{ipq} $(i=1,2,\ldots,k;\ p=1,2,\ldots,U;\ q=1,2,\ldots,T)$ denote the value of the pth element in the qth primary unit of π_i. Let

$$\overline{X}_{ip}=\sum_{q=1}^{T} X_{ipq}/T \qquad \text{and} \qquad \overline{\overline{X}}_{i}=\sum_{p=1}^{U} \overline{X}_{ip}/U.$$

These are, respectively, the population mean per element in the pth primary unit of π_i and the overall population mean per element of π_i. Let the population variance among primary unit means of π_i be denoted by

$$S_{i1}^2=\sum_{p=1}^{U}\frac{\left(\overline{X}_{ip}-\overline{\overline{X}}_{i}\right)^2}{(U-1)}$$

and the population variance among elements within primary units of π_i by

$$S_{i2}^2 = \sum_{p=1}^{U} \sum_{q=1}^{T} \frac{\left(X_{ipq} - \overline{X}_{ip}\right)^2}{(T-1)U}.$$

It is assumed that $S_{i1}^2 = S_1^2$ and $S_{i2}^2 = S_2^2$, $i = 1, \ldots, k$ and that S_1^2 and S_2^2 are known. The values of $\overline{X}_{ip}$ and $\overline{\overline{X}}_i$ are unknown. The goal is to select the population associated with $\overline{\overline{X}}_{[k]}$ with the requirement that the PCS $\geqslant P^*$ whenever $\overline{\overline{X}}_{[k]} - \overline{\overline{X}}_{[k-1]} \geqslant \delta^*$. For this problem, Bechhofer (1967) proposed the following two-state subsampling procedure R based on a random sample of t elements from each primary unit in a random sample of u primary units from each population. Let $\overline{\overline{x}}_i = \Sigma_{p=1}^{u} \Sigma_{q=1}^{t} x_{ipq}$, where x_{ipq} is a sample observation. Then R selects the population associated with $\overline{\overline{x}}_{[k]}$.

For given $\{k, T, U\}$, known $\{S_1^2, S_2^2\}$, and specified $\{\delta^*, P^*\}$, the basic probability requirement can be approximately guaranteed if t and u are chosen such that

$$\left(\frac{U-u}{U}\right)\frac{S_1^2}{u} + \left(\frac{T-t}{T}\right)\frac{S_2^2}{tu} \leqslant \left(\frac{\delta^*}{z}\right)^2 \tag{5.1}$$

where $z = z(k, P^*)$ is given by

$$\int_{-\infty}^{\infty} \Phi^{k-1}(y+z)\, d\Phi(y) = P^* \tag{5.2}$$

and hence can be obtained from the tables of Bechhofer (1954), Gupta [1963a], and Gupta, Nagel, and Panchapakesan [1973].

Let c_1 and c_2 be the cost of sampling for a primary unit and for an element from a primary unit, respectively. Then the total cost of sampling is

$$c(t,u) = u(c_1 + c_2 t). \tag{5.3}$$

The problem is one of integer programming, namely, to find a pair of integers (t, u) which minimizes $u(c_1 + c_2 t)$ subject to (5.1) and the additional constraints: $1 \leqslant t \leqslant T$, $1 \leqslant u \leqslant U$.

Bechhofer (1967) has proposed the following simple method for obtaining a solution to the problem. Let $c_3 = (\delta^*/z)^2$ and L be the smallest integer $\geqslant US_1^2/(Uc_3 + S_1^2)$. Define $u_i = i$, $i = L, L+1, \ldots, U$. For each u_i, let t_i be the smallest integer $\geqslant S_2^2[(S_2^2/T) - S_1^2 + u_i(c_3 + S_1^2/U)]^{-1}$. Determine the costs $u_i(c_1 + c_2 t_i)$, $i = L, L+1, \ldots, U$, and choose the pair

(t^*, u^*) for which

$$\min_{L \leqslant i \leqslant U} u_i(c_1 + c_2 t_i) = u^*(c_1 + c_2 t^*).$$

Then (t^*, u^*) is a solution to the problem. The i-value(s) at which the minimum is achieved clearly depend(s) on c_1, c_2 only through c_1/c_2. One could use the k-population analog of the method given in Section 10.6 of Cochran [1963]; however, as pointed out by Bechhofer (1967), this method need not yield the correct solution (when the solution is unique) nor even an admissible solution.

5.1.2 Selection from Exponential Populations

Let $\pi_1, \ldots, \pi_k$ be k independent exponential populations where π_i has the density

$$f(x; \gamma_i, \theta) = \frac{1}{\theta} e^{-(x-\gamma_i)/\theta}, \qquad x \geqslant \gamma_i, \qquad \theta > 0.$$

The parameter γ_i is called the guaranteed life of the population π_i.

For the case of known θ, assuming it to be unity, Barr and Rizvi (1966a) and Raghavachari and Starr (1970) considered the goal of selecting the t best populations, that is, the populations associated with $\gamma_{[k-t+1]}, \ldots, \gamma_{[k]}$. The P^*-requirement for PCS is to be met whenever $\gamma_{[k-t+1]} - \gamma_{[k-t]} \geqslant \delta^*$, where $\delta^* > 0$ is specified. Let Y_i be the minimum of n independent observations from π_i, $i = 1, \ldots, k$. The preceding authors proposed the natural rule R_1 which selects the populations that yielded the t largest values of the Y_i. For this rule R_1, the LFC is given by $\gamma_{[1]} = \cdots = \gamma_{[k-t]} = \gamma$; $\gamma_{[k-t+1]} = \cdots = \gamma_{[k]} = \gamma + \delta^*$. Equating the PCS corresponding to the LFC to P^*, we get

$$(1 - e^{-n\delta^*})^{k-t} + (k-t)e^{n\delta^* t} I(e^{-n\delta^*}; t+1, k-t) = P^*, \tag{5.4}$$

where $I(x; a, b)$ is the incomplete beta function. Raghavachari and Starr (1970) have tabulated the values of $\nu = e^{-n\delta^*}$ satisfying (5.4) for $k = 1(1)15$, $t = 1(1)k-1$, and $P^* = 0.90$, 0.95, 0.99.

For the case of unknown θ and the goal of selecting the best population (i.e., $t = 1$), Desu, Narula, and Villarreal (1977) have proposed a two-stage procedure R_2 similar to the one considered by Bechhofer, Dunnett, and Sobel (1954) which we discussed in §2.1.2.

The procedure R_2 consists of the following steps:

1. Take an initial sample of size n_0 from each population; let $X_{i[1]} \leqslant \cdots \leqslant X_{i[n_0]}$ be the ordered observations from π_i, $i = 1, \ldots, k$.

2. Let $Y_{i0}=X_{i[1]}$, $i=1,\dots,k$, and estimate θ by

$$\hat{\theta}=\sum_{i=1}^{k}\sum_{j=1}^{n_0}\frac{(X_{i[j]}-Y_{i0})}{k(n_0-1)};$$

3. Choose h satisfying the equation

$$\int_0^\infty\left\{\int_0^\infty(1-e^{-z-ht})^{k-1}e^{-z}\,dz\right\}g(t)\,dt=P^*,$$

where $g(t)$ is the density of the chi-square distribution with $2k(n_0-1)$ degrees of freedom.

4. Compute $\eta=2hk(n_0-1)\hat{\theta}/\delta^*$. The total sample size N is given by

$$N=\begin{cases} n_0 & \text{if} \quad \eta\leqslant n_0 \\ [\eta]+1 & \text{if} \quad \eta>n_0 \end{cases}$$

where $[\cdot]$ is the largest integer function;

5. If necessary, take $(N-n_0)$ additional observations from each population, and compute Y_i the minimum of all the N observations from π_i, $i=1,\dots,k$.

6. Select the population that yielded the largest Y_i value.

The procedure R_2 satisfies the P^*-requirement. The values of h are tabulated for $P^*=0.95$, $k=2(1)6$, and $n_0=2(1)30$. Desu, Narula, and Villarreal have also discussed several criteria for the choice of the initial sample size n_0. This aspect of the problem has not been explored fully.

5.2 SOME PROBLEMS OF OPTIMUM SAMPLING

Let $\pi_1,\dots,\pi_k$ be k populations with distribution functions $F_{\theta_i}(x)$, $i=1,\dots,k$, where the θ_i are unknown real-valued parameters. The goal is to select the population associated with $\theta_{[k]}$. The problem is to determine the optimum sample size for each population balancing the cost of sampling and the PCS. Let us assume without any loss of generality that $\theta_1\leqslant\theta_2\leqslant\cdots\leqslant\theta_k$. Let $W(\theta_i,\theta_k)$ be the loss incurred in selecting the population π_i, where W is a real-valued function with continuous second derivative. Let $W(\theta_k,\theta_k)=0$ and $W(\theta_i,\theta_k)\geqslant 0$. Let a total sample of size kn be taken by an arbitrary procedure: one-stage, two-stage, and so on. Let $C(n)$ be the cost of a sample of size kn, where $C(n)$ is a real-valued function. Let p_i be the probability that π_i is chosen. Then, the expected loss is

$$L = \sum_{i=1}^{k-1} W(\theta_i, \theta_k) p_i + C(n). \tag{5.5}$$

Somerville (1954) has shown the following: Given θ_k, a maximum of L occurs, under certain conditions, when $\theta_1 = \cdots = \theta_{k-1} = \theta'$, where θ' will, in general, depend on θ_k. If, $W(\theta_i, \theta_k)$ depends on a function $g_i = g(\theta_i, \theta_0)$ and each p_i can be expressed explicitly in terms of these g_i, then a maximum of L occurs (again under certain conditions) when $g_1 = \cdots = g_{k-1} = g'$, where g' is independent of θ_k.

5.2.1 Selection of the Population with the Largest Mean

Suppose that π_i is normal with mean θ_i and variance σ^2. While selecting the population with the largest mean, it is desired to use this population for N occasions. A preliminary sample of size n is taken from each population. The N individuals are taken from the population yielding the largest preliminary sample mean. Thus the expected loss L is given by (5.5) where $p_i = \Pr(\bar{X}_i > \bar{X}_j, j = 1, \ldots, k; j \neq i)$. If our object is to make the expected value of the N individuals as large as possible, then it is reasonable to take

$$W(\theta_i, \theta_k) = N(\theta_k - \theta_i). \tag{5.6}$$

Under the preceding setup, Somerville (1954) has shown that a local maximum of L occurs at a point $\boldsymbol{\theta}$ for which $\theta_1 = \cdots = \theta_{k-1} = \theta_k - \delta^*$. This gives a unique maximum in the case of two normal populations. For $k > 2$, such a result is not shown to be true.

Suppose we define $W(\theta_i, \theta_k)$ as follows:

$$W(\theta_i, \theta_k) = \begin{cases} 1 & \text{if } \theta_k - \theta_i \geqslant \delta^*, \\ 0 & \text{otherwise.} \end{cases} \tag{5.7}$$

If we assume that the populations are normally distributed, the expected loss from choosing the wrong population is

$$L^* = \sum_{i=1}^{k-1} W(\theta_i, \theta_k) p_i. \tag{5.8}$$

It can be proved that L^* is maximized when $\theta_1 = \theta_2 = \cdots = \theta_{k-1} = \theta_k - \delta^*$. The optimal choice of n is found by minimizing $L^*_{\max}$.

Somerville (1954) has also discussed a two-stage procedure in the case of three normal populations. As before, it is desired to choose N individuals from π_3. We are allowed a preliminary sample of total size $3n$. The procedure is as follows.

Take a sample of size n_1 from each of the three populations. Discard the population having the smallest sample mean in the first stage, and take a further sample of size n_2 from each of the remaining populations, where $3n_1+2n_2=3n$. Choose the N individuals from the population having the largest cumulative sample mean.

Let p_i be the probability that the N individuals are selected from π_i, $i=1,2,3$. The expected loss is

$$L=\sum_{i=1}^{2} W(\theta_i,\theta_3)p_i+C(n), \tag{5.9}$$

where $W(\theta_i,\theta_3)=N(\theta_3-\theta_i)$.

The computational results of Somerville (1954) show that the two-stage sampling procedure has a smaller maximum expected loss than the one-stage procedure using the same total sample size, provided that $1\leqslant\alpha\leqslant 1.92$, where $\alpha=[(n_1+n_2)/n_1]^{1/2}$.

Somerville's formulation of the two-stage procedure for three normal populations is generalized by Fairweather (1968) to an arbitrary finite number of populations as follows. Fix an integer t, $1\leqslant t\leqslant k-1$. Take n_1 independent observations from each population; compute the k sample means; discard the populations that produced the t smallest sample means; take n_2 additional, independent observations from each of the remaining $k-t$ populations and select as the population associated with θ_k that population which entered the second stage and produced the largest sample mean based on all n_1+n_2 observations.

Let n be such that the total sample size for the procedure is $kn=kn_1+(k-t)n_2$. The problem is to find the allocation of sample sizes that minimizes the maximum of L over the parametric configurations. Let $\alpha^2=(n_1+n_2)/n_1$. For numerical comparisons, Fairweather (1968) has considered the case of $k=4$ with $t=1$ and 2. For each case the two-stage procedure has a smaller maximum expected loss than the associated one-stage procedure. Also, the smallest maximum expected loss can be achieved by taking $t=2$; however, this result is not uniform in α.

For a wide range of losses due to incorrect selection and for a wide range of procedures, one-stage, two-stage, and so on, the expected loss is maximized for a configuration of the type $\theta_1=\cdots=\theta_{k-1}=\theta_k-\Delta$ ($\Delta>0$), and to find the optimal allocation we need to compute p_0, the probability of a correct selection for the preceding configuration. As we remarked earlier, the computations of Somerville (1954) and Fairweather (1968) are only for $k=3$ and 4, respectively. An algorithm for calculating p_0 in the case of k normal populations is given by Somerville (1971b).

Somerville (1970) considered another two-stage problem. Let $\pi_1,\ldots,\pi_k$ be k populations. It is required to choose N individuals in all from these k populations. In the first stage we choose n individuals from each population. In the second stage, we choose the remaining $N-kn$ individuals from the best population as determined from the first stage, that is, the population with the largest first-stage sample mean.

We assume that $\theta_1 \leqslant \theta_2 \leqslant \cdots \leqslant \theta_k$ are the unknown means and σ^2 is the common known variance. The cost unique to the first stage is given by $C(n)=c_0+c_1 n$, where c_0 is the fixed overhead cost, and c_1 is the cost of taking one individual from each of the k populations. The loss in choosing an individual from π_i (common to both stages) is taken to be $c_2(\theta_k-\theta_i)$ where $c_2>0$ is a constant. Therefore the expected loss is given by

$$L=nc_2\sum_{i=1}^{k-1}(\theta_k-\theta_i)+\left[N-nk\right]c_2\sum_{i=1}^{k-1}(\theta_k-\theta_i)p_i+C(n), \qquad (5.10)$$

where p_i is the probability that π_i is chosen as the best from the first stage.

Somerville (1970) obtained a minimax solution for the optimum n by minimizing the maximum expected loss over the parametric configurations. It is obvious (see also Ofosu (1974a)) that the loss function considered by Somerville is not bounded. Thus the minimax solution does not exist. However, if we have prior information regarding bounds on $\theta_k-\theta_i$, then we can find a minimax solution in the restricted range. Somerville (1974a) has given some computations to show that there are situations in which his procedure can give "fairly good" results.

Ofosu (1975c) has considered a Bayes single sample solution to the problem of selecting the largest of k normal means. Let π_i $(1\leqslant i\leqslant k)$ be a normal population with unknown mean μ_i and known variance σ_i^2. The μ_i have independent normal prior distributions with known means θ_i $(i=1,\ldots,k)$ and a common known variance σ_0^2. We select N individuals in two stages as follows. In the first stage, n_i individuals are selected from π_i and in the second stage $(N-n_1-\cdots-n_k)$ individuals are selected from the population that yielded the largest posterior mean in the first stage. Ofosu (1975c) has discussed the problem of how large the n_i should be.

Similar to his (1954) one-stage procedure, Somerville (1957) has considered the problem of optimal sampling in the case of two binomial populations.

5.2.2 Selection of the Gamma Population with the Largest (Smallest) Scale Parameter

Let π_i have the density

$$f(x,\theta_i)=\{\theta_i\Gamma(r)\}^{-1}\left(\frac{x}{\theta_i}\right)^{r-1}\exp\left\{\frac{-x}{\theta_i}\right\}, \qquad x>0, \tag{5.11}$$

with unknown scale parameter $\theta_i(>0)$ and a known shape parameter $r>0$. It is desired to select the population associated with $\theta_{[k]}$ based on a sample of size n from each population and to use the selected population for N occasions. We assume that $\theta_1 \leqslant \cdots \leqslant \theta_k$. Ofosu (1972) has discussed a minimax procedure for this problem when the expected loss L is given by (5.5) with

$$C(n)=kc_1n^d+c_0, \tag{5.12}$$

where c_0 is the fixed overhead cost, c_1 is a known positive constant measured in the same units as c_0, and d is a positive constant. Some possible loss functions are

$$W_1(\theta_i,\theta_k)=Nc_2h_i, \tag{5.13a}$$

$$W_2(\theta_i,\theta_k)=Nc_2h_i^{\alpha}, \qquad \alpha>0, \tag{5.13b}$$

$$W_3(\theta_i,\theta_k)=Nc_2(1-h_i^{-\beta}), \qquad \beta>0, \tag{5.13c}$$

where $h_i=\log(\theta_k/\theta_i)$ and c_2 is a positive constant measured in the same units as c_0 and c_1. Ofosu (1972) considers the loss function in (5.13a). The problem is to determine n which minimizes the maximum of L over all parametric configurations $(\theta_1,\ldots,\theta_k)$. To evaluate this maximum of L, Ofosu (1972) appeals to the sufficient condition of Somerville (1954, Theorem 2) which is usually verified numerically. If this condition holds, then we need to evaluate $\max_\delta L(n,\delta)$, where $L(n,\delta)$ is the expected loss when $\theta_k/\theta_i=\delta$, $i=1,\ldots,k-1$. Let δ^* be the value of δ such that $L(n,\delta^*)=\max_\delta L(n,\delta)$. It is not analytically feasible to minimize $L(n,\delta^*)$ with respect to n. One has to evaluate $L(n,\delta^*)$ over a grid of n values or $L(n,\delta)$ over a grid of n,δ values and obtain the solution by inspection. The numerical study of Ofosu (1972) indicates that for all realistic values of c_1, c_2, d, r, k, and n, the function $L(n,\delta)$ which takes the value $C(n)$ for $\delta=1$, increases to a unique maximum as δ increases, and then decreases approaching $C(n)$ as $\delta\to\infty$. Asymptotic ($\nu=nr\to\infty$) minimax solution is given by

$$n=\left[Nc_2M_{k-1}\{(2r)^{1/2}c_1dk\}^{-1}\right]^{2/(2d+1)}, \tag{5.14}$$

where $M_{k-1}=\max_u u\{1-F_{k-1,1/2}(u,\ldots,u)\}$ and $F_{k-1,1/2}(\cdot)$ denotes the cdf of the $(k-1)$-variate normal distribution with zero means, unit variances, and correlation coefficients $\frac{1}{2}$. Applications of these results are considered by Ofosu (1972) to the scale parameter problem for the Weibull distributions and to the normal variances problem.

5.2.3 Optimal Sampling Using Γ-Minimax Criterion

Gupta and D. Y. Huang (1977b) have discussed the problem of optimal sampling using the Γ-minimax criterion. In other words, they assume that the parameters are random variables and determine the sample size that minimizes the maximum expected risk over Γ, a class of prior distributions. If Γ consists of a single prior, then the Γ-minimax criterion is the Bayes criterion for that prior. At the other extreme, if Γ consists of all possible priors then the Γ-minimax criterion is the usual minimax criterion. Gupta and Huang (1977b) have discussed both the location and the scale parameter cases.

Let us first consider the location parameter case. The population π_i has the associated distribution function $F(x-\theta_i)$, where θ_i is unknown. A sample of size n is taken from each population. Let $G_n(y|F)$ be the cdf of $\bar{X}_i-\theta_i$, where $\bar{X}_i$ is the sample mean from π_i. We note that $G_n(y|F)$ is independent of θ_i. The population that yields the largest sample mean is declared as the population associated with $\theta_{[k]}$. Let p_i be the probability that π_i is selected. Clearly,

$$p_i=\int_{-\infty}^{\infty}\prod_{\substack{j=1\\ j\neq i}}^{k} G_n(x+\theta_i-\theta_j|F)\,dG_n(x|F). \tag{5.15}$$

Let $\Omega=\{\boldsymbol{\theta}|\boldsymbol{\theta}=(\theta_1,\ldots,\theta_k)\}$ and $\Omega_i=\{\boldsymbol{\theta}|\theta_i\geqslant\max_{j\neq i}\theta_j+\Delta\}$, where Δ is a given positive constant. Then $\Omega=\cup_{i=0}^{k}\Omega_i$, where Ω_0 is the indifference zone. Define

$$\delta_i(\mathbf{x})=\begin{cases}1 & \text{if } \bar{x}_i=\bar{x}_{[k]},\\ 0 & \text{otherwise.}\end{cases} \tag{5.16}$$

If ρ denotes a distribution over Ω, the expected loss is given by

$$\gamma_n(\rho)=ckn+\sum_{i=1}^{k}\sum_{j=1}^{k}\int_{\Omega_i}\int_{E^k}L^{(i)}(\boldsymbol{\theta},\delta_j(\mathbf{x}))\,dF_{\boldsymbol{\theta}}(\mathbf{x})\,d\rho(\boldsymbol{\theta}), \tag{5.17}$$

where c is the cost per observation and $L^{(i)}(\boldsymbol{\theta},\delta_j(\mathbf{x}))$ is the loss incurred by selecting π_j when $\boldsymbol{\theta}\in\Omega_i$. We consider the loss function $L^{(i)}(\boldsymbol{\theta},\delta_j(\mathbf{x}))=c'(\theta_i-\theta_j)\delta_j(\mathbf{x})$, where c' is a known positive constant.

It is assumed that partial information is available so that we are able to specify $\lambda_i = \Pr(0 \in \Omega_i), \Sigma_{i=0}^k \lambda_i = 1$. Define

$$\Gamma = \left\{ \rho(\boldsymbol{\theta}) | \int_{\Omega_i} d\rho(\boldsymbol{\theta}) = \lambda_i,\ i = 1, \ldots, k \right\}. \tag{5.18}$$

Gupta and Huang (1977b) have shown that

$$\sup_{\rho \in \Gamma} \gamma_n(\rho) = ckn + c'(1-\lambda_0)R_M, \tag{5.19}$$

where

$$R_M = \sup_{g \geq \Delta} \left[g - g\int_{-\infty}^{\infty} G_n^{k-1}(x+g)\,dG_n(x) \right]. \tag{5.20}$$

When π_i is the normal population $N(\theta_i, 1)$, we get

$$\sup_{\rho \in \Gamma} \gamma_n(\rho) = ckn + \sqrt{\frac{2}{n}}\ c'(1-\lambda_0)M_{k-1}, \tag{5.21}$$

where M_{k-1} is as defined in (5.14). The Γ-minimax solution is given by n^*, the smallest integer greater than or equal to $[(2c^2k^2)^{-1}c'^2(1-\lambda_0)^2 M_{k-1}^2]^{1/3}$. Evaluation of M_{k-1} has been discussed by Dunnet (1960). Gupta and Huang (1977b) have tabulated n^* for $k = 2(1)6$, $\lambda_0 = 0.1, 0.3, 0.5$, and $c'/kc = 5, 10, 15, 30, 50, 100$.

Gupta and Huang (1977b) have also discussed the scale parameter problem with $L^{(i)}(\theta, \delta_j(\mathbf{x})) = c'\delta_j(\mathbf{x})\log(\theta_i/\theta_j)$. The evaluation of $\sup_{\rho \in \Gamma}\gamma_n(\rho)$ is analogous to the earlier case. If we consider the special case of gamma populations (density as in (5.11)), it is not analytically feasible to minimize the maximum risk with respect to n by differentiation. One can determine n^* numerically as explained by Ofosu (1972). Asymptotic Γ-minimax solution can be obtained by usual considerations and it is given by the smallest integer greater than or equal to $(2r)^{-1}[1 + \{2rc'(1-\lambda_0) M_{k-1}/ck\}^{2/3}]$. Similar applications can be made in the case of the Weibull scale parameters problem and the normal variances problem.

5.3 ESTIMATION OF THE PROBABILITY OF CORRECT SELECTION

So far we have approached the problem of selecting the best of k populations from the point of view of designing an experiment that will guarantee a certain probability of a correct selection. However, once the sample has been taken and the decision made, one may be interested in estimating the true PCS based on the sample data. Olkin, Sobel, and Tong

(1976) have studied the point estimation problem for the location and scale parameter families. The interval estimation aspect is discussed by Anderson, Bishop, and Dudewicz (1977).

5.3.1 Point Estimation: Location Parameter Case

Let π_i, $i=1,\dots,k$, be independent populations, where π_i has the associated distribution function $F(x,\theta_i)$. We are interested in selecting the population associated with $\theta_{[k]}$ and we use the rule R which selects the population that yields the largest Y_i, where Y_i is an appropriate statistic based on n independent observations $X_{i1},\dots,X_{in}$ from π_i, $i=1,\dots,k$. For this rule R, the PCS is given by

$$\alpha(\delta_1,\dots,\delta_{k-1})=\int_{-\infty}^{\infty}\left[\prod_{i=1}^{k-1} G(y-\theta_i)\right]dG(y), \tag{5.22}$$

where $G(y-\theta_i)$ is the cdf of Y_i, and $\delta_i=\theta_{[k]}-\theta_{[i]}$.

Suppose that Y_i is a consistent estimator of θ_i. Then $\alpha(\hat{\delta}_1,\dots,\hat{\delta}_{k-1})$ will be a consistent estimator of $\alpha(\delta_1,\dots,\delta_{k-1})$, where $\hat{\delta}_i=Y_{[k]}-Y_{[i]}$, $i=1,\dots,k-1$. However, the evaluation of the estimate is often difficult. Olkin, Sobel, and Tong (1976) have given upper and lower bounds for the PCS that hold for any true configuration, with no regard to any least favorable configuration.

For any fixed $r\leqslant p$, we let $a_1,\dots,a_r$ be *distinct* positive integers such that

$$0\equiv a_0<a_1<\cdots<a_r\equiv p$$

and let $p_j=a_j-a_{j-1}$, $j=1,\dots,r$. Now define the partition

$$\delta(A)=\Big(\underbrace{\delta_1,\dots,\delta_{a_1}}_{p_1},\underbrace{\delta_{a_1+1},\dots,\delta_{a_2}}_{p_2},\dots,\underbrace{\delta_{a_{r-1}+1},\dots,\delta_p}_{p_r}\Big),$$

where $\sum_{j=1}^{r}p_j=p=k-1$. Thus $\delta(A)$ is the vector $\boldsymbol{\delta}=(\delta_1,\dots,\delta_{k-1})$ in which the components are partitioned into r sets. Let $\bar{\delta}_j(j=1,\dots,r)$ be the average of the δ-values in the jth set in the partition. Define $\bar{\delta}(A)$ as the vector obtained from $\delta(A)$ by replacing the δ-values in each set of the partition by the average value of that set. In other words,

$$\begin{aligned}\bar{\delta}(A)&\equiv\big(\bar{\delta}_1(A),\dots,\bar{\delta}_r(A)\big)\\&=\Big(\underbrace{\bar{\delta}_1,\dots,\bar{\delta}_1}_{p_1},\underbrace{\bar{\delta}_2,\dots,\bar{\delta}_2}_{p_2},\dots,\underbrace{\bar{\delta}_r,\dots,\bar{\delta}_r}_{p_r}\Big).\end{aligned} \tag{5.23}$$

Here A denotes the set $A=\{a_1,\dots,a_r\}$ which induces the partition.

Let $A \subset B$ mean that the partition $\delta(B)$ is a refinement of $\delta(A)$, that is, every partition set of $\delta(B)$ is contained in some partition set of $\delta(A)$. Olkin, Sobel, and Tong (1976) have obtained the following results.

Theorem 5.1. If $\log G(y)$ is concave, and $A \subset B$, then

$$\alpha(\delta_1,\ldots,\delta_p) \leqslant \alpha\big(\bar{\delta}_1(B),\ldots,\bar{\delta}_s(B)\big) \leqslant \alpha\big(\bar{\delta}_1(A),\ldots,\bar{\delta}_r(A)\big). \tag{5.24}$$

We can obtain bounds for $\alpha(\delta_1,\ldots,\delta_p)$ that reduce the dimensionality of the integrals involved. This is achieved in terms of the size of the intervals $a_j - a_{j-1}$ $(j=1,\ldots,r)$. Define

$$\beta(\delta_1(A),\ldots,\delta_r(A)) = \int_{-\infty}^{\infty} \prod_{j=1}^{a_1} G(y+\delta_j)\,dG(y) \times \int_{-\infty}^{\infty} \prod_{j=a_1+1}^{a_2} G(y+\delta_j)\,dG(y)$$

$$\times \cdots \times \int_{-\infty}^{\infty} \prod_{j=a_{r-1}+1}^{p} G(y+\delta_j)\,dG(y). \tag{5.25}$$

In particular, for $r=p$, we have $\beta(\delta_1,\ldots,\delta_p) = \alpha(\delta_1,\ldots,\delta_p)$. The following theorem is due to Kimball [1951].

Theorem 5.2. If $A \subset B$, then

$$\beta(\delta_1(B),\ldots,\delta_s(B)) \leqslant \beta(\delta_1(A),\ldots,\delta_r(A)) \leqslant \alpha(\delta_1,\ldots,\delta_p). \tag{5.26}$$

Of particular interest are the results for the partition $A = \{1,\ldots,p\}$, namely,

$$\prod_{j=1}^{p} \int_{-\infty}^{\infty} G(y+\delta_j)\,dG(y) \leqslant \alpha(\delta_1,\ldots,\delta_p)$$

$$= \int_{-\infty}^{\infty} \prod_{j=1}^{p} G(y+\delta_j)\,dG(y). \tag{5.27}$$

If G is normal, then the left-hand side of (5.27) is equal to $\prod_{j=1}^{p} G(\delta_j/\sqrt{2}\,)$, and the evaluation of the right-hand side integral has been discussed by Milton [1970, 1972].

Consider the case of normal distribution under configurations having equally spaced means with common spacing d. Then the right-hand side of (5.27) becomes

$$\int_{-\infty}^{\infty} \prod_{j=1}^{p} \Phi(y+jd)\,d\Phi(y) = \Pr\left\{ Z_j \leqslant \frac{jd}{\sqrt{2}}, j=1,\ldots,p \right\} \tag{5.28}$$

where the Z_i are equally correlated standard normal variables with correlation $\frac{1}{2}$. Olkin, Sobel, and Tong (1976) have tabulated the values of the integral in (5.28) for $d=0.00(0.02)0.20(0.10)4.00$ and $k=2(1)10$, 12, 15, and 20.

Consider again the case of normal distributions with known common variance which is taken to be unity. Here Y_i is the sample mean, and the PCS is given by

$$\alpha(\delta)\equiv\alpha(\delta_1,\ldots,\delta_p)=\int_{-\infty}^{\infty}\prod_{j=1}^{p}\Phi\left(t+\sqrt{n}\,\delta_i\right)d\Phi(t) \tag{5.29}$$

and its estimator $\alpha(\hat{\delta})=\alpha(\hat{\delta}_1,\ldots,\hat{\delta}_p)$ where $\hat{\delta}_i=Y_{[i]}$, $i=1,\ldots,k$. The following theorem of Olkin, Sobel, and Tong (1976) gives the asymptotic distribution of $\alpha(\hat{\delta})$.

Theorem 5.3. $\mathcal{L}(\sqrt{n}\,[\alpha(\hat{\delta})-\alpha(\delta)])\xrightarrow{d}N(0,v_n)$, where $v_n(\delta)=\Sigma_{i=1}^{p}d_i^2+(\Sigma_{i=1}^{p}d_i)^2$, and

$$d_i=\sqrt{\frac{n}{2}}\;\varphi\left(\sqrt{\frac{n}{2}}\;\delta_i\right)\int_{-\infty}^{\infty}\prod_{j\neq i}\Phi\left(\frac{t}{\sqrt{2}}+\sqrt{n}\,\delta_j-\frac{\sqrt{n}\,\delta_i}{2}\right)d\Phi(t). \tag{5.30}$$

The asymptotic variance v_n defined in Theorem 5.3 is complicated. One can use the estimate $\hat{\delta}_i$ in $v_n(\delta)$ and obtain a confidence interval for the PCS by using the fact that

$$\frac{\sqrt{n}\,(\alpha(\hat{\delta})-\alpha(\delta))}{\sqrt{v_n(\hat{\delta})}}\sim N(0,1).$$

5.3.2 Point Estimation: Scale Parameter Case

The handling of this case is analogous to the location parameter case with a few changes. If we are interested in the largest scale parameter, we define $\delta_i=\theta_{[k]}/\theta_{[i]}$, $i=1,\ldots,k-1$. For the smallest scale parameter, $\delta_{k-i+1}=\theta_{[i]}/\theta_{[1]}$. In either case $\delta_1\geqslant\cdots\geqslant\delta_{k-1}\geqslant 1$ and the PCS is given by

$$\alpha(\delta_1,\ldots,\delta_{k-1})=\int_{-\infty}^{\infty}\left[\prod_{i=1}^{k-1}G(\delta_i y)\right]dG(y). \tag{5.31}$$

If $G(y,\theta)=G(y/\theta)$ with $G(0)=0$, we take the $\bar{\delta}_j$ in (5.23) as the geometric means instead of arithmetic means. The results of Theorem 5.1 and 5.2 hold with obvious changes.

5.3.3 Interval Estimation

Anderson, Bishop, and Dudewicz (1977) considered one-sided (lower) confidence bound for the true PCS in the cases of selection from normal and exponential populations. We briefly discuss their approach in the case of normal populations in this section.

Let $\pi_i (1 \leqslant i \leqslant k)$ be a normal population with mean μ_i and variance σ^2. We consider the natural selection rule of Bechhofer (1954) for the goal of selecting the population associated with $\mu_{[k]}$. The rule selects the population that yields the largest sample mean based on samples of size n from the k populations. The approach of Anderson, Bishop, and Dudewicz is to obtain first a lower confidence bound on Δ/σ where $\Delta = \mu_{[k]} - \mu_{[k-1]}$, and then use it to obtain a lower confidence bound for PCS. Suppose that $\Pr\{\Delta/\sigma \geqslant L\} = \gamma$, then with $100\gamma\%$ confidence we can say that

$$\mathrm{PCS} \geqslant \int_{-\infty}^{\infty} \Phi^{k-1}(x + L\sqrt{n})\, d\Phi(x). \tag{5.32}$$

5.4 NOTES AND REMARKS

The idea of estimating the PCS has been used by Tong (1978) to define an adaptive sequential procedure for selecting the best population. The point estimate of PCS is used for defining the stopping rule. The idea here is to have the PCS approximately equal to P^* under the true configuration. This procedure is discussed in §9.3.

CHAPTER 6

Sequential Selection Procedures

6.1 INTRODUCTION

Early contributions to sequential ranking procedures under the indifference zone formulation date back to mid-fifties and these are attributed to Bechhofer and Sobel (1953, 1954b, 1956a, 1956b, 1956c, 1958a). Many of the results of these authors along with the contributions of Kiefer have taken the final form of a monograph by the three authors (1968). Their book lays special emphasis on selection from Koopman–Darmois family which includes many known particular cases. We do not attempt here to survey all their results; however, some of these results are referred to at appropriate places. Also, a brief discussion of the general approach of Bechhofer, Kiefer, and Sobel is presented in §6.8.

6.2 SELECTION OF THE LARGEST NORMAL MEAN

Stein (1948) considered the following problem. Let X_{ij}, $i=1,\ldots,k$; $j=1,2,\ldots,$ be independent random variables where X_{ij} is distributed normally with mean $\xi_i+\eta_j$ and variance σ_j^2. It is assumed that the ξ_i and η_j are unknown and the σ_j^2 are known. Let ϵ and α be fixed such that $\epsilon>0$ and $0<\alpha<1$. The problem of Stein can be described as follows. For every $l=1,2,\ldots,k$, and $\xi_1,\ldots,\xi_k,\eta_1,\eta_2,\ldots$ satisfying $\xi_l \geqslant \xi_i+\epsilon$ for $i\neq l$, we want to choose an $M\in\{1,2,\ldots,k\}$ such that $\Pr(M=l)\geqslant 1-\alpha$. For this problem Stein proposed the following procedure. For stage $n(n=1,2,\ldots)$, we compute

$$A_i=\sum_{j=1}^{n}\frac{1}{\sigma_j^2}\left[X_{ij}-\bar{X}_j-\frac{\epsilon(t_j-1)}{t_j}\right],\qquad i=1,\ldots,k,. \tag{6.1}$$

where $\bar{X}_j$ is the average of the observations with the same second subscript j and t_j is the number of such observations. We continue taking observations $X_{i,n+1},\ldots$ only for those i for which

$$A_i>\frac{\log\alpha}{\epsilon}. \tag{6.2}$$

Eventually there will be at most one subscript $i=1,\ldots,k$, for which one continues to take observations. If there is one, this is chosen to be M. If there is none, the subscript i for which A_i is maximum is chosen to be M. This procedure is a straightforward application of a lemma of Wald [1947, p. 146].

Bechhofer and Sobel (1954b) considered a sequential procedure for selecting the population with the largest mean from a set of k normal populations with unknown means $\mu_1,\ldots,\mu_k$ and a known common variance σ^2. A modification of this procedure is given by Bechhofer and Sobel (1953). This latter procedure is given in the monograph of Bechhofer, Kiefer, and Sobel (1968, pp. 258–259). We denote this procedure by R_1 and describe it in following text. The requirement is that $P(CS|R_1)\geqslant P^*$ whenever $\mu_{[k]}-\mu_{[k-1]}\geqslant\delta^*$.

Let X_{ij}, $j=1,2,\ldots,$ be a sequence of independent observations from π_i, $i=1,\ldots,k$. At the mth stage of the experiment, $m=1,2,\ldots,$ we take an observation from each of the k populations and compute the sample totals $Y_{im}=\Sigma_{j=1}^{m}X_{ij}$, $i=1,\ldots,k$, based on the first m observations. Let

$$W_m=\sum_{i=1}^{k-1}\exp\left\{-\frac{\delta^* D_{im}}{\sigma^2}\right\} \tag{6.3}$$

where $D_{im}=Y_{[k]m}-Y_{[i]m}$, $i=1,\ldots,k-1$ and $Y_{[1]m}\leqslant\cdots\leqslant Y_{[k]m}$ denote the ranked values of the Y_{im}. If $W_m\leqslant(1-P^*)/P^*$, stop experimentation and choose the population yielding $Y_{[k]m}$. If $W_m>(1-P^*)/P^*$, take another observation from each of the k populations and compute W_{m+1}. Continue in this manner until the rule calls for stopping. It has been shown by Bechhofer, Kiefer, and Sobel ((1968) Theorem 6.1.1.) that the P^*-condition is satisfied by R_1. Further the following result is true regarding $E(N|R_1)$, the expected number of stages to terminate the experiment using R_1.

Theorem 6.1. [Bechhofer, Kiefer, and Sobel (1968, p. 161)]. If $\mu_1 \leqslant \mu_2 \leqslant \cdots \leqslant \mu_{k-1} = \mu_k - \delta^*$, then

$$E(N|R_1) = \sigma^2(\delta^*)^{-2}\log(1-P^*)^{-1} + o(\log(1-P^*)) \text{ as } P^* \to 1. \quad (6.4)$$

Huang (1976) has obtained an upper bound for $E(T|R_1)$, the expected sample size using the procedure R_1. Let $a = \max(b/4, b^2/4\delta^{*2})$ where $b = (k-1)P^*(1-P^*)^{-1}$. Then Huang's bound is given by

$$E(T|R_1) \leqslant (k-1)\{(8+4a)(\delta^*)^{-2}+1\}. \quad (6.5)$$

Bechhofer (1958) proposed a procedure for the case of unknown σ^2; however, some difficulties arise with this procedure [see Bechhofer (1970)]. It is still of interest to note that the Monte Carlo robustness investigations of Kleijnen (1977) indicate that the procedure of Bechhofer (1958) works often though not always.

In the remaining part of this section, we discuss the procedures of Paulson (1964a), and Robbins, Sobel, and Starr (1968). In the next section, we discuss the procedures of Srivastava (1966), and Srivastava and Ogilvie (1968). The procedures of Paulson (1964a) satisfy the probability requirement for any fixed P^* and $\delta^* > 0$. The other procedures meet the requirement for a fixed P^* asymptotically as $\delta^* \to 0$ and no definitive result is available in the case of an arbitrarily fixed $\delta^* > 0$. Also, none of these procedures is shown to capitalize on favorable configurations of the population means; that is, the expected number of stages to terminate the experiment is not shown to decrease when the configuration of the population means becomes more favorable.

Paulson (1964a) considers both the cases where σ^2 is known and unknown. His procedure modifies that of Paulson (1962) and differs from the procedures of Bechhofer, Kiefer, and Sobel (1968) and Stein (1948). Let $X_{ij}, j = 1,2,\ldots,$ be a sequence of independent observations from π_i which is $N(\mu_i, \sigma^2)$. Let us first consider *the case of known* σ^2. Under this case, Paulson (1964a) defines a class $\{S_\lambda\}$ of sequential procedures indexed by a parameter λ $(0 < \lambda < \delta^*)$, using triangular stopping regions. Such regions were earlier considered among others by Armitage [1957], Donelly [1957], and Anderson [1960]. Given P^* and δ^*, define

$$a_\lambda = \frac{\sigma^2}{\delta^* - \lambda}\log\left(\frac{k-1}{1-P^*}\right),$$

$$W_\lambda = \text{the largest integer less than } \frac{a_\lambda}{\lambda}. \tag{6.6}$$

The procedure S_λ is defined as follows.

Let $X_{11},\ldots,X_{k1}$ be a set of one observation from each population. Eliminate from consideration any population π_i for which

$$X_{i1} < \max_{1 \leqslant r \leqslant k} X_{r1} - a_\lambda + \lambda. \tag{6.7}$$

If all but one population are eliminated, then stop the experiment and select the remaining population. Otherwise take one observation from each population not eliminated. At the mth stage, $m = 2,3,\ldots, W_\lambda$, eliminate any of the remaining population π_i for which

$$\sum_{j=1}^{m} X_{ij} < \max\left(\sum_{j=1}^{m} X_{rj}\right) - a_\lambda + m\lambda \tag{6.8}$$

where the maximum is taken over all oppulations left after the $(m-1)$st stage. If only one population is left after the mth stage, stop the experiment and select it. Otherwise, go to $(m+1)$st stage. If more than one population remains after the W_λth state, the experiment is terminated at the $(W_\lambda + 1)$st state by selecting the population for which the sum of $(W_\lambda + 1)$ observations ia a maximum. It is shown that

$$P(CS|S_\lambda) \geqslant P^* \quad \text{whenever} \quad \mu_{[k]} - \mu_{[k-1]} \geqslant \delta^*. \tag{6.9}$$

The optimum choice of λ is not known. Calculations of Paulson (1964a) indicate $\lambda = \delta^*/4$ to be a reasonably good choice. For some discussion on analytic choice of λ, see Fabian (1974a) who has also given an improved bound on the probability of an incorrect selection. Some Monte Carlo comparisons of the Bechhofer–Kiefer–Sobel procedures and S_λ are made by Ramberg (1966) and Ramberg and Bechhofer (1973).

For Paulson's procedure S_λ, let N_i denote the number of observations needed to eliminate π_i, $i = 1,\ldots,k$. The following theorem has been proved by Perng (1969).

Theorem 6.2. If $\mu_1 \leqslant \mu_2 \leqslant \cdots \leqslant \mu_{k-1} = \mu_k - \delta^*$, then

$$E(N_i|S_\lambda) = \sigma^2(\lambda + \mu_k - \mu_i)^{-1}(\delta^* - \lambda)^{-1}\log\frac{(k-1)}{(1-P^*)} + o\left(\log\frac{(k-1)}{(1-P^*)}\right) \tag{6.10}$$

as $P^* \to 1$, for $i = 1, \ldots, k-1$.

As a corollary to Theorem 6.2, we get

$$\Pr\{N_1 < N_2 < \cdots < N_{k-1}\} \to 1 \tag{6.11}$$

as $P^* \to 1$ when $\mu_1 < \mu_2 < \cdots < \mu_k$.

Perng (1969) has compared procedures R_1 and S_λ in terms of the asymptotic ($P^* \to 1$) expected total number of observations. Denoting the total number of observations by T, we have

$$\lim_{P^* \to 1}\left[\frac{E(T|R_1)}{E(T|S_\lambda)}\right] = k(\delta^* - \lambda)(\delta^*)^{-2}\left[\sum_{i=1}^{k-2}(\lambda + \mu_k - \mu_i)^{-1} + 2(\lambda + \delta^*)^{-1}\right]^{-1}. \tag{6.12}$$

When $k = 2$ or $\mu_1 = \mu_2 = \cdots = \mu_{k-1} < \mu_k$, (6.12) becomes

$$\lim_{P^* \to 1}\left[\frac{E(T|R_1)}{E(T|S_\lambda)}\right] = \left[(\delta^*)^2 - \lambda^2\right](\delta^*)^{-2} < 1. \tag{6.13}$$

From (6.13) one can conclude that if $k = 2$, or $\mu_1 = \cdots = \mu_{k-1} < \mu_k$, the procedure R_1 is more efficient asymptotically. From (6.12) we see that

a. if $\Sigma_{i=1}^{k-2}(\lambda + \mu_k - \mu_i)^{-1} < k(\delta^* - \lambda)(\delta^*)^{-2} - 2(\lambda + \delta^*)^{-1}$ then the procedure S_λ is more efficient asymptotically, and that

b. $\lim_{P^* \to 1}[E(T|R_1)/E(T|S_\lambda)]$ is a decreasing function of λ.

Intuitively speaking, this means that (1) S_λ is better when k and the differences $(\mu_k - \mu_i)$ are large, and (2) to maximize the asymptotic relative efficiency of S_λ, we should take a small value of λ.

When the common variance σ^2 is *unknown*, Paulson (1964a) considers two situations, namely, (A) one-way classification and (B) randomized block design. Let $X_{ij} \sim N(m_{ij}, \sigma^2)$. Then we have $m_{ij} = \mu_i$ and $m_{ij} = \mu_i + \nu_j$, respectively, in the two situations. For n_0 observations from each population, let

$$\bar{X}_{i.}=n_0^{-1}\sum_{j=1}^{n_0} X_{ij}, \bar{X}_{.j}=k^{-1}\sum_{i=1}^{k} X_{ij}, \quad \text{and} \quad \bar{X}=(kn_0)^{-1}\sum_{i=1}^{k}\sum_{j=1}^{n_0} X_{ij}.$$

Let s^2 be the usual estimate of σ^2 (depending on whether we have the situation A or B) so that $\nu s^2/\sigma^2$ is distributed as a chi-square with ν degrees of freedom where $\nu=k(n_0-1)$ or $(k-1)(n_0-1)$ depending on the situation. Define

$$g=\left[\left(\frac{k-1}{1-P^*}\right)^{2/\nu}-1\right]\frac{\nu}{2}, \qquad a^*=\frac{4s^2g}{3\delta^*}, \tag{6.14}$$

and let W^* be the largest integer less than $4a^*/\delta^*$. The procedure S' of Paulson (1964a) is defined as follows.

Take n_0 observations from each population. If $n_0>W^*$, stop the experiment and select the population with the largest cumulative sum $\Sigma_{j=1}^{n_0}X_{ij}$. If $n_0\leqslant W^*$, then eliminate π_i for which

$$\sum_{j=1}^{n_0} X_{ij}<\max_{1\leqslant r\leqslant k}\left(\sum_{j=1}^{n_0} X_{rj}\right)-a^*+4^{-1}n_0\delta^*. \tag{6.15}$$

If only one population is left, stop the experiment and select it. Otherwise, go to (n_0+1)st stage taking one observation from all the populations that are not eliminated. At the (n_0+t)th stage $(t=1,2,\ldots,W^*-n_0)$, we eliminate π_i for which

$$\sum_{j=1}^{n_0+t} X_{ij}>\max\left(\sum_{j=1}^{n_0+t} X_{rj}\right)-a^*+\frac{(n_0+t)\delta^*}{4} \tag{6.16}$$

where the maximum is taken over all populations not eliminated after the (n_0+t-1)st stage. If more than one population is left after the W*th stage, we terminate the experiment at the (W* + 1)st stage and select the population with the largest cumulative sum.

This procedure S' is shown to meet the probability requirement. The choice of n_0 is very important in determining the efficiency of the sequential procedure. A very small value of n_0 results in a large g which makes the procedure inefficient. A very large value of n_0 also makes it inefficient because there is no elimination until the n_0th stage. An optimum procedure for selecting n_0 is unknown. Paulson (1964a) suggests the following tentative procedure. Two cases arise: (1) No information about σ is available; (2) it is known from past experience that σ' is an approximate value of σ. Under Case (1), we select ν (which determines n_0) so that g shall not exceed its limiting ($\nu \to \infty$) value by more than a specified amount, say, 10 percent. Under Case (2), let $N(\sigma')$ = the number of stages required by the fixed sample size procedure corresponding to $\sigma = \sigma'$. We then take $n_0 = \min\{N(\sigma')/3, \bar{n}_0\}$, where $\bar{n}_0$ is the value of n_0 corresponding to ν of Case (1).

Another sequential procedure has been proposed and investigated by Robbins, Sobel, and Starr (1968) for selecting the largest normal mean when the common variance σ^2 is unknown. Let X_{ij}, $j = 1, 2, \ldots$ denote a sequence of independent observations from π_i, $i = 1, \ldots, k$, and define for $r \geqslant 2$

$$u_r = [k(r-1)]^{-1} \sum_{i=1}^{k} \sum_{j=1}^{r} \left(X_{ij} - \bar{X}_i(r)\right)^2 \tag{6.17}$$

where $\bar{X}_i(r)$ is the sample mean of r observations from π_i. For the procedure R_3 of Robbins, Sobel, and Starr (1968), sampling proceeds sequentially where at the rth stage we take a single observation from each of the populations and compute a fresh estimate u_r of σ^2. We terminate sampling at the Nth stage, where the random variable N is the least odd integer $n \geqslant 5$ such that

$$u_n \leqslant n\delta^{*2}h^{-2}, \tag{6.18}$$

where $h = h(k, P^*)$ is the solution of

$$\int_{-\infty}^{\infty} \Phi^{k-1}(y+h)\, d\Phi(y) = P^*. \tag{6.19}$$

If α is the smallest i such that $\bar{X}_i(N) \geqslant \bar{X}_l(N)$ for all $l = 1, \ldots, k$, we select π_α. The sample is terminated only with an odd number of k-tuples of observations to simplify the computations of the probability distribution of N. It is shown that

$$\liminf_{\delta^* \to 0} P(CS|R_3) \geqslant P^* \quad \text{for} \quad \boldsymbol{\mu} \in \Omega_{\delta^*} \tag{6.20}$$

where $\Omega_{\delta^*} = \{\boldsymbol{\mu}\colon\, \mu_{[k]} - \mu_{[k-1]} \geqslant \delta^*\}$.

As a measure of the cost of ignorance of σ^2 we define

$$I = I(\lambda) = E(N) - n^*, \tag{6.21}$$

where $\lambda = \sigma/\delta^*$ and $n^* = (\sigma h/\delta^*)^2$.

The following theorem is due to Robbins, Sobel, and Starr (1968).

Theorem 6.3. For all $\delta^*>0$, $0<\sigma^2<\infty$,

$$I \leqslant 5. \tag{6.22}$$

REMARK 6.1. The result in (6.20) holds whatever the distribution of the X_{ij} provided only that $\sigma^2<\infty$. Also, the inequality (6.22) cannot be sharpened since the procedure requires that $N \geqslant 5$, and $I \rightarrow 5$ as $\delta^* \rightarrow \infty$.
□

It should be noted that Paulson's procedure combines two properties that other indifference zone procedures do not typically possess; namely, the procedure is truncated and fully sequential. Truncation is self-explanatory, and by fully sequential we mean that the procedure is able to discontinue taking observations from a particular population before the procedure terminates sampling altogether. A fully sequential procedure will often make a substantial savings in the total number of observations taken by quickly eliminating "bad" populations if they are present among the populations under consideration.

In the next section we discuss sequential procedures for selecting the population with the largest mean from k populations having fixed distributions F_i (possibly unknown) with finite but unknown variance σ^2. These procedures could be called asymptotically nonparametric in the sense that the optimum properties of these procedures are large sample properties and the asymptotic normality of chosen statistics hold.

6.3 ASYMPTOTICALLY NONPARAMETRIC PROCEDURES FOR SELECTING THE LARGEST MEAN

Let π_i $(i=1,\ldots,k)$ be a population having an unknown distribution F_i with an unknown mean μ_i and an unknown but finite variance σ^2. The problem is to design a sequential procedure R for selecting the population with the largest mean so that

$$\lim_{\delta^* \rightarrow 0} \inf_{\Omega(\delta^*)} P(CS|R) \geqslant P^* \tag{6.23}$$

where $\Omega(\delta^*)=\{\boldsymbol{\mu}: \mu_{[k]}-\mu_{[k-1]} \geqslant \delta^*\}$.

The procedure of Robbins, Sobel, and Starr (1968) discussed in §6.2 is applicable here (see Remark 6.1). For this problem Srivastava (1966) has investigated two classes of sequential procedures which are discussed in the following text.

Let X_{ij} denote the jth observation from population π_i, $i=1,2,\ldots,k$; $j=1,2,\ldots$. The observations are all assumed to be independent. Our attention is restricted to two situations: (A) one-way classification and (B) randomized block design. If we let $E(X_{ij})=\theta_{ij}$, then $\theta_{ij}=\mu_i$ in the case of (A) and $\theta_{ij}=\mu_i+\nu_j$ in the case of (B). The ranking of the populations is in terms of the μ_i.

Define

$$v_n=(nk)^{-1}\left\{\sum_{i=1}^{k}\sum_{j=1}^{n}\left(X_{ij}-\bar{X}_i(n)\right)^2+1\right\},\qquad n>1, \tag{6.24}$$

where $\bar{X}_i(n)=n^{-1}\Sigma_{j=1}^{n}X_{ij}$. Let $a_1,a_2,\ldots$ be any sequence of positive constants such that

$$\lim a_n=a \tag{6.25}$$

and

$$\Pr(Z\leqslant -a)=\frac{(1-P^*)}{(k-1)} \tag{6.26}$$

where Z is a standard normal variable.

This sequence $\{a_n\}$ determines a member of the class $\mathcal{C}$ of sequential procedures defined by Srivastava (1966) and described as follows.

$\mathcal{C}$: Sample one observation at a time from each population and stop according to the stopping variable N defined by

$$N=\text{smallest } n>1 \text{ such that } v_n\leqslant n\left(\frac{\delta^*}{\sqrt{2}\,a_n}\right)^2. \tag{6.27}$$

When sampling is stopped at $N=n$, select the population with the largest sample mean.

The following theorem of Srivastava (1966) states the asymptotic properties of the class $\mathcal{C}$.

Theorem 6.4. Under the assumption that $0<\sigma^2<\infty$,

$$\lim_{\delta^*\to 0}\frac{\delta^{*2}N}{2a^2\sigma^2}=1\quad\text{a.s.}, \tag{6.28}$$

$$\lim_{\delta^*\to 0}\inf_{\Omega(\delta^*)}P(CS|\mathcal{C})\geqslant P^*, \tag{6.29}$$

$$\lim_{\delta^*\to 0}\frac{\delta^{*2}E(N)}{2a^2\sigma^2}=1. \tag{6.30}$$

Any procedure of the class $\mathcal{C}$ is said to be *asymptotically consistent* if (6.29) holds and *asymptotically efficient* if (6.30) holds. The asymptotic efficiency here implies that the procedure has expected sample size equal (asymptotically as $\delta^* \to 0$) to the smallest sample size one could use if σ^2 were known.

The property (6.28) is proved by using the following lemma of Chow and Robbins [1965].

Lemma 6.1. Let y_n $(n=1,2,\ldots)$ be a sequence of random variables and $f(n)$ be any sequence of constants such that $y_n > 0$ a.s., $\lim_{n\to\infty} y_n = 1$ a.s., $f(n) > 0$, $\lim_{n\to\infty} f(n) = \infty$, and $\lim_{n\to\infty} f(n)/f(n-1) = 1$. For each $t > 0$, define

$$N = N(t) = \text{smallest } m \geqslant 1 \text{ such that } y_m \leqslant \frac{f(m)}{t}. \tag{6.31}$$

Then N is well defined and nondecreasing as a function of t, and

$$\lim_{t\to\infty} N = \infty \text{ a.s.}, \quad \lim_{t\to\infty} E(N) = \infty, \quad \lim_{t\to\infty} \frac{f(N)}{t} = 1 \text{ a.s.} \tag{6.32}$$

An alternative class $\mathcal{C}'$ of sequential procedures proposed by Srivastava (1966) is as follows. Let $b_1, b_2, \ldots$ be any sequence of positive constants such that

$$\lim_{n\to\infty} b_n = b \tag{6.33}$$

and

$$\Pr\left\{ \max_{1 \leqslant j \leqslant k-1} (Z_j - Z_k) \leqslant \sqrt{2}\, b \right\} = P^* \tag{6.34}$$

where $Z_1, \ldots, Z_k$ are independent $N(0,1)$ variables. The class $\mathcal{C}'$ is defined similar to $\mathcal{C}$ with b_n instead of a_n in (6.27). This class $\mathcal{C}'$ also enjoys those properties of $\mathcal{C}$ given in Theorem 6.4. These two classes of procedures with some slight changes have been compared by Srivastava and Ogilvie (1968) in the case of normal populations. Let

$$V_n = (km)^{-1} \sum_{i=1}^{k} \sum_{j=1}^{n} \left(X_{ij} - \bar{X}_i(n) \right)^2, \quad n > 1, \tag{6.35}$$

where $m = n-1$. Let $H_\nu(x)$ denote the cdf of a Student's t random variable with ν degrees of freedom. Suppose a and a_{km} are chosen so that

$$\Phi(-a) = H_{km}(-a_{km}) = (1-P^*)(k-1)^{-1}. \tag{6.36}$$

Then we have $\lim_{m\to\infty} a_{km} = a$ and the sequence $\{a_{km}\}$ determines a procedure A of the class $\mathcal{C}$.

In the case of $\mathcal{C}'$, we choose b given by (6.34) and b_{km} such that $G_{km}(b_{km}) = P^*$ where $G_{km}(x)$ is an appropriate multivariate t-distribution in order to yield $\lim_{m\to\infty} b_{km} = b$. Then the sequence $\{b_{km}\}$ determines a procedure B of the class $\mathcal{C}'$.

The stopping variable N is defined to be the smallest integer $n > 1$ such that $V_n \leqslant n(\delta^*/\sqrt{2}\, a_{km})^2$ in the case of $\mathcal{C}$, and $V_n \leqslant n(\delta^*/\sqrt{2}\, b_{km})^2$ in the case of $\mathcal{C}'$.

We note that for $k=2$ the two procedures are identical. Let $\lambda = \delta^*/(\sqrt{2}\,\sigma)$. Further let $e_A(\lambda)$ and $e_B(\lambda)$ denote the probability of not selecting the best population using the procedures A and B, respectively, when $\boldsymbol{\mu} \in \Omega(\delta^*)$. Computations of these error probabilities and the average sample sizes for $k=4$ and 6 and several values of λ indicate that the average sample size for the procedure B is less than that for the procedure A, whereas the error probabilities are almost the same.

6.4 SELECTION OF THE LARGEST CENTER OF SYMMETRY

Let $\pi_1, \ldots, \pi_k$ be k populations with distribution functions $F(x-\theta_1), \ldots, F(x-\theta_k)$ where F is continuous and symmetric about the origin, so that $\theta_1, \ldots, \theta_k$ are the centers of symmetry of the respective distributions. The function F and the θ_i are unknown. Our goal is to select the population associated with the largest θ_i with probability at least P^* whenever $\theta_{[k]} - \theta_{[k-1]} \geqslant \delta$, where P^* and δ are preassigned constants with $k^{-1} < P^* < 1$ and $\delta > 0$. For this problem we describe a general method of constructing sequential procedures discussed by Geertsema (1972).

At each stage of sampling we take one independent observation from each population. Let X_{ij} denote the jth observation from π_i. Let $L_i(n) = L(X_{i1}, \ldots, X_{in})$ be an estimator of θ_i such that as $n \to \infty$, $L_i(n)$ has the normal distribution $N(\theta_i, (nA^2)^{-1})$, $i = 1, \ldots, k$, where A is a constant depending upon F. The stopping variable $N = N(\delta)$ is defined to be the first $n \geqslant n_0$ such that $n \geqslant h^2 S_n^2/\delta^2$ where n_0 is a suitable positive integer, S_n^2 is an estimator of A^{-2} based on all k populations, and h is given by (6.19). Upon termination of sampling, select π_r where r is the smallest i such that $L_i(N) \geqslant L_\alpha(N)$ for all $\alpha = 1, \ldots, k$.

The procedure of Robbins, Sobel, and Starr (1968) discussed in §6.2 is of this type. For the asymptotic analysis of the procedure just described we need a set of assumptions. In other words, we restrict F to a subclass Λ of the continuous and symmetric distributions for which the following conditions hold.

Conditions 6.1. For all $\boldsymbol{\theta}=(\theta_1,\ldots,\theta_k)\in\Omega$, and $F\in\Lambda$,

a.

$$\sqrt{n}\,(L_i(n)-\theta_i)=A^{-1}Z_i(n)+o(1) \text{ a.s.} \qquad \text{as } n\to\infty,$$

where $Z_i(n)$ is a standardized average of independent and identically distributed random variables having finite second moment and $A>0$ is a constant depending only on F.

b. $\lim S_n^2=A^{-2}$ a.s. as $n\to\infty$.

c. The set $\{\delta^2N(\delta)\}_{\delta>0}$ is uniformly integrable.

Then the following results hold for the sequential procedure (call it R_2) discussed previously.

Theorem 6.5. Under the Conditions 6.1 the following results hold for all $\boldsymbol{\theta}\in\Omega$ and all $F\in\Lambda$:

(1) $N(\delta)\uparrow\infty$ as $\delta\to0$ a.s., $EN(\delta)\to\infty$ as $\delta\to0$,
(2) $\lim_{\delta\to0}\delta^2N(\delta)=h^2A^{-2}$ a.s.,
(3) $\lim_{\delta\to0}\delta^2E(N(\delta))=h^2A^{-2}$,
(4) $\lim_{\delta\to0}\inf_{\Omega(\delta)}P(CS|R_2)\geqslant P^*$,

where $\Omega(\delta)=\{\boldsymbol{\theta}\colon \theta_{[k]}-\theta_{[k-1]}\geqslant\delta\}$.

These results can be used when we employ statistics other than the sample means. Two examples considered by Geertsema (1972) are procedures based on sample medians and estimators of Hodges and Lehmann [1963].

When $L_i(n)$ are the sample medians we choose

$$S_n^2=n\left(4kz_\alpha^2\right)^{-1}\sum_{i=1}^{k}\left(X_{n,a(n)}(i)-X_{n,b(n)}(i)\right)^2, \tag{6.37}$$

where $X_{n,1}(i)\leqslant\cdots\leqslant X_{n,n}(i)$ are the order statistics of the sample from π_i, $i=1,\ldots,k$,

$$b(n)=\max\left\{1,\left[\frac{n}{2}-\frac{z_\alpha\sqrt{n}}{2}\right]\right\},$$

$$a(n)=n-b(n)+1 \tag{6.38}$$

$$\Phi(z_\alpha)=1-\alpha, \qquad \tfrac{1}{2}<\alpha<1,$$

and $[x]$ is the largest integer less than or equal to x.

The Hodges–Lehmann estimator $L_i(n)$ for θ_i is the sample median of the $n(n+1)/2$ quantities $(X_{ij}+X_{ij'})/2$ for $1\leqslant j\leqslant j'\leqslant n$, $i=1,\ldots,k$. In this case choose

$$S_n^2 = n(4kz_\alpha^2)^{-1} \sum_{i=1}^{k} (W_{n,a(n)}(i) - W_{n,b(n)}(i))^2 \tag{6.39}$$

where $W_{n,1}(i) \leqslant \cdots \leqslant W_{n,n(n+1)/2}(i)$ are the ordered $(X_{ij}+X_{ij'})/2$ for $1 \leqslant j \leqslant j' \leqslant n$,

$$b(n) = \max\left\{1, \left[\frac{n(n+1)}{4} - z_\alpha\left(\frac{n(n+1)(2n+1)}{24}\right)^{1/2}\right]\right\},$$

$$a(n) = \frac{n(n+1)}{2} - b(n) + 1. \tag{6.40}$$

In both the cases some regularity conditions are assumed so that the Conditions 6.1 are satisfied.

Carroll (1974a) has considered several generalizations of the results of Geertsema (1972), Robbins, Sobel, and Starr (1968), and Srivastava (1966). Suppose $\pi_1,\ldots,\pi_k$ are k populations with distribution functions $F_{\theta_i}(x)$, $i=1,\ldots,k$, where $\theta_i \in \Theta$, a compact set. Let $X_{i1},\ldots,X_{in}$ be n independent observations from π_i and $\{w_n\}$ be a sequence of constants such that

$$w_n > 0, \quad \lim_{n\to\infty} w_n = \infty, \quad \text{and} \quad \lim_{n\to\infty}\left(\frac{w_n}{w_{n-1}}\right) = 1. \tag{6.41}$$

We assume that for each π_i there are statistics $L_i(n) = L(X_{i1},\ldots,X_{in})$ and $g_{in} = g_n(X_{i1},\ldots,X_{in})$ and constants μ_{θ_i} and A_{θ_i} such that

(i) $g_{in} \to A_{\theta_i}$ a.s. and $g_{in} > 0$ a.s.

(ii) $w_n(L_i(n) - \mu_{\theta_i})/A_{\theta_i} \xrightarrow{\mathcal{L}} N(0,1)$.

(iii) $\{L_i(n)\}$ are uniformly continuous in probability (see Anscombe [1952]).

(iv) $k_n, l_n \to \infty$ and $w_{k_n}/w_{l_n} \to 1$ imply $k_n/l_n \to 1$.

Let $N_i(\delta)$ be the first integer n such that $w_n \geqslant h g_{in}/\delta$, where h is given by (6.19). The procedure R selects the population with the largest $L_i(N_i(\delta))$ as the best. The main theorem of Carroll (1974a) states (under a set of sufficient conditions) that

$$\lim_{\delta\downarrow 0} \inf_{\substack{\boldsymbol{\mu}\in\Omega(\delta) \\ \boldsymbol{\theta}\in\Theta^k}} P(CS|R) \geqslant P^*, \tag{6.42}$$

where $\Theta^k = \Theta \times \ldots \times \Theta$ and

$$\Omega(\delta) = \{\boldsymbol{\mu} = (\mu_1,\ldots,\mu_k) \mid \mu_i = \mu_{\theta_i}, \mu_{[k]} - \mu_{[k-1]} \geqslant \delta\}.$$

6.5 A PAULSON-TYPE PROCEDURE BASED ON LIKELIHOOD RATIOS

As we remarked in §6.2, the procedure of Paulson (1964) for selecting the largest normal mean is both truncated and fully sequential unlike other procedures. By Paulson-type we mean a procedure having these two features. Hoel (1971a) has discussed a method of constructing such sequential procedures based on log-likelihood ratios.

Let $\pi_1,\ldots,\pi_k$ be k populations whose distribution functions are not completely known. Let X_{ij} denote the jth observation from π_i. It is assumed that the observations $\{X_{ij}\}$ are independent for all i and j. Let $T_{ij}(n)$, $n=1,2,\ldots,$ be a sequence of statistics defined by

$$T_{ij}(n)=f_n(X_{i1},\ldots,X_{in};\ X_{j1},\ldots,X_{jn}) \tag{6.43}$$

as functions of the first n observations from π_i and π_j. In a given problem, the function f_n is chosen so that it indicates the differences between the populations in a reasonable way. We assume that $\boldsymbol{T}_{ij}(n)=(T_{ij}(1),\ldots,T_{ij}(n))$ has a joint probability density function $g_{n,\tau_{ij}}(\boldsymbol{t}_{ij}(n))$ depending on the parameter τ_{ij}. Usually $T_{ij}(n)$ is a sufficient and transitive sequence and also invariantly sufficient for τ_{ij} (see Hall, Wijsman, and Ghosh [1965]). The parameter τ_{ij} is used as a measure of distance between the populations π_i and π_j. Define

$$\tau=\max_{i}\min_{j\neq i}\tau_{ij}. \tag{6.44}$$

The best population is π_i for which $\min_{j\neq i}\tau_{ij}=\tau$. The problem is to design a sequential procedure for which PCS $\geqslant P^*$ whenever $\tau\geqslant\tau^*$, where $k^{-1}<P^*<1$ and $\tau^*>\tau_{ii}$ are preassigned constants. We further assume that the functions $\{f_n\}$ have been chosen so that there exists a monotonically decreasing function h such that

$$\tau_{ji}=h(\tau_{ij}). \tag{6.45}$$

Let $\tau_0=h(\tau^*)$ and define the log-likelihood ratios

$$l_n(\boldsymbol{T}_{ij}(n))=\log g_{n,\tau_1}(\boldsymbol{T}_{ij}(n))-\log g_{n,\tau_0}(\boldsymbol{T}_{ij}(n)). \tag{6.46}$$

In (6.46) we assume that τ_1 is a fixed constant greater than τ_0.

The procedure (denoted by R_3) of Hoel (1971a) is defined as follows:

$\boldsymbol{R_3}$: Begin at stage one by taking a single observation from each of the k populations. Calculate the values of the $k(k-1)$ log-likelihood ratios

$l_1(T_{ij}(1))$, $i=1,\ldots,k$; $j=1,\ldots,k$; $i\neq j$. If for some i

$$l_1(T_{ij}(1)) \geqslant \log\{(k-1)(1-P^*)^{-1}\}, \tag{6.47}$$

eliminate population π_j from further consideration. If only one population remains we terminate sampling and declare this population as the best. If two or more populations remain we proceed to the next stage and take one observation from each of the remaining populations. The log-likelihood ratios for the contending populations are again computed and the same elimination rule is used except that $l_2(T_{ij}(2))$ replaces $l_1(T_{ij}(1))$ everywhere. We continue in this manner until only one population is left at which time the procedure is terminated and this population is selected as the best. If at any stage (n say) all the remaining populations are eliminated, we then select as the best from among these populations sampled at stage n that π_i for which $\min_{j\neq i} l_n(T_{ij}(n))$ is a maximum.

By varying the choice of $\tau_1 > \tau_0$, we have a family of selection procedures. If we pick $\tau_1=\tau^*$, the comparison made by the procedure between any two populations becomes a sequential probability ratio test (SPRT) with stopping boundaries $(-a,a)$ where $a=\log\{(k-1)(1-P^*)^{-1}\}$. Thus the procedure can be considered as consisting of $k(k-1)/2$ simultaneous SPRTs when $\tau_1=\tau^*$. The following theorem of Hoel (1971a) states the sufficient conditions for R_3 to satisfy the P^*-condition.

Theorem 6.6. Assume that for each i and j

$$P_{\tau_{ij}}\left[l_n(T_{ij}(n)) \geqslant \log\{(k-1)(1-P^*)^{-1}\}\right] \tag{6.48}$$

is a nondecreasing function of τ_{ij}. Then the sequential procedure satisfies the requirement:

$$P(CS|R_3) \geqslant P^* \quad \text{whenever} \quad \tau \geqslant \tau^*, \tag{6.49}$$

provided that the procedure terminates with probability one.

In all applications considered, the procedure R_3 is truncated whenever τ_1 is chosen in the open interval (τ_0, τ^*). Also, it has been noted that there is less over protection (PCS $-\,P^*$) when $\tau_1 \geqslant \tau_*$ where $\tau_* = h(\tau_{ii})$. Thus, one will generally choose τ_1 in the interval $[\tau_*, \tau^*]$ and a choice recommended by Hoel (1971a) is $\tau_1 = (\tau_* + \tau^*)/2$.

The procedure R_3 can be applied to selecting the population with the largest mean from a set of k normal populations with unknown means $\theta_1,\ldots,\theta_k$ and known common variance σ^2. Choose $T_{ij}(n)=\Sigma_{r=1}^n(X_{ir}-X_{jr})$.

Here $\tau_{ij}=\theta_i-\theta_j$ and $\tau=\theta_{[k]}-\theta_{[k-1]}$. The rule for eliminating populations at any stage n is: Eliminate π_j if

$$\sum_{r=1}^{n} X_{jr} \leqslant \max \sum_{r=1}^{n} X_{ir} - 2\sigma^2 a(\tau_1-\tau_0)^{-1} - 2^{-1}n(\tau_0+\tau_1), \qquad (6.50)$$

where $a=-\log[(k-1)(1-P^*)^{-1}]$ and the maximum is taken over those populations which remain under consideration at stage n. This procedure is truncated if $\tau_1<-\tau_0=\tau^*$ and it is precisely the procedure S_λ of Paulson (1964) [discussed in §6.2], who requires that $\tau_0<\tau_1<-\tau_0$ and recommends $\tau_1=-\tau_0/2=(\tau_*+\tau^*)/2$.

6.6 SELECTION OF THE NORMAL POPULATION WITH THE SMALLEST VARIANCE

Let $\pi_1,\dots,\pi_k$ be k normal populations $N(\mu_i,\sigma_i^2)$, $i=1,\dots,k$. The μ_i and σ_i are all unknown. For selecting the population associated with $\sigma_{[1]}^2$, Hoel (1971a) has applied his sequential procedure R_3 which was described in the previous section. Define

$$\bar{X}_{im}=m^{-1}\sum_{r=1}^{m} X_{ir},$$

$$S_m^2=(m-1)^{-1}\sum_{r=1}^{m}\left(X_{ij}-\bar{X}_{im}\right)^2, \qquad (6.51)$$

and let $T_{ij}(n)=s_{in}^2/s_{jn}^2$, $n=1,2,\dots$. Thus

$$l_n(T_{ij}(n))=(n-1)\log\left\{\frac{\left(\tau_0^{-1}+T_{ij}(n)\right)}{\left(\tau_1^{-1}+T_{ij}(n)\right)}\left(\frac{\tau_0}{\tau_1}\right)^{1/2}\right\} \qquad (6.52)$$

with $\tau_{ij}=\sigma_j^2/\sigma_i^2$ and $\tau=\sigma_{[2]}^2/\sigma_{[1]}^2$. It is shown by Hoel (1971a) that the procedure can continue for at most

$$1+\log\{(k-1)(1-P^*)^{-1}\}\log^{-1}\left\{\frac{\tau_0^{-1}+1}{\tau_1^{-1}+1}\left(\frac{\tau_0}{\tau_1}\right)^{1/2}\right\}$$

stages provided $\tau_1<\tau_0^{-1}$. The procedure terminates with probability 1 if $\tau_1=\tau_0^{-1}$. Thus the probability requirement is satisfied for $\tau_1\leqslant\tau_0^{-1}$. The procedure is both location and scale invariant.

Empirical results of Hoel (1971a) indicate that truncated procedures can often make considerable savings in undesirable cases without giving up too much when the parameters are in the least favorable configuration.

Bechhofer, Kiefer, and Sobel (1968) have considered a different measure of distance between π_i and π_j, namely $\sigma_{[j]}^{-2} - \sigma_{[i]}^{-2}$ for $i \geqslant j$. They have shown that in this case there does not exist a truncated procedure; and have given a procedure [p. 117] for the case of the largest variance. One can also use an analog of Paulson's procedure for normal means by eliminating population π_i at stage m if $s_{im}^2 > \min_{j \neq i} s_{jm}^2 + 2a\Delta^{-1}$. For this procedure, PCS $\geqslant P^*$ whenever $\sigma_{[1]}^{-2} - \sigma_{[2]}^{-2} \geqslant \Delta$.

6.7 SELECTION FROM POPULATIONS BELONGING TO KOOPMAN–DARMOIS FAMILY

As we mentioned in the beginning of this chapter, the monograph of Bechhofer, Kiefer, and Sobel (1968) deals with sequential identification and ranking procedures with special reference to Koopman–Darmois family. For all the selection problems discussed earlier, Bechhofer, Kiefer, and Sobel have given sequential procedures. As stated earlier, we explain in the next section the general approach of their monograph. In this section we are concerned with the procedures of Hoel and Mazumdar (1968) and Hoel and Weiss (1974).

Let $\pi_1, \ldots, \pi_k$ be independent populations with associated density functions $f_{\theta_i}(x)$, $i = 1, \ldots, k$, where

$$f_{\theta_i}(x) = \exp\{T(x)Q(\theta_i) + R(x) + S(\theta_i)\}. \tag{6.53}$$

The function $Q(\cdot)$ is assumed to be a continuous, strictly increasing function of θ. The functions T, Q, R, and S are assumed to satisfy usual regularity conditions. Let $\tau_i = Q(\theta_i)$. Then the τ_i are called the Koopman–Darmois parameters. Hoel and Mazumdar (1968) have investigated a procedure to select the population associated with $\tau_{[k]}$ satisfying the requirement that PCS $\geqslant P^*$ whenever $\tau_{[k]} - \tau_{[k-1]} \geqslant \delta^* > 0$. Their procedure is an extension of the procedure of Paulson (1964a) discussed in §6.2. However, the present procedure is not truncated whereas Paulson's is. As a matter of fact, Bechhofer, Kiefer, and Sobel (1968) have shown that in general no procedure using a finite number of stages can guarantee the probability requirement for all the members of the Koopman–Darmois family when the measure of distance between two populations π_i and π_j is taken to be $|\tau_i - \tau_j|$. The Hoel–Mazumdar procedure, of course, terminates with probability one.

Let X_{ij} denote the jth observation from π_i, $i=1,\ldots,k$. The X_{ij} are mutually independent. Let $Y_{im}=\Sigma_{j=1}^{m}T(X_{ij})$, $i=1,\ldots,k$; $m=1,2,\ldots$. The procedure of Hoel and Mazumdar (1968) (denoted by R_4) is defined as follows.

R_4: Take one observation from each population and compute Y_{i1}, $i=1,\ldots,k$. Eliminate π_i if

$$Y_{i1} \leqslant \max_r Y_{r1} - (\delta^*)^{-1}\log\{(k-1)(1-P^*)^{-1}\}, \tag{6.54}$$

$i=1,\ldots,k$. For the next stage, take one observation from every population not eliminated at the first stage. At stage m, we eliminate any population π_i not eliminated until $(m-1)$st stage if

$$Y_{im} \leqslant \max_r Y_{rm} - (\delta^*)^{-1}\log\{(k-1)(1-P^*)^{-1}\}, \tag{6.55}$$

where the maximum is taken over those populations not eliminated until $(m-1)$st stage. This procedure is terminated when only one population has not been eliminated and this population is selected.

It has been shown by Hoel and Mazumdar that R_4 guarantees the P^*-requirement. However, the number of inequalities used in its construction makes the exact PCS always exceed the nominal requirement P^*. The amount of overprotection for low values of P^* may be considerable. Some improvements have been achieved by Hoel and Mazumdar (1968) for $k \leqslant 5$ by strengthening one of the inequalities used.

Hoel and Weiss (1974) have discussed the problem of choosing the better of two negative exponential lifetime distributions. In terms of clinical trials, we assume that the success or failure of a treatment is measured by the lifetime following it, and that this lifetime has a negative exponential distribution. Let the parameters of the two distributions π_1 and π_2 be λ_1 and λ_2 (i.e., the mean survival times are λ_1^{-1} and λ_2^{-1}), respectively, with $\lambda_2 > \lambda_1$. Let Δ^* be a constant satisfying $0<\Delta^*<1$, and T^* be a fixed time, both chosen in advance of the trial. Most of the investigation of Hoel and Weiss is concerned with the two-decision problem that tests H_1: $\exp(-\lambda_1 T^*) = \Delta^* + \exp(-\lambda_2 T^*)$ against H_2: $\exp(-\lambda_2 T^*) = \Delta^* + \exp(-\lambda_1 T^*)$. In other words, the survival probabilities at time T^* are compared. They consider two stopping rules. One makes use of the likelihood function and the other is an SPRT type stopping rule proposed by Flehinger and Louis [1972]. The purpose of their investigation is to compare several methods for adaptively allocating treatments to patients who arrive sequentially. Based on results obtained by simulation, the technique of Flehinger and Louis appears to be effective in reducing the number of tests on the poorer treatment.

Discrete analogs of designs proposed by Flehinger and Louis [1972], and Hoel and Weiss (1974) have been used by Simon, Weiss, and Hoel (1975) for selecting the better of two binomial populations. This has been discussed in §4.5.2.

6.8 OTHER CONTRIBUTIONS

As we have mentioned earlier, substantial contributions to the theory of sequential selection and ranking problems have been made by Bechhofer, Kiefer, and Sobel (referred to in this section as BKS for short) in their 1968 monograph which began as a short paper more than 12 years before it was published. The extent of this work is large and therefore we will not attempt to survey the results in detail. We have, in fact, referred to some of the results in the earlier sections. Our aim here is to provide a brief summary explaining the general approach of the authors. The reader is referred to the reviews of this monograph by Dudewicz (1970b), Fabian (1970b), and Gupta (1972).

In the beginning of their momograph, BKS develop a sequential procedure for multiple decision problems. This procedure is a generalization of Wald's Sequential Probability Ratio Test (WSPRT) and it coincides with WSRPT in the special case of $k=2$ populations. For a certain class of sequential procedures which includes their basic ranking procedure, they have developed useful bounds for the expected loss and ASN.

The emphasis is on untruncated procedures without elimination which facilitates obtaining of important quantitative properties. The basic approach of BKS is to obtain the solution of a ranking problem through that of an identification problem (discussed in §6.8.1). Though the identification problems are of interest in themselves, the emphasis is on the insight they provide into the solution of a class of ranking problems.

6.8.1 Identification Problems

The basic identification problem can be described in simple terms as follows. Suppose that $\mathbf{X}=(X_1,\ldots,X_k)$ is a vector of observations having the frequency function $f(\mathbf{x}|\boldsymbol{\theta};\psi)=f(x_1,\ldots,x_k|\theta_1,\ldots,\theta_k;\psi)$ where $\theta_1,\theta_2,\ldots,\theta_k$ are the unknown scalar parameters that are of interest in our identification or ranking problem, and ψ is an unknown vector of nuisance parameters. The random variable X_i is paired with θ_i in the sense that, for every subset $\{i_1,i_2,\ldots,i_s\}$ of $\{1,2,\ldots,k\}$, the joint distribution of $\{X_{i_1},\ldots,X_{i_s}\}$ depends on the set $\{\theta_1,\ldots,\theta_k;\psi\}$ only through the subset $\{\theta_{i_1},\ldots,\theta_{i_s};\psi\}$. The pair

(X_i, θ_i) is said to be associated with the component π_i of the frequency function f. When the components are independent, they are regarded as k populations.

The ranked values of the θ_i are denoted by $\theta^0_{[1]} \leqslant \theta^0_{[2]} \leqslant \cdots \leqslant \theta^0_{[k]}$. In the identification problem these values are assumed to be known a priori; it is further assumed that the true pairing of the π_i with the $\theta^0_{[j]}$ is unknown to the experimenter. Suppose that $\theta^0_{[k-1]} < \theta^0_{[k]}$. Then an identification goal would be "to identify the component π_i associated with $\theta^0_{[k]}$." The probability requirement for any decision procedure for this problem is that it identifies the correct population with a minimum guaranteed probability P^* for specified $\boldsymbol{\theta}^0 = (\theta^0_{[1]}, \ldots, \theta^0_{[k]})$ with $\theta^0_{[k-1]} < \theta^0_{[k]}$.

Definition 6.1. A frequency function f is said to satisfy the *Symmetry Condition* if it is invariant with respect to interchanging $(x_\alpha, \theta_\alpha)$ with (x_β, θ_β) for $\alpha, \beta = 1, 2, \ldots, k$.

Let $f^{(0)}$ denote the frequency function $f(\mathbf{x}|\boldsymbol{\theta}; \psi)$. For $1 \leqslant \alpha, \beta \leqslant k$, let $\boldsymbol{\theta}^{(\alpha,\beta)}$ denote the vector obtained from $\boldsymbol{\theta} = (\theta_1, \ldots, \theta_k)$ by exchanging θ_α and θ_β. Define $f^{(\alpha,\beta)}(\mathbf{x}|\boldsymbol{\theta}; \psi) = f(\mathbf{x}|\boldsymbol{\theta}^{(\alpha,\beta)}; \psi)$. If f satisfies the Symmetry Condition, then $f^{(\alpha,\beta)}(\mathbf{x}|\boldsymbol{\theta}; \psi) = f(\mathbf{x}^{(\alpha,\beta)}|\boldsymbol{\theta}, \psi)$.

Definition 6.2. A frequency function f is said to satisfy the *Rankability Condition* if it satisfies the Symmetry Condition and for fixed ψ, there exists a strictly increasing real-valued function $\tau = \tau(\theta; \psi)$ on the domain of θ, and a real-valued function y on the space $\mathcal{X}_1$ of X_1, such that for any pair (α, β), $\alpha, \beta = 1, 2, \ldots, k$, we have

$$\frac{f^{(\alpha,\beta)}(\mathbf{x}|\boldsymbol{\theta}; \psi)}{f^{(0)}(\mathbf{x}|\boldsymbol{\theta}; \psi)} = \exp\{-(\tau_\beta - \tau_\alpha)(y_\beta - y_\alpha)\} \tag{6.56}$$

where $\tau_i = \tau(\theta_i; \psi)$ and $y_i = y(x_i)$, $i = 1, \ldots, k$.

BKS have concentrated their attention on two important goals that are special cases of the general ranking goal of Bechhofer (1954). These are: Goal I of selecting an unordered set of components associated with the set $\{\theta_{[k-t+1]}, \ldots, \theta_{[k]}\}$, and Goal II of selecting the population associated with $\theta_{[k-t+1]}$, the one associated with $\theta_{[k-t+2]}, \ldots,$ and the one associated with $\theta_{[k]}$. They have shown (Theorem 6.1.3) that the Rankability Condition provides a sufficient condition for their basic procedure for solving identification problems to lead to a solution of the corresponding ranking problem for the particular choice δ of the distance function, given by $\delta(\theta_i, \theta_j) = \tau_{[j]} - \tau_{[i]}$. Of course, this result holds for much more general goals. In the case of $k = 2$ populations, they have shown (Theorem 3.6.1) that, under certain regularity conditions, the identification problem leads to a solution of the ranking problem if and only if the frequency function $f(\mathbf{x}|\theta)=f(\mathbf{x}|\theta,\theta+\delta^*)$ is Koopman–Darmois with connected natural parameter, that is, $\log f(\mathbf{x}|\theta)=P(\mathbf{x})Q(\theta)+R(\mathbf{x})+S(\theta)$ with Q continuous and strictly monotone.

In their monograph, BKS have discussed in detail applications of their general results to ranking Koopman–Darmois parameters with special reference to normal means. For selecting the best population under Goal I, they have discussed several special cases (Section 12.6), which include (a) normal populations with unknown means and known variances, (b) normal populations with unknown variances and known means, (c) exponential populations with unknown scale parameters and known location parameters, (d) Bernoulli populations with unknown probabilities of success on a single trial, (e) Poisson populations with unknown means, and (f) negative-binomial populations with unknown probabilities of success on a single trial and common known number of successes to be achieved per stage. They have also given some Monte Carlo sampling results for the problem of ranking means of normal populations with a common known variance.

CHAPTER 7

Selection from Multivariate Populations

7.1 INTRODUCTION

In many practical situations, when comparing systems or groups, we are interested simultaneously in two or more characteristics of each component or individual. In other words, our observations are vector-valued and the components of the vector observations are correlated random variables. For example, we may be interested in comparing the achievements of two or more groups of students. The achievement is defined in terms of verbal and quantitative skills. Thus we will have tests designed to give two scores for each student, one for each type of skill. The most common multivariate model is the multivariate normal distribution.

Let $\pi_1, \pi_2, \ldots, \pi_k$ be k independent p-variate normal populations with mean vectors $\boldsymbol{\mu}_i$ and covariance matrices Σ_i, $i=1,\ldots,k$, respectively. All our vectors are column vectors and the Σ_i are assumed to be positive definite. In general, one may define a partial order relation on the set of multivariate normal distributions. This would naturally mean that any two of these normal populations may not be ordered in terms of a vector-valued parameter. However, one may define a real-valued function θ_i of the parameters $\boldsymbol{\mu}_i$ and Σ_i and use the θ_i to rank the populations. This has been the approach in most of the investigations. In the subsequent sections we are concerned with ranking of several multivariate normal distributions as well as ranking of components of a single multivariate normal distribution.

7.2 RANKING OF SEVERAL MULTIVARIATE NORMAL POPULATIONS

Here we consider several scalar functions θ_i of the mean vector $\boldsymbol{\mu}_i$ and covariance matrix Σ_i, namely, (a) Mahalanobis distance, (b) generalized variance, (c) multiple correlation coefficient, (d) sum of bivariate product–moment correlations, and (e) coefficient of alienation.

7.2.1 Selection in Terms of Mahalanobis Distance

The ranking parameter here is $\theta_i = \boldsymbol{\mu}_i' \Sigma_i^{-1} \boldsymbol{\mu}_i$, which is Mahalanobis' distance function. This parameter can also be looked upon as the multivariate analog of the reciprocal of the square of the coefficient of variation. The best population is the one associated with $\theta_{[k]}$. Let $\overline{X}_i$ and S_i denote the sample mean vector and covariance matrix based on a sample of size n from π_i, $i=1,\ldots,k$. Let $\Omega_1 = \{\boldsymbol{\theta}\colon \theta_{[k-t+1]} - \theta_{[k-t]} \geqslant \delta_1\}$ and $\Omega_2 = \{\boldsymbol{\theta}\colon \theta_{[k-t+1]}/\theta_{[k-t]} \geqslant \delta_2\}$ for some $\delta_1 > 0$, $\delta_2 > 1$. Define $\Omega_3 = \Omega_1 \cap \Omega_2$. For the cases of Σ_i known and unknown, Alam and Rizvi (1966) have proposed rules for selecting the t best populations such that $PCS \geqslant P^*$ whenever $\boldsymbol{\theta} \in \Omega_3$.

Case I:* Σ_i *known. Let $U_i = \overline{\mathbf{X}}_i' \Sigma_i^{-1} \overline{\mathbf{X}}_i$, $i=1,\ldots,k$. Then the procedure R_1 of Alam and Rizvi is to select the populations associated with $U_{[k-t+1]}, \ldots, U_{[k]}$. One can write the expression for the PCS using the fact that nU_i has a noncentral chi-square distribution with p degrees of freedom and noncentrality parameter $n\boldsymbol{\mu}_i' \Sigma_i^{-1} \boldsymbol{\mu}_i$. The LFC is given by

$$\begin{aligned} \theta_{[1]} = \cdots = \theta_{[k-t]} &= \delta_1(\delta_2 - 1)^{-1}; \\ \theta_{[k-t+1]} = \cdots = \theta_{[k]} &= \delta_1\delta_2(\delta_2 - 1)^{-1}. \end{aligned} \tag{7.1}$$

The smallest value of n required to satisfy the P^* condition is obtained from

$$t\int_0^\infty F_p^{k-t}\left(x,\frac{n\delta_1}{\delta_2-1}\right)\left\{1-F_p\left(x,\frac{n\delta_1\delta_2}{\delta_2-1}\right)\right\}^{t-1} f_p\left(x,\frac{n\delta_1\delta_2}{\delta_2-1}\right)dx = P^*, \tag{7.2}$$

where $f_p(x,\theta)$ and $F_p(x,\theta)$ denote the density and the distribution function, respectively, of a noncentral chi-square random variable with p degrees of freedom and noncentrality parameter θ.

Case II:* Σ_i *unknown. In this case, the procedure R_2 of Alam and Rizvi (1966) selects the t populations associated with $V_{[k-t+1]},\ldots,V_{[k]}$, where $V_i=[(n-p)/np](\overline{\mathbf{X}}_i'S_i^{-1}\overline{\mathbf{X}}_i)$. The random variable nV_i has a noncentral F distribution with p and $(n-p)$ degrees of freedom and noncentrality parameter $n\theta_i$. The LFC is again given by (7.1). The smallest value of n required to satisfy the P^*-condition is obtained from (7.2) with $F_p(x,\lambda)$ and $f_p(x,\lambda)$ replaced by $G_{p,n-p}(x,\lambda)$ and $g_{p,n-p}(x,\lambda)$ where the last two represent, respectively, the distribution function and the density of a noncentral F random variable with p and $(n-p)$ degrees of freedom and noncentrality parameter λ.

Sequential Procedures. Srivastava and Taneja (1972) have investigated sequential procedures for selecting the best population in terms of (i) the Euclidean distance function $\delta_i^2=\boldsymbol{\mu}_i'\boldsymbol{\mu}_i$, and (ii) the Mahalanobis distance function $\theta_i=\boldsymbol{\mu}_i'\Sigma^{-1}\boldsymbol{\mu}_i$, where Σ is the common covariance matrix.

Let us first consider the Euclidean distance. For selecting the population associated with $\delta_{[k]}$, Srivastava and Taneja have proposed a class $\mathcal{R}$ of rules based on the sequential theory of Chow and Robbins [1965] so that

$$\lim_{\Delta\to 0} P(CS|\mathcal{R}) \geqslant P^* \quad \text{whenever} \quad \delta_{[k]}-\delta_{[k-1]} \geqslant \Delta, \tag{7.3}$$

where P^* and Δ are preassigned constants.

We first assume that Σ is known. Let $\overline{\mathbf{X}}_{in}$ be the sample mean vector based on n independent observations from π_i, $i=1,\ldots,k$. The procedure R_3 of Srivastava and Taneja is defined as follows.

Take a sample of size n from each population where n is given by

$$n \geqslant a^2\lambda_1\Delta^{-2} \tag{7.4}$$

where

$$\lambda_1 = \max_{\mathbf{c}:\,\mathbf{c}'\mathbf{c}=1} \mathbf{c}'\Sigma\mathbf{c} \tag{7.5}$$

and a is given by

$$\Phi\left(-\frac{a}{\sqrt{2}}\right)=(1-P^*)(k-1)^{-1}. \tag{7.6}$$

Note that λ_1 defined by (7.5) is the maximum characteristic root of Σ. Now select the population associated with the largest $\overline{\mathbf{X}}'_{in}\overline{\mathbf{X}}_{in}$.

Suppose Σ is not known. Let

$$S_n=(nk)^{-1}\sum_{i=1}^{k}\sum_{j=1}^{n}\left(\mathbf{X}_{ij}-\overline{\mathbf{X}}_{in}\right)\left(\mathbf{X}_{ij}-\overline{\mathbf{X}}_{in}\right)' \tag{7.7}$$

and

$$\lambda_{1n}=\max_{\mathbf{b}:\,\mathbf{b}'\mathbf{b}=1}\mathbf{b}'S_n\mathbf{b}. \tag{7.8}$$

Note that $\lim_{n\to\infty}\lambda_{1n}=\lambda_1$ a.s. Let $\{a_n\}$ be a sequence of positive constants such that $\lim_{n\to\infty}a_n=a$, where a is defined by (7.6). Then the sequence $\{a_n\}$ determines a member of the class $\mathcal{R}$ of sequential procedures defined as follows:

$\mathcal{R}$: Start taking $n_0>p$ observations from each population and then one observation at a time from each population and stop according to the stopping rule defined by

$$N=\text{smallest } n\geqslant n_0 \text{ such that } \lambda_{1n}\leqslant n\Delta^2a_n^{-2}. \tag{7.9}$$

When the sampling is stopped at $N=n$, select the population with the largest $\overline{\mathbf{X}}'_{in}\overline{\mathbf{X}}_{in}$ as the best population.

The random sample size N depends on Δ and $N(\Delta)\to\infty$ a.s. as $\Delta\to0$. It can be shown that

$$E(N)\leqslant a^2\lambda_1\Delta^{-2}+0(1). \tag{7.10}$$

Thus the sequential procedure is asymptotically efficient and $E(N)-a^2\lambda_1\Delta^{-2}$ is uniformly bounded.

Now we consider the problem in terms of $\theta_i=\boldsymbol{\mu}_i'\Sigma^{-1}\boldsymbol{\mu}_i$, $i=1,\ldots,k$. For selecting the best population, Srivastava and Taneja (1972) have investigated truncated and nontruncated sequential procedures similar to those of Paulson (1964a) and Hoel and Mazumdar (1968).

A useful lemma in this context is a version of a lemma of Bechhofer, Kiefer, and Sobel (1968, p. 164) and is given as follows.

Lemma 7.1. Let $Z_1,Z_2,\ldots$ be a sequence of independently distributed random variables having the same distribution as $Z=X-Y$ where X and

Y are independent noncentral chi-square random variables with p degrees of freedom and noncentrality parameter θ and ϕ, respectively. Let $\theta<\phi$, $0\leqslant\lambda<\phi-\theta$, and $b>0$. Then

$$\Pr\left\{\sup_{m}\sum_{i=1}^{m}(Z_i+\lambda)>b\right\}\leqslant e^{-t_0 b} \tag{7.11}$$

where $t_0>0$ is the solution of

$$\max\left\{t:\,(1-4t^2)^{-p/2}e^{t[\lambda-4\lambda t^2+2t(\phi+\theta)-(\phi-\theta)](1-4t^2)^{-1}}\leqslant 1\right\},\qquad 0<t<\tfrac{1}{2}. \tag{7.12}$$

To define the nontruncated procedure R_4 let $U_{ij}=\mathbf{X}'_{ij}\Sigma^{-1}\mathbf{X}_{ij}$, $i=1,\ldots,k$; $j=1,2,\ldots$, and $Z_{in}=\Sigma_{j=1}^{n}U_{ij}$. Now, for $\tau^*>0$, define

$$C_{\tau^*}=t_0^{-1}\log\{(k-1)(1-P^*)^{-1}\} \tag{7.13}$$

where t_0 is the solution of (7.12) with $\lambda=0$. The value of t_0 is tabulated by Srivastava and Taneja for selected values of $\phi-\theta$ and $\phi+\theta$ for $p=2,3,4$.

$\boldsymbol{R_4}$: Take one observation from each population. Eliminate from further consideration any population π_s for which

$$Z_{s1}\leqslant\max_{r}Z_{r1}-C_{\tau^*}. \tag{7.14}$$

If all but one population are eliminated, terminate the experiment and select the remaining population as the best. Otherwise we go on to the next stage taking one observation from each population not eliminated. Proceeding this way, at the mth stage ($m=2,3,\ldots$), take one observation from each population not eliminated until $(m-1)$st stage, and then eliminate any population π_s for which

$$Z_{sm}\leqslant\max_{r}Z_{rm}-C_{\tau^*}, \tag{7.15}$$

where the maximum is taken over all the populations left after the $(m-1)$st stage. We terminate when there is only one population left out and select it as the best.

It has been shown that the procedure R_4 terminates with probability one and that $P(CS|R_4)\geqslant P^*$ whenever $\tau_{[k]}-\tau_{[k-1]}\geqslant\tau^*$. However, no upper bound for the expected number of observations is available. For this reason, Srivastava and Taneja (1972) have considered a class of truncated procedures similar to that of Paulson (1964a). Let η be chosen such that

$0<\eta<\tau^*$ and define

$$C_\eta = t_\eta^{-1}\log\{(k-1)(1-P^*)^{-1}\}, \tag{7.16}$$

where t_η is the solution of (7.12) with $\lambda=\eta$. Let W_η be the largest integer less than $\eta^{-1}C_\eta$. The class S_η of sequential procedures defined by Srivastava and Taneja (1972) is as follows.

$\boldsymbol{S_\eta}$: Start sampling as in the case of the nontruncated procedure R_4 described above with (7.14) and (7.15) replaced, respectively, by

$$Z_{s1} \leqslant \max_r Z_{r1} - C_\eta + \eta \tag{7.17}$$

and

$$Z_{sm} \leqslant \max_r Z_{rm} - C_\eta + m\eta. \tag{7.18}$$

If more than one population remain after the W_ηth stage, the experiment is terminated at the next stage by selecting the population with the largest Z-value. For this procedure with $\eta\in(0,\tau^*)$ we have

$$P(CS|S_\eta) \geqslant P^* \quad \text{whenever} \quad \tau_{[k]}-\tau_{[k-1]} \geqslant \tau^*. \tag{7.19}$$

As in the case of Paulson's procedure, the optimum choice of η is not known. One may take $\eta=\tau^*/4$ as recommended by Paulson (1964a). The remarks of Fabian (1974a) about improving the bound of the probability of an incorrect selection for the procedure S_λ of Paulson (1964a) in §6.2 apply here, too.

7.2.2 Selection in Terms of the Generalized Variance

Eaton (1967b) has considered the ranking of the multivariate normal populations in terms of the generalized variances $\Lambda_i=|\Sigma_i|$, $i=1,\ldots,k$, where $|\cdot|$ denotes the determinant of the appropriate matrix. His results in this case follow from the theorems of his earlier paper (1967a) discussed in §§3.2 and 3.3 and the property (proved in his latter paper) that the density function of the sample generalized variance possesses a monotone likelihood ratio. It is assumed that the loss function for the ranking problem depends only on $(\Lambda_1,\ldots,\Lambda_k)$ and satisfies (3.13) and (3.14). Let φ^* be the decision rule which ranks the populations according to the values of the observed sample generalized variances V_i, $i=1,\ldots,k$. Then the following theorem holds.

Theorem 7.1. Within the class of decision rules that depend only on $V=(V_1,\ldots,V_k)$, φ^* is (i) minimax, (ii) admissible, and (iii) the uniformly best decision rule within the class of rules which are invariant under permutations of the vector $V=(V_1,\ldots,V_k)$.

It also follows from a general minimax theorem that the rule φ^* is minimax within the class of all decision rules. This in turn implies that φ^* is a most economical decision rule (see Eaton (1967a) and Hall (1959)).

7.2.3 Selection in Terms of Multiple Correlation Coefficients

Let $\lambda_i=\rho^2_{i:\,1.2\ldots p}$ be the squared population multiple correlation coefficient for π_i between the first and the remaining set of variables. Let

$$\Omega_1=\{\boldsymbol{\lambda}: \boldsymbol{\lambda}=(\lambda_1,\ldots,\lambda_k), \lambda_{[k-t+1]}-\lambda_{[k-t]}\geqslant\delta_1\} \tag{7.20}$$

and

$$\Omega_2=\{\boldsymbol{\lambda}: \boldsymbol{\lambda}=(\lambda_1,\ldots,\lambda_k), \lambda_{[k-t+1]}\geqslant\delta_2\lambda_{[k-t]}\} \tag{7.21}$$

where $0<\delta_1<1$ and $\delta_2>1$ are specified constants. Define $\Omega_3=\Omega_1\cap\Omega_2$. Rizvi and Solomon (1973) have investigated the following procedure (denoted by R_5) for selecting the t best populations so that

$$P(CS|R_5)\geqslant P^* \quad \text{whenever} \quad \boldsymbol{\lambda}\in\Omega_3. \tag{7.22}$$

Consider a random sample of size n $(n>p)$ from each of the k populations and let $Y_i=R^2_{i:\,1.2\ldots p}$, the squared sample multiple correlation coefficient. Then the procedure R_5 selects the population associated with the t largest Y_i's. Rizvi and Solomon consider only the asymptotic case. We can compute the PCS using the result that, for fixed p as $n\to\infty$, nY_i is asymptotically distributed as a noncentral chi-square random variable with $q=(p-1)$ degrees of freedom and noncentrality parameter $n\lambda_i$. The LFC is given by

$$\lambda_{[1]}=\cdots=\lambda_{[k-t]}=\delta_1(\delta_2-1)^{-1};\lambda_{[k-t+1]}=\cdots=\lambda_{[k]}=\delta_1\delta_2(\delta_2-1)^{-1}. \tag{7.23}$$

From the results of Rizvi (1963), it follows that the procedure R_5 is, for large n, unique Bayes, unique minimax, and hence also admissible in the class of invariant rules.

The previous procedure R_5 has been investigated subsequently in the exact sample case by Alam, Rizvi, and Solomon (1975) except for the difference that the probability requirement in (7.22) is now met whenever $\boldsymbol{\lambda}\in\Omega_1'\cap\Omega_2$, where

$$\Omega_1'=\{\boldsymbol{\lambda}: \boldsymbol{\lambda}=(\lambda_1,\ldots,\lambda_k), (1-\lambda_{[k-t]})\geqslant\delta_1'(1-\lambda_{[k-t+1]}), \delta_1'>1\}. \tag{7.24}$$

It is shown that the LFC is given by

$$\lambda_{[i]}=\begin{cases}(\delta_1'-1)(\delta_1'\delta_2-1)^{-1}, & i=1,\ldots,k-t,\\ \delta_2(\delta_1'-1)(\delta_1'\delta_2-1)^{-1}, & i=k-t+1,\ldots,k.\end{cases} \tag{7.25}$$

7.2.4 Selection in Terms of the Sum of Bivariate Product–Moment Correlations

Let ν_i be a measure associated with the population π_i defined by

$$\nu_i=\sum_{m=1}^{p}\sum_{\substack{l=1\\ m\neq l}}^{p}\frac{\rho_{ml}^{(i)}}{p(p-1)}, \qquad i=1,\ldots,k, \tag{7.26}$$

where $\rho_{ml}^{(i)}$ is the bivariate product–moment correlation coefficient between the mth and the lth coordinates of a vector $\mathbf{X}$ from π_i.

Govindarajulu and Gore (1971) have studied the procedure R_6 described as follows for selecting the population associated with $\nu_{[k]}$ so that the P^*-condition is asymptotically ($n\to\infty$) satisfied. Define V_i similar to (7.26) with sample correlation coefficient $r_{ml}^{(i)}$ in the place of $\rho_{ml}^{(i)}$. Then R_6 selects the population associated with the largest V_i. It is shown by Govindarajulu and Gore that, for large n,

$$P(CS|R_6)\geqslant \Pr\left[U_i\leqslant\sqrt{n}\,d\left\{\frac{2(p+3)(p-3)}{p(p-1)}\right\}^{-1/2}, \qquad i=1,\ldots,k-1\right] \tag{7.27}$$

whenever $\nu_{[k]}-\nu_{[k-1]}\geqslant d$, where $U_1,\ldots,U_{k-1}$ are standard normal random variables with equal correlation $\frac{1}{2}$.

In the bivariate case, ν_i will be the product–moment correlation and V_i will be the sample correlation coefficient. In this case,

$$P(CS|R_6)\geqslant \Pr\left[U_i\leqslant\frac{\sqrt{n}\,d}{\sqrt{2}}, \qquad i=1,\ldots,k-1\right]. \tag{7.28}$$

7.2.5 Selection in Terms of the Coefficient of Alienation

Let $\mathbf{X}_i=(\mathbf{Y}_i,\mathbf{Z}_i)'$ be a $(q+p)$-dimensional random variable with covariance matrix

$$\Sigma_i=\begin{bmatrix}\Sigma_{yy}^i & \Sigma_{yz}^i\\ \Sigma_{zy}^i & \Sigma_{zz}^i\end{bmatrix}.$$

We assume that $q \leqslant p$ and let $\nu_{i1}^2, \ldots, \nu_{iq}^2$ be the canonical correlations associated with $\mathbf{Y}_i$ and $\mathbf{Z}_i$. The *coefficient of alienation* between $\mathbf{Y}_i$ and $\mathbf{Z}_i$ is γ_i defined by

$$\gamma_i^2 = \frac{|\Sigma_i|}{|\Sigma_{yy}^i||\Sigma_{zz}^i|}. \tag{7.29}$$

The coefficient of alienation is a measure of association between the two sets of variables. In particular, for $q=1$ it is equal to $1-\rho^2$, where ρ is the multiple correlation coefficient between Y and $(Z_1, \ldots, Z_p)$.

Frischtak (1973) considered selection of the population associated with $\gamma_{[1]}$ subject to the P^*-requirement whenever $\gamma_{[2]}^2 \geqslant \theta^* \gamma_{[1]}^2$, $\theta^* > 1$. Let U_i be defined by

$$U_i^2 = \frac{|S_i|}{|S_{yy}^i||S_{zz}^i|}. \tag{7.30}$$

where S_i is the sample covariance matrix based on N independent vector observations from π_i and, S_{yy}^i and S_{zz}^i are the appropirate components of the partitioned S_i matrix. Frischtak proposed the natural rule R which selects the population that gives the smallest U_i. An asymptotic ($N \to \infty$) lower bound on the PCS is given by

$$\Pr\left\{ V_j \leqslant \frac{n^{\frac{1}{2}} \log \theta^*}{2(2q)^{\frac{1}{2}}}, \qquad j=2,\ldots,k \right\}, \tag{7.31}$$

where $V_2, \ldots, V_k$ are standard normal random variables with equal correlation $\frac{1}{2}$.

7.3 PROCEDURES RELATING TO A SINGLE MULTIVARIATE NORMAL POPULATION

In this section we briefly discuss several selection procedures relating to a single k-variate normal population $N_k(\boldsymbol{\mu}, \Sigma)$. Let $\Gamma = (\rho_{ij})$ denote the correlation matrix. If we assume that $\Sigma = I$, the identity matrix, the problem reduces to selection from k independent normal populations which has been discussed extensively in Chapters 2 and 6. When we do not have independence, we refer to the marginal normal populations as "components" and talk about selecting the best component. This distinction is not crucial but sometimes convenient.

7.3.1 Selection of the Component with the Largest Population Mean (Equal Known Variances)

Here we assume that $\Sigma=\sigma^2\Gamma$, where σ^2 is known. For selecting the variate associated with $\mu_{[k]}$, Frischtak (1973) investigated the procedure R_1 that selects the component corresponding to the largest sample mean $\bar{X}_i$ based on N vector observations from the population. The problem is to determine the smallest integer N so that $\text{PCS}\geqslant P^*$ whenever $\mu_{[k]}-\mu_{[k-1]}\geqslant\delta^*>0$.

If the correlations ρ_{ij} are all equal to ρ, then it is easy to show that

$$\inf_{\Omega_1} \text{PCS}=\Pr\left\{Z_i>-a(N)\left[\frac{(k-1)}{k}\right]^{\frac{1}{2}}, \quad i=1,\dots,k-1\right\}, \tag{7.32}$$

where $a(N)=(\delta^*/\sigma)(N/2)^{\frac{1}{2}}$, the Z_i are equicorrelated $N(0,1)$ variables with correlations $\frac{1}{2}$, and $\Omega_1=\{(\boldsymbol{\mu},\rho): \mu_{[k]}-\mu_{[k-1]}\geqslant\delta^*\}$. The smallest integer N satisfying the P^*-condition can be obtained using the tables of Gupta [1963a] and Gupta, Nagel, and Panchapkesan [1973].

For $k=2$, the PCS is minimized when $\mu_{[2]}-\mu_{[1]}=\delta^*$ and $\rho_{12}=-1$. Thus we get $a(N)=\sqrt{2}\,\Phi^{-1}(P^*)$. In the general case, let us assume, without loss of generality, that $\mu_k\geqslant\mu_j$ $(j\neq k)$. For $\mu_k-\mu_j\geqslant\delta^*$, we have

$$\text{PCS}\geqslant\Pr\left\{Y_j>-a(N)(1-\rho_{jk})^{-\frac{1}{2}}, \quad j=1,\dots,k-1\right\}, \tag{7.33}$$

where $a(N)=(\delta^*/\sigma)(N/2)^{\frac{1}{2}}$ and the Y_j are standard normal random variables with

$$\text{corr}(Y_i,Y_j)=\left[1-\rho_{ki}-\rho_{kj}+\rho_{ij}\right]/2(1-\rho_{ki})^{\frac{1}{2}}(1-\rho_{kj})^{\frac{1}{2}}. \tag{7.34}$$

For $k=3$, the infimum of P_1, the right-hand side of (7.33), is shown to occur when $|\Gamma|=0$. But there is no analytical result available on the configuration of the ρ_{ij} which gives the global minimum of P_1. Fewer small sample results are available in the case of $k>3$. Of course, one can get a conservative approximation to the sample size by using Bonferroni's inequality to obtain a lower bound of P_1.

7.3.2 Selection of the Component with the Smallest Variance

We write the covariance matrix as $\Sigma=\Lambda\,\Gamma\,\Lambda$ where $\Lambda=\text{diag}(\sigma_1,\dots,\sigma_k)$ and $\Gamma=(\rho_{ij})$. We want to select the component associated with $\sigma_{[1]}$ such that $\text{PCS}\geqslant P^*$ whenever $\sigma_{[2]}^2\geqslant\theta^*\sigma_{[1]}^2$, $\theta^*>1$. Frischtak (1973) has investigated the "natural" rule which selects the population associated with the smallest

$S_{ii}=\Sigma_{\alpha=1}^{N}(X_{i\alpha}-\bar{X}_i)^2$. For $k=2$, the infimum of the PCS occurs when $\sigma_{[2]}^2=\theta^*\sigma_{[1]}^2$ and $\rho=0$. For $k>2$, only a conservative approximation to the sample size has been obtained. Of course, one can obtain large sample results which are special cases of the results of the following section.

7.3.3 Selection of a Subclass of Components with the Smallest Population Generalized Variance

Let us consider a kp-variate normal population with unknown mean vector $\boldsymbol{\mu}$ and unknown covariance matrix

$$\Sigma=\begin{bmatrix} \Sigma_{11} & \Sigma_{12} & \cdots & \Sigma_{1k} \\ \Sigma_{21} & \Sigma_{22} & \cdots & \Sigma_{2k} \\ \vdots & \vdots & \vdots & \vdots \\ \Sigma_{k1} & \Sigma_{k2} & \cdots & \Sigma_{kk} \end{bmatrix}$$

where the Σ_{ii} $(1\leqslant i\leqslant k)$ are $p\times p$ symmetric positive definite matrices. Let $\lambda_i=|\Sigma_{ii}|$. Thus the variates are partitioned into k subclasses and λ_i is the generalized variance of the ith subclass. We want to select the subclass associated with $\lambda_{[1]}$. It is required that PCS $\geqslant P^*$ whenever $\lambda_{[2]}\geqslant\theta^*\lambda_{[1]}$, $\theta^*>1$. This problem is a generalization of the problem discussed in §7.3.2 as well as the problem in §7.2.2. Frischtak (1973) considered the natural procedure which selects the subclass associated with the smallest $|S_{ii}|$, where S_{ii} is the submatrix of the sample covariance matrix S appropriate to the ith subclass. However, he has obtained only an asymptotic solution. Asymptotically $(N\to\infty)$ the LFC is given by $\lambda_{[1]}\theta^*=\lambda_{[2]}=\cdots=\lambda_{[k]}$ and $\Sigma_{ij}=0$ $(i\neq j)$. Using the asymptotic normality of $(N-1)^{\frac{1}{2}}(\log|S_1|-\log\lambda_1,\ldots,\log|S_k|-\log\lambda_k)$, we get

$$\inf_{\Omega(\theta^*)} \text{PCS}\approx\Pr\left\{Y_j\leqslant 2^{-1}\left[\frac{(N-1)}{p}\right]^{\frac{1}{2}}\log\theta^*,\qquad j=1,\ldots,k-1\right\} \tag{7.35}$$

where $\Omega(\theta^*)=\{\boldsymbol{\lambda}:\ \lambda_{[2]}\geqslant\theta^*\lambda_{[1]}\}$ and the Y_j are equicorrelated standard normal variables with correlations equal to $\frac{1}{2}$.

The reader can easily see that there are not enough satisfactory results available in many of the cases discussed previously.

7.4 NOTES AND REMARKS

In all the problems discussed so far, we used a scalar function of the unknown parameters to rank the multivariate normal populations. It is necessary to consider other partial ordering relations. For example, if we consider two bivariate normal populations with mean vectors $\boldsymbol{\mu}_1 = (\mu_{11}, \mu_{12})$ and $\boldsymbol{\mu}_2 = (\mu_{21}, \mu_{22})$ and a common known covariance matrix Σ. The components of the mean vector may be the population averages for two characteristics for both of which high values are desirable. In this case, we will naturally prefer the first population if $\mu_{1j} \geqslant \mu_{2j}, j = 1, 2$. However, if $\mu_{11} > \mu_{21}$ and $\mu_{12} < \mu_{22}$, the two are not comparable.

There is practically no result available in this direction. Also, there has been no work done for distributions other than multivariate normal populations.

CHAPTER 8

Nonparametric Selection Procedures

8.1 INTRODUCTION

In this chapter we discuss several procedures for ranking populations whose distribution functions are not known. In some cases we may justifiably assume that distributions have location or scale parameters but the form of the distribution functions may not be known. If no parameter in the usual sense (such as mean and variance) is involved, one can rank the populations according to quantile of a given order. We consider procedures based on sample quantiles, and rank order statistics. We also consider procedures based on paired comparisons with special reference to ranking players in a tournament. Finally, a brief discussion is given on selection in terms of rank correlation coefficients for bivariate populations and a multivariate analog of this.

8.2 SELECTION IN TERMS OF QUANTILES

Let $\pi_1,\ldots,\pi_k$ be k populations with continuous distribution functions $F_i(x)$, $i=1,\ldots,k$, respectively. Given $0<\alpha<1$, let $x_\alpha(F)$ denote the αth quantile of F. For simplicity, we assume that the α-quantiles of the k populations are unique. Let $F_{[i]}=F_{[i]}(x)$ denote the cdf associated with the population having the ith smallest quantile. Let $F_{[i]}\lesssim F_{[j]}$ (π_j better than π_i) denote the fact that $x_\alpha(F_{[i]})\leqslant x_\alpha(F_{[j]})$. It is not required that the distributions have the same support, nor is it assumed that they differ only in a location parameter.

Sobel (1967) has investigated the problem of selecting the populations with the t-largest α-quantiles under the indifference zone formulation. In order to specify the preference zone in which the probability requirement is to be satisfied, Sobel has adopted two basic formulations and two slight modifications of them.

FORMULATION 1A. Let $\epsilon_\gamma^*>0$ $(\gamma=1,2)$ be two specified numbers such that $\epsilon_1^*\leqslant\alpha\leqslant 1-\epsilon_2^*$. Define

$$\underline{x}_\beta(F)=\inf\{x\colon F(x)=\beta\},$$

$$\bar{x}_\beta(F)=\sup\{x\colon F(x)=\beta\}. \tag{8.1}$$

In this section, unless otherwise stated, the subscripts i and j run over the ranges $i=1,\ldots,k-t$; $j=k-t+1,\ldots,k$. Let

$$\underline{F}\equiv\underline{F}(x)=\min_i F_{[i]}(x),$$

$$\bar{F}\equiv\bar{F}(x)=\max_j F_{[j]}(x), \tag{8.2}$$

and let I be the closed interval $[\bar{x}_{\alpha-\epsilon_1^*}(\bar{F}), \underline{x}_{\alpha+\epsilon_2^*}(\bar{F})]$. We now define

$$d = \min_{(i,j)} d_{ij} \tag{8.3}$$

where $d_{ij} = \inf_{x \in I}(F_{[i]}(x) - F_{[j]}(x))$ so that we have

$$d = \inf_{x \in I} \big(\underline{F}(x) - \bar{F}(x)\big). \tag{8.4}$$

The goal is to select the t best populations with the guarantee that

$$\text{PCS} \geqslant P^* \quad \text{whenever } d \geqslant d^* \tag{8.5}$$

where $d^* > 0$ and $\binom{k}{t}^{-1} < P^* < 1$ are specified in advance.

Let n be the common size of independent samples from the k populations. We assume n to be sufficiently large so that $1 \leqslant (n+1)\alpha \leqslant n$. We define r to be a positive integer such that $r \leqslant (n+1)\alpha < r+1$ so that $1 \leqslant r \leqslant n$. Let

$$W_i = \big[r+1-(n+1)\alpha\big]Y_{ri} + \big[(n+1)\alpha - r\big]Y_{r+1,i}, \tag{8.6}$$

where Y_{ji} denotes the jth order statistic from F_i, $i = 1,\ldots,k$. Then Sobel's rule R is:

$$\text{Select } \pi_i \text{ if and only if } W_i \geqslant W_{[k-t+1]}. \tag{8.7}$$

We assume α to be rational so that $(n+1)\alpha$ will be an integer for some arithmetic progression N of n-values and $r = (n+1)\alpha$. In this case $W_i = Y_{ri}$. Find the smallest n in N for which the P^*-condition is satisfied. The inefficiency due to the restriction of n to the set N is negligible. With the additional assumption that for all x

$$\begin{aligned} F_{[i]}(x) &\geqslant F_{[k-t+1]}(x) = \bar{F}(x), \qquad i = 1,\ldots,k-t, \\ F_{[j]}(x) &\leqslant F_{[k-t]}(x) = \underline{F}(x), \qquad j = k-t+1,\ldots,k, \end{aligned} \tag{8.8}$$

we will refer to the formulation as 1B. Of course, for the special cases of $t = 1$ and $t = k-1$ we omit the appropriate part of (8.8).

One could assume that for all pairs (γ, δ) with $\gamma < \delta$

$$F_{[\delta]}(x) \leqslant F_{[\gamma]}(x) \quad \text{for all } x; \tag{8.9}$$

but the weaker assumption (8.8) is sufficient for our purpose.

FORMULATION 2A. Let $\epsilon_\gamma^* > 0$ $(\gamma = 1,2)$ be two specified numbers such that $\epsilon_1^* \leqslant \alpha \leqslant 1 - \epsilon_2^*$. Let

$$d_{ij}' = \underline{x}_{\alpha - \epsilon_1^*}(F_{[j]}) - \bar{x}_{\alpha + \epsilon_2^*}(F_{[i]}) \tag{8.10}$$

and define

$$d' = \min_{(i,j)} d_{ij}' = \underline{x}_{\alpha - \epsilon_1^*}(\bar{F}) - \bar{x}_{\alpha + \epsilon_2^*}(\underline{F}). \tag{8.11}$$

Under this formulation it is required that

$$\text{PCS} \geqslant P^* \quad \text{whenever} \quad d' \geqslant 0. \tag{8.12}$$

The form of the selection procedure is the same as in (8.7). As before, we assume α to be rational and find the smallest $n \in N$ such that (8.12) is satisfied. With the additional assumption stated in (8.8), this is referred to as Formulation 2B.

Sobel (1967) has obtained a lower bound for PCS under all the four formulations. Under Formulation 1A, the LFC is given by $F_{[i]} = \underline{F}$, $F_{[j]} = \bar{F}$, and $\underline{F}(y) = \bar{F}(y) + d^*$. This gives

$$P(CS|R) \geqslant t \int_{\alpha - \epsilon_1^*}^{\alpha + \epsilon_2^*} G^{k-t}(u + d^*)\left[1 - G(u)\right]^{t-1} dG(u)$$

$$+ G^{k-t}(\alpha + \epsilon_2^* + d^*)\left[1 - G(\alpha + \epsilon_2^*)\right]^t, \tag{8.13}$$

where $G(p) = G_{r,n-r+1}(p)$ denotes the incomplete beta function

$$G(p) = \left[\frac{n!}{(r-1)!(n-r)!}\right] \int_0^p x^{r-1}(1-x)^{n-r}\,dx. \tag{8.14}$$

With fixed n and $d^* \to 0$, the right side of (8.13) approaches a lower bound that is less than or equal to $\binom{k}{t}^{-1}$, with the equality holding only if $\alpha + \epsilon_2^* = 1$ and $\alpha - \epsilon_1^* = 0$. For $n \to \infty$ and $r/n \to \alpha$, it is shown that the right side of (8.13) approaches unity for any fixed $d^* > 0$. If simultaneously $d^* \to 0$, then the limit must be at least $\binom{k}{t}^{-1}$.

Corresponding to (8.13), we have under Formulation 1B a lower bound that is greater than the right side of (8.13), and hence the n-value for Formulation 1B is not larger than that for 1A. The lower bounds obtained in both cases are asymptotically equivalent for large n (or, equivalently, for d^* close to zero). Similar remarks hold in the cases of Formulations 2A and 2B. The table values show that the difference in the sample size because of the assumption (8.8) is quite small for $P^* \geqslant 0.75$ and that it decreases with increasing P^* and also with increasing k. The tables of Sobel (1967) relate to selecting the population with the largest median ($\alpha = \frac{1}{2}$). He has tabulated the (exact) smallest n required under all the four formulations for $k = 2(1)10$, $P^* = 0.550(0.050)0.950$, 0.975, 0.990, $\epsilon_1^* = \epsilon_2^* = \epsilon^* = 0.05(0.05)0.20$, and $d^* = \epsilon^*$ (under Formulations 1A and 1B).

Sobel (1977) has discussed a procedure R' for selecting the population associated with the smallest inter (α,β)-range, $Q_i = x_\beta(F_i) - x_\alpha(F_i)$, where α and β are specified numbers such that $0 < \alpha < \frac{1}{2} < \beta < 1$, with neither being close to $\frac{1}{2}$. The quantity Q_i is a measure of dispersion.

For any fixed number n of observations, let $S_{i\alpha}$ and $S_{i\beta}$ denote the αth and βth sample quantiles from F_i and let $S_i = S_{i\beta} - S_{i\alpha}$. The procedure R' selects the population associated with the smallest S_i. The preference zone for meeting the P^*-requirement is specified in a way similar to the case of the procedure R.

8.3 SELECTION PROCEDURES BASED ON RANK ORDER STATISTICS

Selection procedures based on ranks were first proposed and investigated by Lehmann (1963) for location and scale parameter families of distributions. The basic techniques of this paper have been used by others and are discussed in this section.

8.3.1 A Class of Procedures Based on Ranks

Let X_{ij}, $j = 1,\ldots,n$, be independent observations from the population π_i which has the associated cdf $F(x - \theta_i)$, $i = 1,\ldots,k$. The goal is to select the population associated with $\theta_{[k]}$. Toward this end, the following approach is made by Lehmann (1963).

Definition 8.1. $\pi_i (i = 1,\ldots,k)$ is called a *good population* if

$$\theta_i \geqslant \theta_{\max} - \Delta^*, \tag{8.15}$$

where $\theta_{\max} = \max(\theta_1,\ldots,\theta_k)$ and $\Delta^* > 0$ is given. It is required that

$$\Pr(\text{selected population is good}) \geqslant P^*. \tag{8.16}$$

The selection of a good population is a special case of Δ-correct ranking of Fabian (1962) discussed in §2.7.

When F is normal, the natural procedure is the means procedure R_1 which selects the population associated with the largest $\overline{X}_i$, where $\overline{X}_i$, $i = 1,\ldots,k$, are the sample means. Without the assumption of normality, we can find a large sample solution which depends on σ^2, the common variance of the populations. Consider a sequence of situations for increasing n and define the population π_i to be *good* if

$$\theta_i \geqslant \theta_{\max} - \Delta^{(n)}, \tag{8.17}$$

where $\Delta^{(n)}$ is defined in Lemma 8.1. This definition seems rather opportunistic since the acceptable θ-values change with n, the sample size. The sequence is really improvised as a mathematical device for approximating the actual situation. If in a practical situation, the definition (8.15) applies with a given value of Δ^*, then $\Delta^{(n)}$ will be identified with Δ^*.

In what follows we assume without loss of generality that $\theta_k = \theta_{\max}$. For the means procedure R_1, the LFC for selecting a good population is given by

$$\theta_1 = \cdots = \theta_{k-1} = \theta_k - \Delta^{(n)} \tag{8.18}$$

and the sample size is therefore determined by

$$\Pr\left(\bar{X}_k = \bar{X}_{\max}\right) = P^* \tag{8.19}$$

when $(\theta_1, \ldots, \theta_k)$ satisfies (8.18).

The large sample solution follows from the following lemma of Lehmann (1963).

Lemma 8.1. For fixed P^*, and with "goodness" of a population defined by (8.17), let n be determined so that (8.18) and (8.19) hold. Then, as $n \to \infty$,

$$\Delta^{(n)} = \Delta\sigma n^{-1/2} + o(n^{-1/2}), \tag{8.20}$$

where σ^2 is the variance of F, and Δ is determined by

$$\Pr\left(U_i \leqslant \Delta/\sqrt{2}\,, \qquad i = 1, \ldots, k-1\right) = P^*. \tag{8.21}$$

Here the U_i are equicorrelated $N(0,1)$ random variables with correlation $\frac{1}{2}$.

For a given Δ^*, the large sample solution of n for which (8.15) and (8.16) hold is obtained by putting $\Delta^* = \Delta\sigma n^{-1/2}$. Thus,

$$n = \left(\frac{\Delta\sigma}{\Delta^*}\right)^2. \tag{8.22}$$

To define the procedure R_2 of Lehmann (1963) based on scores, let $R_{ij}=R(X_{ij})$ denote the rank of X_{ij} in the set of pooled observations. Define

$$h(r)=E_{F_0}(Z_{r,N}), \qquad r=1,\dots,N \tag{8.23}$$

where $Z_{r,N}$ is the rth order statistic based on $N(=kn)$ independent observations from the distribution F_0 and E_{F_0} denotes the expectation with respect to F_0.

The procedure R_2 is:

$$\text{Select } \pi_i \text{ if and only if } V_i=\max_{\alpha} V_\alpha, \tag{8.24}$$

where

$$V_i=\sum_{j=1}^{n} h(R_{ij}). \tag{8.25}$$

For R_2, the claim of Lehmann (1963) that (8.18) is the LFC is not true in general. This has been shown by Rizvi and Woodworth (1970) through counterexamples. But the approach of Lehmann has been adopted with some modifications by other authors.

Let $m=g(n)$ be the sample size required by R_2 if it is to satisfy (8.16) with "goodness" given by (8.17) and (8.20). Lehmann's interest was to study the asymptotic efficiency of R_2 relative to R_1 defined by $\lim_{n\to\infty} n/m$.

When F_0 is a uniform distribution over $(0,1)$, the procedure R_2 is equivalent to selecting the population π_i for which $R_i=\Sigma_{j=1}^{n}R_{ij}$ is the largest. This is the procedure considered by Alam and Thompson (1971) who assume that the populations $\pi_1,\dots,\pi_k$ are stochastically ordered and the best population is $F_{[k]}$ which is stochastically the largest. They have shown that, for $k=3$, when the distributions are normal, the slippage configuration is the least favorable for large n. Monte Carlo studies for $k=3,4,5$ in the normal case indicate that the preceding result may hold for all n in these cases.

8.3.2 More Procedures Based on Ranks

Puri and Puri (1968, 1969) have investigated procedures based on scores for location and scale parameter cases. Their approach and results are similar to those of Lehmann (1963); however, they get by the difficulty relating to the least favorable configuration by restricting themselves to that part of the parameter space where $\theta_{[k]}-\theta_{[1]}=0(n^{-1/2})$ in the location parameter case and to that part where $\theta_{[k]}/\theta_{[1]}=1+0(n^{-1/2})$ in the scale parameter case. It seems difficult to translate this idea into practical terms.

Puri and Puri (1969) have considered the goals of selecting the t best populations with and without reference to ordering, from populations belonging to a location parameter family. In this case, it is required that

$$\text{PCS} \geqslant P^* \quad \text{whenever} \quad \theta_{[k-t+1]} - \theta_{[k-t]} \geqslant \Delta^*. \tag{8.26}$$

Under nonnormality, one can use the means procedure R_3 which selects the populations associated with the set $\{\bar{X}_{[k-t+1]}, \ldots, \bar{X}_{[k]}\}$, where the $\bar{X}_i$ are the sample means. The result obtained by Puri and Puri (1969) is similar to that of Lehmann (1963). We state the result in the following lemma.

Lemma 8.2. For fixed P^*, under the condition that $\theta_{[k-t+1]} - \theta_{[k-t]} > \Delta^{(n)}$, let n be the smallest integer such that $P(CS|R_3) \geqslant P^*$ for the least favorable configuration $\theta_{[1]} = \cdots = \theta_{[k-t]} = \theta_{[k-t+1]} - \Delta^{(n)}$; $\theta_{[k]} - \theta_{[k-t+1]} = 0$. Then, as $n \to \infty$,

$$\Delta^{(n)} = \Delta \sigma n^{-1/2} + o(n^{-1/2}). \tag{8.27}$$

Here σ^2 is the variance of F and Δ is determined by

$$\Pr\left(U_i \leqslant 2^{-1/2}\Delta \text{ and } W_j \leqslant 0;\ i = 1, \ldots, k-t;\ j = 1, \ldots, t-1\right) = \frac{P^*}{t}, \tag{8.28}$$

where the U_i and the W_j are $N(0,1)$ random variables such that $\text{cov}(U_i, U_{i'}) = \text{cov}(W_j, W_{j'}) = \frac{1}{2}$ for $i \neq i'$, $j \neq j'$, and $\text{cov}(U_i, W_j) = -\frac{1}{2}$.

Given Δ^*, a large sample solution for n is given by

$$n = \left(\frac{\Delta\sigma}{\Delta^*}\right)^2. \tag{8.29}$$

To define the F_0-scores procedure R_4 of Puri and Puri (1969), let X_{ij}, $j = 1, \ldots, n$ be independent observations from π_i, $i = 1, \ldots, k$. For the pooled sample of $N = kn$ observations, define

$$Z_{j,N}^{(i)} = \begin{cases} 1, & \text{if the } j\text{th smallest observation is from } \pi_i, \\ 0, & \text{otherwise.} \end{cases} \tag{8.30}$$

Let

$$T_i = n^{-1} \sum_{j=1}^{N} E(V_{j,N}) Z_{j,N}^{(i)}, \tag{8.31}$$

where $V_{1,N} < V_{2,N} < \cdots < V_{N,N}$ is an ordered sample from a given continuous distribution F_0. The procedure R_4 selects the populations associated with the set $\{T_{[k-t+1]}, \ldots, T_{[k]}\}$.

Puri and Puri (1969) have obtained the sample size m required to guarantee (8.26) subject to the condition that the differences of any two θ's is of order $m^{-1/2}$. The asymptotic efficiency of the F_0-scores procedure R_4 relative to the means procedure R_3 is given by

$$e_{R_4, R_3}(F) = \sigma^2 A^{-2}\left[\int \frac{d}{dx}\{J(F(x))\}\, dF(x)\right]^2, \tag{8.32}$$

where

$$J = F_0^{-1},$$

$$A^2 = \int_0^1 J^2(x)\,dx - \left(\int_0^1 J(x)\,dx\right)^2. \tag{8.33}$$

If F_0 is a uniform distribution over $(0,1)$, $e_{R_4, R_3} \geqslant 0.864$ for all F, and if F_0 is an $N(0,1)$ distribution, $e_{R_4, R_3} > 1$ for all nonnormal F.

For the problem of selecting the t best populations with order, the means procedure R_5 and the F_0-scores procedure R_6 select the same set as R_3 and R_4, respectively, ordering them in that order starting with $\bar{X}_{[k]}$ and $T_{[k]}$ as the best, respectively. The sample size n for R_5 is chosen so that

$$\Pr(CS \text{ with regard to order}) \geqslant P^*, \tag{8.34}$$

where the θ_i are subject to the conditions

$$\theta_{[i+1]} - \theta_{[i]} \geqslant \delta^{(n)}, \qquad i = k-t, \ldots, k-1, \tag{8.35}$$

with

$$\delta^{(n)} = \delta\sigma n^{-1/2} + o(n^{-1/2}). \tag{8.36}$$

The constant δ is determined by

$$(k-t)Q_{k-1}(\underbrace{0,\ldots,0}_{k-t-1}, \underbrace{\delta 2^{-1/2},\ldots,\delta 2^{-1/2}}_{t}) \geqslant P^*, \qquad 1 \leqslant t \leqslant k-1; \tag{8.37}$$

where Q_{k-1} stands for the cdf of a normal random vector

$(U_1,\ldots,U_{k-t-1},W_1,\ldots,W_t)$ satisfying

$$\left.\begin{aligned}
&E(U_i)=E(W_j)=0;\\
&\operatorname{cov}(U_i,U_{i'})=\tfrac{1}{2}(1+\delta_{ii'});\\
&\operatorname{cov}(W_j,W_{j'})=\begin{cases} 1 & \text{if } j=j',\\ -\tfrac{1}{2} & \text{if } |j-j'|=1,\\ 0 & \text{otherwise;}\end{cases}\\
&\operatorname{cov}(U_i,W_j)=\begin{cases} -\tfrac{1}{2} & \text{if } j=1,\\ 0 & \text{otherwise;}\end{cases}\\
&i,i'=1,\ldots,k-t-1;\quad j,j'=1,\ldots,t.
\end{aligned}\right\} \tag{8.38}$$

The $\delta_{ii'}$ in (8.38) are Kronecker deltas.

Let m be the sample sizes needed for R_6 subject, as before, to the condition that the difference of any two θ's is of order $m^{-1/2}$. A comparison of R_5 and R_6 similar to that of R_3 and R_4 gives

$$e_{R_6,R_5}(F)=e_{R_4,R_3}(F). \tag{8.39}$$

In another paper, Puri and Puri (1968) have investigated the selection problem with reference to a scale parameter. The general trend of the investigation is the same and we briefly indicate the formulation and the results as follows.

Let X_{ij}, $j=1,\ldots,n$, be independent observations from $\pi_i(i=1,\ldots,k)$ whose associated distribution function is $F(x/\sigma_i)$, $\sigma_i>0$. In this case, we define the population with $\sigma_{[1]}$ as the best. Let $S_i^2=n^{-1}\sum_{j=1}^{n}X_{ij}^2$. For selecting the t best populations without regard to order, it is required that $\text{PCS}\geqslant P^*$ whenever $\theta_{t+1,t}\geqslant\theta^*>1$, where $\theta_{ij}=\sigma_{[i]}^2/\sigma_{[j]}^2$ and θ^* is specified in advance. The normal theory procedure R_7 selects the populations associated with $\{S_{[1]}^2,\ldots,S_{[t]}^2\}$. The large sample solution in this case is given by $n=(\gamma_2+2)\Delta^2(\theta^*-1)^{-2}$, where γ_2 is the kurtosis of F.

To define the F_0-scores procedure R_8, consider the $N=kn$ absolute values $|X_{ij}|$ and define

$$Z_{j,N}^{(i)}=\begin{cases} 1 & \text{if the } j\text{th smallest is from } \pi_i,\\ 0 & \text{otherwise.}\end{cases} \tag{8.40}$$

The procedure R_8 selects the populations associated with the set

$\{T_{[1]},\ldots,T_{[t]}\}$, where

$$T_i = n^{-1}\sum_{j=1}^{N} E_{j,N} Z_{j,N}^{(i)} \tag{8.41}$$

and $E_{j,N}$ denotes the expected value of the square of the jth order statistic based on N independent observations from F_0. Same procedures with the natural ordering of the populations of the selected set are used if t best populations are to be selected with regard to order. The asymptotic efficiency of R_8 relative to R_7 (for selection with or without reference to ordering) is given by

$$e_{R_8,R_7}(F) = \frac{A^{-2}(\gamma_2+2)I^2(J,G)}{4}, \tag{8.42}$$

where $G(x/\sigma_i)$ is the cdf of $|X_{ij}|$, $J=(F_0^{-1})^2$, A^2 is given by (8.33), and

$$I(J,G) = \int_0^\infty x\frac{d}{dx}\{J[G(x)]\}\,dG(x). \tag{8.43}$$

Randomized and nonrandomized procedures based on rank-sum statistics have been studied by Bartlett and Govindarajulu (1968) for selecting the best population; however, there are difficulties regarding the least favorable configuration. Investigations based on techniques similar to Lehmann (1963) and Puri and Puri (1968, 1969) have been made by Bhapkar and Gore (1971) for procedures based on generalized U-statistics.

8.3.3 Selection Procedures Based on Hodges–Lehmann Estimators

In this section we discuss some selection procedures for location parameters based on one-sample and two-sample estimators of Hodges and Lehmann [1963]. Let X_{ij}, $j=1,\ldots,n$, $i=1,\ldots,k$ be independent observations from k populations $\pi_1,\ldots,\pi_k$, where π_i has the cdf $F(x-\theta_i)$. Assume that F is absolutely continuous. The goal is to select a good population which is any π_i for which $\theta_i > \theta_{\max} - \Delta$ for a specified $\Delta > 0$.

Let $X' = (\mathbf{X}_1',\ldots,\mathbf{X}_k')$ where $\mathbf{X}_i = (X_{i1},\ldots,X_{in})'$, $i=1,\ldots,k$. Suppose $\Psi(\mathbf{X}_i,\mathbf{X}_t)$ is an estimator of $\theta_i - \theta_t$ with the following properties for each $\mathbf{x}_i$ and $\mathbf{x}_t$:

F $\mathbf{c}=(c,\ldots,c)'$ is an n-vector, then

$$\Psi(\mathbf{x}_i+\mathbf{c},\mathbf{x}_t) = c+\Psi(\mathbf{x}_i,\mathbf{x}_t); \tag{8.44a}$$

$$\Psi(\mathbf{x}_i,\mathbf{x}_t) = -\Psi(\mathbf{x}_t,\mathbf{x}_i); \tag{8.44b}$$

$$\text{If } x_{ij} > x_{tj} \quad \text{for} \quad j=1,\ldots,n,\ \Psi(\mathbf{x}_i,\mathbf{x}_t) > 0. \tag{8.44c}$$

Some examples of estimates with these properties are:

(a) $$\bar{x}_i - \bar{x}_t,$$

(b) $$\operatorname*{median}_{\alpha,\beta}(x_{i\alpha} - x_{t\beta}),$$

(c) $$\operatorname*{median}_{\alpha}(x_{i\alpha}) - \operatorname*{median}_{\beta}(x_{t\beta}).$$

For testing the equality of $F(x-\theta_i)$ and $F(x-\theta_t)$, consider the F_0-scores test based on the statistic $\Sigma_{j=1}^{n} E_{F_0}(V_{r_j,2n})$ where $V_{1,2n} < \cdots < V_{2n,2n}$ are the order statistics of a random sample of size $2n$ from the distribution F_0, and r_j is the rank of x_{ij} in the joint ranking of $x_{i1},\dots,x_{in}, x_{t1},\dots,x_{tn}$. Hodges and Lehmann [1963] have proposed estimates of $\theta_i - \theta_t$ based on these rank tests. It can be shown that, if $\Psi(\mathbf{x}_i,\mathbf{x}_t)$ is the Hodges–Lehmann (H-L) estimate corresponding to the two-sample F_0-scores test, $\Psi(\mathbf{x}_i,\mathbf{x}_t)$ satisfies properties (8.44).

We now estimate $\theta_i - \theta_t$ by

$$Z_{it}(\mathbf{x}) = \Psi_{i.}(\mathbf{x}) - \Psi_{t.}(\mathbf{x}), \tag{8.45}$$

where $\Psi_{i.}(\mathbf{x}) = k^{-1}\Sigma_{s=1}^{k}\Psi(\mathbf{x}_i,\mathbf{x}_s)$. The procedure R investigated by Randles (1970) is:

$$\text{Select } \pi_i \text{ if } Z_{it}(\mathbf{x}) > 0 \text{ for all } t \neq i. \tag{8.46}$$

In other words, the rule R selects the population corresponding to $\max\{\Psi_{1.}(\mathbf{x}),\dots,\Psi_{k.}(\mathbf{x})\}$. Here it is assumed that $\Psi(\mathbf{x}_i,\mathbf{x}_t)$ satisfies properties (8.44).

REMARK 8.1. If $\Psi(\mathbf{x}_i,\mathbf{x}_t) = \varphi(\mathbf{x}_i) - \varphi(\mathbf{x}_t)$, the procedure R selects the population corresponding to $\max\{\varphi(\mathbf{x}_1),\dots,\varphi(\mathbf{x}_k)\}$. In particular, if $\Psi(\mathbf{x}_i,\mathbf{x}_t) = \bar{x}_i - \bar{x}_t$, the resulting procedure is one studied by Bechhofer (1954). □

The following theorem of Randles (1970) gives the result regarding the infimum of PCS. Here a correct selection refers to selection of any good population.

Theorem 8.1. For any procedure R in the class of rules (8.46),

$$\inf_{\Omega} P(CS|R) = \Pr(\text{selecting } \pi_k | \theta_1 = \cdots = \theta_{k-1} = \theta_k - \Delta) \tag{8.47}$$

where Ω is the parameter space and the probability on the right side of (8.47) is independent of θ_k.

To compare two procedures S_1 and S_2 in the class of rules (8.46), we determine the corresponding sample sizes $n_1 = n_1(\Delta, P^*, F)$ and $n_2 = n_2(\Delta, P^*, F)$ so that

$$\inf_{\Omega} P(CS|R) = P^*. \tag{8.48}$$

Now we assume that the sample size n_j and its corresponding Δ are such that

$$\Delta^{(n_j)} = K_j n_j^{-1/2} + o\left(n_j^{-1/2}\right) \quad \text{as } n_j \to \infty, \tag{8.49}$$

where K_j is a positive constant. An asymptotic comparison is made by examining the ratio of the sample sizes under a sequence of Δ values approaching zero.

Randles (1970) has compared the asymptotic efficiency of the procedure S_1 for which $\Psi(\mathbf{x}_i, \mathbf{x}_t)$ is the H-L estimate of $\theta_i - \theta_t$ corresponding to the two-sample F_0-scores test relative to the procedure S_2 for which $\Psi(\mathbf{x}_i, \mathbf{x}_t) = \bar{x}_i - \bar{x}_t$. If $F(x)$ is the distribution function of an absolutely continuous random variable with a finite variance σ^2, then under certain regularity conditions, the asymptotic efficiency of S_1 relative to S_2 is same as the Pitman efficiency of the two-sample F_0-scores test relative to the t-test.

As one can see, the procedure R of Randles (1970) requires the computation of $\binom{k}{2}$ two-sample H-L estimates each based on ranks of $2n$ observations. Ghosh (1973) has proposed an alternative procedure R' which uses only one-sample H-L estimates each based on ranks of n observations. However, Ghosh assumes F to be symmetric.

Assume that $\theta_1 \leqslant \cdots \leqslant \theta_k$ and define, as before, π_i to be a good population if $\theta_i > \theta_k - \Delta$, where $\Delta > 0$. Ghosh (1973) considered three goals, namely,

I. Select a good population.
II. Select the t best populations without regard to order.
III. Select the t best with regard to order.

Let

$$R_{ij} = \tfrac{1}{2} + \sum_{j=1}^{n} u(|X_{ij}| - |X_{ij'}|), \qquad j = 1, \ldots, n; \quad i = 1, \ldots, k; \tag{8.50}$$

where

$$u(t)=\begin{cases} 1 & \text{if } t>0, \\ \frac{1}{2} & \text{if } t=0, \\ 0 & \text{if } t<0. \end{cases} \tag{8.51}$$

Thus R_{ij} is the rank of $|X_{ij}|$ among $|X_{i1}|,\ldots,|X_{in}|$. Define

$$h(\mathbf{X}_i)=\sum_{j=1}^{n}\operatorname{sgn}(X_{ij})E\big(J(U_{n,R_{ij}})\big), \qquad i=1,\ldots,k, \tag{8.52}$$

where

$$\operatorname{sgn}(t)=\begin{cases} 1 & \text{if } t>0, \\ 0 & \text{if } t=0, \\ -1 & \text{if } t<0, \end{cases} \tag{8.53}$$

and $U_{n,1}\leqslant\cdots\leqslant U_{n,n}$ are the order statistics of a sample of n independent observations from a uniform distribution over $(0,1)$. The function J in (8.52) is given by

$$J(u)=\Psi^{-1}\left(1+\frac{u}{2}\right), \tag{8.54}$$

where $\Psi(x)$ is the cdf of a symmetric random variable.

The one-sample H-L estimators are given by

$$\hat{\theta}_i(\mathbf{X}_i)=2^{-1}\left[\hat{\theta}_{i1}(\mathbf{X}_i)+\hat{\theta}_{i2}(\mathbf{X}_i)\right], \qquad i=1,\ldots,k, \tag{8.55}$$

where

$$\left.\begin{aligned} \hat{\theta}_{i1}(\mathbf{X}_i)&=\sup\{a\colon h(\mathbf{X}_i-a\mathbf{1}_n)>0\},\\ \hat{\theta}_{i2}(\mathbf{X}_i)&=\inf\{a\colon h(\mathbf{X}_i-a\mathbf{1}_n)<0\},\\ \mathbf{1}_n&=(1,\ldots,1),\text{ an } n\text{-vector.} \end{aligned}\right\} \tag{8.56}$$

In particular, when $J(u)=u$ or $\chi_1^{-1}(u)$ (the inverse of the cdf of a chi random variable with 1 degree of freedom), the statistic $\hat{\theta}_i$ becomes the Wilcoxon signed-rank statistic or normal scores statistic.

The procedures of Ghosh (1973) corresponding to the three goals are:

I. Select the population associated with $\hat{\theta}_{[k]}$.

II. Select the populations associated with the set $\{\hat{\theta}_{[k-t+1]},\ldots,\hat{\theta}_{[k]}\}$.
III. Select the t populations in II and order them according to the order of the $\hat{\theta}_i$.

In each of these cases, let Ω' denote the subset of the parameter space over which we require PCS $\geqslant P^*$ for any goal. For Goal I, $\Omega' = R^k$, the k-dimensional Euclidean space. For Goals II and III, Ω' is $\{\boldsymbol{\theta}: \theta_{k-t+1} - \theta_{k-t} \geqslant \Delta\}$ and $\{\boldsymbol{\theta}: \theta_{i+1} - \theta_i \geqslant \Delta, i = k-t,\ldots,k-1\}$, respectively. It has been shown that the least favorable configurations for these goals are:

I. $\theta_1 = \cdots = \theta_{k-1} = \theta_k - \Delta$.
II. $\theta_1 = \cdots = \theta_{k-t} = \theta_{k-t+1} - \Delta = \cdots = \theta_k - \Delta$.
III. $\theta_1 = \cdots = \theta_{k-t} = \theta_{k-t+1} - \Delta = \cdots = \theta_k - t\Delta$.

In each of the three cases, the asymptotic efficiency of the proposed procedure relative to Bechhofer's means procedure turns out to be the same as the Pitman efficiency of a Chernoff–Savage test with respect to Student's t-test.

8.3.4 An Optimal Procedure Based on Relative Rank—The Secretary Problem

In this section, we discuss an optimal procedure investigated by Chow, Moriguti, Robbins, and Samuels (1964) for the problem of selecting the best person from among n rankable persons who appear sequentially in a random order. This problem is called *Secretary Problem* or *Marriage Problem*. Here rank 1 will mean the best and rank n will indicate the worst.

For every person we can observe only the rank relative to the predecessors. So, for the ith person, the relative rank = 1 + the number of predecessors better than the ith person. If we select the ith person, the process terminates; otherwise we reject the ith person and go to the $(i+1)$st person. The ith person cannot be recalled.

Let X denote the absolute rank of the person selected. Then X can take on values 1 through n with probabilities determined by the strategy used. The problem is to find a stopping rule for which $E(X)$ is minimized.

Let $(X_1,\ldots,X_n)$ denote a random permutation of the integers $(1,\ldots,n)$. All permutations are equally likely. The integer 1 corresponds to the best person. For any $i = 1,\ldots,n$, let

$$Y_i = 1 + \text{number of terms } X_1,\ldots,X_{i-1} \text{ which are } < X_i. \qquad (8.57)$$

Then Y_i is the relative rank of the ith person among the first i persons to appear. A simple but interesting result is that $Y_1,\ldots,Y_n$ are independent.

For any stopping rule τ,

$$E(X)=E\left[\frac{n+1}{\tau+1}Y_\tau\right]. \tag{8.58}$$

We want to minimize this by an optimal choice of τ. The technique used here is the usual method of backward induction. Define, for $i=0,\ldots,n-1$,

$$c_i=c_i(n)=\text{minimal possible expected absolute rank of the person selected if we must confine ourselves to stopping rules } \tau \text{ such that } \tau\geqslant i+1. \tag{8.59}$$

It is easy to see that $c_{n-1}=(n+1)/2$ and for $i=n-1,\ldots,1$,

$$c_{i-1}=i^{-1}\sum_{j=1}^{i}\min\left(\frac{n+1}{i+1}j,c_i\right). \tag{8.60}$$

Now, define

$$\begin{cases} s_i=\left[\dfrac{i+1}{n+1}c_i\right], & i=n-1,\ldots,1,\\ s_n=n, \end{cases} \tag{8.61}$$

where $[x]$ denotes the greatest integer $\leqslant x$.

It has been shown by Chow et al. (1964) that the optimal stopping rule is: Stop with the first $i\geqslant 1$ such that $Y_i\leqslant s_i$. The expected absolute rank of the person selected is c_0.

Lindley (1961) has treated this problem heuristically for large n. His results indicate that for $n\to\infty$, $c_0\to$a finite limit. But his method is too rough to give a value of this limit. A more adequate, but still heuristic approach indicates that $\lim_{n\to\infty}c_0=\prod_{j=1}^{\infty}((j+2)/j)^{1/(j+1)}\cong 3.8695$. Chow et al. prove this result by an alternative method involving several inequalities.

One of the interesting results is that $c_0=c_0(n)$ is strictly increasing with n. Thus we find that $c_0(n)<3.87$, $n=1,2,\ldots$.

There are many mathematically interesting results relating to several variations of the secretary problem; however, we do not discuss them here. For more information the reader is referred to Deely and Smith [1975], Gilbert and Mosteller [1966], Mucci [1973], and Samuels and Gianini [1976].

8.4 AN ADAPTIVE PROCEDURE FOR THE LARGEST LOCATION PARAMETER

Let $X_{i1},\ldots,X_{in}$ be independent observations from a population π_i ($i=1,\ldots,k$) which has the density $f(x-\theta_i)$, where the θ_i are unknown. It is assumed that $f(x)$ is symmetric about the origin. The goal is to select the population associated with $\theta_{[k]}$. It is conceivable that by using different estimators one may arrive at different conclusions about which is the best population. In this section we are concerned with three kinds of estimators, namely, (a) sample mean, (b) a trimmed mean, and (c) a mean of trimmings. The choice of the estimator depends on whether the distributions are light, or medium, or heavy tailed. The selection procedure investigated by Randles, Ramberg, and Hogg (1973) involves two stages. The first stage is to classify the underlying distribution as light tailed (LT), or medium-tailed (MT), or heavy-tailed (HT). The second stage uses an estimator appropriate to this classification and selects the best population.

For details regarding adaptive estimators and related results, see Hogg [1967, 1972] and Davenport [1971].

Let U_β and L_β denote, respectively, the sum of the largest $n\beta$ order statistics and the sum of the smallest $n\beta$ order statistics. If $n\beta$ is not an integer, fractional items are used. For example, if $n=30$ and $\beta=0.05$, we take $U_{0.05}=y_{30}+0.5y_{29}$ and $L_{0.05}=y_1+0.5y_2$, where $y_1\leqslant\cdots\leqslant y_{30}$ are the ordered observations.

The selection procedure R of Randles, Ramberg, and Hogg (1974) uses, as the estimator of θ_i, $\overline{X}^c_{i,1/8}$, or $\overline{X}_{i.}$, or $\overline{X}_{i,3/8}$ (defined below) according as the underlying distribution is classified as LT, or MT, or HT on the basis of the sample data.

The estimator $\overline{X}^c_{i,1/8}$ is a mean of trimmings; it is the mean of $n/8$ smallest and $n/8$ largest order statistics. The estimator $\overline{X}_{i,3/8}$ is a trimmed mean; it is the mean of the middle $\frac{1}{4}$ of these order statistics. The estimator $\overline{X}_{i.}$ is the usual sample mean. These estimators can also be written as

$$\begin{aligned}\overline{X}^c_{i,1/8}&=4n^{-1}\left(U^i_{0.125}+L^i_{0.125}\right),\\ \overline{X}_{i.}&=n^{-1}\left(U^i_{0.5}+L^i_{0.5}\right),\\ \overline{X}_{i,3/8}&=4n^{-1}\left(U^i_{0.5}-U^i_{0.375}+L^i_{0.5}-L^i_{0.375}\right).\end{aligned}\tag{8.62}$$

Let Y_i, $i=1,\ldots,k$, denote the appropriate estimators of the θ_i. Then R selects the population associated with $Y_{[k]}$.

Randles, Ramberg, and Hogg (1973) have made Monte Carlo studies for comparing the performance of the adaptive rule (which we will call R_1

now) with Bechhofer's means procedure R_2 and Lehmann's procedure R_3 which selects the population with the largest rank-sum. The criterion used for the comparison is the percentage of samples for which the procedure correctly selected the population with the largest location parameter. The results of their study can be summarized as follows.

1. All the three procedures perform equally well in the case of normal populations (medium tailed).
2. For uniform populations (light tailed), R_2 and R_3 perform equally well, and R_1 is considerably better than either of the other two if $n \geqslant 10$.
3. In the case of Cauchy distributions (heavy tailed) R_2 is poor. If $n \geqslant 10$, R_1 seems to be the best.
4. The observed percent of correct selection using R_3 was fairly constant over a wide range of underlying distributions.

Since the performances of R_1 and R_2 are approximately the same for normal, one can use the n-value of R_2 as an ad hoc choice of the sample size for R_1.

8.5 SELECTION PROCEDURES USING PAIRED COMPARISONS

In this section we are concerned with selection procedures in an experiment involving k treatments and a design with paired comparisons. If every possible pair of treatments is compared a fixed number of times the design is called the *completely balanced paired-comparison design*. For example, if we have 5 brands of toothpastes, we can present for comparison all 10 possible pairs to 10 judges. This balanced design is similar to the scheme of round robin tournament. In comparing a pair of treatments the possibility of a tie arises. We discuss problems with the possibility of ties as well as with the assumption of no ties. Some procedures for selecting the best player in a tournament with two or three players are discussed in the next section.

Trawinski and David (1963) investigated the problem of comparing k treatments by the method of paired comparisons based on n replications of all $k(k-1)/2$ pairs. Let $T_i \to T_j$ denote the fact that the treatment T_i is preferred to T_j and define

$$x_{ij\gamma} = \begin{cases} 1 & \text{if } T_i \to T_j, \\ 0 & \text{if } T_j \to T_i, \end{cases} \tag{8.63}$$

$$i,j = 1,\ldots,k; \quad i \neq j, \quad \gamma = 1,\ldots,n.$$

It is assumed that there are no ties and no replication effect. It is also assumed that all the comparisons are independent. Let

$$\Pr(x_{ij\gamma}=1)=\pi_{ij},$$
$$\Pr(x_{ij\gamma}=0)=\pi_{ji}=1-\pi_{ij}. \tag{8.64}$$

The score for treatment T_i is given by

$$a_i=\sum_{\gamma=1}^{n}\sum_{j\neq i}x_{ij\gamma} \tag{8.65}$$

which is the number of times the treatment T_i is preferred in the total of $n(k-1)$ comparisons that it was involved in. The π_{ij} are called the preference probabilities. Trawinski and David (1963) have assumed that the preference probabilities satisfy a linear model defined as follows.

Suppose the treatment T_i has the true value or "merit value" V_i when judged on some characteristic. Its observed merit Y_i may be represented by a continuous random variable on the same scale. If it is possible to construct a merit scale such that the probability of preference $\pi_{ij}=\Pr(Y_i-Y_j>0)$ can be expressed for all i and j as $H(V_i-V_j)$, where $H(x)$ is a nondecreasing function such that $H(-\infty)=0$, $H(\infty)=1$, and $H(-x)=1-H(x)$, then the preference probabilities are said to satisfy a *linear model*.

The treatments are ranked according to the V_i. The goal is to select the population associated with $V_{[k]}$. The procedure R proposed by Trawinski and David (1963) selects the population associated with $a_{[k]}$, using randomization to break the ties, if any.

Without loss of generality, we let $V_1\leqslant\cdots\leqslant V_k$. It is assumed that

$$\pi_{kj}=p>\tfrac{1}{2},\qquad j=1,\ldots,k-1,$$
$$\pi_{ij}=\tfrac{1}{2},\qquad i,j=1,\ldots,k-1;\quad i\neq j. \tag{8.66}$$

The model (8.66) has not been shown to be least favorable. Trawinski and David (1963) have tabulated the smallest n needed to guarantee that $PCS\geqslant P^*$ under the assumption (8.66) for $k=2(1)10(2)20$, $p=0.55(0.05)0.95$, and $P^*=0.75$, 0.90, 0.95, 0.99.

Under the assumption in (8.66), let each pair be compared once and the best treatment be selected by the procedure R. It is shown by Huber (1963a) that $\lim_{k\to\infty}\text{PCS}=1$ for all fixed $p>\frac{1}{2}$.

Bühlmann and Huber (1963) considered the following problem of comparing $k \geqslant 3$ items pairwise. It is convenient to describe the problem in terms of tournaments. For the game between Player i and Player j, let

$$a_{ij} = \begin{cases} 1, & \text{if } i \text{ wins,} \\ 0, & \text{if } j \text{ wins.} \end{cases} \tag{8.67}$$

We assume that there are no draws. Let $\Pr(a_{ij}=1)=p_{ij}$, and $\Pr(a_{ij}=0)=p_{ji}$. Obviously, $p_{ij}+p_{ji}=1$. It is also assumed that all games are independent. The matrix $P=((p_{ij}))$ describes the underlying probability structure of the preference matrix $A=((a_{ij}))$. The a_{ij} can be considered as follows.

(i) $a_{ij}(i<j)$ are independent zero–one trials with $p_{ij}=\Pr(a_{ij}=1)$.
(ii) $a_{ij}(i>j)=1-a_{ji}$.
(iii) $a_{ii}=\frac{1}{2}$ for all i.

The players could be ranked in several ways, namely,

a. in the descending order of m_i, where $m_i=\min_{j\neq i} p_{ij}$;
b. in the descending order of p_i, where $p_i=\sum_j p_{ij}$; (8.68)
c. in the descending order of $p_i^{(N)}$, where $\mathbf{p}=(p_1,\ldots,p_k)$ with p_i as defined in (b) and $\mathbf{p}^{(N)}=(p_1^{(N)},\ldots,p_k^{(N)\prime})$ is given by $\mathbf{p}^{(N)}=P^{N-1}\mathbf{p}$.

Now, if $p_{ij}=F(\theta_i-\theta_j)$, where the θ_i are parameters associated with the players and $F(t)$ is the distribution function of a symmetric random variable, then they should obviously be ranked in the descending order of θ_i. This ranking satisfies all three principles of ranking in (8.68). This special case arises if the performance of Player i is of the form $\theta_i+\Delta_i$ and Player i wins over Player j if $\theta_i+\Delta_i>\theta_j+\Delta_j$, the fluctuation terms Δ_i being independent and identically distributed random variables.

The problem is to hit the correct ranking (defined in terms of P) given the outcome A. If there are more than one correct ranking, we arbitrarily distinguish one of them as the best. Let $\mathbf{d}=(d_1,\ldots,d_k)$ denote the ranking vector, where $d_i=l$ means that Player i has been assigned rank l. Let $\mathcal{P}=\{P\}$ be the space of probability matrices. If $p_{ij}=F(\theta_i-\theta_j)$ with fixed F, $\mathcal{P}$ is in one–one correspondence with $\Theta=\{\boldsymbol{\theta}\}$. Let $\mathcal{Q}$ and $\mathcal{D}$ denote, respectively, the space of the tournament outcomes and the space of rankings. Thus a ranking procedure δ is the mapping $\delta: \mathcal{Q} \to \mathcal{D}$. The risk function is given by

$$R(\delta, P) = E_P[L(\delta(A), P)], \tag{8.69}$$

where E_P denotes the expectation if P is the true underlying probability matrix. The ranking procedure δ is *uniformly better than* the procedure δ' (or δ *dominates* δ') for the class $\mathcal{P}$ if

$$R(\delta, P) \leqslant R(\delta', P) \quad \text{for all } P \in \mathcal{P}.$$

Let $\mathcal{P}_0$ be the class of a given matrix P, and all matrices P^σ obtained from P by permutations of the players, that is, $p_{ij}^\sigma = p_{\sigma(i)\sigma(j)}$, where σ stands for permutation of the integers 1 through n. In other words, $\mathcal{P}_0$ is the orbit of P under the group of all permutations of the players. We define

$$L(\mathbf{d}, P) = \begin{cases} 0 & \text{if } \mathbf{d} \text{ is the correct ranking,} \\ 1 & \text{otherwise.} \end{cases} \tag{8.70}$$

The problem is to find $\delta(A)$ which, among all permutation invariant procedures, is uniformly best for $\mathcal{P}_0$. Let $W_P(A)$ denote the probability that the outcome of the tournament is A, given P is the true underlying probability matrix. Then the following procedure δ is shown to be the best permutation invariant procedure in the class $\mathcal{P}_0$.

We determine the permutation σ for which $W_{P^\sigma}(A)$ is maximum. The procedure δ ranks according to the true ranking for P^σ. In other words $\delta(A) = (\sigma(1), \sigma(2), \ldots, \sigma(n))$. If the maximum is attained for more than one σ, then one of them is chosen randomly.

Thus the preceding procedure δ chooses a ranking with maximum a posteriori probability given the outcome, for the uniform a priori distribution on $\mathcal{P}_0$.

The preceding problem where the class $\mathcal{P}$ is restricted to $\mathcal{P}_0$, the orbit of a given matrix P under the permutation group, is referred to as the *reduced ranking problem*. The *general ranking problem* will be to find a ranking procedure δ which is uniformly best among permutation invariant procedures, given a class $\mathcal{P}$, and an invariant (under σ) loss function $L(\mathbf{d}, P)$. An example of Bühlmann and Huber (1963) shows that no such procedure exists unless the class $\mathcal{P}$ is quite restricted or the loss function is very unrealistic. So, one can ask the question: What is the class $\mathcal{P}_u$ for which the usual ranking procedure by the scores $s_i = \Sigma_j a_{ij}$ turns out to be uniformly best? The following theorems of Bühlmann and Huber (1963) provide answers to this question.

Theorem 8.2. Assume that P is a probability matrix such that $p_{ij} \neq 0$ for all i,j. Then $W_P(A)$ depends on A only through the score vector $\mathbf{s} = (s_1, \dots, s_n)$ if and only if $P = (p_{ij})$ is of the form

$$p_{ij} = F(\theta_i - \theta_j) \tag{8.71}$$

where $\{\theta_i\}$ are constants and $F(t) = (1 + e^{-t})^{-1}$ (logistic function).

The preceding theorem reduces the class $\mathcal{P}_u$ that we are seeking to a class of P of the form (8.71). Let us denote this reduced class by $\mathcal{P}_l$.

COROLLARY 8.1. Let $\mathcal{P}$ be a class of matrices for which $p_{ij} \neq 0$ for all i and j, and which, with each element P, contains all matrices P^{σ}. Let $L(\mathbf{d}, P)$ be defined by (8.70). Then, if $\mathcal{P}$ contains at least one element $P_0 \notin \mathcal{P}_l$, the usual ranking procedure using the scores s_i is not uniformly best for $\mathcal{P}$ among the invariant procedures.

Theorem 8.3. Assume that $\mathcal{P} = \mathcal{P}_l$ and $L(\mathbf{d}, P) = L(\mathbf{d}, \boldsymbol{\theta})$ is invariant under permutations and

$$L(\mathbf{d}, \theta) \leqslant L(\mathbf{d}', \theta) \tag{8.72}$$

whenever $d_k < d_l$, $d'_k = d_l$, $d'_l = d_k$, $d'_i = d_i$ for the remaining i, and $\theta_k \geqslant \theta_l$. Then the usual ranking procedure based on the scores is uniformly best among all permutation invariant procedures for the class $\mathcal{P}$.

REMARK 8.2. The assumption (8.72) means that the loss does not decrease if two players are "wrongly interchanged" in the ranking. □

REMARK 8.3. Theorems 8.2 and 8.3 remain true if a fixed number $k \geqslant 1$ of games is played between each pair of players and if a_{ij} is the number of games Player i has won against Player j. Theorem 8.2 remains true (with obvious modifications) if each player does not play against every other one. □

Huber (1963b) considered the problem of comparing k players when draws are possible. Let x_{ij} be a real number that is a measure of the comparison made between Player i and Player j such that

$$x_{ij} + x_{ji} = 0 \quad \text{for all } i \text{ and } j. \tag{8.73}$$

The number x_{ij} might be some measured difference between i and j, or x_{ij} could be defined by

$$x_{ij} = \begin{cases} 1 & \text{if } i \text{ wins,} \\ 0 & \text{if the game ends in a draw,} \\ -1 & \text{if } j \text{ wins,} \end{cases} \tag{8.74}$$

or x_{ij} could be a statistic summarizing several comparisons between i and j. Let $s_i = \sum_{j=1}^{n} x_{ij}$.

The ranking procedure which ranks the players in the descending order (rank 1 means the best) of the scores s_i is termed the *row sum procedure* by Huber (1963b). The following theorem of Huber (1963b) gives an optimal property of this procedure.

Theorem 8.4. Assume that the joint distribution of the s_i is of the form

$$Q_{\boldsymbol{\theta}}(d\mathbf{s}) = c(\boldsymbol{\theta}) f(\boldsymbol{\theta}, \mathbf{s}) \lambda(d\mathbf{s}), \tag{8.75}$$

where λ is some permutation invariant measure on $\mathcal{Y}$, the space of statistics, and f is a density function satisfying

$$\begin{aligned} &f(\boldsymbol{\theta}^{\sigma}, \mathbf{s}^{\sigma}) = f(\boldsymbol{\theta}, \mathbf{s}), \\ &f(\boldsymbol{\theta}, \mathbf{s}) \geqslant f(\boldsymbol{\theta}^{\alpha}, \mathbf{s}), \text{ whenever } \theta_i \geqslant \theta_j,\ s_i \geqslant s_j, \text{ and } \alpha \text{ is the} \\ &\qquad \text{transposition interchanging Players } i \text{ and } j. \end{aligned} \tag{8.76}$$

If $L(\mathbf{d}, \boldsymbol{\theta})$ is a loss function invariant under permutation and satisfying (8.72), then ranking in descending order of the s_i, breaking ties randomly, has minimal risk among all invariant ranking procedures that depend on $\mathbf{x}$ only through $\mathbf{s}$.

It is also shown by Huber (1963b) that, for independent x_{ij}, the exponential distributions given by

$$\Pr(x_{ij} \leqslant t) = c(\theta_i - \theta_j) \int_{-\infty}^{t} e^{(\theta_i - \theta_j)y} \mu(dy) \tag{8.77}$$

are essentially the only distributions for which $\mathbf{s}$ is sufficient.

8.6 INVERSE SAMPLING AND OTHER PROCEDURES FOR TOURNAMENTS WITH TWO OR THREE PLAYERS

In this section we discuss mainly some inverse sampling and related procedures for ranking two or three players in a tournament, which have been investigated by Sobel and Weiss (1970b) and Bechhofer (1970b).

It is assumed that the comparisons are independent and there are no draws. The inverse sampling procedure of Sobel and Weiss uses a Drop-the-Loser (DL) sampling rule. This rule is a 3-player analog of the Play-the-Winner (PW) rule discussed in Chapter 4. Let p_1, p_2, and p_3

denote the single-game probability that A beats B, B beats C, and C beats A, respectively, and let $q_i = 1 - p_i$, $i = 1,2,3$. Let p_0 and P^* denote specified constants with $\frac{1}{2} < p_0 < 1$ and $\frac{1}{3} < P^* < 1$. Player A is defined to be the best player if and only if $\min(p_1, q_3) > \frac{1}{2}$. According to this definition, it is possible that none of the players is best. However, Sobel and Weiss assume that there are only three possible decisions available to us, that is, one of the players has to be selected and declared the best player. We seek a procedure R such that

$$P(CS|R) = \Pr(\text{selecting } A \text{ using } R) \geqslant P^* \text{ whenever } \min(p_1, q_3) \geqslant p_0, \tag{8.78}$$

regardless of the value of p_2. Some of the procedures investigated by Sobel and Weiss do not satisfy (8.78). So a weaker form of (8.78) is required in these cases. Letting θ to stand for p_2, it is required that (8.78) hold for specific values of θ and for $p_0 > p > \frac{1}{2}$, where $p = p(\theta)$ may depend on θ as well as the procedure R.

8.6.1 Inverse Sampling Procedure R_I

The sampling rule of R_I consists of deciding the first game randomly out of the three possible games (A vs. B, B vs. C, C vs. A), and for the succeeding games, using the DL-sampling rule, that is, the loser of any game sits out the next game. Sampling is terminated as soon as any one player has a total of r wins and this player is selected as the best.

PROBABILITY OF A CORRECT SELECTION. Let w_A denote the number of games won by A and let $w'_A = r - w_A$; similar quantities for B and C are denoted by w_B, w'_B, w_C, and w'_C. Let $W' = (w'_A, w'_B, w'_C)$. We use "wt" to denote "wins the tournament" and "PNG=" to denote "the players of the next game are." Define

$$\left.\begin{aligned} R_{k,m,n} &= \Pr(A \text{ wt}|\mathbf{W}' = (k,m,n) \text{ and PNG} = A \text{ vs. } B), \\ S_{k,m,n} &= \Pr(A \text{ wt}|\mathbf{W}' = (k,m,n) \text{ and PNG} = B \text{ vs. } C), \\ T_{k,m,n} &= \Pr(A \text{ wt}|\mathbf{W}' = (k,m,n) \text{ and PNG} = C \text{ vs. } A). \end{aligned}\right\} \tag{8.79}$$

The DL-sampling rule gives recursive relations for $R_{k,m,n}$, $S_{k,m,n}$, and $T_{k,m,n}$, and the stopping rule of R_I gives the boundary conditions. Sobel and Weiss (1970b) have solved the recursive relations numerically and calculated the PCS given by

$$P(CS|R_I) = \tfrac{1}{3}(R_{r,r,r} + S_{r,r,r} + T_{r,r,r}). \tag{8.80}$$

It has been shown by Elliot and Liu in an appendix to the paper by Sobel and Weiss (1970b) that the LFC for R_I is given by

$$\begin{aligned} p_1 = q_3 = p_0 \quad & \left(\tfrac{1}{2} \leqslant p \leqslant 1\right), \\ p_2 = \theta \quad & \left(\tfrac{1}{2} \leqslant \theta \leqslant 1\right). \end{aligned} \tag{8.81}$$

The configuration in (8.81) is least favorable in the sense that $P(CS|R_I)$ is minimum for this configuration over all triples (p_1, p_2, p_3) with $\min(p_1, q_3) \geqslant p_0$ and fixed $p_2 = \theta$.

The expected number of games required to reach a decision is denoted by $E(N|R_I)$ and can be evaluated by a technique similar to that employed for $P(CS|R_I)$. Since R_I depends only on the total number of wins for each player an inefficiency arises which is affected by the sampling rule used. For certain points in the parameter space where A is the best player, $P(CS|R_I)$ does not approach 1 as the number of games grows indefinitely. Under the LFC given in (8.81), the PCS approaches 1 if

$$p > \tfrac{1}{2}\left(-\theta + (\theta^2 + 4\theta)^{1/2}\right). \tag{8.82}$$

If we use the vector-at-a-time (VT) sampling rule, in the place of (8.82) we have

$$p > \tfrac{1}{3}(1 + \theta). \tag{8.83}$$

Since the right members of (8.82) and (8.83) are both increasing in θ, and equal to $(\sqrt{5} - 1)/2$ and $\frac{2}{3}$, respectively, for $\theta = 1$, we see that PCS$\to$1 regardless of how close θ is to 1, when $p > (\sqrt{5} - 1)/2$ under the DL-sampling rule and when $p > \frac{2}{3}$ under the VT-sampling rule. An upper bound and approximations for both $P(CS|\text{LFC})$ and $E(N|\text{LFC})$ have been obtained for R_I by Sobel and Weiss.

REMARK 8.4. An alternative definition of best player is to define A as the best player if the average of his probabilities of beating each of the other contestants (in a single game) is larger than the corresponding averages for B and C, that is, if and only if

$$\frac{p_1 + q_3}{2} > \max\left\{\frac{q_1 + \theta}{2}, \frac{p_3 + 1 - \theta}{2}\right\}. \tag{8.84}$$

For this definition of the best player, it can be shown that (8.78) is satisfied, regardless of the value of θ, under both DL and VT sampling rules by taking r sufficiently large. □

8.6.2 Inverse Sampling with Elimination: Procedure R_E

The procedure R_E, based on inverse sampling and the DL-rule, eliminates one of the three players at or before the termination of the tournament. As in the case of R_I the recurrence formulae provide an algorithm for obtaining the exact values for $P(CS|\text{LFC})$ and $E(N|\text{LFC})$. The first game is decided by randomization. Using the DL-rule, a player is eliminated as soon as he accumulates a total of r losses.

The PCS and $E(N)$ have been evaluated for the configuration (8.81) but this configuration has not been shown to be the least favorable for R_E even when θ is fixed as in the case of R_I.

8.6.3 Other Sequential Procedures

For the purposes of comparisons, Sobel and Weiss (1970b) have considered several sequential procedures and obtained Monte Carlo results. Let W_A, W_B, and W_C denote the number of games won by A, B, and C, respectively. The ordered values are denoted by $W_1 \leqslant W_2 \leqslant W_3$, where ties are clearly possible.

PROCEDURE R_S. This procedure uses the DL-sampling rule. The sampling is terminated as soon as $W_3 - W_2 \geqslant d$ for a specified $d > 0$, and selects the player with W_3 wins as the best. The constant d is to be determined subject to the probability requirement.

The three other sequential procedures use a certain likelihood ratio statistic W. To define W, let W_{AB} be the number of games in which A beat B and let W_{BA}, W_{AC}, W_{CA}, W_{BC}, and W_{CB} denote similar quantities. Six likelihoods are defined for the LFC given by (8.81). These are denoted by L_{ABC}, L_{ACB} and so on. For example,

$$L_{ABC} = p^{W_{AB}} q^{W_{BA}} \theta^{W_{BC}} (1-\theta)^{W_{CB}} p^{W_{AC}} q^{W_{CA}}. \tag{8.85}$$

The sum of L_{ABC} and L_{ACB} represents the "total likelihood" that the common $p(=p_1 = q_3)$ is associated with Player A, that is, that A is the best player. Now, let $\mathcal{L}_A$, $\mathcal{L}_B$, and $\mathcal{L}_C$ denote the total likelihood for A, B, and C, respectively. Then W is defined by

$$W = \frac{\max(\mathcal{L}_A, \mathcal{L}_B, \mathcal{L}_C)}{\mathcal{L}_A + \mathcal{L}_B + \mathcal{L}_C}. \tag{8.86}$$

For all the other three procedures, sampling is terminated as soon as $W \geqslant P^*$ and the player with the largest $\mathcal{L}$-value is selected. For $P^* > \frac{1}{2}$ we cannot have ties for the first place. For $\frac{1}{3} < P^* \leqslant \frac{1}{2}$, we may have ties at termination, in which case randomization is used to break the ties. The three procedures differ only in the sampling rule. One of them uses the

DL-sampling rule and is denoted by R_{DL}. Another one called the "Double Duty" procedure and denoted by R_{DD} uses the sampling rule: Find the two largest $\mathcal{L}$'s and play the corresponding two players in the next game. In the case of ties and at the outset use appropriate randomization.

For the third procedure denoted by R_{DDR}, we define

$$\left.\begin{aligned} s_1 &= \frac{(\mathcal{L}_A + \mathcal{L}_B)}{2(\mathcal{L}_A + \mathcal{L}_B + \mathcal{L}_C)}, \\ s_2 &= \frac{(\mathcal{L}_B + \mathcal{L}_C)}{2(\mathcal{L}_A + \mathcal{L}_B + \mathcal{L}_C)}, \\ s_3 &= \frac{(\mathcal{L}_C + \mathcal{L}_A)}{2(\mathcal{L}_A + \mathcal{L}_B + \mathcal{L}_C)}. \end{aligned}\right\} \tag{8.87}$$

At the outset we randomize by taking a uniform variate $U(0 \leqslant U < 1)$ and play A vs. B if $0 \leqslant U < s_1$, play B vs. C if $s_1 \leqslant U < s_1 + s_2$, and play C vs. A if $s_1 + s_2 \leqslant U < 1$.

The procedures R_{DD} and R_{DDR} eliminate noncontenders in a more natural way by gradually lowering the probability of putting them into the next game rather than by a sudden complete withdrawal from the tournament.

All the six procedures, R_I, R_E, R_S, R_{DL}, R_{DD}, and R_{DDR} have been compared using Monte Carlo estimates of $P(CS|\mathrm{LFC})$ and $E(N|\mathrm{LFC})$ for the last four.

The procedures R_{DL}, R_{DD}, and R_{DDR} require an arbitrarily large number of observations to terminate sampling, when the true state of affairs is such that the probabilities that i beats j, that j beats k, and that k beats i ($i \neq j$, $j \neq k$, $k \neq i$, $i,j,k = 1,2,3$) all approach one. As pointed out by Bechhofer (1970b), this is due to the fact that the Sobel–Weiss formulation does not allow as a possible decision not selecting any one as the best player. Bechhofer considers the problem with the addition of the preceding decision.

8.6.4 The Modified Formulation of Bechhofer

One modification introduced by Bechhofer (1970b), as described in §8.6.3, is to permit a fourth decision: "No player is best." His formulation also includes an additional probability requirement.

Let $p_{ij} = \Pr(\text{Player } i \text{ beats Player } j)$, $i \neq j$; $i,j = 1,2,3$. The p_{ij} are unknown and it is assumed that we have no knowledge about whether $p_{ij} < \frac{1}{2}$ or $p_{ij} > \frac{1}{2}$. (We assume that the possibility $p_{ij} = \frac{1}{2}$ cannot occur). If $p_{ij} > \frac{1}{2}$, then

Player i is better than Player j. Let $w_{ab,cd,ef}$ be the State of Nature in which $p_{ab} > \frac{1}{2}$, $p_{cd} > \frac{1}{2}$, $p_{ef} > \frac{1}{2}$. Then there are eight possible States of Nature: $\omega_{12,23,31}, \ldots, \omega_{21,32,31}$, which are grouped into sets $\Omega_0, \ldots, \Omega_3$ as follows.

$$\left.\begin{aligned} \Omega_0 &= \{\omega_{12,23,31}, \omega_{21,32,13}\}, \\ \Omega_1 &= \{\omega_{12,23,13}, \omega_{12,32,13}\}, \\ \Omega_2 &= \{\omega_{21,23,13}, \omega_{21,23,31}\}, \\ \Omega_3 &= \{\omega_{12,32,31}, \omega_{21,32,31}\}. \end{aligned}\right\} \tag{8.88}$$

Clearly, Ω_0 comprises the states corresponding to "no player is best," and Ω_i $(i=1,2,3)$ includes the states for which Player i is best. The goal is to determine which one of the Ω_i $(i=0,1,2,3)$ represents the true State of Nature. Let d_i $(i=0,1,2,3)$ denote the decision that the true State of Nature is Ω_i. Prior to the start of the play, four constants $\{p_0, P_0^*; p_1, P_1^*\}$ are specified with $\frac{1}{2} < p_0, p_1 < 1$, and $\frac{1}{4} < P_0^*, P_1^* < 1$, and it is required that

$$C_0: \quad \begin{aligned} P_0 = \; & \Pr(\text{making decision } d_0) \geqslant P_0^* \\ & \text{when } \min\{p_{12}, p_{23}, p_{31}\} \geqslant p_0, \text{ or} \\ & \text{when } \min\{p_{21}, p_{32}, p_{13}\} \geqslant p_0, \end{aligned} \tag{8.89}$$

and

$$C_1: \quad \begin{aligned} P_i = \; & \Pr(\text{making decision } d_i) \geqslant P_1^* \; (i=1,2,3) \\ & \text{when } \min\{p_{ij}, p_{ik}\} \geqslant p_1 \\ & (j \neq i;\; k \neq i;\; j \neq k;\; j,k = 1,2,3). \end{aligned} \tag{8.90}$$

8.6.5 A Single-Stage Procedure R_B

A c-cycle tournament is conducted, where c is a predetermined number, and a cycle consists of three games (Player 1 vs. Player 2, Player 2 vs. Player 3, Player 3 vs. Player 1). Let $\hat{p}_{ij}$ denote the proportion of games in which Player i beat Player j in the c-cycle tournament. The procedure R_B of Bechhofer (1970b) compares each of these $\hat{p}_{ij}$ to $\frac{1}{2}$ and makes the obvious terminal decision. If ties occur, randomization is used to choose between the decisions. For example, we randomize between decisions d_0 and d_1 with probability $\frac{1}{2}$ if $\hat{p}_{12} > \frac{1}{2}$, $\hat{p}_{23} > \frac{1}{2}$, and $\hat{p}_{13} = \frac{1}{2}$. Of course, ties cannot occur if c is odd.

PROBABILITIES OF TERMINAL DECISIONS. Assuming c to be odd for sake of simplicity, and letting $B(x|n,p)=\sum_{y=x}^{n}\binom{n}{y}p^y(1-p)^{n-y}$, we have

$$P_0=B\left(\frac{c+1}{2}|c,p_{12}\right)B\left(\frac{c+1}{2}|c,p_{23}\right)B\left(\frac{c+1}{2}|c,p_{31}\right)$$
$$+\left[1-B\left(\frac{c+1}{2}|c,p_{12}\right)\right]\left[1-B\left(\frac{c+1}{2}|c,p_{23}\right)\right]\left[1-B\left(\frac{c+1}{2}|c,p_{31}\right)\right],\tag{8.91}$$

and for $j\neq i$, $k\neq i$, $j\neq k$, and $j,k=1,2,3$, we have

$$P_i=B\left(\frac{c+1}{2}|c,p_{ij}\right)B\left(\frac{c+1}{2}|c,p_{ik}\right),\qquad i=1,2,3.\tag{8.92}$$

It has been shown by Bechhofer (1970b) that the LFC for P_0 is given by $p_{12}=p_{23}=p_{31}=p_0$ when $\min\{p_{12},p_{23},p_{31}\}\geqslant p_0$ and by $p_{12}=p_{23}=p_{31}=1-p_0$ when $\min\{p_{21},p_{32},p_{13}\}\geqslant p_0$, and that the LFC for P_i is given by $p_{ij}=p_{ik}=p_1$ when $\min\{p_{ij},p_{ik}\}\geqslant p_1$. The minimum value of P_0 is same in both cases. The smallest value c^* of c that will guarantee the probability requirements (8.89) and (8.90) is given by

$$c^*=\max\{c_0,c_1\},\tag{8.93}$$

where c_0 and c_1 are the smallest (odd) integers c such that

$$B^3\left(\frac{c+1}{2}|c,p_0\right)+\left[1-B\left(\frac{c+1}{2}|c,p_0\right)\right]^3\geqslant P_0^*\tag{8.94}$$

and

$$B^2\left(\frac{c+1}{2}|c,p_1\right)\geqslant P_1^*,\tag{8.95}$$

respectively.

8.6.6 A Sequential Identification Procedure R_0

Let p_{ij} be defined as in §8.6.4. We assume that p_0,p_1,θ are three *given* probabilities with $\frac{1}{2}<p_0,p_1,\theta\leqslant 1$, and that there are eight possible States of Nature: $\omega^0_{12,23,31},\ldots,\omega^0_{21,32,31}$, which represent certain possible pairings of the triple (p_{12},p_{23},p_{31}) with the probabilities p_0,p_1,θ. The pairings and the groupings of the ω^0's into sets $\Omega^0_0,\ldots,\Omega^0_3$ are given in the following table.

State of Nature		*Pairings of* (p_{12},p_{23},p_{31}) *with* p_0,p_1,θ		
		p_{12}	p_{23}	p_{31}
Ω_0^0	$\omega^0_{12,23,31}$	p_0	p_0	p_0
	$\omega^0_{21,32,13}$	$1-p_0$	$1-p_0$	$1-p_0$
Ω_1^0	$\omega^0_{12,23,13}$	p_1	θ	$1-p_1$
	$\omega^0_{12,32,13}$	p_1	$1-\theta$	$1-p_1$
Ω_2^0	$\omega^0_{21,23,31}$	$1-p_1$	p_1	θ
	$\omega^0_{21,23,13}$	$1-p_1$	p_1	$1-\theta$
Ω_3^0	$\omega^0_{12,32,31}$	θ	$1-p_1$	p_1
	$\omega^0_{21,32,31}$	$1-\theta$	$1-p_1$	p_1

We further assume a *known* prior probability $\xi, \xi(\omega^0) \geqslant 0 (\Sigma_{\Omega^0}\xi(\omega^0)=1)$, associated with each one of the eight States of Nature in $\Omega^0 = \cup_{i=0}^{3}\Omega_i^0$. It seems reasonable, because of the symmetries, to set $\xi(\omega^0)=(1-3\lambda)/2$ for each of the two ω^0's in Ω_0^0, and $\xi(\omega^0)=\lambda/2$ for each of the six ω^0's in $\Omega_1^0 \cup \Omega_2^0 \cup \Omega_3^0$. Then $\xi_0 = \xi(\Omega_0^0) = 1-3\lambda$ and $\xi_i = \xi(\Omega_i^0) = \lambda (i=1,2,3)$.

The goal is to determine which one of the Ω_i^0 represents the true State of Nature. Let d_i be the decision that the true State of Nature is in Ω_i^0, $i=1,2,3$. Two constants $\{P_0^*, P_1^*\}$ with $\frac{1}{4} < P_0^*, P_1^* < 1$ are specified in advance and it is required that

$$C_0^0: P_0^0 = \Pr(\text{making decision } d_0 \text{ when } \Omega_0^0 \text{ is true}) \geqslant P_0^*, \tag{8.96}$$

and

$$C_1^0: P_i^0 = \Pr(\text{making decision } d_i \text{ when } \Omega_i^0 \text{ is true}) \geqslant P_1^*, \ i=1,2,3. \tag{8.97}$$

For simplicity, only symmetric procedures are considered so that the probability expressions of (8.96) and (8.97) are the same for both members of Ω_i^0 $(i=0,1,2,3)$.

For this problem, Bechhofer (1970b) has proposed a sequential procedure R_0 which is based on eight likelihoods, each one of which is associated with a different one of the eight States of Nature. Letting $L_m^{(12,23,31)},\ldots,L_m^{(21,32,31)}$ denote the eight likelihoods based on m games, one can define the four likelihood sums $\mathfrak{L}_m(d_i)$, $i=0,1,2,3$. For example, $\mathfrak{L}_m(d_1)=L_m^{(12,23,31)}+L_m^{(21,32,13)}$. Let

$$P_{j,m}^0=\frac{\xi_j\mathfrak{L}_m(d_j)}{\sum_{r=0}^{3}\xi_r\mathfrak{L}_m(d_r)}, \qquad j=0,1,2,3. \tag{8.98}$$

The stopping and terminal decision rules of the procedure R_0 are based on the $P_{j,m}^0$ and a constant $\hat{P}^*(\frac{1}{4}<\hat{P}^*<1)$ which is chosen to guarantee (8.96) and (8.97). As stated earlier, we let $\xi_0=1-3\lambda$ and $\xi_i=\lambda$ $(i=1,2,3)$, where λ is assumed to be known.

THE PROCEDURE R_0. Run the tournament in cycles. Stop the tournament at the first cycle n for which

$$\max_{0\leqslant j\leqslant 3} P_{jn}^0\geqslant\hat{P}^*, \tag{8.99}$$

where

$$\hat{P}^*=\begin{cases} P_0^* & \text{for } \lambda=0,\\ \max\{P_0^*+3\lambda(1-P_0^*),1-3\lambda(1-P_1^*)\} & \text{for } 0<\lambda<\frac{1}{3},\\ P_1^* & \text{for } \lambda=\frac{1}{3}.\end{cases} \tag{8.100}$$

Make the decision that has associated with it the largest of P_{jn}^0, ties being resolved by randomization.

The identification problem discussed above is the first step toward a sequential ranking procedure. The probability requirement (8.96) and (8.97) corresponds to protecting against the least favorable configuration of (8.89) and (8.90) with the addition of the nuisance parameter θ. However, Bechhofer (1970b) has not defined a sequential ranking procedure that reduces to the sequential identification procedure R_0.

The procedure R_0 which uses $\hat{P}^*$ of (8.100) is very conservative for $0<\lambda<\frac{1}{3}$ because of the crude inequalities employed in deriving it.

8.6.7 Some Remarks on Sobel–Weiss and Bechhofer Procedures

As stated earlier, the Sobel–Weiss formulation of the problem is similar to that of Bechhofer except that the decision d_0 is never permitted. Thus, Sobel and Weiss consider only the probability requirements C_1 of (8.90) and C_1^0 of (8.97). Ideally they would like all their sequential procedures for tournaments with three players to guarantee C_1 of (8.90). However, this has not been proved for any of their procedures. They have shown that their procedure R_I of §8.6.1 does so if the value of the nuisance parameter $\theta = p_{jk}$ in (8.90) is *known*. The procedure R_E of §8.6.2 has been shown by Sobel and Weiss to guarantee the probability requirement C_1^0 of (8.97). The procedure R_S of §8.6.3 is expected to guarantee C_1^0 but there are no analytic results. The other procedures of §8.6.3, namely, R_{DL}, R_{DD} and R_{DDR} guarantee C_1^0.

The important advantage of Bechhofer's formulation is that his procedure will tend to terminate early by making the decision d_0 when either $\min\{p_{12}, p_{23}, p_{31}\} \to 1$ or $\min\{p_{21}, p_{13}, p_{31}\} \to 1$.

8.7 SELECTION IN TERMS OF RANK CORRELATION

In this section, we discuss some selection procedures for choosing the best of k p-variate populations which are ranked according to the usual rank correlation coefficient when $p = 2$ and according to a p-variate analog of the rank correlation coefficient when $p > 2$. We first discuss the bivariate case.

8.7.1 Selection from Bivariate Populations

Let $F_i(x,y)$ denote the continuous distribution function of the bivariate population π_i with rank correlation τ_i $(i = 1,\ldots,k)$. Let (X_{ij}, Y_{ij}), $j = 1,\ldots,n$ and $i = 1,\ldots,k$ denote n independent observations from each of the k populations. Let $R_{i1},\ldots,R_{in}$ denote the ranks of the Y_{ij} $(j = 1,\ldots,n)$ when the X_{ij} $(j = 1,\ldots,n)$ are such that $X_{i1} < \cdots < X_{in}$. Then the rank correlation coefficient is defined by

$$T_i = \binom{n}{2}^{-1} \sum_{j<s}^{n} \sum^{n} \operatorname{sgn}(R_{ij} - R_{is}), \qquad i = 1,\ldots,k. \tag{8.101}$$

Govindarajulu and Gore (1971) have investigated a procedure R for selecting the populations associated with $\tau_{[k]}$ which will guarantee that

$$\text{PCS} \geqslant P^* \text{ whenever } \tau_{[k]} - \tau_{[k-1]} \geqslant d^* > 0. \tag{8.102}$$

The procedure R selects the population associated with the largest T_i.

Suppose that

$$\left.\begin{aligned} X_{ij} &= X_{ij}^* + \sqrt{\Delta_i}\, Z_{ij} \\ & \qquad ; i=1,\ldots,k; j=1,\ldots,n; \\ Y_{ij} &= Y_{ij}^* + \sqrt{\Delta_i}\, Z_{ij} \end{aligned}\right\} \tag{8.103}$$

where X_{ij}^*, Y_{ij}^*, and Z_{ij} are independent random variables having absolutely continuous distribution functions F, G, and H, respectively. Under certain conditions on F, G, and H, Govindarajulu and Gore have obtained a lower bound for PCS when n is sufficiently large and the Δ_i are sufficiently small.

8.7.2 Selection from Multivariate Populations

For ranking p-variate populations, Govindarajulu and Gore (1971) consider the measure θ_i given by

$$\theta_i = \Pr\left(\mathbf{X}_l^{(i)} > \mathbf{X}_m^{(i)} \quad \text{or} \quad \mathbf{X}_m^{(i)} > \mathbf{X}_l^{(i)}\right), \tag{8.104}$$

where $\mathbf{X}_l^{(i)} > \mathbf{X}_m^{(i)}$ if and only if $X_{jl}^{(i)} > X_{jm}^{(i)}$, $j=1,\ldots,p$. The parameter θ_i can be looked upon as the probability of concordance for the population π_i, which is taken as a measure of association.

Let

$$\phi_{l,m}^{(i)} = \begin{cases} 1 & \text{if } \mathbf{X}_l^{(i)} > \mathbf{X}_m^{(i)} \text{ or } \mathbf{X}_m^{(i)} > \mathbf{X}_l^{(i)}, \\ 0 & \text{otherwise,} \end{cases} \tag{8.105}$$

and

$$T_i = \binom{n}{2}^{-1} \sum_{l<m}\sum \phi_{l,m}^{(i)}, \qquad l,m=1,\ldots,n. \tag{8.106}$$

Then the procedure of Govindarajulu and Gore (1971) selects the population corresponding to the largest T_i.

8.8 NOTES AND REMARKS

The role of the procedures for selecting the best multinomial cell as nonparametric procedures is obvious. Let $\pi_1,\ldots,\pi_k$ have continuous distributions $F(x;\theta_i)$, $i=1,\ldots,k$, respectively. Assume that the distributions $\{F(x;\theta)\}$ are stochastically increasing in θ. Let X_{ij}, $j=1,\ldots,n$, be independent observations from π_i. Define

$$p_i = \Pr(X_{ij} > X_{rj},\ r \neq i), \qquad i = 1, \ldots, k;\ j = 1, \ldots, n. \tag{8.107}$$

Thus, if $\theta_i > \theta_j$, then $p_i > p_j$. If we let

$$Y_{ij} = \begin{cases} 1 & \text{if } X_{ij} \geqslant X_{rj}, \qquad r = 1, \ldots, k, \\ 0 & \text{otherwise,} \end{cases} \tag{8.108}$$

then $Y_i = \sum_{j=1}^{n} Y_{ij}$ is the number of observations in the cell with probability p_i in n multinomial trials. Thus the procedures for selecting the best cell discussed in §4.9 can be used for selecting the population with the largest (smallest) θ_i.

Dudewicz (1971b) has studied the efficiency of the multinomial procedure of Bechhofer, Elmaghraby, and Morse (1959) when $F(x; \theta_i) = F(x - \theta_i)$, $i = 1, \ldots, k$. For some interesting applications of nonparametric selection procedures, see Lee (1977b), McDonald (1977b), and Sobel (1977).

CHAPTER 9

Fixed-Size Subset Selection: A Generalized Goal and Other Modifications

9.1 INTRODUCTION

Suppose that $\pi_1,\ldots,\pi_k$ are k populations, where the observable random variable X_i from π_i has the cdf $F(x|\theta_i)$, $i=1,\ldots,k$. The parameters θ_i are unknown, and they belong to the interval Θ, of the real line. The t populations with largest θ-values are defined as the t best populations. The problem of selecting the t best populations under the indifference zone formulation was first studied by Bechhofer (1954). One can generalize this basic goal in different ways. For example, one may want to select a subset of size $s(\geqslant t)$ which contains the t best populations. Sometimes it is of interest to select a subset of size $s(\leqslant t)$ which includes any s of the t best populations. These goals can be brought under one generalized goal which has been considered by Mahamunulu (1967) whose procedure is described in the next section.

Let c,s,t be integers such that $\max(1,s+t+1-k) \leqslant c \leqslant \min(s,t)$, which implies that $\max(s,t) \leqslant k-1$. The generalized goal of the experimenter is to select a subset of size s which contains at least c of the t best populations. As in the original formulation of Bechhofer, two positive constants d^* and $P^*(<1)$ are specified. It is required of the selection procedure to guarantee that the probability of realizing the goal (the event which is synonymous with correct selection) is at least P^* whenever the distance (suitably defined) between $\theta_{[k-t]}$ and $\theta_{[k-t+1]}$ is at least d^*. For a nontrivial problem, it is easy to see that

$$P^* \geqslant \binom{k}{s}^{-1} \sum_{i=c}^{\min(s,t)} \binom{t}{i}\binom{k-t}{s-i} = P(c,k,s,t), \text{ say.} \tag{9.1}$$

The constraints on c, s, and t, assure us that the right side of (9.1) is less than 1.

Most of this chapter is devoted to fixed-size subset selection procedures under the generalized goal described previously. We also discuss here a sequential adaptive procedure of Tong (1978) with a formulation free of indifference zone. The central idea here is to use an estimate of the PCS at each stage to define the stopping rule.

9.2 FIXED-SIZE SUBSET SELECTION WITH GENERALIZED GOAL

In this section we describe a fixed-sample procedure of Mahamunulu (1967) and a related inverse problem considered by Desu and Sobel (1968). Further we consider selecting in terms of quantiles.

9.2.1 A Fixed-Sample Procedure

Mahamunulu (1967), known as Desu in later publications, has investigated the procedure R_1 which selects the populations corresponding to the s largest T-values, where T_i is a suitable statistic based on a random sample of size n from π_i, $i=1,\ldots,k$.

The distance function $d(x,y)$ is a continuous nonnegative real-valued function defined for $x \geqslant y$, where x and y are both real, such that $d(x,y)=0$ if and only if $x=y$. It is assumed that $d(x,y)$ is strictly increasing in x for a fixed y, and is strictly decreasing in y for a fixed x. It is also assumed that T_i has an absolutely continuous distribution $G_n(\cdot|\theta)$, and that $\mathcal{G} = \{G_n(\cdot|\theta),\ \theta \in \Theta\}$ is a stochastically increasing family.

Mahamunulu has shown that in $\Omega(d^*)=\{\boldsymbol{\theta}: d(\theta_{[k-t+1]},\theta_{[k-t]})\geqslant d^*\}$ the LFC is given by

$$\theta_{[1]}=\cdots=\theta_{[k-t]}=\theta'; \qquad \theta_{[k-t+1]}=\cdots=\theta_{[k]}=\theta; \qquad d(\theta,\theta')=d^*. \tag{9.2}$$

It has also been shown that the infimum of PCS over the entire parameter space Ω is $P(c,k,s,t)$. Now, the required sample size is given by the smallest n for which

$$\inf_{\Omega_L} P(CS|R_1)\geqslant P^*, \tag{9.3}$$

where Ω_L is the set of parametric points satisfying (9.2). Suppose that the infimum of $P(CS|R_1)$ over Ω_L takes place at the point where $\theta_{[k]}=\theta_0$. Then the existence of the required sample size for the procedure R_1 is assured by the sufficient condition that

$$\lim_{n\to\infty}\int_{-\infty}^{\infty} G_n(x|\theta_0)dG_n(x|\theta_0')=0, \tag{9.4}$$

where $d(\theta_0,\theta_0')=d^*$. If Y_1 and Y_k are independent random variables with cdf $G(x|\theta_0)$ and $G(x|\theta_0')$, respectively, (9.4) is true provided that

$$\lim_{n\to\infty}\frac{V(Y_k)+V(Y_1)}{\left[E(Y_k)-E(Y_1)\right]^2}=0. \tag{9.5}$$

The condition (9.5) is more easily verifiable than (9.4).

More specific results have been obtained by Mahamunulu (1967) in the cases of location and scale parameter, and also for particular cases of the general goal which correspond to (1) $c=t$ when $s\geqslant t$, and (2) $c=s$ when $s\leqslant t$.

9.2.2 An Inverse Problem

In our preceding discussion, the size s of the selected subset is fixed and the problem is the determination of the smallest common sample size needed to meet the probability requirement. Desu and Sobel (1968) are concerned with the inverse problem. For a given common sample size, their problem is to select a subset of the smallest fixed size s that will contain the t best of k populations ($t\leqslant s\leqslant k$). In the notation of §9.2.1, their procedure R_2 selects the populations corresponding to the s largest T-values, where s is the smallest integer for which

$$P(CS|R_2)\geqslant P^* \quad \text{whenever } \boldsymbol{\theta}\in\Omega(d^*). \tag{9.6}$$

The existence of such a value of s is clear from the fact that $P(CS|R_2)\equiv 1$ for all $\boldsymbol{\theta}\in\Omega$ when $s=k$.

It has been shown by Desu and Sobel that, for any $s<k$, the infimum of $P(CS|R_2)$ is attained for a configuration satisfying (9.2). Let $Q(s,t|\theta)$ denote $P(CS|R_2)$ when the parameter point in Ω satisfies (9.2). Define

$$Q(s,t)=\inf_{\theta\in\Theta} Q(s,t|\theta). \tag{9.7}$$

The following theorem of Desu and Sobel (1968) establishes the existence of a unique solution for s.

Theorem 9.1. For $t<s<k$ and any fixed $\theta\in\Theta$,

$$\Delta Q(s,t|\theta)>0 \quad\text{and}\quad \Delta Q(s|t)\geqslant 0, \tag{9.8}$$

where

$$\Delta Q(s,t|\theta)=Q(s,t|\theta)-Q(s-1,t|\theta),$$

$$\Delta Q(s,t)=\inf_{\theta\in\Theta} Q(s,t|\theta)-\inf_{\theta\in\Theta} Q(s-1,t|\theta). \tag{9.9}$$

The latter inequality in (9.8) is strict if, for each s, inf $Q(s,t|\theta)$ is attained for some $\theta\in\Theta$.

For the special case of the normal means (known common variance σ^2), Desu and Sobel (1968) have tabulated correct to four decimal places the values of $\lambda=d^*\sqrt{n}/\sigma$ for which $Q(s,t)=P^*$ for $P^*=0.900$, 0.950, 0.975, 0.990, 0.999; $k=2$, $t=1$, $s=1$; and $k=3(1)9$, $t=1(1)k-2$, $s=t(1)k-1$. For $k\geqslant 3$, the λ-values corresponding to some values of $s=t$ are left out in the table because they can be obtained from earlier entries in the table using the fact that $Q(t,t)=Q(k-t,k-t)$. Since $Q(s,t)$ increases with s and also with λ, the integer s is obtained so that $\lambda(s-1)>\lambda^*\geqslant\lambda(s)$, where λ^* is the computed values of λ. Here we define $\lambda(0)=\infty$ and $\lambda(k)=0$.

The procedure R_2 possesses the usual unbiasedness property. For a loss function proportional to $(t-B)$ where B is the number of best populations included in the subset, the procedure R_2 has the smallest risk among all invariant procedures based on $T_1,\ldots,T_k$.

9.2.3 Nonparametric Procedures

Sobel (1967) investigated nonparametric procedures for selecting the t populations, with the largest α-quantiles, from a given collection of k populations. This has been discussed in §8.2. In this section we discuss the fixed-size subset generalization of this problem. These procedures are the nonparametric analogs of those in §§9.2.1 and 9.2.2.

Let $\pi_1,\ldots,\pi_k$ be k populations with continuous distribution functions $F_i(x)$, $i=1,\ldots,k$. The setup and the notations are the same as in §8.2. There are four formulations as explained earlier. However, P^* must now be greater than $\binom{k-s}{s-t}/\binom{k}{s}$. The procedure R_3 studied by Desu and Sobel (1971) selects the s populations which give rise to the s largest Y_r-values. The least favorable configuration satisfies

$$F_{[1]}=\cdots=F_{[k-t]}=\underline{F};\ F_{[k-t+1]}=\cdots=F_{[k]}=\bar{F}. \tag{9.10}$$

If $\bar{\mathfrak{F}}=(F_{[1]},\ldots,F_{[k]})$ is a configuration satisfying (9.10), then

$$P(CS|R_3,\bar{\mathfrak{F}})\equiv P(\underline{F},\bar{F})=\int_{-\infty}^{\infty}\underline{U}(y)\,d\bar{V}(y)=\int_{-\infty}^{\infty}\left[1-\bar{V}(y)\right]d\underline{U}(y), \tag{9.11}$$

where

$$\left.\begin{aligned}
\underline{U}(y)&=G(\underline{H}(y);k-s,s-t+1),\\
\bar{V}(y)&=1-\left[1-\bar{H}(y)\right]^t,\\
\underline{H}(y)&=G(\underline{F}(y);r,n-r+1),\\
\bar{H}(y)&=G(\bar{F}(y);r,n-r+1),
\end{aligned}\right\} \tag{9.12}$$

and $G(p;r,n-r+1)$ is the incomplete beta function. Desu and Sobel (1971) have obtained lower bounds for $P(CS|R_3,\bar{\mathfrak{F}})$ under the four different formulations. Let $Q(s,t|n)$ denote the lower bound obtained under any particular formulation. Then, for fixed n, k, P^*, d^*, ϵ_1^*, and ϵ_2^*,

$$Q(s,t|n)>Q(s-1,t|n) \tag{9.13}$$

when $s-1\geqslant t$. In particular,

$$Q(s,t|n)>Q(t,t|n)\quad\text{for}\quad s>t. \tag{9.14}$$

This inequality relates the lower bounds of this paper with the corresponding lower bounds for the problem of Sobel (1967), which is discussed in §8.2. This implies that whenever the required n-value for the earlier paper of Sobel exists for a given set of specifications, then the required n-value for the present problem, with the same set of specifications, exists.

Desu and Sobel (1971) have tabulated the smallest odd integer n required for their procedure for selecting the s populations with largest medians ($\alpha=\frac{1}{2}$) under Formulation 1A with $\epsilon_1^*=\epsilon_2^*=d^*$, and under Formulation 2A with $\epsilon_1^*=\epsilon_2^*$. The values of n are tabulated for $k=2(1)6$, $\epsilon_1^*=\epsilon_2^*=0.15$, 0.20, $P^*=0.75(0.05)0.95$, 0.975, 0.99, and all combinations of (s,t) such that $t<s$, and $t=s=1$. Under Formulation 2A, when $k\geqslant 3$, the n-values required for $s=t\geqslant 2$ are the same as the n-values for $s=t=1$ (that is, when $\alpha=1/2$ and $\epsilon_1^*=\epsilon_2^*$).

The related problem of determining the smallest integer $s\geqslant t$ when the sample size n is specified can be treated in a manner similar to Desu and Sobel (1968).

9.2.4 Selection from Restricted Families of Distributions

In many situations, even though we may not know the underlying distributions, we may be able to make reasonable assumptions regarding some partial ordering relation that they may have with a *known* distribution G. Such situations occur very naturally in reliability studies. The *increasing failure rate* (IFR) and *increasing failure rate average* (IFRA) distributions are well-known examples of such ordered families. These are called *restricted families of probability distributions*. In fact, IFR and IFRA class of distributions are special cases of families of distributions which are *convex* and *star-shaped*, respectively, with respect to a known G, this being exponential in these special cases.

Contributions to ranking and selection problems for restricted families of distributions under (random-sized) subset selection approach are due to Barlow and Gupta (1969), Barlow, Gupta, and Panchapakesan (1969), Panchapakesan (1969), Gupta and Panchapakesan (1974, 1975), Lu (1976), Gupta and Lu (1977), Hooper (1977), and Hooper and Santner (1979). The fixed-size subset selection problems (either with an indifference zone or a preferred population formulation) have been discussed by Hooper (1977), Hooper and Santner (1979), and Gupta and Lu (1977).

We have dealt with the order relations in some detail in Chapter 16 dealing with subset selection procedures. Here we deal with two order relations and some associated selection problems.

Let I be a given interval (finite or infinite) of the real line and let $\mathcal{F}_I$ be the class of distributions F which satisfy:

1. F is absolutely continuous with respect to Lebesgue measure,
2. $S(F)$, the support of F, is a convex set contained in I.

For convenience, let $\mathcal{F}$ and $\mathcal{F}_0$ denote the set $\mathcal{F}_I$ when I=real line and $I=[0,\infty)$, respectively. For F, $G\in\mathcal{F}_0$, (1) F is *convex ordered* w.r.t. G $(F\prec_c G)$ if and only if $G^{-1}F(x)$ is convex on $S(F)$, and F is *star-ordered* w.r.t. $G(F\prec_* G)$ if and only if $G^{-1}F(\lambda x)\leqslant\lambda G^{-1}F(x)$ whenever $x\in S(F)$, $\lambda x\in S(F)$, and $0\leqslant\lambda\leqslant 1$. For F, $G\in\mathcal{F}$, F is said to be *tail ordered* w.r.t. G $(G\prec_t G)$ if and only if $G^{-1}F(x)-x$ is nondecreasing on $S(F)$.

Hooper and Santner (1979) studied fixed-size subset selection procedures for selecting good populations in terms of α-quantiles from star- and tail-ordered distributions; in other words, from distributions $F_1,\ldots,F_k$ belonging to $\mathcal{F}$ or $\mathcal{F}_0$. Let $x_\alpha(F_i)$ denote the (unique) α-quantile of F_i. Also, let $F_{[i]}$ be the cdf with the ith smallest α-quantile and let its α-quantile be denoted by $x_\alpha(F_{[i]})$.

Any population π_i is *preferred* in the star-ordered case if and only if $x_\alpha(F_i)>c^*x_\alpha(F_{[k-t+1]})$ where $0<c^*<1$. In the case of tail-ordered distributions, π_i is preferred if and only if $x_\alpha(F_i)>x_\alpha(F_{[k-t+1]})-d^*$ where $d^*>0$.

Given $t(1\leqslant t<k)$, $r(1\leqslant r\leqslant t)$, $s(r\leqslant s<k-t+r)$ and $\alpha(0<\alpha<1)$, the goal of Hooper and Santner (1979) is to select a subset of size s containing at least r preferred populations. Let $N_\alpha=\min\{n\geqslant 1|1\leqslant(n+1)\alpha\leqslant n\}$. For each $n\geqslant N_\alpha$, Hooper and Santner (1979) have proposed the following procedure R_4 based on sample α-quantiles $T_1,\ldots,T_k$ from samples of size n from each population.

***R*$_4$:** Select π_i if and only if

$$T_i\geqslant T_{[k-s+1]}. \tag{9.15}$$

The problem is to determine the smallest $n\geqslant N_\alpha$ so that

$$P(CS|R_4)\geqslant P^* \tag{9.16}$$

for all $\mathbf{F}=(F_1,\ldots,F_k)\in\Omega_*$ in the star-ordered case and $\mathbf{F}\in\Omega_t$ in the tail-ordered case, where

$$\begin{aligned}\Omega_*&=\{\mathbf{F}=(F_1,\ldots,F_k)|F_i\prec_* G\ \forall\ i\},\\ \Omega_t&=\{\mathbf{F}=(F_1,\ldots,F_k)|F_i\prec_t G\ \forall\ i\},\end{aligned} \tag{9.17}$$

and G is a known distribution. Recall that for a nontrivial problem, $P^*\geqslant P(r,k,s,t)$ defined in (9.1).

Let $B(p;q,n)$ be the incomplete beta function given by

$$B(p;q,n)=\frac{n!}{(n-q)!(q-1)!}\int_0^p x^{q-1}(1-x)^{n-q}\,dx$$

and define

$$H(y)=B(B(y);k-t-s+r,k-t),$$

$$J(y)=B(B(y):t-r+1,t),$$

$$F_n^* \equiv F_n^*(r,t,s,k,\alpha,c^*)$$

$$=\int_{G(c^*)}^{\alpha} H\left(G\left(\frac{G^{-1}(y)}{c^*}\right)\right)dJ(y)+H\left(G\left(\frac{1}{c^*}\right)\right)(1-J(\alpha)),$$

$$F_n^t \equiv F_n^t(r,t,s,k,\alpha,d^*)$$

$$=\int_{G(-d^*)}^{\alpha} H(G(G^{-1}(y))+d^*))dJ(y)+H(G(d^*))[1-J(\alpha)].$$

Hooper and Santner (1979) have shown that the infimum of $P(CS|R_4)$ over Ω_* in the star-ordered case is given by F_n^* and that the infimum over Ω_t in the tail-ordered case is given by F_n^t. Actually in the latter case, the probability requirement can be guaranteed for any $P^*<1$.

Consider the case where $G(x)=1-e^{x\log(1-\alpha)}$, $x \geqslant 0$. Then Ω_* is the set of all configurations of IFRA distributions. For the goal of selecting at least $r=t$ preferred populations, Hooper and Santner (1979) have tabulated the smallest odd sample sizes required for $\alpha=0.25(0.25)0.75, c^*=0.65, 0.70, P^*=0.75, 0.95, k=2(1)9, t=1(1)[k/2], r=t,\ s=t(1)[k/2]$. Here $[k/2]$ denotes the largest integer contained in $k/2$.

Gupta and Lu (1977) have considered the problem of selecting the best population using the indifference zone approach from k populations which are convex ordered w.r.t. G. The best population is the one which is stochastically larger than any other. It is assumed that there exists an $F_{[k]}(x)$ such that $F_{[i]}(x) \geqslant F_{[k]}(x/\delta)$ for all x, $i=1,\ldots,k-1$, where $\delta \in (0,1)$ is specified.

9.3 AN ADAPTIVE SEQUENTIAL PROCEDURE

In Chapter 5, we discussed the problem of estimating the PCS and the results of Olkin, Sobel, and Tong (1976). Any procedure R in the classical indifference zone approach guarantees that the PCS $\geqslant P^*$ whenever the true parametric configuration belongs to a specified preference zone. The idea of estimating the PCS is important because we do not know whether the true parametric configuration belongs to the preference zone or the indifference zone. Tong (1978) proposed an adaptive procedure which is designed to yield a PCS that is approximately P^* under the true configuration.

Let $\pi_1,\ldots,\pi_k$ be k independent populations. Let $X_{i1},X_{i2},\ldots$ be a sequence of independent observations from π_i with distribution function $F(x,\theta_i)$, $\theta_i\in\Theta$, a subset of the real line. We are interested in selecting the population associated with $\theta_{[k]}$. The observations are taken sequentially vector-at-a-time. For n vector observations, let $T_i=T_i^{(n)}=T(X_{i1},\ldots,X_{in})$, $i=1,\ldots,k$, be appropriately chosen statistics. For our discussion we assume that T_i has a continuous distribution with cdf $G_n(y-\theta_i)$. The decision rule R selects the population that yields the largest T_i. For any fixed n, the PCS is given by

$$P_{\boldsymbol{\delta}}(CS|R)=\int_{-\infty}^{\infty}\prod_{i=1}^{k-1}G_n(y+\delta_i)\,dG_n(y)$$

$$=\gamma_1^{(n)},\text{ say,} \tag{9.18}$$

where $\boldsymbol{\delta}=(\delta_1,\ldots,\delta_{k-1})$ and $\delta_i=\theta_{[k]}-\theta_{[i]}$. Let $\hat{\delta}_i=\hat{\delta}_i^{(n)}=T_{[k]}^{(n)}-T_{[i]}^{(n)}$ and $\hat{\gamma}_1^{(n)}$ is the estimate of $\gamma_1^{(n)}$ obtained by substituting $\hat{\delta}_i$ for δ_i in (9.18).

For a predetermined P^*, Tong (1978) proposed the following procedure R_1.

$\boldsymbol{R_1}$: (a) Observe sequence of vector observations. Stop with $N_1=n$ where

$$n=\text{the smallest integer satisfying } \hat{\gamma}_1^{(n)}\geqslant P^*. \tag{9.19}$$

(b) When sampling terminates, select the population corresponding to $T_{[k]}^{(n)}$.

Tong (1978) has also proposed two more procedures R_2 and R_3 for the purpose of obtaining bounds on N_1. Define

$$\hat{\Delta}=\hat{\Delta}_n=\hat{\delta}_{k-1}^{(n)},\ \hat{\bar{\delta}}=\hat{\bar{\delta}}_n=\frac{1}{k-1}\sum_{i=1}^{k-1}\hat{\delta}_i^{(n)}; \tag{9.20}$$

and evaluate

$$\hat{\gamma}_2^{(n)}=\int_{-\infty}^{\infty}G_n^{k-1}(y+\hat{\Delta})\,dG_n(y),$$

$$\hat{\gamma}_3^{(n)}=\int_{-\infty}^{\infty}G_n^{k-1}(y+\hat{\bar{\delta}})\,dG_n(y). \tag{9.21}$$

The procedures R_2 and R_3 are defined by using $\hat{\gamma}_2^{(n)}$ and $\hat{\gamma}_3^{(n)}$, respectively, in the place of $\hat{\gamma}_1^{(n)}$ in (9.19).

For $k=2$, all three procedures are identical. The following theorem of Tong (1978) compares the three procedures.

Theorem 9.2. If $\log G_n(y)$ is a concave function of y for every n, then

(a)
$$\hat{\gamma}_2^{(n)} \leqslant \hat{\gamma}_1^{(n)} \leqslant \hat{\gamma}_3^{(n)} \text{ a.s.}$$

for every $\boldsymbol{\theta}$ and n, and hence

(b)
$$N_3 \leqslant N_1 \leqslant N_2 \text{ a.s.}$$

for every $\boldsymbol{\theta}$.

It can be easily seen that the first inequality in the statement (a) of the preceding theorem does not require the log concavity of $G_n(y)$. The second inequality of this statement follows from a theorem of Olkin, Sobel, and Tong (1976).

Under certain regularity-type conditions, Tong (1978) has shown that, for any arbitrary but fixed $\epsilon > 0$, and $\boldsymbol{\theta}$ for which $\theta_{[k]} - \theta_{[k-1]} > 0$, there exists a $\gamma_0^{(t)} = \gamma_0^{(t)}(\epsilon, k, \boldsymbol{\theta})$ such that

$$|P_\delta(CS|R_t) - P^*| < \epsilon \tag{9.22}$$

for all $P^* > \gamma_0^{(t)}$, $t = 1, 2, 3$. In particular, we have

$$P_\delta(CS|R_t) \to 1 \text{ as } P^* \to 1, \qquad t = 1, 2, 3. \tag{9.23}$$

Tong (1978) performed Monte Carlo studies in the case of $k(=2, 4, 5, 8,$ and $10)$ normal populations with a common unit variance under two configurations: (i) $\boldsymbol{\theta} = (0, 0, \ldots, 0, 0.8)$, and (ii) $\boldsymbol{\theta} = (0.0, 0.4, 0.8, 1.2, \ldots)$, for $P^* = 0.75$, 0.90, 0.95, and 0.99. The performances of R_1, R_2, and R_3 were studied with regard to observed probability of correct selection and the average sample number (ASN). The results of the study can be summarized as follows:

1. The procedure R_2 yields an observed PCS in excess of P^* in all the cases considered, but it is too conservative in the sense of ASN.
2. The procedure R_1 seems to perform satisfactorily with regard to ASN, but very poorly in terms of the observed PCS.
3. The procedure R_1 performs very well under all circumstances.

9.4 NOTES AND REMARKS

The formulation of Desu and Sobel (1968) is useful if the experimenter wants to take a subset of smallest possible fixed size when the common sample size is given. The natural appeal in the random subset size approach of Gupta (1956) is that the experimenter need not be tied down to selection of a predetermined number of populations unless warranted by the evidence of the sample.

One can restrict the size of the random subset by requiring that it may not exceed a specified integer $m(<k)$. Procedures of this type have been proposed by Panchapakesan (1969), Santner (1973), and Gupta and Santner (1973). A generalization of the subset selection goal of Gupta (1956) in a different direction has been investigated by Deverman (1969), and Deverman and Gupta (1969). Their generalized goal is to select a random subcollection of the collection of all subsets of a given size s from a set of k populations such that at least one such selected subset contains at least c of the t best populations. This generalization can be thought of as the counterpart of Mahamunulu's generalization of the Bechhofer formulation. These variations of the basic goal have been discussed in Chapter 11.

CHAPTER 10

Bayesian Selection and Ranking Procedures

10.1 INTRODUCTION

When we rank k populations according to the values of their parameters $\theta_i, i=1,\ldots,k$, it is assumed that no a priori information about the true configuration of the k unknown parameters is available to the experimenter. Hence the experimenter would adopt a procedure that protects him against all possible configurations of the θ_i, including the least favorable one. Situations occur in practice, as for example in the slippage problems, in which the experimenter feels fairly confident of facing the least favorable configurations.

On the other hand, there also exist situations in which the least favorable configuration of the population parameters is a priori so unlikely to occur that the experimenter may not consider it necessary to design his procedure to cover this eventuality. One such situation is that in which the population parameters have been chosen at random from a priori distributions.

Bross (1949) considered the problem of selecting the best of k normal population means, under the assumption that the k means were a priori normally distributed with unequal expected values, and using a linear loss function. His main concern was the criterion to be used for the selection rather than the optimum sample size. In another paper (1950), he determined the optimum sample size in the case of $k=2$ populations with equal prior estimates of the means. His result is a special case of that of Grundy, Healy, and Rees (1956) who considered the problem of selecting from two normal populations the one with the greater mean. They considered the situation in which prior estimates of μ_1 and μ_2, the two population means under consideration, are available from the results of a preliminary experiment. Assuming a linear loss function, they integrated the expected loss or the risk over the fiducial distribution of μ_1 and μ_2 defined by these preliminary estimates, and then determined the sample size n which minimized this integrated risk. Formally, one can obtain the same results by assuming that the parameters have a priori distributions over which the expected loss can be integrated. This, in fact, was done by Dunnett (1960) who considered several different formulations of the problem of selecting the largest of k normal population means, when each population mean has an a priori normal distribution. Dunnett is the first author to investigate in detail Bayesian procedures. In this chapter we review the results of Dunnett (1960), Raiffa and Schlaifer (1961), Guttman and Tiao (1964), Bland and Bratcher (1968), and Govindarajulu and Harvey (1974).

10.2 SELECTING THE LARGEST OF k NORMAL POPULATION MEANS: DETERMINATION OF SAMPLE SIZE TO ACHIEVE A SPECIFIED PROBABILITY OF A CORRECT SELECTION

We have k normal populations with a common known variance σ^2. It is assumed that the mean μ_i of the population π_i is distributed a priori normally with mean U_i and variance σ_0^2. It is also assumed that the μ_i are independent, that σ_0^2 is known, and that the values of U_i are known at least up to an unknown additive constant. The goal is to select the population associated with $\mu_{[k]}$. The procedure R_1 described in this section is due to Dunnett (1960) and is based on the adjusted sample means $Z_i=\bar{X}_i+U_i/N$, $i=1,\dots,k$, where $\bar{X}_i$ is the mean of a sample of size n from π_i, and

$N = n\sigma_0^2/\sigma^2$. The effect of considering the adjusted sample mean is to weight U_i and $\bar{X}_i$ in inverse proportion to their variances.

$\boldsymbol{R_1}$: Select the population that corresponds to the largest Z_i. Note that the procedure R_1 reduces to that of Bechhofer (1954) when the U_i are equal. Let P_i denote the probability that π_i is selected. Then

$$P_i = F_{k-1,\frac{1}{2}}\{\alpha_{1i},\ldots,\alpha_{i-1,i},\alpha_{i+1,i},\ldots,\alpha_{k,i}\} \tag{10.1}$$

where

$$\alpha_{j,i} = -\frac{\mu_j - \mu_i + (U_j - U_i)/N}{\sigma\sqrt{2/n}}, \tag{10.2}$$

and $F_{k-1,\frac{1}{2}}\{z_1,\ldots,z_{k-1}\}$ denotes the cdf of the $(k-1)$-variate normal distribution with zero means, unit variances, and correlation coefficients $\frac{1}{2}$. Integrating the expression for P_i with respect to the joint $(k-1)$-variate a priori distribution of $\alpha_{1,i},\ldots,\alpha_{k,i}$ over negative values of $\mu_j - \mu_i (j = 1,\ldots, i-1, i+1,\ldots,k)$, we get $\bar{P}_i$, the joint probability that μ_i is the largest population mean and π_i is selected, which can be written as

$$\bar{P}_i = \int \cdots \int f(v_1,\ldots,v_{2k-2})\,dv_1\ldots dv_{2k-2} \tag{10.3}$$

the range of integration being from $-\infty$ to $\sqrt{N+1}\,(U_i - U_s)/\sigma_0\sqrt{2N}$ for the first $k-1$ variables and from $-\infty$ to $(U_i - U_s)/\sigma_0\sqrt{2}$ for the last $k-1$ variables $(s = 1,\ldots,i-1,i+1,\ldots,k)$. The function $f(v_1,\ldots,v_{2k-2})$ represents the $(2k-2)$-variate normal density function with zero means, unit variances and correlation matrix

$$\begin{bmatrix} A & B \\ B & A \end{bmatrix},$$

where $A = ((a_{ij}))$ and $B = ((b_{ij}))$ are submatrices of order $k-1$ in which

$$a_{ij} = \begin{cases} 1 & \text{for } i = j \\ \frac{1}{2} & \text{for } i \neq j \end{cases} \tag{10.4}$$

$$b_{ij} = \begin{cases} \rho & \text{for } i = j \\ \rho/2 & \text{for } i \neq j \end{cases} \tag{10.5}$$

and $\rho = (N/(N+1))^{\frac{1}{2}}$. Then the probability of a correct selection is given by

$$P(CS|R_1) = \sum_{i=1}^{k} \bar{P}_i\,. \tag{10.6}$$

Dunnett (1960) has compared his method with that of Bechhofer (1954) in the case of $k=2$ populations, by equating δ^* (the width of the indifference zone) to the expected separation of μ_1 and μ_2, which is $1.1284\sigma_0$. The values of n required by Bechhofer's procedure to discriminate between two μ's differing by an amount δ^* are always smaller than the corresponding n required to discriminate between two μ's whose expected separation is δ^* (except for $P^*=0.50$ where $n=0$ by both methods). As one might expect, the difference in the n required by the two procedures is considerable when P^* is near unity.

Dunnett has also considered the conditional probability of a correct selection given that the largest mean exceeds the others by an amount δ^* or more. This probability is given by

$$P(CS|R;\delta^*)=\frac{\sum_{i=1}^{k}\int\cdots\int f(v_1,\ldots,v_{2k-2})\,dv_1\cdots dv_{2k-2}}{\sum_{i=1}^{k}\int\cdots\int f_{k-1,1/2}(x_1,\ldots,x_{k-1})\,dx_1\cdots dx_{k-1}}, \qquad (10.7)$$

where the range of integration is from $-\infty$ to $(U_i-U_s-\delta^*)/(\sigma_0\sqrt{2}\,)$ for the variables in the denominator and for the last $k-1$ variables in the numerator, and from $-\infty$ to $\sqrt{N+1}\,(U_i-U_s)/\sigma_0\sqrt{2N}$ for the first $k-1$ variables in the numerator $(s=1,\ldots,i-1,i+1,\ldots,k)$. The function $f(v_1,\ldots,v_{2k-2})$ is the same multivariate normal density that appears in (10.3) and $f_{k-1,\frac{1}{2}}(x_1,\ldots,x_{k-1})$ is the $(k-1)$-variate normal density corresponding to the cdf $F_{k-1,\frac{1}{2}}(x_1,\ldots,x_{k-1})$ defined earlier.

Dunnett (1960) has also considered the probability that the selected mean is within a specified amount of the largest mean. This probability (denoted by P) can also be obtained by defining a loss function that is zero if the selected mean is within an amount δ^* of the largest, and unity otherwise. If $\bar{R}$ is the risk integrated with respect to the a priori distribution of the μ_i, then $P=1-\bar{R}$.

10.3 SELECTION OF THE LARGEST NORMAL MEAN USING A LINEAR LOSS FUNCTION

Consider the loss function

$$L_i=c'(\mu_{[k]}-\mu_i), \qquad i=1,\ldots,k, \qquad (10.8)$$

where L_i is the loss if π_i is selected, and c' is a positive constant. The cost of performing an experiment involving n observations from each population is cn, where c is a positive constant. Overhead and other fixed costs are omitted here. The risk function is

$$R_n(\mu_1,\ldots,\mu_k)=cn+c'\left(\mu_{[k]}-\sum_{i=1}^{k}\mu_i P_i\right), \tag{10.9}$$

where P_i, the probability of selecting π_i, is given by (10.1).

The sample size was determined by the minimax risk method by Bross (1950) when $k=2$ and by Somerville (1954) for general k. Somerville shows that the risk is maximized over the parameter space when all the μ_i other than $\mu_{[k]}$ are equal. If we denote by M_{k-1} the maximum, with respect to y, of the function $y[1-F_{k-1,\frac{1}{2}}(y,\ldots,y)]$, then the minimax solution of n is given by

$$n=\left\{\frac{1}{2}M_{k-1}^2\frac{c'^2\sigma^2}{c^2}\right\}^{1/3}. \tag{10.10}$$

Dunnett (1960) has investigated the optimum sample size for the procedure R_1 when the μ_i have a priori normal distributions with known means U_i $(i=1,\ldots,k)$, and a common known variance σ_0^2. In some applications, the number k of populations from which a selection is to be made is at our disposal, as in plant selection where k is the number of varieties, the potential number of which may be large. Even in situations where k is fixed, one may wish to consider the desirability of discarding some of these without test. Motivated by these considerations, Dunnett (1960) has considered the optimum value of k as well as that of n.

When k is not fixed, the loss function defined in (10.8) is not satisfactory since the expected value of $c'\mu_{[k]}$ varies with k. Dunnett defines an *yield* function

$$Y_n(\mu_1,\ldots,\mu_k)=c'\sum_{i=1}^{k}\mu_i P_i-cnk-c''k, \tag{10.11}$$

where $c'>0$, $c>0$, and $c''\geqslant 0$. The worth of π_i is $c'\mu_i-cn$. Thus the yield in (10.11) represents the conditional expected worth, of the selected population for fixed values of the μ_i. As far as the optimum sample size is concerned, the problem is not affected by shifting our consideration from (10.8) to (10.11), because, for fixed k, minimizing (10.8) is equivalent to maximizing (10.11).

It has been shown by Dunnett that the expected yield, $\overline{Y}$, is given by

$$\frac{\sigma_0^2}{c\sigma^2}\overline{Y}=\lambda'\left(\frac{N}{2(N+1)}\right)^{1/2}\chi_k'\left(\frac{U_i(N+1)}{N};\frac{\sigma_0^2(N+1)}{N}\right)-k(N+\nu), \tag{10.12}$$

where $\lambda' = c'\sigma_0^3\sqrt{2}\,/(c\sigma^2)$, $\nu = c''\sigma_0^2/(c\sigma^2)$, and $\sigma\chi_k'(\theta_i,\sigma^2)$ is the expected value of the maximum of k independent normal variates with means θ_i, $i=1,\ldots,k$, and a common variance σ^2. The optimum choice of N for which the right-hand side of (10.12) is a maximum has been investigated by Dunnett. For the special case in which the U_i are all equal, the equation (10.12) reduces to

$$\frac{\sigma_0^2}{c\sigma^2}\bar{Y} = \frac{\lambda' U}{\sigma_0\sqrt{2}} + \frac{\lambda'}{\sqrt{2}}\chi_k'\sqrt{\frac{N}{N+1}} - k(N+\nu), \tag{10.13}$$

where U is the common value of the U_i, and χ_k' is the expected value of the largest of k standard normal variates.

The quantity $\chi_k'\sigma_0[N/(N+1)]^{1/2}$ is the expected increase due to selection, measured from the mean U of the parent population from which the μ_i are drawn. It is equivalent to the so-called "genetic advance" in plant selection.

The expression for $\bar{Y}$ in (10.13) is maximized with respect to N if N is chosen as the unique positive root of

$$N^{1/2}(N+1)^{3/2} = \frac{\lambda'\chi_k'}{2\sqrt{2}\,k}. \tag{10.14}$$

This optimum choice of N is for a fixed number k of populations. When k is at our disposal, we choose the particular combination of k and N for which $\bar{Y}$, given by (10.13), is as large as possible. If $\nu=0$, it is interesting to see that it never pays to select from fewer than 5 populations no matter how small λ' is. This result has also been noted by Curnow [1958] in considering the problem of maximizing the average genetic advance with a fixed total number of test plots.

The selection of the best normal population has also been studied by Raiffa and Schlaifer (1961) whose terminal utility is linear in the mean of the chosen population. Their analysis is based on differential utilities and their general approach for selecting the best of several processes is described in the next section.

10.4 SELECTION OF THE BEST OF SEVERAL PROCESSES: ANALYSIS IN TERMS OF DIFFERENTIAL UTILITY

Let $\pi_i(i=1,\ldots,k)$ be a process that generates independent random variables $X_{i1}, X_{i2}, \ldots$ with identical densities $f(x|\theta_i)$. Let $u(e,\mathbf{z},a_i,\boldsymbol{\theta})$ denote the utility of performing experiment e, observing $\mathbf{z}$, and choosing the action a_i

(which corresponds to choosing the process π_i), when $\boldsymbol{\theta}=(\theta_1,\ldots,\theta_k)$ is the true state of nature. The following assumptions A1 through A5 are made.

A1: The utilities of terminal action and experimentation are additive.

A2: The terminal utility of choosing the action a_i is linear in θ_i and is independent of θ_j, $j\neq i$.

Formally, we assume that

$$u(e,\mathbf{z},a_i,\boldsymbol{\theta})=u_t(a_i,\boldsymbol{\theta})-c_s(e,\mathbf{z}), \qquad i=1,\ldots,k, \tag{10.15}$$

and that there exists a mapping W from the state space Θ into a new space Ω carrying $\boldsymbol{\theta}$ into $\boldsymbol{\omega}=W(\boldsymbol{\theta})$, such that

$$u_t(a_i,\boldsymbol{\theta})=A_i+l_i\omega_i, \qquad i=1,\ldots,k, \quad \text{and all } \boldsymbol{\theta}\in\Theta. \tag{10.16}$$

Here $u_t(a_i,\boldsymbol{\theta})$ denotes the utility of a terminal decision and $c_s(e,\mathbf{z})$ is the sampling cost for the experiment.

A3: The experimental outcome $\mathbf{x}_i=(x_{i1},\ldots,x_{in})$ from the process π_i can be described by a sufficient statistic $\mathbf{y}_i=(y_{i1},\ldots,y_{im})$ of *fixed dimensionality* [which means that m is independent of n], where $y_{ir}=y_{ir}(x_{i1},\ldots,x_{in})$.

A4: Given any n and any sample $(x_{i1},\ldots,x_{in})$ from π_i, there exists a function h and an m-tuple of real numbers

$$\mathbf{y}_i(x_{i1},\ldots,x_{in})=\mathbf{y}_i=(y_{i1},\ldots,y_{im})$$

where m does not depend on n, such that the likelihood of the sample is proportional to $h(\mathbf{y}_i,\theta_i)$.

A5: It is assumed that the prior measure $P_{\boldsymbol{\theta}}$ on Θ is a member of the conjugate family with respect to the conditional measure $P_{\mathbf{z}|e,\boldsymbol{\theta}}$ on the sample space Z.

Let $U_i=u_t(a_i,\boldsymbol{\theta})$ and $\mathbf{U}=(U_1,\ldots,U_k)$. The measure $P_{\boldsymbol{\theta}}$ induces a measure P_U on the space $\mathfrak{U}=\{\mathbf{U}\}$ with respect to which we can evaluate the expectation $E(\mathbf{U})=(E(U_1),\ldots,E(U_k))$. The experimenter will obviously choose the action a^* which is a member of $\{a_1,\ldots,a_k\}$ for which the expected utility is maximum under the prior distribution. Let us assume that a_{i*} is an optimal action and denote the associated utility by U_{i*}.

Now, for an experiment e with outcome $\mathbf{z}$, we can substitute for $P_{\boldsymbol{\theta}}$ the posterior measure $P'_{\boldsymbol{\theta}|\mathbf{z}}$ which induces a posterior measure $P'_{\mathbf{U}|\mathbf{z}}$ which permits us to evaluate $E'_{\mathbf{U}|\mathbf{z}}(\mathbf{U})=(E'(U_1),\ldots,E'(U_k))$ and select an action that is optimal under the posterior distribution.

The prior expected terminal utility of a_{i*} is at least as large as that of any other action. However, its actual terminal utility may be exceeded by that of one or more other actions. If the decision maker were to be given perfect information on $\boldsymbol{\theta}$ and thus on $\mathbf{U}$, he would choose an action for which the terminal utility is maximum. The *value of this information* is

$$\begin{aligned} v_t(e_\infty,\boldsymbol{\theta}) &= \max\{U_1,\ldots,U_k\} - U_{i*} \\ &= \max\{\delta_1,\ldots,\delta_k\}, \end{aligned} \tag{10.17}$$

where

$$\delta_i = U_i - U_{i*}. \tag{10.18}$$

Before the perfect information is received, $v_t(e_\infty,\boldsymbol{\theta})$ is a random variable, and one can compute the *expected value of perfect information*

$$v_t^*(e_\infty) = E_{\boldsymbol{\delta}}\max\{\delta_1,\ldots,\delta_k\}, \tag{10.19}$$

where the expectation is with respect to the measure $P'_{\boldsymbol{\delta}}$ induced by the measure $P_{\boldsymbol{\theta}}$ via $P_{\mathbf{U}}$.

Let a_{i*} be the action optimal under the posterior distribution. The resulting increase in utility or value of the information $\mathbf{z}$ is obviously

$$\begin{aligned} v_t(e,\mathbf{z}) &= \max(E(U_1),\ldots,E(U_k)) - E(U_{i*}) \\ &= \max\{\delta'_1,\ldots,\delta'_k\}, \end{aligned} \tag{10.20}$$

where

$$\delta'_i = E(U_i) - E(U_{i*}).$$

When the experiment e is being considered but has not been performed, $\boldsymbol{\delta}'$ is a random variable with a measure $P_{\boldsymbol{\delta}'|e}$ induced by the marginal measure $P_{\mathbf{z}|e}$. One can compute for any e the *expected value of sample information*

$$v_t^*(e) = E_{\boldsymbol{\delta}'|e}v_t(e,\mathbf{z}). \tag{10.21}$$

The *net gain of sampling* is defined by

$$v^*(e) = v_t^*(e) - c_s^*(e), \tag{10.22}$$

where $c_s^*(e) = E_{\boldsymbol{\delta}'|e}(c_s(e,\mathbf{z}))$.

On the basis of this, one can find an optimal e. Specific results for normal distributions under various cases are given in Chapter 5B of Raiffa and Schlaifer (1961).

10.5 SELECTION OF THE BEST POPULATION USING UTILITY FUNCTIONS INVOLVING COVERAGES

Let π_i $(i=1,\ldots,k)$ have the associated density function $f(y|\theta_i)$, where θ_i may be vector-valued. Associated with π_i is a utility function $U(\theta_i)$. The best population is the one with the largest value among the $U(\theta_i)$. Usually the experimenter is interested in a specific *criterion* $h_i = g(\theta_i)$ where g is known. It would then be natural for the experimenter to regard his utility function U as a function of h_i, that is, $U = U(h_i)$. In this section, we are concerned with the case in which $g(\theta)$ is defined as

$$h = g(\theta) = \int_{a_1}^{a_2} f(y|\theta)\,dy, \tag{10.23}$$

for specified a_1 and a_2. The interval (a_1, a_2) is referred to as a tolerance interval and the quantity h is called the *coverage* of this interval.

Guttman and Tiao (1964) have discussed the problem by assuming that $U(h_i) = h_i$ for all i and comparing the expected values of the posterior distribution of the coverages. Their results relate to normal and exponential populations and are discussed in the next two subsections.

The prior distribution of the parameters of the k populations under consideration is denoted by $p(\theta_1,\ldots,\theta_k)$. Given the samples $(\mathbf{y}_1,\ldots,\mathbf{y}_k)$ from these populations one can in principle determine the joint posterior distribution of $(h_1,\ldots,h_k)$. It is assumed that the population parameters $(\theta_1,\ldots,\theta_k)$ are locally independent a priori. This means that in the region in which the likelihood is appreciable, the joint distribution $p(\theta_1,\ldots,\theta_k)$ can be written approximately as the product $p_1(\theta_1)\ldots p_k(\theta_k)$. Such an assumption will be appropriate in situations where the prior distribution of the parameters is diffuse and gently changing over a wide range. This independence assumption together with the assumption about the independence of samples implies that the h_i are locally independent a posteriori and hence can be analyzed individually.

10.5.1 The Case of Normal Populations

It is assumed that little is known a priori about the values of (μ, σ). In other words, the information we have about (μ, σ) comes primarily from the sample. We may then assume (see, for example, Box and Tiao [1962]) that μ and $\log\sigma$ are independent and locally uniformly distributed a priori.

It has been shown by Guttman and Tiao (1964) that the posterior expectation is given by

$$E(h|y_1,\ldots,y_n) = F_{n-1}(t_2) - F_{n-1}(t_1), \tag{10.24}$$

where F_{n-1} is the cdf of a Student t-variable with $(n-1)$ degrees of freedom,

$$t_j = \left(\frac{n-1}{n+1}\right)^{1/2}\left(\frac{a_j-\bar{y}}{s}\right), \qquad j=1,2, \tag{10.25}$$

and $\bar{y}$ and s^2 are the maximum likelihood estimators of μ and σ^2, respectively.

In the special case where $a_1 = -\infty$, the selection procedure is equivalent to deciding that the best population is the one with the largest value of $(a_2-\bar{y})/s$. This is an intuitively pleasing result because the coverage $h = g(\mu,\sigma) = \Phi[(a_2-\mu)/\sigma]$, where Φ is the standard normal cdf.

10.5.2 The Case of Exponential Populations

Here the density is

$$f(y|\eta,\sigma) = \begin{cases} \dfrac{1}{\sigma}\exp\{-(y-\eta)/\sigma\}, & y \geqslant \eta, \\ 0 & \text{otherwise.} \end{cases} \tag{10.26}$$

We assume, as before, that the prior distributions for η and $\log\sigma$ are locally uniform. The posterior expectation of h is

$$E(h|y_1,\ldots,y_n) = \int_{a_1}^{a_2} p(x|y_1,\ldots,y_n)\,dx, \tag{10.27}$$

where

$$p(x|y_1,\ldots,y_n) = \begin{cases} n[(n+1)w]^{-1}[1+n(y^*-x)/(n-1)w]^{-n} & \text{if } x<y^*, \\ n[(n+1)w]^{-1}[1+(x-y^*)/(n-1)w]^{-n} & \text{if } x>y^*, \end{cases} \tag{10.28}$$

$$w = \Sigma\frac{(y_j-y^*)}{(n-1)}, \tag{10.29}$$

and y^* is the smallest observation in the sample.

Guttman and Tiao (1964) have discussed the problem when the utility function is some function of h. They have also discussed the cases where $U(h)=h$ and the criterion h is either a location or a scale parameter of the population distribution.

10.6 BAYESIAN PROCEDURES FOR COMPLETE RANKING

Govindarajulu and Harvey (1974) have considered Bayesian procedures for deciding on the complete ranking of k populations. Let $f(x|\theta,\omega)$ denote the likelihood function for these distributions, where θ is the single parameter of interest and ω denotes a vector of nuisance parameters. Let θ_i denote the true value of θ associated with π_i and $h_i(\theta)$ denote the marginal posterior density concerning θ_i. Probabilities and expectations taken with respect to posterior distributions are denoted by P'' and E'', respectively.

A typical decision about the ranking of the k populations in terms of their θ-values will be denoted by R: $\theta_{i_1}<\theta_{i_2}<\cdots<\theta_{i_k}$, where $\{i_1,i_2,\ldots,i_k\}$ is a permutation of the integers $\{1,2,\ldots,k\}$. The decision R will be denoted by the ordered sequence $(i_1,\ldots,i_k)$ itself. Let $l(R,\boldsymbol{\theta})$ denote the loss in making decision R when $\boldsymbol{\theta}$ is the true parametric vector. Then the procedure is to choose an R for which

$$L(R)=\int_{\Omega}\cdots\int_{\Omega}l(R,\boldsymbol{\theta})h(\boldsymbol{\theta})\,d\theta_1\cdots d\theta_k \tag{10.30}$$

is minimized.

The simplest loss function is, of course, the one that assigns loss 1 to any incorrect ranking, and loss zero to the correct ranking. This loss function, simple as it is, does not differentiate between a slightly incorrect ranking and a drastically incorrect one. We can define a loss function by considering a distance function $d(R,R')$, which is defined as the minimum number of interchanges to go from R to R', where an *interchange* means a transposition of two adjacent indices in a ranking. Denoting the correct ranking by $R^0(\boldsymbol{\theta})$, we have

$$l(R,\boldsymbol{\theta})=d(R,R^0(\boldsymbol{\theta})). \tag{10.31}$$

If two rankings R and R' differ only by an interchange of $\theta_i<\theta_j$ in R to $\theta_j<\theta_i$ in R', then it has been shown by Govindarajulu and Harvey that

$$L(R)-L(R')=P''(\theta_j<\theta_i)-P''(\theta_i<\theta_j). \tag{10.32}$$

10.6.1 Loss Functions Depending on θ_i.

Instead of defining the loss function as in (10.31), we can make it depend on the true values θ_i in situations where the ranking is for the specific purpose of selecting later one or more populations according to a given procedure. It is somewhat more convenient to consider utility rather than loss. Let $u(R,\boldsymbol{\theta})$ denote the utility function chosen and $U(R)$ the posterior expected utility. It is assumed that a utility $w_i(\theta_i)$ is realized whenever π_i is selected, where the w_i are preassigned functions.

Govindarajulu and Harvey (1974) have considered four types of situations, namely, (i) selection of a single population from the k ranked populations according to a probability distribution depending only on the ranks of the populations, (ii) selection of a specified number r $(1 \leqslant r < k)$ of the ranked populations, again according to a probability distribution depending only on the ranks of the populations, (iii) selection of r populations, where r is not specified in advance, and q_i $(i=1,\ldots,k)$ is the imputed probability that the i populations with the highest ranking are selected, and (iv) selection of a single or no population from the ranked populations. The selection probabilities in the first three cases are based entirely on the ranks of the populations. In the fourth case, the populations are considered one at a time starting with the highest ranked population. The decision to accept or reject each population depends on the population itself rather than its ranking. A utility u_0 is associated with the decision of not selecting any population. It has been shown that, in all these cases, the optimal ranking based on the posterior expected utility is given by

$$W_{i_1}'' < W_{i_2}'' < \cdots < W_{i_k}'', \tag{10.33}$$

where

$$W_j'' = E''\big(w_j(\theta_j)\big).$$

10.6.2 Comparison of Ranking Procedures

In the beginning of Section 10.6, we discussed optimal ranking which minimizes $L(R)$ defined by (10.30), where the loss function $l(R,\boldsymbol{\theta})$ is given by (10.31). In §10.6.1, optimal ranking was discussed in terms of maximizing the posterior expected utility $U(R)$ for different subsequent selection goals. One may be interested in finding out whether these procedures can lead to the same optimal ranking. The fact that such is the case has been proved by Govindarajulu and Harvey (1974) under the following assumptions.

A$_1$: The likelihoods $f(x|\theta_i)$, $i=1,\ldots,k$, associated with the populations belong to the exponential family,

$$f(x|\theta_i)=c(\theta_i)\exp\{u(x)Q(\theta_i)\}r(x) \tag{10.34}$$

where $Q(\theta_i)$ is a strictly increasing function of θ_i.

A$_2$: For each population π_i $(i=1,\ldots,k)$, an independent prior density $g_i(\theta)$ has been assigned from the natural conjugate family, that is, for some constants η_i and τ_i, and a function K independent of θ,

$$g_i(\theta)=[c(\theta)]^{\eta_i}\exp\{\tau_i Q(\theta_i)\}K, \tag{10.35}$$

and an independent sample $\mathbf{x}_i=(x_{i1},\ldots,x_{in_i})$ of size n_i has then been taken.

A$_3$: The posterior densities, which are of the form

$$h_i(\theta)=[c(\theta)]^{\eta_i+n_i}\exp\{(\tau_i+T(\mathbf{x}_i))Q(\theta)\}r(\mathbf{x}_i), \tag{10.36}$$

where $T(\mathbf{x}_i)=\sum_{j=1}^{n_i}u(x_{ij})$ have all the same value η_0'' for η_i+n_i.

Under these assumptions, the optimal ranking in terms of minimizing $l(R,\boldsymbol{\theta})$ or maximizing $U(R)$ is given by

$$\tau_{i_1}+T(\mathbf{x}_{i_1})<\cdots<\tau_{i_k}+T(\mathbf{x}_{i_k}). \tag{10.37}$$

There are cases where the assumptions A$_1$–A$_3$ are not satisfied. It has been shown by Govindarajulu and Harvey (1974) that the ranking procedures become identical if each posterior density $h_i(\theta)$, $-\infty<\theta<\infty$, is symmetric about some point, τ_i'', $i=1,\ldots,k$, and $w(\theta)=\alpha\theta+\beta$, $\alpha>0$. The optimal ranking is given by $\tau_{i_1}''<\tau_{i_2}''<\cdots<\tau_{i_k}''$.

10.7 PARTIAL RANKING OF BINOMIAL PROBABILITIES

Bland and Bratcher (1968) have considered a partial ranking problem for binomial success probabilities where the goal is to divide the set of k binomial populations into s ranked groups. Here s is not predetermined. Their procedure is based on comparison of the populations in pairs. In other words, the ranking decision is based on several component two-decision problems.

Let p_i be the success probability of π_i. It is assumed that each p_i has the same prior distribution, that the loss functions are the same for each component problem, and that the loss for a decision in the ranking problem is the sum of the losses for the generating decisions in the component two-decision problems. With these assumptions, it is necessary to find a Bayes rule for only one of the component two-decision problems.

Consider the component two-decision problem with decisions

$$d_1 : p_1 \leqslant p_2 \qquad \text{and} \qquad d_2 : p_1 > p_2.$$

The loss functions considered are

$$L_1(p_1, p_2) = \begin{cases} 0, & p_1 \leqslant p_2; \\ m_1(p_1 - p_2), & p_1 > p_2, \end{cases} \tag{10.38}$$

and

$$L_2(p_1, p_2) = \begin{cases} m_2(p_2 - p_1), & p_1 \leqslant p_2, \\ 0, & p_1 > p_2, \end{cases} \tag{10.39}$$

where L_i is the loss corresponding to the decision d_i, and m_1 and m_2 are positive constants (the same for each pair of p's). The prior distribution $\phi(p)$ for each p is taken to be a beta distribution. The Bayes rule for the two-decision problem considered previously is:

Make decision d_1 or d_2 according as $h(t) >$ or < 0;

make an arbitrary decision if $h(t) = 0$,

where $t = X_1 - X_2$ (X_i is the number of successes in n trials of π_i),

$$h(t) = \int_0^1 \int_0^1 [L_2(p_1, p_2) - L_1(p_1, p_2)] \phi(p_1) \phi(p_2) f(t | p_1, p_2) \, dp_1 \, dp_2, \tag{10.40}$$

and $f(t | p_1, p_2)$ is the probability function of $X_1 - X_2$.

For the component two-decision rules to be compatible, it is necessary to impose the additional restriction that $m = m_1 / m_2$ be not less than unity. The ranking procedure is now easily carried out in the obvious manner.

10.8 NOTES AND REMARKS

In §3.3 we discussed the results of Eaton (1967a). Theorem 3.6 establishes optimal properties of his ranking procedure for a class of populations. His procedure is a Bayes procedure. Tiao and Afonja (1976) have studied the choice of design with respect to a variety of utility functions for ranking.

Bayesian procedures under subset selection formulation have been discussed by Deely (1965), Deely and Gupta (1968), Govindarajulu and Harvey (1974), Bickel and Yahav (1977), Chernoff and Yahav (1977), Goel and Rubin (1977), Gupta and Hsu (1978), and Hsu (1977). These procedures are discussed in Chapter 18.

PART II

SUBSET SELECTION FORMULATION

CHAPTER 11

Subset Selection: General Theory

11.1 INTRODUCTION

In Chapter 1, we briefly discussed the basic problem of selection and ranking formulations using either the indifference zone approach or the random subset size approach. The several succeeding chapters dealt with the investigations carried out under the indifference zone approach and some modifications of the basic formulation. In the present chapter as well as in a few following ones, we are concerned with the random subset size approach of Gupta (1956).

The early investigations of Gupta (1956, 1963, 1965), and Gupta and Sobel (1957, 1960, 1962a, 1962b) were concerned with ranking of normal means and variances, gamma scale parameters, and binomial success probabilities. General results were obtained for the cases of location and scale parameters. In view of the subsequent developments in the literature, we supersede the chronological order and present in this chapter some general theory of subset selection.

11.2 BASIC GENERAL THEORY

11.2.1 Formulation of the Problem

Let $\pi_1,\ldots,\pi_k$ be k independent populations. Let Λ be an interval on the real line. Associated with $\pi_i (i=1,\ldots,k)$ is a real-valued random variable X_i with an absolutely continuous distribution $F_i \equiv F_{\lambda_i}, \lambda_i \in \Lambda$, and density function $f_i \equiv f_{\lambda_i}$, It is assumed that the functional form of F_{λ_i} is known, but not the value of λ_i. It is also assumed that $\{F_\lambda\}, \lambda \in \Lambda$, is a stochastically increasing family of distributions, all having the same support. The distribution associated with $\lambda_{[k]}$ is defined to be the best. (The population associated with $\lambda_{[1]}$ could be defined as the best and an analogous theory could be developed.) If there are more than one population with $\lambda_i = \lambda_{[k]}$, then we assume that one of them is tagged as the best. It should be noted that this assumption is made only for the purpose of evaluating the infimum of the PCS. In practice, if $\lambda_i = \lambda_{[k]}$ for more than one population, then they are all equally good. As described in Chapter 1, the goal is to select a subset of the k populations to include the best population with probability at least equal to $P^*(k^{-1} < P^* < 1)$. We want to define a rule R so that

$$P(CS|R) \geq P^* \quad \text{for all } \boldsymbol{\lambda} \in \Omega, \tag{11.1}$$

where $\Omega = \{\boldsymbol{\lambda} : \boldsymbol{\lambda} = (\lambda_1,\ldots,\lambda_k)\}$.

11.2.2 The Class of Procedures R_h

Let $h \equiv h_{c,d}, c \in [1, \infty), d \in [0, \infty)$ be a class of real-valued functions defined on the real line satisfying the following conditions (A): For every x belonging to the support of F_λ,

(i) $h_{c,d}(x) \geqslant x$,
(ii) $h_{1,0}(x) = x$,
(iii) $h_{c,d}(x)$ is continuous in c and d, and
(iv) $\lim_{d\to\infty} h_{c,d}(x) = \infty$, c fixed, and/or
$\lim_{c\to\infty} h_{c,d}(x) = \infty$, d fixed, $x \neq 0$.

Using the class of functions h defined previously, Gupta and Panchapakesan (1972a) have considered the following class of procedures whose typical member is denoted by R_h.

R_h: Include the population π_i in the selected subset if and only if

$$h(x_i) \geqslant \max_{1 \leqslant r \leqslant k} x_r, \tag{11.2}$$

where x_i is an observation from $\pi_i (i = 1, \ldots, k)$. For this procedure we have

$$P(CS|R_h) = \int \prod_{r=1}^{k-1} F_{[r]}(h(x)) f_{[k]}(x)\, dx, \tag{11.3}$$

where $f_{[r]}$ denotes the density corresponding to $F_{[r]}$, and the integral is taken over the common support of the distributions. Because of the stochastic ordering,

$$P(CS|R_h) \geqslant \int F_{[k]}^{k-1}(h(x)) f_{[k]}(x)\, dx. \tag{11.4}$$

Define

$$\psi(\lambda; c, d, t+1) = \int F_\lambda^t(h(x)) f_\lambda(x)\, dx. \tag{11.5}$$

Then

$$\inf_{\Omega} P(CS|R_h) = \inf_{\lambda \in \Lambda} \psi(\lambda; c, d, k). \tag{11.6}$$

Because of the set of conditions (A) imposed on h, we have for any $\lambda \in \Lambda$,

(i) $\psi(\lambda; c, d, k) \geqslant \frac{1}{k}$,
(ii) $\psi(\lambda; 1, 0, k) = \frac{1}{k}$,
(iii) $\lim_{d\to\infty} \psi(\lambda; c, d, k) = 1$, c fixed, and/or
$\lim_{c\to\infty} \psi(\lambda; c, d, k) = 1$, d fixed. (11.7)

In general, the preceding conditions are not enough to ensure the existence of constants c and d so that the P^*-condition is satisfied. If the condition (11.8) (stated in Theorem 11.1) is satisfied then $\psi(\lambda; c, d, k)$ is nondecreasing in λ and $\inf_{\lambda\in\Lambda} \psi(\lambda; c, d, k) = \psi(\lambda_0; c, d, k)$, say. Then, we can evaluate the constants c and d so that the P^*-condition is satisfied, provided the set of conditions (A) hold for λ_0 in case $\lambda_0 \notin \Lambda$. Under these conditions, there can be obviously several choices of c and d satisfying the P^*-condition. Some additional conditions can be imposed to determine the choice of a pair (c, d).

Now we state a theorem of Gupta and Panchapakesan (1972a) which gives a sufficient condition for the monotonicity of $\psi(\lambda; c, d, k)$. The proof of the theorem assumes regularity conditions not explicitly stated. The theorem is stated as follows in completeness.

Theorem 11.1. Let $\{F_\lambda(\cdot)\}, \lambda \in \Lambda$, be family of absolutely continuous distributions on the real line with continuous densities $f_\lambda(\cdot)$. Let $\psi(x, \lambda)$ be a bounded real-valued function possessing first partial derivatives ψ_x and ψ_λ with respect to x and λ respectively and satisfying regularity conditions (11.9). Then $E_\lambda[\psi(x, \lambda)]$ is nondecreasing in λ provided for all $\lambda \in \Lambda$,

$$A(x,\lambda) \equiv f_\lambda(x) \frac{\partial}{\partial\lambda} \psi(x,\lambda) - \frac{\partial}{\partial\lambda} F_\lambda(x) \frac{\partial}{\partial x} \psi(x,\lambda) \geqslant 0 \quad \text{for a.e. } x, \qquad (11.8)$$

and

(i) for all $\lambda \in \Lambda, (\partial/\partial x)\, \psi(x, \lambda)$ is Lebesgue integrable in R;

(ii) for every $[\lambda_1, \lambda_2] \subset \Lambda$ and $\lambda_3 \in \Lambda$ there exists $M(x)$ depending only on λ_1, λ_2, and λ_3 such that

$$\left| f_{\lambda_3}(x) \frac{\partial}{\partial\lambda} \psi(x,\lambda) - \frac{\partial}{\partial\lambda} F_\lambda(x) \frac{\partial}{\partial x} \psi(x,\lambda_3) \right| \leqslant M(x)$$

$$\text{for all } \lambda \in [\lambda_1, \lambda_2] \qquad (11.9)$$

and $M(x)$ is Lebesgue integrable on R.

REMARK 11.1. If $\psi(x,\lambda)=\psi(x)$ for all $\lambda\in\Lambda$, then (11.8) reduces to $(\partial/\partial\lambda)F_\lambda(x)\cdot(d/dx)\psi(x)\leqslant 0$. This is satisfied if $\{F_\lambda\}$ is a stochastically increasing family of distributions and $\psi(x)$ is nondecreasing in x, in which case, $E_\lambda\psi(x)$ is nondecreasing in λ. This special case is a result of Lehmann [1959, p. 112]. A generalization of Lehmann's result has been given by Alam and Rizvi (1966) and Mahamunulu (1967) for the case of independent random variables with distribution functions $F_{\lambda_i}(i=1,\ldots,k)$ and $\psi(x_1,\ldots,x_k)$ nondecreasing in each argument. □

REMARK 11.2. If we assume that $\psi(x,\lambda)\geqslant 0$, then, for any positive integer t, we can let $\phi(x,\lambda)=\psi^t(x,\lambda)$ play the role of $\psi(x,\lambda)$ of Theorem 11.1. It follows immediately that $E_\lambda\psi^t(x,\lambda)$ is nondecreasing in λ if (11.8) holds. □

In view of Theorem 11.1 and Remark 11.2, we have the following theorem of Gupta and Panchapakesan (1972a).

Theorem 11.2. For the procedure R_h defined by (11.2), $\psi(\lambda;c,d,k)=P(CS|R_h;\lambda_1=\cdots=\lambda_k=\lambda)$ is nondecreasing in λ provided that for all λ and x,

$$f_\lambda(x)\frac{\partial}{\partial\lambda}F_\lambda(h(x))-h'(x)f_\lambda(h(x))\frac{\partial}{\partial\lambda}F_\lambda(x)\geqslant 0, \tag{11.10}$$

where $h'(x)=(d/dx)h(x)$. The monotonicity of $\psi(\lambda;c,d,k)$ is strict if strict inequality holds in (11.10) on a set of positive Lebesgue measure.

11.2.3 Some Properties of R_h

The procedure R_h is *monotone*; that is,

$$\Pr(\text{selecting } \pi_i)\leqslant \Pr(\text{selecting } \pi_j) \quad \text{when } \lambda_i<\lambda_j, \tag{11.11}$$

provided that $h(x)$ is nondecreasing in x. This implies that the procedure R_h is *unbiased*, that is, $p_k\geqslant p_i, i=1,\ldots,k-1$, where p_i is the probability that the population associated with $\lambda_{[i]}$ is included in the selected subset.

An obvious criterion for evaluating the performance of a subset selection procedure is the expected value of S, the size of the selected subset. It is easy to see that, for any rule $R, E(S)=p_1+\cdots+p_k$. We are interested in the supremum of $E(S|R_h)$ over the parameter space Ω and the following theorem of Gupta and Panchapakesan (1969b) gives a sufficient condition under which this supremum is attained for $\boldsymbol{\lambda}\in\Omega_0$ where

$$\Omega_0=\{\boldsymbol{\lambda}:\lambda_1=\cdots=\lambda_k\}. \tag{11.12}$$

Theorem 11.3. For the procedure R_h defined by (11.2), the supremum of $E(S|R_h)$ takes place for a $\boldsymbol{\lambda}$ in Ω_0 provided that

$$\frac{\partial}{\partial\lambda_1}F_{\lambda_1}(h(x))f_{\lambda_2}(x)-h'(x)\frac{\partial}{\partial\lambda_1}F_{\lambda_1}(h(x))f_{\lambda_2}(h(x))\geqslant 0$$

$$\text{for all } \lambda_1\leqslant\lambda_2 \quad \text{and all } x. \tag{11.13}$$

REMARK 11.3. The condition (11.13) with $h(x)=x+d, d\geqslant 0$, is essentially the assumption made by Sobel (1969) so that the supremum of $E(S)$ for his rule is attained when the distributions are identical. □

It should be noted that the condition (11.13) implies condition (11.10). Thus the condition (11.13) which ensures that the the worst configuration for $\sup E(S)$ lies in Ω_0 also establishes the monotonicity of $P(CS|R_h)$ in Ω_0, a fact observed by Gupta and Panchapakesan (1972a). This incidentally shows that, if $P(CS|R_h)$ is constant ($=P^*$) in Ω_0 (as is seen to be the case for location and scale parameter families with appropriate choices of $h(x)$), then the $\sup E(S)=kP^*$ when (11.13) holds. For monotone rules $E(S)\leqslant kp_k$.

Instead of $E(S)$, one can consider $E(S')$ where S' denotes the number of nonbest populations included in the selected subset. Gupta and Panchapakesan (1969b) have shown that the supremum of $E(S'|R_h)$ takes place in Ω_0 provided that (11.13) holds.

Nagel (1970) has defined a just rule. In our case, this could be stated as follows.

Definition 11.1. Let $\mathbf{x}=(x_1,\ldots,x_k)$ and $\mathbf{y}=(y_1,\ldots,y_k)$ be two sets of observations from the populations such that $x_i\leqslant y_i$ and $x_j\geqslant y_j, j\neq i$. Then a rule R is said to be *just* if the probability of selecting π_i based on $\mathbf{y}$ is at least as large as the probability of selecting π_i based on $\mathbf{x}$.

It is easy to see that the procedure R_h is just if $h(x)$ is nondecreasing in x.

11.2.4 Some Special Cases

We now discuss applications of the preceding general results to selection problems concerning parameters of location, scale, and convex mixtures of distributions. These three broad categories cover most of the selection procedures investigated under this formulation.

(a) ***Location Parameter Case.*** In this case, $F_\lambda(x)=F(x-\lambda), \Lambda=$real line, and we see that the condition (11.13) reduces to

$$h'(x)f_{\lambda_1}(x)f_{\lambda_2}(h(x))-f_{\lambda_2}(x)f_{\lambda_1}(h(x))\geqslant 0. \tag{11.14}$$

Since $h(x) \geqslant x$, (11.14) is satisfied if $h'(x) \geqslant 1$ and $f_\lambda(x)$ has a monotone likelihood ratio in x. The location parameter case has been discussed by Gupta (1965) using the obvious choice, $h(x) = x + d, d \geqslant 0$. It can be seen that $\psi(\lambda; c, d, k)$ is independent of λ so that the construction of the constants d is accomplished by solving

$$\int_{-\infty}^{\infty} F^{k-1}(x+d)\, dF(x) = P^*. \tag{11.15}$$

An important special problem in this case is, of course, selection from normal population in terms of means. This is discussed in the next chapter in detail.

(b) *Scale Parameter Case.* Here we have $F_\lambda(x) = F(x/\lambda), x \geqslant 0, \Lambda = [0, \infty)$. The condition (11.13) becomes

$$xh'(x) f_{\lambda_1}(x) f_{\lambda_2}(x) - h(x) f_{\lambda_1}(h(x)) f_{\lambda_2}(x) \geqslant 0. \tag{11.16}$$

If $xh'(x) \geqslant h(x) \geqslant 0$ and $f_\lambda(x)$ has a monotone likelihood ratio in x, then (11.16) is satisfied. The usual choice in this case has been $h(x) = cx, c \geqslant 1$. Procedure R_h with this choice of $h(x)$ has been studied by Gupta (1965). For this choice, $\psi(\lambda; c, d, k)$ becomes independent of λ and thus the constant c is given by

$$\int_{0}^{\infty} F^{k-1}(cx)\, dF(x) = P^*. \tag{11.17}$$

Of particular interest under this case are the problems of selection of normal populations in terms of variances and selection from gamma populations. These are discussed in detail in Chapter 13.

In the preceding two cases, it is clear from our discussion in §11.2.3 that the supremum of $E(S) = kP^*$ for the appropriate choices of $h(x)$ provided that the density $f_\lambda(x)$ has a monotone likelihood ratio in x, a result proved earlier by Gupta (1965).

(c) *Convex Mixtures.* In this case, we assume that

$$f_\lambda(x) = \sum_{j=0}^{\infty} w(\lambda, j) g_j(x) \tag{11.18}$$

where $g_j(x) = 0, 1, \ldots$ is a sequence of density functions and $w(\lambda, j)$ are nonnegative weights such that $\sum_{j=0}^{\infty} w(\lambda, j) = 1$. We consider weights given by

$$\left.\begin{aligned} w(\lambda,j) &= \frac{a_j\lambda^j}{A(\lambda)j!}, \qquad A(\lambda) \geqslant 0, \qquad \lambda \geqslant 0,\\ a_{j+1} &= (m+lj)a_j, \qquad j=0,1,\ldots;l,m>0. \end{aligned}\right\} \tag{11.19}$$

From (11.19), we have

$$A(\lambda) = a_0(1-\lambda l)^{-m/l} \tag{11.20}$$

provided that $\lambda < l^{-1}$.

This special case is of interest. If we set $m=1, l=0$, and $a_0=1$, we get $w(\lambda,j)=e^{-\lambda}\lambda^j/j!$. Thus the densities $g_j(x)$ are weighted by Poisson probabilities. Familiar examples of $f_\lambda(x)$ in this case are the densities of noncentral chi-square and noncentral F variables with noncentrality parameter λ. Again, if we set $l=1$ and $a_0=1$, we get densities $g_j(x)$ with negative binomial weights. The distribution of R^2, where R is the multiple correlation coefficient, in the so-called unconditional case is an example of the preceding. Subset selection procedures in these cases have been discussed by Gupta and Panchapakesan (1969a), and Gupta and Studden (1970) with $h(x)=cx, c \geqslant 1$. These procedures have been developed in the context of selection from multivariate normal populations and as such are discussed in a later chapter. The preceding authors have obtained sufficient conditions for the monotonicity of $\psi(\lambda;c,d,k)$ in λ for those specific cases.

For the case of the densities $f_{\lambda_i}(x)$ given by (11.18) with weight functions specified by (11.19) it has been shown by Gupta and Panchapakesan (1972a) that the sufficient condition (11.13) is equivalent to

$$\sum_{i=0}^{\infty} \frac{\lambda^i}{i!} \sum_{\alpha=0}^{i} \binom{i}{\alpha} a_\alpha a_{i-\alpha} T_\alpha(x) \geqslant 0, \tag{11.21}$$

where

$$\begin{aligned} T_\alpha(x) = {} & b^{i-\alpha}(m+l\alpha)g_{i-\alpha}(x)\Delta G_\alpha(h(x))\\ & - h'(x)b^\alpha(m+l(i-\alpha))g_\alpha(h(x))\Delta G_{i-\alpha}(x), \end{aligned} \tag{11.22}$$

and $\Delta G_j(x) = G_{j+1}(x) - G_j(x)$.

One can obtain more stringent sufficient condition by saying that the coefficient of λ^i for every i in (11.21) is nonnegative. Thus, (11.21) is implied by the condition that, for every integer $i \geqslant 0$,

$$\sum_{\alpha=0}^{i} \binom{i}{\alpha} a_\alpha a_{i-\alpha} T_\alpha(x) \geqslant 0. \tag{11.23}$$

Grouping the terms corresponding to α and $i-\alpha$ in the summation of (11.23), we see that (11.23) is in its turn implied by the condition that, for $\alpha = 0, 1, \ldots, [i/2]$,

$$T_\alpha(x) + T_{i-\alpha}(x) \geqslant 0 \tag{11.24}$$

where $[s]$ denotes the largest integer $\leqslant s$. Though (11.24) is a much stronger condition than (11.21), it is satisfied in the cases investigated by Gupta and Panchapakesan (1969a), and Gupta and Studden (1970).

In discussing these special cases we find that the choice of $h(x)$ has been either $h(x) = cx, c \geqslant 1$, or $h(x) = x + d, d \geqslant 0$. A third choice of interest is $h(x) = cx + d, c \geqslant 1, d \geqslant 0$. This has been used to advantage by Gupta and D. Y. Huang (1975a) in discussing selection from Poisson populations. For this problem, there is difficulty in meeting the P^*-requirement by using the usual choices of $h(x)$. We examine this in detail while discussing selection problems for discrete distributions. There remains to be answered, however, the general question: How does one determine a "reasonable" choice of $h(x)$ for a given class of distributions?

McDonald (1977a) has discussed subset selection procedures based on sample ranges when π_i is $F[(x-\mu_i)/\sigma_i]$ and the ranking is in terms of the σ_i. The general results of this section can be applied to this problem.

11.2.5 A Decision-Theoretic Formulation and Best Invariant Rules

Let $\pi_i (i = 1, \ldots, k)$ be described by the probability space $(\mathcal{X}, \mathcal{B}, P_i)$, where P_i belongs to some family $\mathcal{P}$ of probability measures. We assume that there is a partial order relation $\succsim$ defined in $\mathcal{P}$. To be specific, $P_i \succsim P_j$ means that P_i is better than or equal to P_j; in other words, P_i is preferred over P_j. In many problems, $\succsim$ denotes stochastic ordering. Other partial orderings that have been considered are: convex ordering, star-shaped ordering, and tail ordering. In the preceding setup, we assume that there exists a population π_j such that $\pi_j \succsim \pi_i$ for all i. This population will be called the best population. As usual, in the case of more than one contender we assume that one of them is tagged as the best.

From each population we observe a random element X_i. The space of the observations is: $\mathcal{X}^k = \{\mathbf{x} = (x_1, \ldots, x_k), x_i \in \mathcal{X}, i = 1, \ldots, k\}$. In most applications $\mathcal{X}^k$ will be a real vector space. The decision space $\mathcal{D}$ consists of the 2^k subsets d of the set $\{1,2,\ldots,k\}$: to put it formally

$$\mathfrak{D}=\{d \mid d\subseteq\{1,2,\ldots,k\}\}. \tag{11.25}$$

Thus a decision d corresponds to selection of a subset (which could be $\varnothing$, the empty set) of the k given populations. Any decision $d\in\mathfrak{D}$ is called a *correct selection* if $j\in d$, where π_j is the best population.

Definition 11.2. A measurable function δ defined on $\mathfrak{X}^k\times\mathfrak{D}$ is called a *selection procedure* provided that for each $\mathbf{x}\in\mathfrak{X}^k$, we have

$$\delta(\mathbf{x},d)\geqslant 0 \quad \text{for any } d\in\mathfrak{D}, \quad \text{and}$$

$$\sum_{d\in\mathfrak{D}}\delta(\mathbf{x},d)=1 \tag{11.26}$$

where $\delta(\mathbf{x},\mathrm{d})$ denotes the probability that the subset d is selected when $\mathbf{x}$ is observed.

The *individual selection probability* $p_i(\mathbf{x})$ for the population π_i is then given by

$$p_i(\mathbf{x})=\sum_{d\ni i}\delta(\mathbf{x},d) \tag{11.27}$$

the summation being over all subsets that contain i. If the selection probabilities $p_i(\mathbf{x}), i=1,\ldots,k$, take on only values 0 and 1, then the selection procedure $\delta(\mathbf{x},d)$ is completely specified by them. The following example shows that this is not true in general.

EXAMPLE 11.1. [Nagel (1970), p. 9]. For $k=2$, and some fixed $\mathbf{x}$, selection rules δ and δ' are defined by

$$\delta(\mathbf{x},\varnothing)=\delta(\mathbf{x},\{1\})=\delta(\mathbf{x},\{2\})=\delta(\mathbf{x},\{1,2\})=\tfrac{1}{4};$$

$$\delta'(\mathbf{x},\varnothing)=\delta'(\mathbf{x},\{1,2\})=0;$$

$$\delta'(\mathbf{x},\{1\})=\delta'(\mathbf{x},\{2\})=\tfrac{1}{2}.$$

Both rules have the same individual selection probabilities $(\frac{1}{2},\frac{1}{2})$.

Let us now assume that the probability measures P_i are members of the family $\{P_\theta\}$. We say that $P_{\theta_i}\succsim P_{\theta_j}$, if $\theta_i\geqslant\theta_j$. Let $f(\mathbf{x};\boldsymbol{\theta})$ denote the density of the vector of observations $\mathbf{x}=(x_1,\ldots,x_k)$, where $\boldsymbol{\theta}=(\theta_1,\ldots,\theta_k)$. Studden (1967) has investigated the problem of defining "optimal" subset selection rules for the case where we have k *fixed* density functions and only the correct pairing of the densities and the populations is unknown.

Let Ω be the set of k-vectors consisting of permutations of the k fixed θ-values. We shall assume that a selection of the subset $d \in \mathfrak{D}$ results in a loss $L(\boldsymbol{\theta},d)=\Sigma_{i\in d}L_i(\boldsymbol{\theta})$ where $L_i(\boldsymbol{\theta})$ is the loss whenever π_i is selected. An additional loss L is imposed if a correct selection is not made. It is easy to see that

$$\sum_{\mathfrak{D}} L(\boldsymbol{\theta},d)E_{\boldsymbol{\theta}}\delta(\mathbf{x},d)=\sum_{i=1}^{k} L_i(\boldsymbol{\theta})E_{\boldsymbol{\theta}}p_i. \tag{11.28}$$

The problem is to minimize the risk

$$R(\boldsymbol{\theta},\mathbf{p})=\sum_{i=1}^{k} L_i(\boldsymbol{\theta})E_{\boldsymbol{\theta}}p_i+L[1-P_{\boldsymbol{\theta}}(CS|\mathbf{p})]. \tag{11.29}$$

This minimization is done under the usual symmetry conditions imposed by the group G of permutations $g(1,\ldots,k)\rightarrow(g1,\ldots,gk)$. If $h=g^{-1}$ denotes the inverse of g, let $g\mathbf{x}=(x_{h1},\ldots,x_{hk})$ and $g\boldsymbol{\theta}=(\theta_{h1},\ldots,\theta_{hk})$. All quantities involved are assumed to be invariant under G, that is, the measure μ (with respect to which $f(\mathbf{x};\boldsymbol{\theta})$ is the density) is invariant, $f(\mathbf{x};\theta)=f(g\mathbf{x};g\boldsymbol{\theta}), L_i(\boldsymbol{\theta}) = L_{gi}(g\boldsymbol{\theta})$, and $\delta(\mathbf{x},d)=\delta(g\mathbf{x},gd)$, where $gd=\{gi|i\in d\}$. The invariance of the selection rule δ implies that the individual selection probabilities are also invariant, that is, $p_i(\mathbf{x})=p_{gi}(g\mathbf{x}), i=1,\ldots,k$. An invariant selection rule δ or a set of invariant individual selection probabilities that minimizes (11.29) is said to be *best invariant*.

Since G acts transitively on Ω it is easily seen that for invariant δ both $R_1(\boldsymbol{\theta},\mathbf{p})=\Sigma_{i=1}^{k}L_i(\boldsymbol{\theta})E_{\boldsymbol{\theta}}p_i$ and $P_{\boldsymbol{\theta}}(CS|\mathbf{p})$ are invariant under G; that is, they are independent of $\boldsymbol{\theta}\in\Omega$. The class of invariant loss functions includes the following cases.

(i) $L_i(\boldsymbol{\theta})\equiv 1$. In this case $R_1(\boldsymbol{\theta},\mathbf{p})=\Sigma_{i=1}^{k}E_{\boldsymbol{\theta}}p_i$ is the expected size of the selected subset.

(ii) $$L_i(\boldsymbol{\theta})=\begin{cases}1 & \text{if} \quad \theta_i\neq\theta_{[k]}\\ 0 & \text{if} \quad \theta_i=\theta_{[k]}\end{cases} \qquad i=1,\ldots,k.$$

In this situation $R_1(\boldsymbol{\theta},\mathbf{p})$ is the expected subset size excluding the best population.

(iii) $L_i(\boldsymbol{\theta})=$rank of θ_i in the sequence $\theta_{[k]},\ldots,\theta_{[1]}$. Here we assume that $\theta_{[k]}$ has rank one and $\theta_{[1]}$ has rank k. Then $R_1(\boldsymbol{\theta},\mathbf{p})$ is the expected sum of ranks of the populations in the selected subset.

Let Ω_k be the set of permutations of $(\theta_1,\ldots,\theta_k)$ such that $\theta_{[k]}$ (or the tagged parameter) is the last component. In addition, let

$$u_i(\mathbf{x};\boldsymbol{\theta})=[(k-1)!]^{-1}\sum_{G_i} f(\mathbf{x};g\boldsymbol{\theta}), \qquad i=1,\ldots,k \tag{11.30}$$

where $G_i = \{g \mid g^{-1}k = i\}$. Studden (1967) has proved the following theorem.

Theorem 11.4. A selection rule δ is best invariant if and only if

$$p_k(\mathbf{x}) = \begin{cases} 1 & \text{if} \quad Lu_k(\mathbf{x};\boldsymbol{\theta}) > \sum_{i=1}^{k} L_i(\boldsymbol{\theta})u_i(\mathbf{x};\boldsymbol{\theta}) \\ 0 & \text{if} \quad Lu_k(\mathbf{x};\boldsymbol{\theta}) < \sum_{i=1}^{k} L_i(\boldsymbol{\theta})u_i(\mathbf{x};\boldsymbol{\theta}) \end{cases} \tag{11.31}$$

for $\boldsymbol{\theta} \in \Omega_k$, almost everywhere μ. The function $p_i, i \neq k$, is defined by the invariant conditions on $\mathbf{p}$.

Let $\gamma(L) = P(CS|\mathbf{p})$ where $\mathbf{p}$ satisfies (11.31) for a given value of L. Then the following corollary of Studden is immediate.

COROLLARY 11.1. An invariant selection procedure minimizes $\Sigma_{i=1}^{k} L_i(\boldsymbol{\theta}) E_{\boldsymbol{\theta}} p_i$ subject to the condition

$$P_{\boldsymbol{\theta}}(CS|\mathbf{p}) \geq \gamma(L) \quad \text{for all } \boldsymbol{\theta} \in \Omega \tag{11.32}$$

if and only if the individual selection probabilities are determined by (11.31).

The function $\gamma(L)$ is nondecreasing in L so that for a fixed value of $\gamma \in (0,1)$ the minimum of $\Sigma_{i=1}^{k} L_i(\boldsymbol{\theta}) E_{\boldsymbol{\theta}} p_i$ subject to $P_{\boldsymbol{\theta}}(CS|\mathbf{p}) \geq \gamma$ will be attained by the $\mathbf{p}$ in (11.31) where L is the minimal value of L for which $P_{\boldsymbol{\theta}}(CS|\mathbf{p}) \geq \gamma$.

The expression given in (11.31) defining the selection probabilities that minimize the risk $R(\boldsymbol{\theta};\mathbf{p})$ is rather complicated when written down explicitly in terms of the original densities. However, for the slippage situation when the underlying densities are from an exponential family and $L_i(\boldsymbol{\theta}) \equiv 1$, the expressions simplify considerably. Let $f(x,\theta_i) = C(\theta_i)e^{\theta_i x}$ with respect to some σ-finite measure μ on the real line, and let $\theta_{[1]} = \cdots = \theta_{[k-1]} = \theta_{[k]} - \Delta$ for some fixed $\Delta > 0$. It has been shown by Studden (1967) that an invariant selection rule δ minimizes the expected subset size subject to the condition that $P(CS|\mathbf{p}) \geq \gamma$ if and only if for almost all $\mathbf{x}$

$$p_k(\mathbf{x}) = \begin{cases} 1 & \text{if} \quad \sum_{i=1}^{k-1} e^{\Delta x_i} < Ce^{\Delta x_k}, \\ 0 & \text{if} \quad \sum_{i=1}^{k-1} e^{\Delta x_i} > Ce^{\Delta x_k}. \end{cases} \tag{11.33}$$

Let us define

$$A_k = \left\{ \mathbf{x} \middle| \sum_{i=1}^{k-1} e^{\Delta x_i} > Ce^{\Delta x_k} \right\}. \tag{11.34}$$

This is the region for which we select the kth population with probability one. This region is shown by Studden to have nice properties. The region is (i) convex, (ii) translation invariant, that is, if $(x_1,\ldots,x_k)\in A_k$ then $(x_1+b,\ldots,x_k+b)\in A_k$ for all b, and (iii) symmetric in $x_1,\ldots,x_{k-1}$. If the constant C in (11.33) is $>k-1$ then $\Sigma_{i=1}^k p_i(\mathbf{x})\geqslant 1$ for all $\mathbf{x}$; that is, we select at least one population with probability one. On the other hand, if $C<k-1$, then $\Sigma_{i=1}^k p_i(\mathbf{x})<1$ at least for $\mathbf{x}$ in a neighborhood of the equiangular line. The following example shows that situations where $\Sigma_{i=1}^k p_i(\mathbf{x})<1$ for some $\mathbf{x}$ do exist.

EXAMPLE 11.2. [Studden (1967)]. Consider the case of $k=2$ with $f(x,\theta)=f(x-\theta)$, where $f(x)$ is the standard normal density. In this case the second population is selected if $x_2>x_1-\Delta^{-1}\log C$. If γ is sufficiently close to $\frac{1}{2}$ and $\theta_2>\theta_1$, then $C<1$. In this situation $p_1(\mathbf{x})+p_2(\mathbf{x})=0$ for $|x_2-x_1|<\Delta^{-1}\log C^{-1}$.

The results described previously have been obtained by Studden (1967) assuming that $\theta_1,\ldots,\theta_k$ are k fixed values. However, he has considered a simple situation concerning the normal populations where the parameters are permitted to vary.

Let $f(\mathbf{x};\boldsymbol{\theta})=\Pi_{i=1}^k f(x_i-\theta_i)$ where $f(x)$ is the standard normal density. For fixed Δ let $\mathbf{p}(\mathbf{x},\Delta)$ denote the selection probabilities defined by (11.33) where C is chosen so that $P_{\boldsymbol{\theta}}(CS|\mathbf{p}(\mathbf{x},\Delta))=\gamma$ for all $\boldsymbol{\theta}=(\theta,\ldots,\theta,\theta+\Delta)$. Let $\Phi(\Delta)$ denote the class of invariant procedures satisfying

$$P_{\boldsymbol{\theta}}(CS|\mathbf{p})\geqslant\gamma \qquad \text{for all } \boldsymbol{\theta}\in\Omega(\Delta)$$

where

$$\Omega(\Delta)=\{\boldsymbol{\theta}|\theta_{[1]}\leqslant\cdots\leqslant\theta_{[k-1]}\leqslant\theta_{[k]}-\Delta\}.$$

Studden (1967) has shown that for any $\boldsymbol{\theta}$ with $\theta_{[1]}=\cdots=\theta_{[k-1]}=\theta_{[k]}-\Delta$ the minimum value of the expected subset size over the class $\Phi(\Delta)$ is attained by $\mathbf{p}(\mathbf{x},\Delta)$.

11.2.6 A Desirable Property of the Probability of a Correct Selection

Here we continue the notations and the formulation introduced in §11.2.5. However, we now do *not* assume the θ_i to be fixed values. We noted in

Example 11.1 that two different selection rules may have the same individual selection probabilities. If we assume that π_k is the best population, then

$$P_\omega(CS|\delta) = E_\omega \sum_{\substack{d \\ d \ni k}} \delta(\mathbf{X}, d) = E_\omega p_k(\mathbf{X}) \tag{11.35}$$

where $\omega = (P_1, P_2, \ldots, P_k)$, $P_i \in \mathcal{P}$, is a typical point of Ω and P_ω denotes the corresponding product measure with respect to $\mathcal{B}^k$. In the case of parameter families $\mathcal{P} = \{P_\theta\}$, Ω denotes the space of parameter vectors. We recall that Ω_0 denotes the subset of Ω where all the P_i are identical.

Suppose that $s(d)$ denotes the number of elements in d. Then

$$\begin{aligned} E_\omega(S|\delta) &= E_\omega \sum_{d \in \mathcal{D}} s(d)\delta(\mathbf{X}, d) \\ &= E_\omega \sum_{d \in \mathcal{D}} \sum_{i \in d} \delta(\mathbf{X}, d) \\ &= E_\omega \sum_{i=1}^{k} \sum_{\substack{d \\ d \ni i}} \delta(\mathbf{X}, d) \\ &= E_\omega \sum_{i=1}^{k} p_i(\mathbf{X}). \end{aligned} \tag{11.36}$$

From (11.35) and (11.36) it is clear that $P_\omega(CS|\delta)$ and $E_\omega(S|\delta)$, the quantities which are of interest to us, depend on δ only through the individual selection probabilities. Therefore we consider two rules with identical selection probabilities to be equivalent and use the following simplified definition of a selection rule.

Definition 11.3. A subset selection rule R is a measurable mapping from $\mathcal{X}^k$ into R^k (the k-dimensional Euclidean space) where R: $\mathbf{x} \to (p_1(\mathbf{x}), \ldots, p_k(\mathbf{x}))$, $0 \leq p_i(\mathbf{x}) \leq 1$.

If the rule R is nonrandomized, that is, the p_i take on only values 0 and 1, then R can equivalently be defined by the sets $A_i = \{\mathbf{x} \in \mathcal{X}^k | p_i(\mathbf{x}) = 1\}$, $i = 1, \ldots, k$. In defining a rule R, we want the basic probability requirement be satisfied. In other words, we require that

$$\inf_{\Omega} P(CS|R) \geq P^*. \tag{11.37}$$

In our discussion of the procedure R_h in §11.2.1, we saw that the infimum on the left side of (11.37) is attained for a parametric point in Ω_0 because of the stochastic ordering.

To define a reasonable property of any rule R, we may require that the rule satisfies the condition

$$\inf_{\Omega} P(CS|R) = \inf_{\Omega_0} P(CS|R) \tag{11.38}$$

and study how to characterize such a rule. This was investigated by Nagel (1970).

Let $\succsim$ define a partial order in $\mathfrak{X}$. We say y is *preferable* to x if $y \succsim x$. A subset $A \subseteq \mathfrak{X}$ is called an *increasing set* if and only if $x \in A$ and $y \succsim x$ imply $y \in A$. For any two members P and Q of $\mathfrak{P}$, we say that P is *stochastically larger* than $Q(P \succ_{\text{st}} Q)$ if and only if $P(A) \geqslant Q(A)$ for all increasing sets $A \in \mathfrak{B}$.

Definition 11.4. A selection rule R, defined by its individual selection probabilities $p_i(\mathbf{x})$, $i = 1, \ldots, k$, is called *just* if and only if

$$x \precsim y_1, x_j \succsim y_j \quad \text{for all } j \neq i \Rightarrow p_i(\mathbf{y}) \geqslant p_i(\mathbf{x}).$$

When $\mathfrak{X} = R^k$ and $\succsim$ stands for "componentwise larger," the increasing sets are those introduced by Lehmann [1955]. In this case, Definitions 11.1 and 11.4 coincide.

Nagel (1970) has shown that (11.38) is satisfied for any R which is just. In addition, if R is also symmetric (i.e., invariant under permutation), then the rule R is monotone. Nagel has shown that many procedures earlier available in the literature are just procedures.

For an arbitrary family of distributions in general one may not have a transformation under which invariance could be required for selection rules. It is reasonable to require that $P(CS|R)$ be constant in Ω_0. The invariance under certain transformation, when reasonably required, turns out to be a sufficient condition for $P(CS|R)$ to be constant over Ω_0. The following lemma of Gupta and Nagel (1971) can be used to construct a rule with constant $P(CS|R)$ in Ω_0.

Lemma 11.1. Let $X_1, \ldots, X_k$ be independent and identically distributed random variables with joint distribution P_θ. Let $T(X_1, \ldots, X_k)$ be a sufficient statistic for θ.

(i) If $E(\delta(X_1, \ldots, X_k)|T) = P^*$ for all T, then $E_\theta \delta = P^*$ for all T.
(ii) If T is complete for the family $\{P_\theta\}$, then $E_\theta \delta = P^*$ for all θ only if $E_\theta(\delta(X_1, \ldots, X_k)|T) = P^*$.

Application of this lemma to several families of distributions such as gamma, binomial, and Poisson is discussed in later chapters.

11.2.7 A Generalization of the Subset Selection Goal

In Chapter 9 we discussed a procedure of Mahamunulu (1967) with a selection goal which is a generalization of Bechhofer's basic formulation. Deverman (1969) and Deverman and Gupta (1969) have discussed a similar generalization of the basic subset selection formulation.

Suppose that there is a binary order relation ($\precsim$) on $\{\pi_1,\ldots,\pi_k\}$. We assume that there exists at least one permutation of subscripts, say, $i_1,\ldots,i_k$, such that $\pi_{i_1}\precsim\pi_{i_2}\precsim\cdots\precsim\pi_{i_k}$. If there exists more than one, that is, if for some i and j, $\pi_i\precsim\pi_j$ and $\pi_j\precsim\pi_i$, we agree to write down a specific one. Let the ranked populations be denoted by $\pi_{(1)}\precsim\cdots\precsim\pi_{(k)}$. Any subset of exactly s populations is called an *s-subset*. The goal is to select a subcollection of the collection of all the $\binom{k}{s}$ s-subsets of the k populations with a minimum guaranteed probability P^* that the chosen subcollection contains at least one s-subset having at least c of the t best populations. Obviously, for a meaningful problem, the integers c, s, t, and k must be such that $2\leqslant k$ and $\max[1,s+t+1-k]\leqslant c\leqslant\min[s,t]$. Selection of any subcollection satisfying the goal is a correct selection. An s-subset that contains at least c of the t best populations will be called a *relevant s-subset*. [Such a subset is called an interesting s-subset by Deverman (1969).] Since there are $\sum_{i=c}^{\min(s,t)}\binom{t}{i}\binom{k-t}{s-i}=N_0(k,t,s,c)$ such sets we will require that

$$P^*\geqslant\binom{k}{s}^{-1}N_0(k,t,s,c). \qquad (11.39)$$

We assume that $T_i=T(X_{i1},\ldots,X_{in})$ is a statistic based on n_i independent observations from π_i (with an absolutely continuous distribution G_i) and that $\pi_i\precsim\pi_j\Rightarrow T_i$ is stochastically smaller than T_j.

Definition 11.5. A continuous real-valued function $\Delta(x,y)$ of real arguments x and y will be called a *generalized difference* if and only if

(i) $\Delta(x,y)=0\Leftrightarrow x=y$,
(ii) for fixed $y=y_0$, $\Delta(x,y_0)$ is increasing in x, and
(iii) for fixed $x=x_0$, $\Delta(x_0,y)$ is decreasing in y.

For any l-set of real numbers $A=\{a_1,a_2,\ldots,a_l\}$, we define the set functions $U_{[i]}(A)$, $i=1,\ldots,l$, by

$$U_{[i]}(A)\equiv a_{[i]},\qquad i=1,\ldots,l. \qquad (11.40)$$

The procedure R_s investigated by Deverman (1969) is based on the observed values of the T_i and is as follows.

***R*$_s$**: The s-subset $\{\pi_{i_1},\pi_{i_2},\ldots,\pi_{i_s}\}$ having the corresponding set of observations $A=\{t_{i_1},\ldots,t_{i_s}\}$ and the complementary set of observations $A^c=\{t_{i_{s+1}},\ldots,t_{i_k}\}$ is included in the collection of selected s-subsets if and only if

$$\Delta[U_{[1]}(A),U_{[k-s]}(A^c)]\geqslant -d^* \tag{11.41}$$

where $d^*>0$ is a constant to be chosen so that the probability of a correct selection is at least P^*.

Deverman (1969) has proved the following theorem.

Theorem 11.5. Let π_i, T_i, and G_i, $i=1,\ldots,k$, be populations, corresponding statistics, and the distribution functions of these statistics, respectively. Let π_i', T_i', and G_i' be similarly defined, where $\pi_{(i)}\precsim\pi_{(i)}'$, $i=1,\ldots,k-t$, and $\pi_{(i)}'\precsim\pi_{(i)}$, $i=k-t+1,\ldots,k$. Then

$$P(CS|R_s)\geqslant P'(CS|R_s) \tag{11.42}$$

where P and P' denote probabilities relative to populations $\pi_1,\ldots,\pi_k$ and $\pi_1',\ldots,\pi_k'$, respectively.

The preceding theorem leads to the fact that the infimum of $P(CS|R_s)$ over all population configurations $\pi_{(1)},\ldots,\pi_{(k)}$ occurs for some configuration(s) wherein all the populations are identical with respect to the binary relation with which they are ordered, that is, $\pi_{(1)}\equiv\pi_{(2)}\equiv\cdots\equiv\pi_{(k)}$.

The procedure R_s has the desirable monotonicity property, namely, if $S_1=\{\pi_{i_1},\ldots,\pi_{i_s}\}$ and $S_2=\{\pi_{j_1},\ldots,\pi_{j_s}\}$ are s-subsets of $\{\pi_1,\ldots,\pi_k\}$ such that $\pi_{i_\alpha}\precsim\pi_{j_\alpha}$, $\alpha=1,\ldots,s$, then

$$\Pr(S_1 \text{ is selected}|R_s)\leqslant \Pr(S_2 \text{ is selected}|R_s). \tag{11.43}$$

Deverman (1969) has also shown that the supremum of the expected number of s-subsets selected occurs for some configuration of the populations wherein all populations are identical provided that certain condition holds. This sufficient condition is a generalized version of (11.13).

The choice of the generalized difference Δ in (11.41) depends upon the nature of populations. In the case of location and scale parameter families we have natural choices of Δ.

Location Parameter Families. Suppose π_i has the associated cdf $F(x-\theta_i)$, $i=1,\ldots,k$. Let the binary order relation $\precsim$ be defined by

$\pi_i \precsim \pi_j \Leftrightarrow \theta_i \leqslant \theta_j$. In this case, we use $T_i = \bar{X}_i$ and $\Delta(a,b) = a - b$. The cdf of T_i is $G(t-\theta_i)$. The constant d^* is the solution of

$$\int_{-\infty}^{\infty} V(y+d^*)dH(y) = P^* \tag{11.44}$$

where $H(\cdot)$ is the cdf of the cth largest of $Z_{(k-t+1)},\ldots,Z_{(k)}$; $V(\cdot)$ is the cdf of the $(s-c+1)$st largest of $Z_{(1)},\ldots,Z_{(k-t)}$; and $Z_{(1)},\ldots,Z_{(k)}$ are independent and identically distributed with cdf $G(t)$. Deverman and Gupta (1969) have applied these results to the normal means problem.

Scale Parameter Families. In this case π_i has the associated distribution function $F(x/\theta_i)$, $\theta_i > 0$. The binary order relation is defined by $\pi_i \precsim \pi_j \Leftrightarrow \theta_j \leqslant \theta_i$. We can use the means procedure with $\Delta(a,b) = (a-b)/b$.

11.3 A RESTRICTED SUBSET SELECTION FORMULATION

11.3.1 Formulation

The basic formulation of the subset selection problem discussed in §11.2.1 has as its goal selection of a nonempty subset that will include the best population with a minimum guaranteed probability. The procedure R_h proposed for this purpose selects a subset of random size. If the data make the choice of the best population less obvious, we are likely to select all the populations. We would expect this to happen if the λ_i are all very close to one another. So it is meaningful to put an additional restriction that the size of the selected subset will not exceed $m(1<m<k)$. Any selection problem with such a restriction is naturally called a *restricted subset selection* problem. Santner (1973, 1975) and Gupta and Santner (1973) have investigated restricted subset selection procedures. In formulating the problem, we need to introduce an indifference zone in the parameter space. As in §11.2.1, we assume that $F_{\lambda_i}, \lambda_i \in \Lambda$, are absolutely continuous distributions. Let $T_{in} = T_n(X_{i1},\ldots,X_{in})$ be a statistic based on n independent observations from π_i. We assume that the sequence $\{T_{in}\}$ is consistent for λ_i, that is, $T_{in} \xrightarrow[p]{} \lambda_i$ as $n \to \infty$. In practice it suffices to assume that T_{in} converges to a monotone function of λ_i so that the resulting selection problem is equivalent to the original one. It is assumed that T_{in} has an absolutely continuous distribution with respect to Lebesgue measure with $G_n(y|\lambda_i)$ as cdf and $g_n(y|\lambda_i)$ as its density. The support $E_n^{\lambda_i}$ of the distribution of T_{in} is assumed to depend on F_i only through λ_i. Also for each n, we assume that $\{G_n(y|\lambda)|\lambda \in \Lambda\}$ forms a stochastically increasing family.

An indifference zone in Ω is defined by means of a real-valued function $p\colon \Lambda \to R$ such that

(i) $p(\cdot)$ is continuous and nondecreasing on Λ,

(ii) $p(\lambda)<\lambda, \lambda\in\Lambda$,

(iii) $p\colon \Lambda' \underset{\text{onto}}{\to} \Lambda$ where $\Lambda'=\{\lambda\in\Lambda|p(\lambda)\in\Lambda\}$. (11.45)

Define

$$\Omega(p)=\{\boldsymbol{\lambda}\in\Omega|\lambda_{[k-1]}\leqslant p(\lambda_{[k]})\}$$

and

$$\Omega^0(p)=\{\boldsymbol{\lambda}\in\Omega|\lambda_{[1]}=\lambda_{[k-1]}=p(\lambda_{[k]})\}.$$

The subspace $\Omega(p)$ is the preference zone and its complement is the indifference zone. The subspace $\Omega^0(p)$ is the set whose elements are the least favorable configurations in $\Omega(p)$.

EXAMPLE 11.3. [Santner (1975)].

(1) $$p_1(\lambda)=\lambda-\delta_1 \quad (\delta_1>0) \quad \Rightarrow \quad \Omega(p_1)=\{\boldsymbol{\lambda}|\lambda_{[k]}-\lambda_{[k-1]}\geqslant\delta_1\},$$

a location-type preference zone.

(2)

$$p_2(\lambda)=\delta_2^{-1}\lambda \quad (\delta_2>1 \text{ and } \Lambda\subset(0,\infty)) \quad \Rightarrow \quad \Omega(p_2)=\{\boldsymbol{\lambda}|\lambda_{[k]}\geqslant\delta_2\lambda_{[k-1]}\},$$

a scale-type preference zone.

(3) $$p_3(\lambda)=\begin{cases}\lambda-\delta_1(\delta_1>0), & 0\leqslant\lambda\leqslant\delta_1\delta_2/(\delta_2-1),\\ \delta_2^{-1}\lambda(\delta_2>1), & \lambda\geqslant\delta_1\delta_2/(\delta_2-1),\end{cases} \qquad (\Lambda=[0,\infty))$$

$$\Rightarrow\Omega(p_3)=\Omega(p_2)\cap\Omega(p_1),$$

a mixed-type preference zone.

REMARK 11.4. The emphasis in the formulation is on the case $1<m<k$ and the strict inequality $p(\lambda)<\lambda$ insures that the indifference zone does not vanish. However, it should be noted that the general theory formally reduces to give the results of Bechhofer (1954) and Gupta (1956, 1965) for the choice of $m=1$ and $m=k$, respectively, if the weaker assumption $p(\lambda)\leqslant\lambda$ is allowed. □

11.3.2 The Class of Procedures $R(n)$

To define a class of procedures for every n, we define a sequence of functions $\{h_n(\cdot)\}$ such that $h_n(\cdot)$: $E_n \to R^1$ (real line) where $\cup_{\lambda\in\Lambda} E_n^{\lambda} \subset E_n$ and the following properties hold.

(i) For each n and x, $h_n(x) > x$.

(ii) For each n, $h_n(x)$ is continuous and strictly increasing in x.

(iii) For each n, $h_n(x) \to x$ as $n \to \infty$. (11.46)

The following class of procedures $R(n)$ was proposed and investigated by Santner (1975).

***R*(*n*)**: Select π_i if and only if

$$T_{in} \geqslant \max\{T_{[k-m+1]n}, h_n^{-1}(T_{[k]n})\} \tag{11.47}$$

where $T_{[1]n} \leqslant \cdots \leqslant T_{[k]n}$ are the ordered T_{in}.

Given P^*, $p(\cdot)$ and the sequence $R(n)$, the goal is to find the common sample size n necessary to achieve

$$P(CS|R(n)) \geqslant P^* \qquad \text{for all } \boldsymbol{\lambda}\in\Omega(p). \tag{11.48}$$

11.3.3 Probability of a Correct Selection and Its Infimum over $\Omega(p)$

For any $\boldsymbol{\lambda}\in\Omega$,

$$P(CS|R(n)) = \sum_{l=k-m}^{k-1} \sum_{\nu=1}^{\binom{k-l}{l}} \int_{-\infty}^{\infty} \prod_{j\in \mathbb{S}_\nu^l(k)} G_n^{(j)}(y) \times \prod_{j\in \overline{\mathbb{S}}_\nu^l(k)} \{G_n^{(j)}(h_n(y)) - G_n^{(j)}(y)\} dG_n^{(k)}(y) \tag{11.49}$$

where $\left\{\mathbb{S}_\nu^l(i) | \nu = 1,\ldots,\binom{k-1}{l}\right\}$ is the collection of all subsets of size l from $U(i) = \{1,\ldots,k\} - \{i\}$,

$$\overline{\mathbb{S}}_\nu^l(i) = U(i) - \mathbb{S}_\nu^l(i),$$

$$G_n^{(j)}(y) = G_n(y|\lambda_{[j]}).$$

The infimum of $P(CS|R(n))$ is shown by Santner to occur at a point in $\Omega^0(p)$. Thus

$$\inf_{\Omega(p)} P(CS|R(n)) = \inf_{\lambda\in\Lambda'} \psi(\lambda,n) \tag{11.50}$$

where

$$\psi(\lambda,n) = \int \{G_n(h_n(y)|p(\lambda))\}^{k-1} \times I\left(\frac{G_n(y|p(\lambda))}{G_n(h_n(y)|p(\lambda))}; k-m, m\right) dG_n(y|\lambda) \tag{11.51}$$

and $I(y;a,b)$ is the usual incomplete beta function with parameters a and b. Sufficient conditions are obtained by Santner (1975) for $\psi(\lambda,n)$ defined in (11.51) to be nondecreasing in λ. His regularity conditions on $G_n(y|\lambda)$ insure that (11.9) holds and hence can be used to obtain the following sufficient conditions.

For all $\lambda\in\Lambda'$

$$g_n(y|\lambda)\frac{\partial G_n(h_n(y)|p(\lambda))}{\partial\lambda} - h_n'(y)g_n(h_n(y)|p(\lambda))\frac{\partial G_n(y|\lambda)}{\partial\lambda} \geqslant 0 \quad \text{a.e. in } E_n, \tag{11.52}$$

$$g_n(y|\lambda)\frac{\partial G_n(y|p(\lambda))}{\partial\lambda} - g_n(y|p(\lambda))\frac{\partial G_n(y|\lambda)}{\partial\lambda} \geqslant 0 \quad \text{a.e. in } E_n. \tag{11.53}$$

11.3.4 Properties of $\{R(n)\}$

Santner (1975) has studied several properties of the sequence $\{R(n)\}$ as well as the individual rules $R(n)$. To describe these properties, we need the following definitions.

Definition 11.6. The sequence of rules $\{R(n)\}$ is said to be *consistent* with respect to Ω' if $\inf_{\Omega'} P(CS|R(n))\rightarrow 1$ as $n\rightarrow\infty$.

Definition 11.7. The rule $R(n)$ is said to be *strongly monotone in* $\pi_{(i)}$ if

$$p_\lambda^n(i) = \Pr(R(n) \text{ selects the population associated with } \lambda_{[i]}) \tag{11.54}$$

is increasing in $\lambda_{[i]}$ when all other components of $\boldsymbol{\lambda}$ are fixed, and is decreasing in $\lambda_{[j]}(j\neq i)$ when all other components of $\boldsymbol{\lambda}$ are fixed.

The results of Santner (1975) are given in the following theorem.

Theorem 11.6. a. If there exists $N\geqslant 1$ and $\lambda_0\in\Lambda'$ such that for all $n\geqslant N, \inf_{\lambda\in\Lambda'}\psi(\lambda,n)=\psi(\lambda_0,n)$, then any sequence $\{R(n)\}$ defined by (11.47) is consistent with respect to $\Omega(p)$.

b. Any rule $R(n)$ of form (11.47) is strongly monotone in $\pi_{(i)}$ for any $i=1,\ldots,k$.

It follows from Theorem 11.6 (b) that the rule $R(n)$ is monotone in the sense of (11.11) and hence is unbiased.

Let $S(n)$ and $S'(n)$ denote, respectively, the number of populations and the number of nonbest populations included in the selected subset. Then the following asymptotic properties of $\{S(n)\}$ and $\{S'(n)\}$ have been established by Santner (1975).

Theorem 11.7. For any $\boldsymbol{\lambda}$ such that $\lambda_{[k]}>\lambda_{[k-1]}$, $p_{\lambda}^{n}(i)\to 1$ for $i=k$ and $p_{\lambda}^{n}(i)\to 0$ for $i<k$ as $n\to\infty$.

Define

$$W_i(n)=\begin{cases}1 & \text{if } \pi_{(i)} \text{ is selected,}\\ 0 & \text{otherwise.}\end{cases} \tag{11.55}$$

COROLLARY 11.2. For any $\boldsymbol{\lambda}\in\Omega$ such that $\lambda_{[k]}>\lambda_{[k-1]}$,

a. $W_i(n)\xrightarrow[p]{}1$ for $i=k$ and $W_i(n)\xrightarrow[p]{}0$ for $i<k$ as $n\to\infty$.

b. $S(n)\xrightarrow[p]{}1$ and $S'(n)\xrightarrow[p]{}0$ as $n\to\infty$.

As a consequence of Corollary 11.2 (b), we see that for $\boldsymbol{\lambda}$ such that $\lambda_{[k]}>\lambda_{[k-1]}$, $E[S(n)]\to 1$ and $E[S'(n)]\to 0$ as $n\to\infty$.

The preceding properties relate to the sequence $\{S(n)\}$. For any fixed n, the supremum of $E(S(n))$ occurs when $\lambda_1=\cdots=\lambda_k$ if certain regularity conditions are satisfied and for all λ_1,λ_2 in Λ with $\lambda_1\leqslant\lambda_2$,

$$\frac{\partial G_n(h_n(y)|\lambda_1)}{\partial\lambda_1}g_n(y|\lambda_2)-\frac{\partial G_n(y|\lambda_1)}{\partial\lambda_1}g_n(h_n(y)|\lambda_2)h_n'(y)\geqslant 0 \qquad \text{a.e. in } E_n \tag{11.56}$$

which is essentially the condition of Gupta and Panchapakesan (1972a) given in (11.13).

Application of these results to the ranking and selection of normal means has been studied by Gupta and Santner (1973) and their results are discussed in the next chapter.

11.3.5 A Generalized Goal for Restricted Subset Selection

The results of Santner (1975) have been generalized by him (1976a) for the goal of selecting at least one of the t best populations. The treatment is similar to the case of $t=1$ but for the natural differences. We briefly indicate the results in this case.

We now define

$$\Omega^t(p) = \{\boldsymbol{\lambda}\in\Omega|\lambda_{[k-t]} \leqslant p(\lambda_{[k-t+1]})\},$$

$$\Omega^t_0(p) = \{\boldsymbol{\lambda}|\lambda_{[1]} = \lambda_{[k-t]} = p(\lambda_{[k-t+1]}),\ \lambda_{[k-t+1]} = \lambda_{[k]}\},$$

$$\Omega_0 = \{\boldsymbol{\lambda}|\lambda_{[1]} = \lambda_{[k]}\}.$$

The sequence of functions $\{h_n(\cdot)\}$ are defined with the properties specified in (11.46). The rule $R_t(n)$ selects π_i if and only if

$$T_{in} \geqslant \max\{T_{[k-m+1]n}, h_n^{-1}(T_{[k]n})\}. \tag{11.57}$$

Given P^*, $p(\lambda)$, t, and $R_t(n)$, we want to find the smallest common sample size n needed to satisfy the requirement

$$P(CS|R_t(n)) \geqslant P^* \quad \text{for all } \boldsymbol{\lambda}\in\Omega^t(p). \tag{11.58}$$

We note that the rule $R_t(n)$ defined by (11.57) is same as $R(n)$ in (11.47). Santner (1976a) has shown that the infimum of $P(CS|R_t(n))$ over $\Omega^t(p)$ occurs when $\boldsymbol{\lambda}\in\Omega^t_0(p)$ for $t \leqslant k-m$ and when $\boldsymbol{\lambda}\in\Omega_0$ for $t > k-m$. The monotonicity of $P(CS|R_t(n))$ in terms of λ when either $\boldsymbol{\lambda} = (p(\lambda),\ldots,p(\lambda),\lambda,\ldots,\lambda)$ or $\boldsymbol{\lambda} = (\lambda,\lambda,\ldots,\lambda)$ is given by the sufficient conditions (11.52) and (11.53). The rule $R_t(n)$ is consistent and it has the other monotonicity properties of $R(n)$.

11.4 A MODIFIED GOAL: ELIMINATION OF NONBEST POPULATIONS

11.4.1 Formulation of the Problem

In the notation of §11.2.1, one may be interested in selecting the populations that are fairly as good as the best populations. The emphasis is on excluding any population that is sufficiently inferior to the best population. Desu (1970) considered such a goal. Let $\delta(\lambda_i,\lambda_j)$ denote a measure of the distance between the populations π_i and π_j. Let δ_1^* and δ_2^* be known constants such that $0<\delta_1^*<\delta_2^*$. Any population π_i is called a *superior* (or *good*) population if $\delta(\lambda_{[k]},\lambda_i)\leqslant\delta_1^*$ and an *inferior* (or *bad*) population if $\delta(\lambda_{[k]},\lambda_i)\geqslant\delta_2^*$. This definition is slightly more general than the one used by Lehmann (1963) [see §8.3]. The goal is to select a nonempty (random-sized) subset that excludes all the inferior populations with a minimum guaranteed probability P^*. Selection of any subset satisfying the goal is called a correct selection. Let T_i be a statistic based on a sample of n independent observations from π_i and let $G_n(t,\theta_i)$ denote the cdf of T_i.

Desu (1970) has investigated the cases when the θ_i are location or scale parameters of the distributions of the T_i. His procedure R selects π_i if and only if

$$d(T_{[k]},T_i)\leqslant d(\delta_2^*,c) \tag{11.59}$$

where $d(a,b)=a-b$ for the location parameter case and $=a/b$ for the scale parameter case, and $c\in(0,\delta_2^*)$ is suitably chosen to satisfy the P^*-requirement.

11.4.2 Location Parameter Case

Let t_1 and t_2 denote the unknown number of inferior and superior populations, respectively, in the given set of k populations. Clearly, $t_1\geqslant 0$, $t_2\geqslant 1$ and $t_1+t_2\leqslant k$. For specified δ_1^* and δ_2^*, let

$$\begin{aligned}\Omega(t_1,t_2)=\{\boldsymbol{\lambda}: \lambda_{[1]}\leqslant\cdots\leqslant\lambda_{[t_1]}\leqslant\lambda_{[k]}-\delta_2^*\\ <\lambda_{[t_1+1]}\leqslant\lambda_{[t_1+2]}\leqslant\cdots\leqslant\lambda_{[k-t_2]}\\ <\lambda_{[k]}-\delta_1^*\leqslant\lambda_{[k-t_2+1]}\leqslant\cdots\leqslant\lambda_{[k]}\}.\end{aligned} \tag{11.60}$$

Then $\Omega=\cup_{t_1,t_2}\{\Omega(t_1,t_2)\}$. To choose c we need $\inf_\Omega P(CS|R)$. Desu (1970) has shown that the infimum of $P(CS|R)$ over $\Omega(t_1,t_2)$ occurs when

$$\begin{aligned}\lambda_{[1]}=\cdots=\lambda_{[t_1]}=m,\\ \lambda_{[t_1+1]}=\cdots=\lambda_{[k-t_2]}=m',\\ \lambda_{[k-t_2+1]}=\cdots=\lambda_{[k-1]}=m'',\\ \lambda_{[k]}=\lambda.\end{aligned} \tag{11.61}$$

Let P' denote $P(CS|R)$ at points $\boldsymbol{\lambda}$ given by (11.61). For fixed λ, P' is minimized by letting m, m', and m'' approach $\lambda-\delta_2^*$, $\lambda-\delta_2^*$, and $\lambda-\delta_1^*$, respectively. Desu has shown that for all possible values of t_1 and t_2, the preceding minimum is least when $t_1 = k$. This gives

$$\inf_{\Omega} P(CS|R) = \int_{-\infty}^{\infty} G_n^{k-1}(t+c) g_n(t)\,dt, \tag{11.62}$$

where $G_n(t,\lambda) = G_n(t-\lambda)$ and $g_n(t-\lambda)$ is the corresponding density. In other words, the infimum in (11.62) is independent of λ. The required c-value is the largest real number c (less than δ_2^*) such that

$$\int_{-\infty}^{\infty} G_n^{k-1}(t+c) g_n(t)\,dt = P^*. \tag{11.63}$$

Even though not explicitly stated by Desu (1970), the solution c_0 of (11.63) will be less than δ_2^* for a proper choice of n. This can be done if the left-hand side of (11.63) tends to 1 as $n \to \infty$.

Let η and ζ denote, respectively, the numbers of inferior and superior populations included in the selected subset. Desu (1970) has investigated the performance of the procedure R by considering $E(\eta)$, $E(\zeta)$, and $E(\eta/t_1)$. He has shown that

$$\sup_{\Omega} E(\eta) = (k-1) \int_{-\infty}^{\infty} \left[G_n(t+\delta_2^* - c) \right]^{k-2} G_n(t-c) g_n(t)\,dt, \tag{11.64}$$

provided that $g_n(t-\lambda)$ has a monotone likelihood ratio in t. If c is chosen to satisfy the P^*-requirement, then

$$\sup_{\Omega} E(\eta) > 1 - P^*. \tag{11.65}$$

With the monotone likelihood ratio assumption, it can also be shown that

$$\sup_{t_1, t_2} \left\{ \sup_{\Omega(t_1,t_2)} E\left(\frac{\eta}{t_1} \right) \right\} = \frac{1}{k-1} \sup_{\Omega} E(\eta) \tag{11.66}$$

and

$$\inf_{\Omega} E(\zeta) = \int_{-\infty}^{\infty} G_n^{k-1}(t+\delta_2^* + \delta_1^* - c) g_n(t)\,dt. \tag{11.67}$$

The treatment of the scale parameter case is similar and hence is omitted. Applications of these results to the ranking of normal means and gamma scale parameters are discussed in the following chapters.

11.4.3 Elimination of Non-t-best Populations

Carroll, Gupta, and Huang (1975) have considered a generalization of the problem of eliminating the nonbest populations discussed in §11.4.2. Any population π_i is called a *strictly non-t-best* population if $\delta(\lambda_{[k-t+1]}, \lambda_i) \geqslant \Delta > 0$. The goal is to select a subset that excludes all the non-t-best populations. Naik (1977) has considered the same problem for $t=1$. The rule R' investigated by Carroll, Gupta, and Huang selects π_i if and only if

$$d(T_{[k-t+1]}, T_i) \leqslant d(\Delta, d_1) \tag{11.68}$$

where $d_1 \in (0, \Delta)$ is to be determined to satisfy the P^*-condition.

In the location case, $d(a,b) = a - b$. It has been shown by Carroll, Gupta, and Huang that, if $t \leqslant k-t$,

$$\inf_{\Omega} P(CS|R') = t\int_{-\infty}^{\infty} G_n^{k-t}(y+d_1)\left[1 - G_n(y)\right]^{t-1} g_n(y)\,dy \tag{11.69}$$

so that for sufficiently large n, a value of d_1 (less than Δ) is given by

$$t\int_{-\infty}^{\infty} G_n^{k-t}(y+d_1)\left[1 - G_n(y)\right]^{t-1} g_n(y)\,dy = P^*. \tag{11.70}$$

The scale parameter case is analogous. Applications to the cases of normal means and gamma scale parameters are obvious.

11.4.4 Selection of Δ_p-Superior Populations

Panchapakesan and Santner (1977) have considered a class of procedures for selecting from a general family of stochastically ordered distributions employing a class of functions to define a good population. Suppose π_i has the distribution function F_{λ_i}, where $\lambda_i \in \Lambda$, a known interval on the real line. Let $X_{i1}, \ldots, X_{in}$ be independent observations from π_i and let $T_{in} = T_n(X_{i1}, \ldots, X_{in})$ be sufficient and consistent for λ_i. Let $G_n(y|\lambda_i)$ denote the cdf of T_{in}. It is assumed that for each n, $\{G_n(y|\lambda)|\lambda \in \Lambda\}$ is a stochastically increasing family and that each member of the family has the same support $S = \Lambda$.

Let $\gamma = \inf\{\lambda|\lambda \in \Lambda\}$ which can possibly be $-\infty$. In order to define a good population we introduce a function p: $\Lambda \to \Lambda$ such that

(i) $p(\lambda)$ is continuous,

(ii) $p(\lambda) < \lambda$ for all $\lambda \in \Lambda' = \{\lambda \in \Lambda | p(\lambda) > \gamma\}$,

(iii) $p(\lambda)$ is strictly increasing on Λ'. (11.71)

For a given $t(1 \leqslant t \leqslant k-1)$, any populations π_i is said to be *good* (relative to the tth best) if $\lambda_i > p(\lambda_{[k-t+1]})$.

Panchapakesan and Santner (1977) have considered two goals: Goal A of selecting a nonempty subset containing only good ones, and Goal B of selecting a subset whose size does not exceed a predetermined level $m(1 \leqslant m < k)$ and which will include at least one good population. In either case, selection of any subset satisfying the goal is called a Δ_p-correct selection ($\Delta_p - CS$).

Define a sequence of functions $h_n(y)$: $S \rightarrow R$ possessing the following properties:

(1) For each n, $h_n(x)$ is continuous and strictly increasing in x.
(2) For each x, $h_n(x) \downarrow x$ as $n \rightarrow \infty$.
(3) For each n and x, $h_n(p(x)) \leqslant x$. (11.72)

Panchapakesan and Santner (1977) have proposed and studied the following classes of procedures $\{R_h(n)\}$ and $\{R_h'(n)\}$ for Goals A and B, respectively:

***R*$_h$(*n*):** Select π_i if and only if

$$T_i \geqslant h_n(p(T_{[k-t+1]})) \tag{11.73}$$

and

***R*$_h'$(*n*):** Select π_i if and only if

$$T_i \geqslant \max\left\{h_n^{-1}(T_{[k]}), T_{[k-m+1]}\right\}. \tag{11.74}$$

The treatment regarding the PCS and the expected number of good populations are similar to that of Gupta and Panchapakesan (1972a) and Santner (1975).

Panchapakesan and Santner have also established a large sample consistency property of $R_h(n)$. If $\lambda_{[1]} \leqslant \cdots \leqslant \lambda_{[k-s]} < p(\lambda_{[k-t+1]}) < \lambda_{[k-s+1]} \leqslant \cdots \leqslant \lambda_{[k]}, t \leqslant s < k$, then the number of good populations selected converges stochastically to s as $n \rightarrow \infty$.

11.5 NOTES AND REMARKS

In this chapter we have confined our attention to the general theory for the basic subset selection formulation and certain important variations of the goal. We have considered only single-stage procedures with equal sample sizes. There are other general results especially for location and scale parameter cases and these results are discussed in the sequel in several chapters devoted specifically to topics such as nonparametric procedures, sequential procedures, Bayes procedures, and procedures for restricted families of distributions. Some minimax properties and construction of Γ-minimax rules are discussed in Chapter 18.

CHAPTER 12

Selection from Univariate Continuous Populations

12.1 INTRODUCTION

In this chapter we are concerned with subset selection procedures for selecting the best population from a set of k populations belonging to several specific families of univariate continuous populations. Of special interest to us are normal, gamma, and uniform populations. We discuss in detail the selection from normal populations in terms of the means, the variances, and the absolute values of the means. As one can see, the problem of selecting the best normal population in terms the variances is essentially selection of chi-square distributions which are special cases of gamma populations. We also discuss selection in terms of the coverages of a fixed interval. Many of the procedures discussed fall under the general type discussed in Chapter 11. However, there are many interesting specific results such as those for procedures based on unequal sample sizes.

Since our attention is mostly on selection from normal populations, let us define the following quantities associated with k independent samples where X_{ij}, $j=1,\ldots,n_i$, are independent observations from π_i $(i=1,\ldots,k)$.

$$N=\sum_{i=1}^{k} n_i$$

$$\bar{X}_i=n_i^{-1}\sum_{j=1}^{n_i} X_{ij}$$

$$s_i^2=(n_i-1)^{-1}\sum_{j=1}^{n_i}\left(X_{ij}-\bar{X}_i\right)^2 \tag{12.1}$$

$$s^2=(N-k)^{-1}\sum_{i=1}^{k}(n_i-1)s_i^2$$

$$v_i^2=n_i^{-1}\sum_{j=1}^{n_i}(X_{ij}-\mu_i)^2.$$

As usual, $\Phi(\cdot)$ and $\varphi(\cdot)$ denote the cdf and the density, respectively, of a standard normal random variable. The cdf of $Y=\bar{X}_i$, the sample mean from $N(\mu_i,\sigma_i^2)$ is denoted by $G_{n_i}(y;\mu_i,\sigma_i^2)$ which equals $\Phi(\sqrt{n_i}\,\sigma_i^{-1}(y-\mu_i))$.

12.2 SELECTION FROM NORMAL POPULATIONS IN TERMS OF THE MEANS

12.2.1 Equal Sample Sizes and Common Known Variance

Let $n_i=n$ and $\sigma_i^2=\sigma^2$ for all $i=1,\dots,k$. For this case Gupta (1956) proposed and studied the rule

$\boldsymbol{R}_1$: Select π_i if and only if

$$\bar{X}_i \geqslant \bar{X}_{[k]} - \frac{d\sigma}{\sqrt{n}} \tag{12.2}$$

where $d=d(k,P^*)>0$ is to be determined so that the P^*-requirement is met. It can be seen that this problem comes under the location parameter case discussed in §11.2.4. The infimum of $P(CS|R_1)$ occurs when $\mu_1=\cdots=\mu_k$ and is independent of the common value of the μ_i. If we let $d_1=d\sigma/\sqrt{n}$, the constant d_1 is given by

$$\int_{-\infty}^{\infty} G_n^{k-1}(y+d_1;0,1)\,dG_n(y;0,1)=P^* \tag{12.3}$$

which can be rewritten as

$$\int_{-\infty}^{\infty} \Phi^{k-1}(y+d)\,d\Phi(y)=P^*. \tag{12.4}$$

Tables giving the values of d have been constructed by several authors in different contexts. Gupta (1956) has tabulated the values of d for $k=2(1)51$ and $P^*=0.75$, 0.90, 0.95, 0.975, and 0.99. Table I of Bechhofer (1954) includes the values of d for $k=2(1)10$ and $P^*=0.10(0.05)\ 0.80(0.02)\ 0.90(.01)\ 0.99$, 0.995, 0.999, 0.9995, the starting value of P^* depending on k. The values of $d/\sqrt{2}$ are given for $k=2(1)51$ and $P^*=0.75$, 0.90, 0.95, 0.975, and 0.99 by Gupta [1963a] as well as by Gupta, Nagel, and Panchapakesan [1973]. The last two papers are concerned with the order statistics from equally corrrelated normal random variables and the d-value required for the procedure R_1 corresponds to the case where the correlation is 0.5.

The monotonicity properties of R_1 and the results relating to the supremum of the expected subset size follow from the discussions in §11.2.4.

Assuming the common variance σ^2 to be unity, Deely and Gupta (1968) have tabulated the PCS, the probability of selecting a fixed nonbest population, and the expected proportion of populations selected for the rule R_1 under the slippage configuration $(\mu,\ldots,\mu,\mu+\delta)$. These values are tabulated by them for $k=2(1)10$, 25, 50; $\delta/\sqrt{n}=0.00(0.50)\ 1.00(0.25)3.00$, 4.00, 5.00, ∞, and $P^*=0.75$, 0.90, 0.95.

12.2.2 Equal Sample Sizes and Common Unknown Variance

In this case, the following rule R_2 was investigated by Gupta (1956).

R_2: Select π_i if and only if

$$\bar{X}_i \geqslant \bar{X}_{[k]} - \frac{d_2 s}{\sqrt{n}} \tag{12.5}$$

where $d_2=d_2(k,n,P^*)>0$ is to be determined subject to the P^*-condition. The infimum of $P(CS|R_2)$ occurs when the μ_i are equal and is independent of the common (unknown) value of the μ_i. Thus the constant d_2 is the solution of

$$\int_0^\infty \int_{-\infty}^\infty \Phi^{k-1}(u+d_2 y)\varphi(u)q_\nu(y)\,du\,dy = P^* \tag{12.6}$$

where $q_\nu(y)$ is the density of $\chi_\nu/\sqrt{\nu}$, $\nu=N-k$. The values of d_2 have been tabulated by Gupta and Sobel (1957) for $k=2$, 5, 10(1)16(2)20(5)40, 50; $\nu=15(1)20$, 24, 30, 36, 40, 48, 60(20)120, 360, ∞; and $P^*=0.75$, 0.90, 0.95, 0.975. It should be noted that d_2 is the $100P^*$-percentage point of $Y=(U_{\max}-U)/s_\nu$, where $U_{\max}=\max(U_1,\ldots,U_{k-1})$, the random variables $U,U_1,\ldots,U_{k-1}$ are independent and identically distributed as $N(\mu,\sigma^2)$, and $\nu s_\nu^2/\sigma^2$ is a χ^2 random variable with ν degrees of freedom which is distributed independently of the set $\{U,U_1,\ldots,U_{k-1}\}$. Moments of Y and the evaluation of the percentage points have been discussed by Gupta and Sobel (1957). See also Carroll (1974c) for some approximations based on a lower bound for PCS.

12.2.3 Unequal Sample Sizes and Common Known Variance

Without loss of generality we can assume that the common variance $\sigma^2=1$. Gupta and W. T. Huang (1974a) have proposed the rule

R_3: Select π_i if and only if

$$\bar{X}_i \geqslant \bar{X}_{[k]} - \frac{d_3}{\sqrt{n_i}}, \tag{12.7}$$

where $d_3=d_3(k,P^*,n_1,\ldots,n_k)>0$ is to be determined subject to P^*-requirement.

Let $n_{(i)}$ denote the sample size associated with the population having the mean $\mu_{[i]}$ and define $r_{ij}^2 = n_{(i)}/n_{(j)}$. Then

$$\inf_{\Omega} P(CS|R_3) = \int_{-\infty}^{\infty} \prod_{j=1}^{k-1} \Phi(r_{jk}(x+d_3))\,d\Phi(x). \tag{12.8}$$

Since the r_{jk} are unknown, the problem reduces to finding $\min_{1 \leqslant i \leqslant k} I(n_i)$, where

$$I(n_i) = \int_{-\infty}^{\infty} \prod_{j \neq i} \Phi\left(\sqrt{\frac{n_j}{n_i}}\,(x+d_3)\right) d\Phi(x). \tag{12.9}$$

Gupta and W. T. Huang (1974a) have shown that

$$I(n_{[k]}) = \min_{1 \leqslant i \leqslant k} I(n_i). \tag{12.10}$$

Thus we have

$$\inf_{\Omega} P(CS|R_3) \geqslant P^* \tag{12.11}$$

for any configuration $(\mu_1, \ldots, \mu_k)$ if d_3 satisfies the equation

$$\int_{-\infty}^{\infty} \prod_{i=1}^{k-1} \Phi(t_i(x+d_3))\,d\Phi(x) = P^* \tag{12.12}$$

where $t_i = n_{[i]}/n_{[k]}$.

One can use a crude lower bound, namely,

$$\inf_{\Omega} P(CS|R_3) \geqslant \int_{-\infty}^{0} \Phi^{k-1}\left(\frac{n_{[k]}}{n_{[1]}}u\right) d\Phi(u-d_3)$$
$$+ \int_{0}^{\infty} \Phi^{k-1}\left(\frac{n_{[1]}}{n_{[k]}}u\right) d\Phi(u-d_3). \tag{12.13}$$

The preceding bound is easier to compute than the left-hand side of (12.12).

An alternative procedure for this problem has been proposed by Gupta and D. Y. Huang (1976a). Their procedure is

R_4: Select π_i if and only if

$$\overline{X}_i \geqslant \max_{1 \leqslant j \leqslant k} \left[\overline{X}_j - d_4\sqrt{\frac{1}{n_i} + \frac{1}{n_j}}\right] \tag{12.14}$$

where $d_4 = d_4(k, P^*, n_1, \ldots, n_k) > 0$ is chosen to satisfy the P^*-condition. For any given association between $(n_1, \ldots, n_k)$ and $(n_{(1)}, \ldots, n_{(k)})$, the infimum of $P(CS|R_4)$ is attained when $\mu_{[1]} = \cdots = \mu_{[k]}$. This gives

$$\min_{n_1,\ldots,n_k} \inf_{\Omega} P(CS|R_4)$$

$$= \min_{1 \leq i \leq k} \int_{-\infty}^{\infty} \prod_{\substack{j=1 \\ j \neq i}}^{k} \Phi\left[\sqrt{\frac{n_{(j)}}{n_{(i)}}}\, y + d_4 \sqrt{1 + \frac{n_{(j)}}{n_{(i)}}}\,\right] d\Phi(y). \tag{12.15}$$

It has been shown by Gupta and Huang (1976a) that the minimum in (12.15) is given by

$$\int_{-\infty}^{\infty} \prod_{j=1}^{k-1} \Phi\left[\frac{d_4 - \alpha_j u}{\left(1 - \alpha_j^2\right)^{1/2}}\right] d\Phi(u). \tag{12.16}$$

where

$$\alpha_i = n_{[i]}^{1/2}\left(n_{[i]} + n_{[k]}\right)^{-1/2}, \; i = 1, \ldots, k-1. \tag{12.17}$$

For this procedure, an upper bound for the expected subset size is given by

$$\sup_{\Omega} E(S|R_4) \leq k\Phi(d_4). \tag{12.18}$$

For $k = 2$ and equal sample sizes, we have equality in (12.18).

12.2.4 Unequal Sample Sizes and Common Unknown Variance

Gupta and W. T. Huang (1974a) proposed the rule R_5 which selects π_i if and only if

$$\bar{X}_i \geq \bar{X}_{[k]} - \frac{d_5 s}{\sqrt{n_i}}, \tag{12.19}$$

where $d_5 = d_5(k, P^*, n_1, \ldots, n_k) > 0$ is to be chosen subject to the P^*-condition. It has been shown by them that

$$\inf_{\Omega} P(CS|R_5) \geq P^* \tag{12.20}$$

provided that d_5 satisfies

$$\int_0^{\infty}\int_{-\infty}^{\infty}\prod_{j=1}^{k-1}\Phi\big(t_j(x+d_5y)\big)\,d\Phi(x)\,dQ_\nu(y)=P^*, \tag{12.21}$$

where $t_j=n_{[j]}/n_{[k]}$, $Q_\nu(y)$ denotes the cdf of $\chi_\nu/\sqrt{\nu}$ and $\nu=N-k$.

An alternative procedure R_6 has been proposed by Gupta and D. Y. Huang (1976a). This procedure selects π_i if and only if

$$\bar{X}_i \geqslant \max_{1\leqslant j\leqslant k}\left[\bar{X}_j-d_6s\sqrt{\frac{1}{n_i}+\frac{1}{n_j}}\right], \tag{12.22}$$

where $d_6=d_6(k,P^*,n_1,\ldots,n_k)>0$ is to be determined so that the P^*-condition is satisfied. In this case

$$\min_{n_1,\ldots,n_k}\inf_{\Omega} P(CS|R_6)=\int_0^{\infty}\int_{-\infty}^{\infty}\prod_{i=1}^{k-1}\Phi\left[\frac{d_6s-\alpha_iu}{\sqrt{1-\alpha_i^2}}\right]d\Phi(u)\,dQ_\nu(s) \tag{12.23}$$

where α_i is defined by (12.17). Also an upper bound for the expected subset size is given by

$$E(S|R_6)\leqslant k\int_0^{\infty}\Phi(d_6x)\,dQ_\nu(x). \tag{12.24}$$

Sitek (1972) proposed the procedure R_7 which selects π_i if and only if

$$\bar{X}_i \geqslant \bar{X}_{[k]}-d_7s\sqrt{\frac{1}{n_i}+\frac{1}{M}} \tag{12.25}$$

where $d_7>0$ is to be chosen approximately and M is the n_j of the population that yielded $\bar{X}_{[k]}$. However, her calculation of the infimum of the PCS is incorrect. Chen, Dudewicz, and Lee (1976) have proposed a modification in the procedure R_7. Their modified rule is

$\boldsymbol{R_8}$: Select π_i if and only if

$$\bar{X}_i \geqslant \bar{X}_{[k]}-d_8s\sqrt{\frac{1}{n_i}+\frac{1}{a}}\,, \tag{12.26}$$

where $a \geqslant 0$ is any fixed constant and $d_8 = d_8(k, P^*, n_1, \ldots, n_k, a) > 0$ is to be determined to satisfy the P^*-condition. For this procedure

$$\inf_{\Omega} P(CS|R_8) \geqslant \Pr\left[T_i \leqslant d_8 \sqrt{\frac{n_{[1]}(a+n_{[k]})}{a(n_{[1]}+n_{[k]})}} \quad, i=1,\ldots,k-1 \right], \tag{12.27}$$

where $(T_1, \ldots, T_{k-1})$ has a $(k-1)$-variate t-distribution with $\nu = N-k$ degrees of freedom and equal correlations $\rho(0 < \rho \leqslant \frac{1}{2})$ where

$$\rho = \left[\left(1 + \frac{n_{[k]}}{n_{[1]}}\right)\left(1 + \frac{n_{[k]}}{n_{[2]}}\right) \right]^{-1/2}.$$

Equality holds in (12.27) if and only if $n_1 = \cdots = n_k$, in which case $\rho = \frac{1}{2}$ and the procedure R_8 reduces to that of Gupta (1956) defined by (12.5). The question of an optimal choice of a in defining R_8 has not been answered. For results of appropriate comparisons among the preceding rules R_3, R_4, R_5, R_6, and R_8, see also Gupta and D. Y. Huang (1977c).

12.2.5 Allocation of Sample Sizes When Variances Are Unknown

Dudewicz and Dalal (1975) have defined a two-stage procedure for selecting the largest mean when the unknown variances need not be equal. The problem here is the determination of the sample sizes n_i so that for the procedure proposed the P^*-condition is satisfied. The sampling procedure is same as that employed by them for this problem under the indifference zone formulation which was discussed in §2.1.2 (procedure R_3). The first stage sample size is n_0 and the total sample size n_i is determined by (2.9) where we replace δ^* by d. The selection procedure of Dudewicz and Dalal selects π_i if and only if

$$\tilde{\bar{X}}_i \geqslant \tilde{\bar{X}}_{[k]} - d, \tag{12.28}$$

where $\tilde{\bar{X}}_i$ is a weighted average defined by (2.10) and (2.11). This procedure has the monotonicity property and it satisfies the P^*-condition irrespective of the choice of d. The choice of d can be made by imposing restrictions on the expected subset size for specific parametric configurations.

Gupta and Wong (1976b) have discussed a single-stage procedure based on unequal sample sizes for selecting the normal population with the largest normal mean when the variances are unknown and possibly unequal. They propose the procedure R_9: Select π_i if and only if

$$\bar{X}_i \geqslant \max_{1 \leqslant j \leqslant k} \bar{X}_j - c \max_{1 \leqslant j \leqslant k} S_j \tag{12.29}$$

where $S_i^2 = n_i^{-1}\Sigma_{j=1}^{n_i}(X_{ij} - \bar{X}_i)^2$. To satisfy the P^*-condition, the constant $c = c(k, P^*, n_1, \ldots, n_k)$ is determined by

$$\int_0^{\infty}\left\{\prod_{i=1}^{k-1}\int_0^{\infty}\Phi\left[\frac{c}{\left(x_i^{-1}+y^{-1}\right)^{1/2}}\right]dG_{[i]}(x_i)\right\}dG_{[i]}(y) = P^*, \qquad (12.30)$$

where $G_{[i]}(\cdot)$ denotes the cdf of a chi-square random variable with $n_{[i]} - 1$ degrees of freedom. Gupta and Wong (1976b) have also shown that an upper bound for the expected subset size is given by

$$E(S|R_9) \leqslant \frac{1}{k-1}\sum_{1 \leqslant i < j \leqslant k}\int_0^{\infty}\int_0^{\infty}\Phi\left[\frac{c}{\left(x^{-1}+y^{-1}\right)^{1/2}}\right]dG_{n_i}(x)\,dG_{n_j}(y), \qquad (12.31)$$

where $G_{n_i}(\cdot)$ is the cdf a chi-square random variable with $n_i - 1$ degrees of freedom.

12.2.6 A Procedure Based on Contrasts of Sample Means

Here it is assumed that the variances are equal but unknown. Seal (1955) defined a procedure based on the ordered sample means and a set of nonnegative constants $c_1, \ldots, c_{k-1}$ such that $\Sigma_{i=1}^{k-1} c_i = 1$. Let $\bar{X}_{[1]}^{(i)} \leqslant \bar{X}_{[2]}^{(i)} \leqslant \cdots \leqslant \bar{X}_{[k-1]}^{(i)}$ denote the ordered sample means after deleting $\bar{X}_i$, $i = 1, \ldots, k$. Then the procedure R_{10} of Seal (1955) selects π_i if and only if

$$\bar{X}_i \geqslant \sum_{r=1}^{k-1} c_r \bar{X}_{[r]}^{(i)} - st, \qquad (12.32)$$

where $t = t(k, P^*, c_1, \ldots, c_{k-1})$ is the upper $100(1-P^*)$ percentage point of the distribution of $\Sigma_{i=1}^{k-1} c_i Y_{[i]} - Y_k$, where $Y_1, \ldots, Y_k$ are i.i.d. $N(0, \sigma^2)$ and $Y_{[1]} \leqslant \cdots \leqslant Y_{[k-1]}$ are the ordered random variables corresponding to the set $\{Y_1, \ldots, Y_{k-1}\}$. It is shown by Seal that

$$P(CS|R_{10}) \geqslant P^*. \qquad (12.33)$$

The procedure is unbiased. Actually, R_{10} defines a class of procedures depending on the various choices of the constants $c_1, \ldots, c_{k-1}$. Let this class be denoted by $\mathcal{C}$. Seal (1955) has considered the optimum choice of the constants by maximizing $P(CS|R_{10})$ when $\mu_{[1]} = \cdots = \mu_{[k-1]} = \mu_{[k]} - \delta$ ($\delta > 0$) and by maximizing the probability of excluding the worst population when $\mu_{[1]} + \delta = \mu_{[2]} = \cdots = \mu_{[k]}$. He has shown that the optimum choice is given by $c_1 = \cdots = c_{k-1} = 1/(k-1)$. This choice approximately (because of normal approximations) maximizes the probabilities considered.

For the determination of the optimal choice of the constants, Seal (1955) did not consider the chance of rejecting nonbest populations other than the worst. Nor did he consider any measure of actual superiority of the rule D_0 (which is R_{10} with $c_1 = \cdots = c_{k-1} = (k-1)^{-1}$) over other members of the class $\mathcal{C}$. However, in another paper (1957), he considered the optimal choice using minimization of the expected subset size as the criterion. Specifically, he considered the case of $k=3$ populations (σ known) and the three types of configurations: (i) $\mu-\delta, \mu, \mu+\delta$; (ii) $\mu, \mu, \mu+\delta$; and (iii) $\mu-\delta, \mu, \mu; \delta>0$. In these situations the optimum choice is seen to be $c_1 = \cdots = c_{k-2} = 0$ and $c_{k-1} = 1$. This rule will be denoted by $D(k-1)$. Under the earlier criterion used by Seal (1955), the superiority of D_0 over $D(k-1)$ is negligible. The conclusions of Seal for the general problem are based on particular cases and not resolved analytically.

It should be noted that the rule $D(k-1)$ of the class $\mathcal{C}$ has been proposed and investigated by Gupta (1956) and is denoted by R_1 (σ known) and R_2 (σ unknown) defined by (12.2) and (12.5), respectively. The rules R_1 and R_2 were derived by Gupta using likelihood test considerations under the slippage alternative hypotheses. This derivation has been generalized by Nagel (1970) for Koopman–Darmois families and more general hypotheses.

Let $\mathcal{C}'$ denote the class of rules $\{D(\alpha): \alpha = 1, \ldots, k-1\}$, where $D(\alpha)$ is defined by $c_\alpha = 1$ and $c_i = 0$ for $i \neq \alpha$. Assume that $\sigma_1^2 = \cdots = \sigma_k^2 = 1$. In this case, the rule $D(\alpha)$ selects π_i if and only if

$$\bar{X}_i \geqslant \bar{X}_{[\alpha]}^{(i)} - \frac{t(k, P^*, \alpha)}{\sqrt{n}} \tag{12.34}$$

and D_0 sleects π_i if and only if

$$\bar{X}_i \geqslant \frac{1}{k-1} \sum_{j \neq i} \bar{X}_j - \frac{t(k, P^*)}{\sqrt{n}}. \tag{12.35}$$

Deely and Gupta (1968) have compared the expected subset sizes for the rules $D(\alpha)$ for the parametric configurations $\boldsymbol{\mu} \in \Omega(\delta)$ where

$$\Omega(\delta) = \{\boldsymbol{\mu} : \mu_{[1]} \leqslant \cdots \leqslant \mu_{[k-1]} \leqslant \mu_{[k]} - \delta,\ \delta > 0\}. \tag{12.36}$$

They have shown that there exists a δ_0 such that

$$\sup_{\Omega(\delta)} E[S \mid D(k-1)] < 2 \text{ for } \delta > \delta_0. \tag{12.37}$$

For P^* sufficiently large,

$$E[S|D(\alpha)] \geqslant 2 \text{ for } \alpha = 1,\ldots,k-2. \tag{12.38}$$

Thus, for P^* and δ sufficiently large,

$$E[S|D(\alpha)] > E[S|D(k-1)] \tag{12.39}$$

which provides a justification for using $D(k-1)$. Deely and Gupta (1968) have also compared the expected subset sizes of the rules D_0 and $D(k-1)$. Under the slippage configuration $\mu_{[1]} = \cdots = \mu_{[k-1]} = \mu_{[k]} - \delta$, they have shown that $E[S|D(k-1)] < E[S|D_0]$ when δ is sufficiently away from zero. Gupta and Miescke (1978) have discussed optimality of $D(k-1)$ in the Seal's class of procedures in terms of $E(S)$ in the case of three normal populations. Numerical work together with asymptotic results show that for every fixed P^* and $\boldsymbol{\mu}$, $D(k-1)$ is optimal for sufficiently large n.

12.2.7 Restricted Subset Size Procedures

Gupta and Santner (1973) have discussed the application of the results of §11.3 to the problem of selecting the normal population with the largest mean when $\sigma_1^2 = \cdots = \sigma_k^2 = \sigma^2$ (known). Corresponding to $\Omega(p)$ and $\Omega^0(p)$ of §11.3 we have

$$\begin{aligned} \Omega(\delta) &= \{\boldsymbol{\mu} : \mu_{[k]} - \mu_{[k-1]} \geqslant \delta\} \\ \Omega^0(\delta) &= \{\boldsymbol{\mu} : \mu_{[1]} = \cdots = \mu_{[k]} - \delta\}, \end{aligned} \qquad \delta > 0. \tag{12.40}$$

The procedure is

$\boldsymbol{R}_{11}$: Select π_i if and only if

$$\bar{X}_i \geqslant \max\left\{ \bar{X}_{[k-m+1]}, \bar{X}_{[k]} - \frac{d\sigma}{\sqrt{n}} \right\}. \tag{12.41}$$

It should be noted that under this formulation, we require the P^*-condition to be satisfied for $\boldsymbol{\mu} \in \Omega(\delta)$. The admissible range of P^* values are given by $k^{-1} < P^* < P_1(k,n,m,\delta)$, where

$$P_1 = (k-m)\binom{k-1}{k-m} \int_{-\infty}^{\infty} \left[1 - \Phi\left(t - \frac{\sqrt{n}\,\delta}{\sigma}\right)\right] \Phi^{k-m-1}(t)[1-\Phi(t)]^{m-1} d\Phi(t). \tag{12.42}$$

The infimum of $P(CS|R_{10})$ over $\Omega(\delta)$ is attained for $\boldsymbol{\mu} \in \Omega^0(\delta)$. The properties of this procedure follows from the general results of §11.3. As one can see, special choices of m, d, and δ lead to known earlier results of Bechhofer (1954; $m=1$, $\delta>0$), Gupta (1956; $m=k$, $0<d<\infty$, $\delta=0$), and Desu and Sobel (1968; $1<m<k$, $d=\infty$, $\delta>0$). For implementation of the procedure R_{11}, Gupta and Santner (1973) have tabulated the values of $\sqrt{n}\,\delta/\sigma$ for $k=3(1)10(5)20$, $m=2(1)\min(5,k-1)$, and $P^*=0.75$, 0.90, 0.975, using rules with $d=0.4$ and 0.7.

For comparing R_{11} with the fixed-size subset rule of Mahamunulu (1967), we define

$$e(P^*,k,n,d)=\frac{n(d)}{n(\infty)} \tag{12.43}$$

where $n(d)$ is the sample size necessary to achieve the basic probability requirement using rule R_{10} with k, m, and d. This ratio has been tabulated by Gupta and Santner (1973) for $k=3(1)9$, $m=2(1)\min(4,k-1)$, $P^*=0.90$, 0.975, and $d=0.4$, 0.7. For larger d-values, this ratio gets closer to 1 indicating that in many cases a slight additional cost will allow us use of a restricted subset procedure and still meet the same probability requirement.

12.2.8 A Two-Stage Procedure for Selecting Good Populations

Santner (1976b) considered the problem of selecting a subset containing only good populations. Any population π_i is said to be good if $\mu_i \geqslant \mu_{[k]}-\delta^*$, where $\delta^*>0$ is specified. It is assumed that the normal populations have a common unknown variance σ^2. The formulation is the same as that of Desu (1970), Carroll, Gupta, and Huang (1975) and Panchapakesan and Santner (1977) discussed in §11.4. The parameter space is $\Omega=\{(\mu_1,\ldots,\mu_k,\sigma^2)|\sigma^2>0,\ -\infty<\mu_i<\infty,\ i=1,\ldots,k\}$ and it is required that $PCS\geqslant P^*$ for all $(\boldsymbol{\mu},\sigma^2)$. A correct selection is selection of any subset which includes only good populations.

Let $\mathbf{X}'=(X_1,\ldots,X_p)$ have a p-variate normal distribution with zero means, unit variances and covariance matrix Σ. Let Z_m be a chi-square random variable with m degrees of freedom, independent of $\mathbf{X}$. Recall that the distribution of $(T_1,\ldots,T_p)$ where $T_i=\sqrt{m}\,X_i/\sqrt{Z_m}$ is said to be the p-variate t distribution with m degrees of freedom and associated matrix Σ. In particular, let $T_m(\cdot)$ denote the cdf of the $(k-1)$-variate t distribution with m degrees of freedom and associated matrix Σ having common covariance $\frac{1}{2}$.

The steps involved in the procedure $R(N)$ of Santner (1976b) are described as follows.

1. Choose n (arbitrary) observations from each population and compute the pooled-estimate s^2 given in (12.1). The choice of n should be made so that s^2 has a reasonable number of degrees of freedom $\nu(=k(n-1))$.
2. Find $d=d(\nu,k,P^*)$, the solution of $T_\nu(d/\sqrt{2},\dots,d/\sqrt{2})=P^*$.
3. Fix N by *any* stopping rule having the form $N=\max(n,[f(s^2)])$, where $[\cdot]$ denotes the greatest integer function and $f(s^2)$ is Borel measurable. Then take an additional $(N-n)$ observations from each population.
4. Compute the sample means, X_i, $i=1,\dots,k$, based on N observations from each population and select all π_i's having $\bar{X}_i \geqslant \bar{X}_{[k]}-\delta^*+ds/\sqrt{n}$.

Santner (1976b) has shown that $R(N)$ satisfies the probability requirement. He has also given a stopping rule which will guarantee a specified minimum proportion p of good populations present in the selected subset. Let h be chosen so that

$$(k-1)T_\nu\left(\delta^*h-\frac{d}{\sqrt{2}},\dots,\delta^*h-\frac{d}{\sqrt{2}},(\delta^*-\Delta)h-\frac{d}{\sqrt{2}}\right)$$
$$+T_\nu\left((\delta^*+\Delta)h-\frac{d}{\sqrt{2}},\dots,(\delta^*+\Delta)h-\frac{d}{\sqrt{2}}\right)$$
$$=kp.$$

Then the special choice of stopping rule is given by $M=\max\{n,[s^2\max\{2h^2,(d/\delta^*)^2\}]\}+1$ where $[\cdot]$ again denotes the greatest integer function.

12.3 SELECTION FROM NORMAL POPULATIONS IN TERMS OF THE ABSOLUTE VALUES OF THE MEANS

In §2.3 we discussed selection from normal populations in terms of the absolute values of their means. Rizvi (1963, 1971) has also investigated a location-type subset selection procedure.

Case (i). We assume that $\sigma_1^2=\cdots=\sigma_k^2=1$. In this case, Rizvi's procedure R_{12} selects π_i if and only if

$$W_i \geqslant W_{[k]}-d, \tag{12.44}$$

where $W_i=|\bar{X}_i|$ and $d=d(k,n,P^*)>0$ is to be determined to satisfy the P^*-condition. Let $\Omega_1=\{\boldsymbol{\theta}:\boldsymbol{\theta}=(\theta_1,\dots,\theta_k),\ \theta_i=|\mu_i|\geqslant 0\}$. The infimum of

$P(CS|R_{12})$ over Ω_1 occurs when $\theta_{[1]}=\cdots=\theta_{[k]}=\theta$ (say). It is shown that $P_\theta(CS|R_{12})=P(CS|R_{12};\theta_1=\cdots=\theta_k=\theta)$ is a monotonic decreasing function of θ. Thus $\inf_{\Omega_1}P(CS|R_{12})=\lim_{\theta\to\infty}P_\theta(CS|R_{12})$. The constant $d=\sqrt{n}\,\gamma$ is given by

$$\int_{-\infty}^{\infty}\Phi^{k-1}(x+\gamma)\,d\Phi(x)=P^*. \qquad (12.45)$$

The value of γ in (12.45) is same as that of d in (12.4) and so can be obtained from the tables mentioned in the other case. Also

$$\sup_{\Omega_1} E(S|R_{12})=2k\int_0^{\infty}\left[2\Phi(u+\gamma)-1\right]^{k-1}d\Phi(u). \qquad (12.46)$$

One can control $\sup E(S|R_{12})$ over

$$\Omega(\xi)=\{\boldsymbol{\theta}:\theta_{[k]}-\theta_{[k-1]}\geqslant\xi>0\}.$$

It is possible to find the smallest n such that $\sup_{\Omega(\xi)}E(S|R_{12})\leqslant 1+\epsilon$ for any arbitrarily preassigned $\epsilon>0$.

Case (ii). $\sigma_1^2,\ldots,\sigma_k^2$ are known. If we can choose n_i such that $\sigma_i^2/n_i=c$, then it reduces to Case (i).

Case (iii). $\sigma_1^2=\cdots=\sigma_k^2=\sigma^2$ (unknown). In this case the procedure R_{12} is modified as

$\boldsymbol{R'_{12}}$: Select π_i if and only if

$$W_i\geqslant W_{[k]}-sd, \qquad (12.47)$$

where $d=d(k,n,P^*)>0$ satisfying the P^*-condition is given by

$$\int_0^{\infty}\int_{-\infty}^{\infty}\Phi^{k-1}\left(u+\frac{d}{\sqrt{n}}y\right)\varphi(u)q_\nu(y)\,du\,dy=P^* \qquad (12.48)$$

where $q_\nu(y)$ is same as in (12.6). The value of d in selected cases can be obtained from the table of Gupta and Sobel (1957). The problem of selecting the population with $\theta_{[1]}$ can be handled analogously.

12.4 SELECTION FROM NORMAL POPULATIONS IN TERMS OF THE VARIANCES

In ranking normal populations in terms of their variances, we define the best population as the one associated with $\sigma_{[1]}^2$. There is really no need to deal with the cases of known and unknown means separately. We first discuss procedures based on the statistics v_i^2 or s_i^2 defined in (12.1) according as the means are known or unknown. To avoid unnecessary duplications, we use s_i^2 in both cases so that $\nu_i s_i^2/\sigma^2$ has a chi-square distribution with ν_i degrees of freedom where $\nu_1 = n_i$ or $n_i - 1$ depending on the case. We also discuss some results for using sample ranges in the decision rule.

12.4.1 Equal Sample Sizes

In this case, Gupta and Sobel (1962a) investigated the rule R_{13} which selects π_i if and only if

$$s_i^2 \leqslant c^{-1} s_{[1]}^2, \tag{12.49}$$

where $0 < c = c(\nu, k, P^*) \leqslant 1$ is determined so that the basic probability requirement is satisfied. Here $\Omega = \{\boldsymbol{\sigma} : \sigma_i \geqslant 0\}$. The infimum of $P(CS|R_{13})$ occurs when $\sigma_{[1]} = \cdots = \sigma_{[k]}$ and is independent of the common value. Thus we have

$$\inf_{\Omega} P(CS|R_{13}) = \int_0^\infty [1 - G_\nu(cx)]^{k-1} dG_\nu(x), \tag{12.50}$$

where $G_\nu(x)$ is the cdf of a chi-square variable with ν degrees of freedom. It should be noted that this problem falls under the scale parameter case discussed in §11.2.4. It can be easily seen that $\sup_\Omega E(S|R_{13}) = kP^*$.

The values of the constant c so that the right-hand side of (12.50) is equal to P^* are tabulated in a companion paper by Gupta and Sobel (1962b) for $k = 2(1)11$, $\nu = 2(2)50$, and $P^* = 0.75$, 0.90, 0.95, and 0.99. These c-values can also be regarded as percentage points of the smallest of several correlated F statistics. For large ν, we can obtain an approximate value of c using the asymptotic normality of $[(\nu-1)/2]^{1/2}\log[s_i^2/\sigma_i^2]$. This approximate value of c is given by (12.4) by setting $d = -[(\nu-1)/2]^{1/2}\log_e c$.

12.4.2 Unequal Sample Sizes

When the degrees of freedom are $\nu_1, \ldots, \nu_k$, let $\nu_{(i)}$ be the degrees of freedom associated with the population with $\sigma_{[i]}^2$. To find the minimum of $P(CS|R_{13})$ over all possible associations between the sets $\{\nu_i\}$ and $\{\nu_{(i)}\}$, Gupta and Sobel (1962a) considered two special cases, namely, (i) $\nu_{[1]} = \cdots = \nu_{[k-1]} = 2 \leqslant \nu_{[k]}$, and (ii) $k = 2$, $\nu_{[1]} \leqslant \nu_{[2]}$. In these cases, the minimum is attained for $\nu_{(1)} = \nu_{[k]}$. It was conjectured that in general $P(CS|R)$ is decreased by an interchange of $\nu_{(1)}$ and $\nu_{(\alpha)}$ $(\alpha > 1)$ if $\nu_{(1)} < \nu_{(\alpha)}$.

For the case of unequal sample sizes, Gupta and D. Y. Huang (1976a) have obtained two different lower bounds for this minimum. They have shown that

$$P(CS|R_{13}) \geqslant \max\{A,B\}, \tag{12.51}$$

where

$$A=\int_0^\infty \prod_{j=1}^{k-1}\left[1-G_{\nu_{[j]}}\left(\frac{c\nu_{[k]}x}{\nu_{[1]}}\right)\right]dG_{\nu_{[k]}}(x),$$

$$B=\int_0^\infty \prod_{j=2}^{k}\left[1-G_{\nu_{[1]}}\left(\frac{c\nu_{[j]}x}{\nu_{[1]}}\right)\right]dG_{\nu_{[k]}}(x).$$

For some large sample approximations, see Gupta and D. Y. Huang (1976c).

12.4.3 Selecting the t Best Populations

Gupta and Sobel (1962a) also considered selecting a subset containing the t best populations. For this problem, they investigated the rule R_{14} which selects π_i if and only if

$$s_i^2 \leqslant c^{-1}s_{[t]}^2. \tag{12.52}$$

In the general case, they conjectured that the least favorable configuration is of the type $0=\sigma_{[1]}^2=\cdots=\sigma_{[s-1]}^2<\sigma_{[s]}^2=\cdots=\sigma_{[k]}^2$, where $1\leqslant s\leqslant k$, and proved the conjecture in the special case of $k=3$, $t=2$, and a common $\nu=2$.

12.4.4 Procedures Based on Sample Ranges

McDonald (1977a) investigated a class of selection procedures based on the sample ranges. Let R_i be the sample range based on n independent observations from π_i, $i=1,\ldots,n$. To select the population associated with the smallest variance, McDonald proposed the rule R_{15} which selects the population π_i if and only if

$$R_i \leqslant \frac{1}{c}\min_{1\leqslant j\leqslant k} R_j \tag{12.53}$$

where $c=c(k,n,P^*)$ is a constant in the interval $(0,1)$ and is chosen as large as possible subject to the probability requirement. The infimum of the PCS occurs when $\sigma_1=\cdots=\sigma_k$ and is independent of the common value. The constant c required in this case is the solution of the equation

$$\int_0^\infty [1-G(cx)]^{k-1}dG(x)=P^* \tag{12.54}$$

where

$$G(x)=n\int_{-\infty}^{\infty}\Phi(y)[\Phi(y+x)-\Phi(y)]^{n-1}dy.$$

McDonald (1977a) has tabulated the c-values for $k=2(1)10$, $n=2(1)10(5)25$, and $P^*=0.75, 0.90, 0.95, 0.975, 0.99$. The values corresponding to $n\geqslant 10$ are approximately determined through Monte Carlo simulations.

As regards the expected subset size, it has been shown that its maximum value is kP^* in the particular cases of $n=2$ and 3. McDonald (1977a) has made Monte Carlo comparisons of his procedure R_{15} with the procedure R_1 (defined in (12.49)) of Gupta and Sobel (1962a) in terms of $E(S)/\text{PCS}$ choosing a slippage and a geometric configuration of the parameters. His studies indicate that both procedures are roughly equivalent for small values of n and k ($n\leqslant 15$ and $k\leqslant 10$) and that R_1 is definitely better otherwise.

12.5 SELECTION FROM NORMAL POPULATIONS IN TERMS OF THE COVERAGES OF A FIXED INTERVAL

Let π_i be a population defined over the sample space $\mathfrak{X}$. Let $(\mathfrak{X},\mathfrak{B},P_i)$ be the probability space associated with π_i. For any set $A\in\mathfrak{B}$, we define

$$\lambda_i=\int_A dP_i. \tag{12.55}$$

The quantity λ_i is the coverage of the set A under the probability measure P_i. Guttman (1961) and Guttman and Milton (1969) have considered selection from normal populations using λ_i as the ranking parameter and the interval $A=(-\infty,a)$ where a is known. In this case we see that

$$\lambda_i=\Phi\left(\frac{a-\mu_i}{\sigma_i}\right). \tag{12.56}$$

It is clear that ranking the populations according to the values of λ_i is equivalent to ranking them according to the values of $(a-\mu_i)/\sigma_i$. If we let $\delta_i=(\mu_i-a)/\sigma_i$, the best population is defined as the one associated with $\delta_{[1]}$. Let us examine several possible cases.

12.5.1 Unknown Means and Common Known Variance

In this case, the best population is the one with the smallest mean and the procedure D_1 is the analog of Gupta's procedure R_1 defined by (12.2). The procedure D_1 selects π_i if and only if

$$\bar{X}_i \leqslant \bar{X}_{[1]} + \frac{\delta_1^* \sigma}{\sqrt{n}} \tag{12.57}$$

where the constant $\delta_1^* = \delta_1^*(n,k,P^*) > 0$ satisfying the P^*-condition is given by

$$\int_{-\infty}^{\infty} [1 - \Phi(x - \delta_1^*)]^{k-1} d\Phi(x) = P^*. \tag{12.58}$$

The results relating to $\inf P(CS|D_1)$ and $\sup E(S|D_1)$ are obtained in a manner quite analogous to the case of R_1 and therefore are omitted. Because of the symmetry of the standard normal distribution, it should be noted that the constant δ_1^* satisfying (12.58) is equal to d defined by (12.4). Thus the constant can be obtained from the tables mentioned earlier.

12.5.2 Unknown Means and Known Variances

Let $Z_i = (\bar{X}_i - a)/\sigma_i$. Then the coverage of $(-\infty, a)$ under π_i is the same as the coverage of $(-\infty, 0)$ under the distribution of Z_i. The Z_i have normal distributions with unit variance. Thus the problem reduces to the case discussed in §12.5.1.

12.5.3 Unknown Means and Common Unknown Variance

The problem in this case is equivalent to selecting the population with the smallest mean; thus we can use the analog of the procedure R_2 of Gupta (1956) defined by (12.5). The analog will be

D_2: Select π_i if and only if

$$\bar{X}_i \leqslant \bar{X}_{[1]} + \frac{\delta_2^* s}{\sqrt{n}}, \tag{12.59}$$

where $\delta_2^* = \delta_2^*(n,k,P^*)$ satisfying the P^*-condition is same as d_2 of (12.6).

12.5.4 Common Known Mean and Unknown Variances

In this case, selecting the population with the largest coverage is equivalent to selecting the population with $\sigma_{[k]}^2$ or $\sigma_{[1]}^2$ according as $\mu > a$ or $\mu < a$. Suppose that $\mu < a$. Then the procedure R_{13} of Gupta and Sobel (1962a) defined by (12.49) is applicable. The values of the constant c are given by Gupta and Sobel (1962b). Guttman and Milton (1969) have given the values of c^{-1} for $k = 2(1)4$, $\nu = 2(2)60(5)100$, and $P^* = 0.75, 0.90, 0.95, 0.99$.

If $\mu > a$, we can define the analogous procedure

D_3: Select π_i if and only if

$$v_i^2 \geqslant \delta_3^* v_{[k]}^2, \tag{12.60}$$

where v_i^2 is defined in (12.1) and $0<\delta_3^*=\delta_3^*(n,k,P^*)<1$ is chosen subject to the P^*-condition. The constant δ_3^* is given by

$$\int_0^\infty G_\nu^{k-1}\left(\frac{y}{\delta_3^*}\right)dG_\nu(y)=P^*, \tag{12.61}$$

where $\nu=n$ and $G_\nu(y)$ is as in (12.50). The values of δ_3^* are given by Gupta (1963) for $k=2(1)11$, $\nu=2(2)50$, and $P^*=0.75$, 0.90, 0.95, 0.99. These values are obtained by Gupta (1963) for his procedure defined for selection from gamma populations. This is discussed in the next section. The values of δ_3^* are also tabulated by Guttman and Milton (1969) for $k=2(1)4$, $\nu=2(2)60(5)100$, and $P^*=0.75$, 0.90, 0.95, 0.99.

12.5.5 Known Means and Unknown Variances

Suppose that $\mu_i>a$ for all $i=1,\ldots,k$. The population with the largest coverage corresponds to the population with $\delta_{[k]}$, where $\delta_i=\sigma_i/(\mu_i-a)$. We define the procedure D_4 which selects π_i if and only if

$$U_i^2 \geqslant \delta_4^* U_{[k]}^2, \tag{12.62}$$

where $U_i=v_i/(\mu_i-a)$ and $0<\delta_4^*=\delta_4^*(n,k,P^*)<1$. It is easy to see that the constant δ_4^* is same as δ_3^* of the procedure D_3 defined by (12.60).

Remark 12.1. Guttman (1961) proposed a procedure that selects π_i if and only if $U_i \geqslant d' U_{[k]}$. His evaluation of the probability of a correct selection can be handled more easily by writing the procedure in terms of U_i^2. Also Guttman and Milton (1969) proposed the procedure (procedure 5.1 in their paper) which selects π_i if and only if $Q_i^2 \geqslant dQ_{[k]}^2$, where $Q_i^2=v_i^2/(\mu_i-a)$. The problem of finding the infimum of the probability of a correct selection for their procedure is complicated and is not established. The difficulty arises in associating the σ_i with the ordered δ_i.

If $\mu_i<a$ for all $i=1,\ldots,k$, then we are selecting the population with the smallest $\Delta_i=\sigma_i/(a-\mu_i)$. We can define an analogous procedure in terms of $U_i^2=[v_i/(a-\mu_i)]^2$.

If some μ_i are less than a and some are greater than a, it is easy to see that we can eliminate the populations with $\mu_i>a$, thereby reducing the problem to the case of all means less than a with a reduced set of populations. □

12.5.6 Unknown Common Mean and Unknown Variances

The best population is the one associated with $\sigma_{[1]}$ or $\sigma_{[k]}$ depending on whether $\mu<a$ or $\mu>a$. Since μ is unknown, Guttman and Milton (1969) proposed a procedure D_5 which first estimates μ and uses the information to decide the type of selection rule. Let $\bar{\bar{X}}=k^{-1}\Sigma\bar{X}_i$, where the means are based on a common sample size.

***D*₅**: If $\bar{\bar{X}}>a$, select π_i if and only if

$$s_i^2 \geqslant \delta_{5,1}^* s_{[k]}^2 \tag{12.63}$$

and if $\bar{\bar{X}}<a$, select π_i if and only if

$$s_i^2 \leqslant \delta_{5,2}^* s_{[1]}^2 \tag{12.64}$$

where $\delta_{5,1}^* \in (0,1)$ and $\delta_{5,2}^* \geqslant 1$.

It is shown by Guttman and Milton that the P^*-condition is satisfied by taking $\delta_{5,1}^*$ and $\delta_{5,2}^*$ to be the constants δ_3^* of the procedure D_3 defined by (12.60) and the reciprocal of the constant c of the procedure R_{13} defined by (12.49), respectively. The degrees of freedom ν must now be taken as $(n-1)$.

REMARK 12.2. Even though Guttman (1961) and Guttman and Milton (1969) considered the coverage of the interval $(-\infty,a)$ as the ranking measure, in several special cases discussed the selection problem is equivalent to selection in terms of means or variances. These procedures have been proposed earlier and more thoroughly investigated by Gupta (1956, 1963, 1965) and Gupta and Sobel (1962a, 1962b). Further the expression for the probability of a correct selection obtained by Guttman (1961) [(2.11) in his paper] for his procedure (2.10) [which is the procedure R_2 of Gupta (1956) defined by (12.5)] is incorrect because the Student -t random variables involved are wrongly assumed to be independent. □

12.6 SELECTION FROM GAMMA POPULATIONS IN TERMS OF THE SCALE PARAMETERS

Let π_i $(i=1,\ldots,k)$ be a population with density

$$f(x,\theta_i)=\begin{cases}\{\Gamma(r)\}^{-1}\theta_i^{-r}\exp(-x/\theta_i)x^{r-1}, & x>0,\ \theta_i>0,\\ 0 & \text{otherwise.}\end{cases} \tag{12.65}$$

As we can see, it is assumed that the populations have the same shape parameter r (>0). Further, r is assumed to be known. The problem of selecting the population with the largest (smallest) θ_i falls under the scale parameter case discussed in §11.2.4. Gupta (1963) investigated the procedure R_{16} which selects π_i if and only if

$$\bar{X}_i \geqslant b\bar{X}_{[k]}, \tag{12.66}$$

where $\bar{X}_i$ is the sample mean based on n_i observations from π_i and $0<b=b(k,P^*,n_1,\ldots,n_k)<1$ is to be determined subject to the P^*-condition. The sample mean $\bar{X}_i$ has a gamma distribution with scale parameter θ_i/n_i and degrees of freedom $\nu_i/2=n_i r$. We have

$$P(CS|R_{16})=\int_0^\infty \prod_{j=1}^{k-1} G_{\nu_{(j)}}\left(\frac{x\nu_{(j)}\theta_{[k]}}{b\nu_{(k)}\theta_{[j]}}\right) dG_{\nu_{(k)}}(x), \tag{12.67}$$

where $\nu_{(i)}$ is the (unknown) ν_i associated with the population having the largest θ_i and $G_\nu(x)$ is the cdf of a standardized gamma random variable (i.e., with $\theta=1$) with parameter $\nu/2$.

Gupta (1963) obtained the infimum of $P(CS|R_{16})$ over $\Omega=\{\boldsymbol{\theta}:\boldsymbol{\theta}=(\theta_1,\ldots,\theta_k),\ \theta_i>0\}$ for the case of equal sample size n. The infimum in this case occurs when $\theta_1=\cdots=\theta_k$ and the constant b is given by

$$\int_0^\infty G_\nu^{k-1}\left(\frac{x}{b}\right) dG_\nu(x)=P^*, \tag{12.68}$$

where $\nu=2nr$. Thus the constant b depends on n and r through ν. We have referred to the solution of b in §12.5.4 and have given the ranges of k, P^*, and ν for which the values of b are tabulated by Gupta (1963). The properties of the procedure R_1 follow from our general discussion of the scale parameter case in §11.2.4.

For the case of unequal sample sizes, Gupta and D. Y. Huang (1976a) have obtained a lower bound for the minimum of $\inf_\Omega P(CS|R_{16})$ over all possible associations between the sets $\{\nu_i\}$ and $\{\nu_{(i)}\}$. They have shown that

$$P(CS|R_{16}) \geqslant \max\left\{ \int_0^\infty \prod_{j=2}^{k} G_{[j]}(b^{-1}\nu'_{[1]}x)\, dG_{[1]}(x),\right.$$

$$\left.\int_0^\infty \prod_{j=1}^{k-1} G_{[k]}(b^{-1}\nu'_{[j]}x)\, dG_{[1]}(x)\right\}, \tag{12.69}$$

where $G_{[j]}(x)=G_{\nu_{[j]}}(x)$ and $\nu'_{[j]}=\nu_{[j]}/\nu_{[k]}$.

To select a subset containing the smallest θ_i, Seal (1958a) investigated a rule which is the scale-type analog of his procedure discussed in §12.2.6. Let $Y_1,\ldots,Y_k$ be a set of observations from the k given populations and $Y_{[j]}^{(i)}$ denote the jth smallest observation in the reduced set obtained by eliminating Y_i. Then the procedure R_{17} of Seal (1958a) selects π_i if and only if

$$Y_i \leqslant q \sum_{j=1}^{k-1} c_j Y_{[j]}^{(i)}, \tag{12.70}$$

where $c_1,\ldots,c_{k-1}$ are nonnegative constants such that $\Sigma_{j=1}^{k-1} c_j = 1$ and $q = q(k, P^*, c_1,\ldots,c_{k-1}) > 1$ is the upper 100 $(1-P^*)$ percentage point of the distribution of $U_1/\Sigma_{j=1}^{k-1} c_j U_{[j]}^{(1)}$, where $U_1,\ldots,U_k$ independent and U_i has the gamma density (12.65) with $\theta_i = 1$. As in the case of ranking normal means, the optimum choice of the c_i is inferred to be $c_1 = 1$ and $c_2 = \cdots = c_k = 0$ based on some special cases and approximations.

12.7 SELECTION FROM EXPONENTIAL POPULATIONS IN TERMS OF THE COVERAGES OF A FIXED INTERVAL

Let π_i $(i = 1,\ldots,k)$ be an exponential population with density

$$f(x; \mu_i, \sigma_i) = \begin{cases} \sigma_i^{-1} \exp\{-(x-\mu_i)\sigma_i^{-1}\}, & x \geqslant \mu_i,\ \sigma_i > 0, \\ 0 & \text{otherwise.} \end{cases} \tag{12.71}$$

Guttman (1961) has considered selection in terms of the coverages of the interval $A = (-\infty, a)$ under these distributions where a is a known constant. The coverage λ_i under π_i is given by

$$\lambda_i = 1 - \exp\{-(a-\mu_i)\sigma_i^{-1}\}. \tag{12.72}$$

Thus selecting the population with the largest λ_i is equivalent to selecting the population with the largest $\delta_i = (a-\mu_i)\sigma_i^{-1}$. As in the normal case discussed in §12.5, here also several cases arise. Let X_{ij}, $j = 1,\ldots,n$, be independent observations from π_i and let $X_{i[1]} \leqslant \cdots \leqslant X_{i[n]}$ denote the ordered observations from π_i $(i = 1,\ldots,k)$. Define

$$Y_{ij} = X_{ij} - u_i,$$

$$\bar{Y}_i = n^{-1} \sum_{j=1}^{n} X_{ij},$$

$$T_i = \min_{1 \leqslant j \leqslant n} X_{ij},$$

$$D_i=(n-1)^{-1}\sum_{j=2}^{n}(X_{i[j]}-X_{i[1]}), \tag{12.73}$$

$$D=k^{-1}\sum_{i=1}^{k}D_i,$$

$$Z_i=(T_i-a)\sigma_i^{-1},$$

$$W_i=D_i(a-\mu_i)^{-1}.$$

Several essential cases and the procedures investigated are listed in the following table.

	Case	Select π_i if and only if
(i)	$\mu_i=\mu$(known), σ_i unknown	$\bar{Y}_i\leqslant c_1\bar{Y}_{[1]}$
(ii)	$\sigma_i=\sigma$(known), μ_i unknown	$T_i\leqslant T_{[1]}+d_1$
(iii)	μ_i unknown, σ_i known	$Z_i\leqslant Z_{[1]}+d_2$
(iv)	All $\mu_i<a$ are known, σ_i unknown	$W_i\leqslant c_2W_{[1]}$
(v)	μ_i and σ_i both unknown	$T_i\leqslant T_{[1]}+d_3D$

The constants $d_i(>0)$ and $c_i(>1)$ are determined subject to the P^*-condition.

12.8 SELECTION FROM UNIFORM POPULATIONS

Barr and Rizvi (1966b) have discussed selection from a set of k uniform populations, where π_i is uniform over $(0,\theta_i)$. The best population is the one associated with $\theta_{[k]}$ (or $\theta_{[1]}$). Their procedure is based on the $X_{i[n]}$, where $X_{i[n]}=\max(X_{i1},\ldots,X_{in})$. They have extended their procedure to a class of nonregular distributions with density,

$$g(x,\theta)=\begin{cases} M(x)Q(\theta), & a(\theta)<x<b(\theta),\\ 0 & \text{otherwise.}\end{cases} \tag{12.74}$$

In (12.74) it is assumed that $a(\theta)$ and $b(\theta)$ are continuously differentiable and that $\sup a(\theta)=\inf b(\theta)$. Further it is assumed that, of the two functions $a(\theta)$ and $b(\theta)$, *either* one is a constant function and the other is strictly increasing in θ, *or* one is strictly increasing and the other strictly decreasing.

McDonald (1976) has considered uniform populations where π_i is uniform over the finite interval (a_i,b_i). Let $h_i=b_i-a_i$. The goal is to select the population associated with $h_{[1]}$. The procedure R of McDonald (1976) selects π_i if and only if

$$Y_i \leqslant c^{-1} Y_{[1]}, \tag{12.75}$$

where Y_i is the range of the sample from π_i and $0 < c = c(k, n, P^*) < 1$ is chosen to satisfy the P^*-condition. The infimum of $P(CS|R)$ is attained when the h_i are equal and the constant c is given by

$$\int_0^1 [1 - F(cx)]^{k-1} dF(x) = P^*, \tag{12.76}$$

where the density $f(x)$ corresponding to the cdf $F(x)$ is given by

$$f(x) = \begin{cases} n(n-1)x^{n-2}(1-x), & 0 \leqslant x \leqslant 1, \\ 0 & \text{elsewhere.} \end{cases} \tag{12.77}$$

12.9 NOTES AND REMARKS

Among other distributions, double exponential is of interest in reliability studies. Subset selection procedures for double exponential populations have been studied by Gupta and Leong (1977). W. T. Huang (1974) deals with Pareto populations. Subset selection procedures for discrete distributions are discussed in Chapter 13. The procedures relating to multivariate normal populations are given in Chapter 14. Chapter 17 deals with sequential procedures.

CHAPTER 13

Selection from Discrete Populations

13.1 INTRODUCTION

Discrete distributions provide theoretical models for several chance mechanisms in a wide variety of disciplines such as biological and medical sciences, social sciences, physical sciences, and engineering. Some examples of how different distributions arise were given in Chapter 4.

The initial investigations of subset selection procedures for discrete distributions were naturally focused on binomial populations. The investigations of Gupta and Sobel (1960) were inconclusive in the sense that the infimum of the probability of a correct selection was shown to occur when the success probabilities are equal but no definite result was obtained, except in the case of $k=2$ populations, about the common value of the success probabilities for which the infimum is attained.

As an alternative, Gupta and Nagel (1971) and Gupta, Huang, and Huang (1976) have investigated conditional procedures where the conditioning is on the number of successes obtained in a fixed number of trials. These and other procedures for binomial populations are discussed in this chapter. We are also concerned with selection procedures for Poisson, negative binomial, hypergeometric, and Fisher's logarithmic distributions. Although one could discuss selection of the best multinomial cell under multivariate distributions, we discuss these procedures also in this chapter.

13.2 SELECTION FROM BINOMIAL POPULATIONS

13.2.1 Selection in Terms of Success Probabilities: Fixed Sample Size Rules

Let θ_i be the probability of a success on a single trial associated with π_i $(i=1,\ldots,k)$. Let n_i be the number of trials from π_i and X_i be the number of successes obtained in these trials. In the case of $n_1=\cdots=n_k=n$, for selecting the population associated with $\theta_{[k]}$, Gupta and Sobel (1960) proposed the rule R_1 which selects π_i if and only if

$$X_i \geqslant X_{[k]}-d, \tag{13.1}$$

where $d=d(n,k,P^*)$ is the smallest nonnegative integer that satisfies the P^*-condition. It is shown by them that $P(CS|R_1)$ is minimized when $\theta_1=\cdots=\theta_k$. Thus the integer d is the smallest nonnegative integer for which

$$\inf_{0<\theta<1} \sum_{\alpha=0}^{n} b(\alpha;n,\theta)\left[B(\alpha+d;n,\theta)\right]^{k-1} \geqslant P^*, \tag{13.2}$$

where

$$b(\alpha;n,\theta)=\binom{n}{\alpha}\theta^{\alpha}(1-\theta)^{n-\alpha},$$
$$B(y;n,\theta)=\sum_{\alpha=0}^{y} b(\alpha;n,\theta). \tag{13.3}$$

In the case of $k=2$ populations, Gupta and Sobel (1960) have shown that the infimum on the left-hand side of (13.2) is attained when $\theta=\frac{1}{2}$. For $k>2$, the common value θ_0 at which the infimum takes place is not known. However, it is shown that this value $\theta_0 \to \frac{1}{2}$ as $n\to\infty$.

For large n, using the normal approximation, the approximate solution $\hat{d}$ is given by

$$\inf_{0<\theta<1}\int_{-\infty}^{\infty}\Phi^{k-1}\left(x+\frac{\hat{d}+0.5}{\sqrt{n\theta(1-\theta)}}\right)d\Phi(x)=P^*. \tag{13.4}$$

The solution $\hat{d}$ need not be an integer. The infimum on the left-hand side of (13.4) occurs when $\theta=\frac{1}{2}$. Thus we get

$$\hat{d}=\frac{(u\sqrt{n}\,-1)}{2}, \tag{13.5}$$

where

$$\int_{-\infty}^{\infty}\Phi^{k-1}(x+u)\,d\Phi(x)=P^*. \tag{13.6}$$

To implement the procedure, we replace d in (13.1) by d', the smallest integer equal to or greater than $\hat{d}$. It should be noted that (13.6) is same as (12.4) and the tables providing the u-values have been described in §12.2.1. The values of d (or d') have been tabulated by Gupta and Sobel (1960) for $k=2(1)20(5)50$, $n=1(1)20(5)50(10)100(25)200(50)500$ and $P^*=0.75$, 0.90, 0.95, 0.99. The values are exact for small values of n. In all cases where both exact and approximate values were computed, it was found that d' was conservative (i.e., $d'\geqslant d$) and never exceeded the exact value by more than unity (i.e., $d'\leqslant d+1$). Gupta and Sobel have also tabulated the expected proportion of the populations retained in the selected subset, $E(S)/k$, under the slippage configuration $\theta_{[1]}=\cdots=\theta_{[k-1]}=\theta=\theta_{[k]}-\delta$ for $\delta=0$, 0.10, 0.25, 0.50; $\theta+\delta=0.50$, 0.75, 0.95, 1.00; $k=2$, 3, 5, 10; $n=5(5)25$; and $P^*=0.75$, 0.90, 0.95. For any fixed n, k, d, and δ, $E(S)$ is the same for $\theta=0$ and $\theta=1-\delta$.

Gupta, Huang, and Huang (1976) have proved the following theorem which gives a conservative value of d for the procedure R_1.

Theorem 13.1. For given P^* and any integer $t(0 \leqslant t \leqslant kn)$, let $d(t)$ be the smallest value such that

$$N(k;d(t),t,n) \geqslant P^*\binom{kn}{t}, \tag{13.7}$$

where $N(k;d(t),t,n)=\Sigma\binom{n}{s_1}\cdots\binom{n}{s_k}$, the summation being over the set of all nonnegative integers s_i such that $\Sigma_1^k s_i = t$ and $s_k \geqslant \max_{1 \leqslant j \leqslant k-1} s_j - d(t)$ for some constant $d(t)$. If $d = \max_{0 \leqslant t \leqslant kn} d(t)$, then

$$\inf_{\Omega} P(CS|R_1) \geqslant P^*. \tag{13.8}$$

The smallest value $d(t)$ satisfying (13.7) has been tabulated by Gupta, Huang, and Huang for $k=2$, 4(1)10, $n=1(1)10$, $t=1(1)20$, and $P^*=0.75$, 0.90, 0.95, 0.99. They have also tabulated the d-values satisfying (13.8) for $P^*=0.75$, 0.90, 0.95, 0.99, and $n=1(1)4$ when $k=3(1)15$, and $n=5(1)10$ when $k=3(1)5$.

When the n_i are not all equal, the procedure R_1 is modified into R_2 which selects π_i if and only if

$$Y_i \geqslant Y_{[k]} - c, \tag{13.9}$$

where $Y_i = X_i/n_i$ and $c = c(k,P^*,n_1,\ldots,n_k) > 0$ is the smallest nonnegative number for which

$$\min_{i=1,\ldots,k}\left\{\inf_{0 \leqslant \theta \leqslant 1} P(CS;\theta,c,n_i)\right\} \geqslant P^*, \tag{13.10}$$

where $P(CS;\,\theta,\,c,\,n_i)$ is the probability of including π_i in the selected subset when $\theta_1 = \cdots = \theta_k$.

For large n, the infimum in (13.10) is attained when $\theta = \frac{1}{2}$. For $k=2$,

$$P(CS;\theta,c,n_i) \approx \Phi\left(c\left\{\theta(1-\theta)\left(\frac{1}{n_1}+\frac{1}{n_2}\right)\right\}^{-1/2}\right), \tag{13.11}$$

which is independent of i. In other words, if n_1 and n_2 are large enough, then (13.11) shows that both possible assignments of sample sizes give approximately the same probability of a correct selection and that the common worst value of θ will be close to $\frac{1}{2}$. Actually, in this case, Gupta and Sobel (1960) have shown a stronger result which states that we need consider only one of the two possible assignments for any n_1 and n_2. They have shown that for any n_1, n_2, c, and θ.

$$P(CS;\theta,c,n_1)=P(CS;1-\theta,c,n_2). \tag{13.12}$$

PROCEDURES BASED ON TRANSFORMATION OF VARIABLES. One can use the arc sine transformation and define a procedure R_3 which selects π_i if and only if

$$Z_i \geqslant Z_{[k]}-b, \tag{13.13}$$

where $Z_i=2 \text{ arc } \sin\sqrt{X_i/n_i}$ and $b=b(k,P^*,n_1,\ldots,n_k)>0$ is to be determined. For $n_1=\cdots=n_k=n$ large, using the asymptotic normal distribution of Z_i, we have

$$\int_{-\infty}^{\infty}\Phi^{k-1}(x+b\sqrt{n}\,)\,d\Phi(x)=P^* \tag{13.14}$$

and b can be obtained from the tables described in §12.2.1. The arc sine transformation can be modified to improve the approach to normality by replacing Z_i with

$$Z_i'=\arcsin\sqrt{\frac{X_i}{n_i+1}}+\arcsin\sqrt{\frac{X_i+1}{n_i+1}}\;. \tag{13.15}$$

13.2.2 Selection in Terms of Success Probabilities: Conditional Rules

Gupta and Nagel (1971) and Gupta, Huang, and Huang (1976) have investigated in the case of equal sample sizes procedures for which the constants defined depend upon $T=\Sigma_{i=1}^{k}X_i$. Since the procedure of Gupta, Huang, and Huang is of the same type as that of R_1 we discuss their rule first. Their rule R_4 selects π_i if and only if

$$X_i \geqslant X_{[k]}-D(t), \qquad \text{given } T=t. \tag{13.16}$$

R_4 actually denotes a class of procedures whose members correspond to the possible values of T. Exact result about $\inf_{\Omega}P(CS|R_4)$ is available only in the case of $k=2$. In this case, it has been shown by Gupta, Huang, and Huang that the infimum is attained when $\theta_1=\theta_2=\theta$ (say) and that this infimum is independent of the common value θ. Their results for $k=2$ and $k>2$ are given in the following theorem.

Theorem 13.2. Given $P^*(k^{-1}<P^*<1)$, let $d(r)$ be the smallest value such that

$$N(2;d(r),r,n) \geqslant \begin{cases} P^*\binom{2n}{r}, & k=2, \\ P_1^*\binom{2n}{r}, & k>2, \end{cases} \tag{13.17}$$

where $P_1^* = 1-(1-P^*)(k-1)^{-1}$, and $r=0,1,\ldots,\min(t,2n)$. Then, for the procedure R_4 using

$$D(t) = \begin{cases} d(t), & k=2, \\ \max\{d(r): r=0,1,\ldots,\min(t,2n)\}, & k>2, \end{cases} \tag{13.18}$$

we have

$$\inf_{\Omega} P(CS|R_4) \geqslant P^* \tag{13.19}$$

Gupta, Huang, and Huang have also obtained upper bounds for $\sup_\Omega E(S|R_4)$ when $k=2$ and $k>2$. It should be noted that for the procedures discussed so far no result is available for the exact evaluation of the infimum of the probability of a correct selection except in the case of $k=2$.

Gupta and Nagel (1971) have defined a rule R_5 which is just in the sense of Definition 11.4 and for which $P(CS|R_5)=P^*$ for all $\boldsymbol{\theta} \in \Omega_0$ where Ω_0 is the set of equal-value parametric configurations. This cannot be achieved with a nonrandomized rule, because when $\boldsymbol{\theta}=(0,\ldots,0)$ or $\boldsymbol{\theta}=(1,\ldots,1)$, the observations will be $\mathbf{x}=(0,\ldots,0)$ or $\mathbf{x}=(n,\ldots,n)$ with probability 1, thus requiring the use of individual selection probabilities $p_i(\mathbf{x})=P^*$. Since we are interested in symmetric rules it is sufficient to know one of the individual selection probabilities, say, p_k. From Lemma 11.1, it follows that

$$E(p_k(\mathbf{X})|T)=P^* \qquad \text{for } T=0,1,\ldots,kn. \tag{13.20}$$

The requirement that R_5 be just leads to

$$\left.\begin{matrix} y_i \leqslant x_i, & i=1,\ldots,k-1 \\ y_k \geqslant x_k & \end{matrix}\right\} \Rightarrow p_k(\mathbf{x}) \leqslant p_k(\mathbf{y}). \tag{13.21}$$

Figure 13.1 shows the partial ordering induced by (13.21) among the observation vectors for the case $k=3$, $n=2$. The individual selection probability $p_3(x_1,x_2,x_3)$ defines a just rule if its values are nondecreasing in

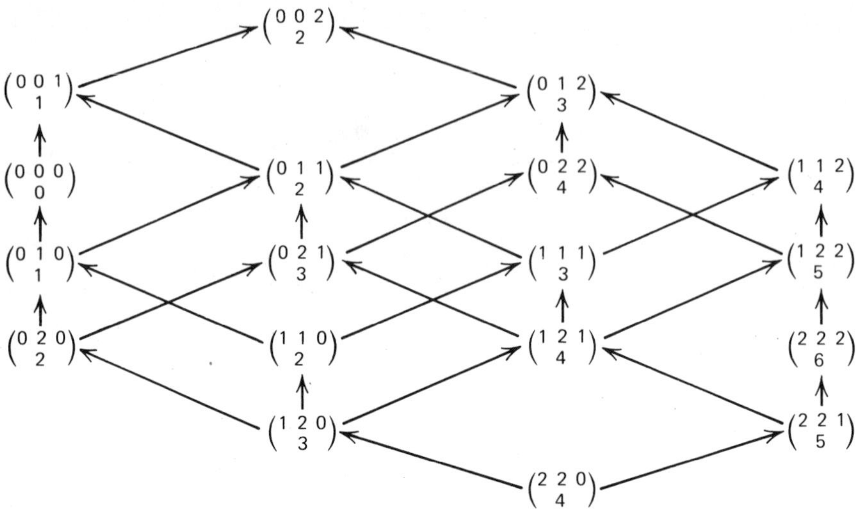

Figure 13-1. Partial ordering for binomial observations. $k=3$, $n=2$.

the direction of the arrows. Because of symmetry only one of the two permutations (x_1, x_2, x_3) and (x_2, x_1, x_3) is plotted. The numbers underneath the observation vectors denote the corresponding T-values.

The conditions (13.20) and (13.21) do not determine a rule uniquely. Gupta and Nagel (1971) have proposed the following rule R_5:

$$p_k(\mathbf{x}) = \begin{cases} 1 & \text{if } x_k > c_T, \\ \rho & \text{if } x_k = c_T, \\ 0 & \text{if } x_k < c_T, \end{cases} \tag{13.22}$$

where $\rho = \rho(T; P^*, k)$ asnd $c_T = c_T(P^*, k)$ are determined to satisfy

$$E(p_k(\mathbf{X})|T) = P^*. \tag{13.23}$$

The conditional distribution of X_k given T is hypergeometric. Let Z_T have the hypergeometric distribution for which

$$\Pr(Z_T = i) = \frac{\binom{n}{i}\binom{(k-1)n}{T-i}}{\binom{kn}{T}}.$$

Then the constant c_T is the smallest integer such that $\Pr(Z_T > c_T) \leqslant P^*$ and

$\Pr(Z_T \geqslant c_T) > P^*$. From (13.23) we get

$$\rho = \frac{P^* - \Pr(Z_T > c_T)}{\Pr(Z_T = c_T)}. \tag{13.24}$$

It has been established by Gupta and Nagel (1971) that the rule R_5 is just and they have tabulated the values of c_T and ρ for $k=2, 3, 5$; $n=5, 10$; and $P^*=0.75, 0.90, 0.95, 0.99$, with T in each case going from 0 to nk.

Since T takes on the values $0, 1, \ldots, kn$, these tables become very extensive for large values of k and n. Therefore it is desirable to find approximations for c_T and ρ. The normal approximation for the hypergeometric distribution gives good results when n is large and T is not extreme (close to 0 or kn). The expectation and the variance of Z_T are $\mu = T/k$, and $\sigma^2 = T(kn-T)(k-1)/(kn-1)k^2$, respectively. Using the fact that Z_T is asymptotically $N(\mu, \sigma^2)$ distributed, we obtain the approximate values $\hat{c}_T$ and $\hat{\rho}$ given by

$$\hat{c}_T = \left[0.5 + \mu - \sigma\Phi^{-1}(P^*)\right],$$

$$\hat{\rho} = \hat{c}_T + 0.5 - (\mu - \sigma\Phi^{-1}(P^*)), \tag{13.25}$$

where $[s]$ denotes the integral part of s. The exact and approximate values of c_T and ρ have been compared by Gupta and Nagel (1971) for $k=2, 3, 5, 10$; $n=5, 10, 20$; and some selected values of T and P^*. Their results show no change in the values of c_T and $\hat{c}_T$, and only small deviations in those of ρ and $\hat{\rho}$.

The nonrandomized version R_5' of R_5 selects π_i if and only if $x_i \geqslant c_T$. This procedure is conservative in the sense of meeting the P^*-requirement. However, R_5' may not be just and it selects large subsets if the θ_i are close to 0 or 1. A comparison of the procedures R_1 and R_5 is difficult because $\inf_\Omega P(CS|R_1)$ is not known. Since it takes place near $\theta = \frac{1}{2}$, the P^*-value for R_5 has been chosen by Gupta and Nagel (1971) to satisfy $P(CS|R_5) = P^*$ for $\boldsymbol{\theta} = (\frac{1}{2}, \ldots, \frac{1}{2})$ which makes the comparison slightly more favorable to R_1. Under the slippage configuration $(\theta, \ldots, \theta, \theta + \delta)$, the numerical computations show that R_5 yields better results for small values of δ, whereas R_1 is better for large δ. It should, however, be noted that R_5, unlike R_1, may select an empty subset.

13.2.3 Selection in Terms of Success Probabilities: Inverse Sampling Rules

In Chapter 4 we discussed procedures for selecting from two or more binomial populations under the indifference zone formulation using vector-at-a-time (VT) and play-the-winner (PW) sampling and stopping rules based on inverse sampling. In this subsection we discuss subset selection rules for these cases.

Stated in terms of treatments and clinical trials, Barron and Mignogna (1977) considered the case of $k=2$ treatments. They have defined two procedures: $R_{I,\mathrm{VT}}$ based on VT sampling rule and $R_{I,\mathrm{PW}}$ based on PW sampling rule. For $R_{I,\mathrm{VT}}$, sampling is terminated whenever there are r successes or c failures for any one of the treatments. For $R_{I,\mathrm{PW}}$, the sampling is terminated when r successes are yielded by one of the treatments or when c failures are yielded by both. Let m be the stopping time for either procedure and S_{im} be the number of successes resulting from π_i, $i=1,2$. The procedures $R_{I,\mathrm{VT}}$ and $R_{I,\mathrm{PW}}$ of Barron and Mignogna (1977) are the following.

$\boldsymbol{R_{I,\mathrm{VT}}}$: Select π_i if and only if

$$S_{im} \geqslant \max(S_{1m}, S_{2m}) - d \tag{13.26}$$

where $d \geqslant 0$, $d=d_1$ if the sampling is terminated with r successes on one treatment and less than c failures on the other, and $d=d_2$ if the termination occurs with c failures on one treatment and less than r successes on the other.

$\boldsymbol{R_{I,\mathrm{PW}}}$: Select π_i if and only if

$$S_{im} \geqslant \max(S_{1m}, S_{2m}) - f \tag{13.27}$$

where $f \geqslant 0$, and $f=f_1$ or f_2 according as the termination occurs with r successes on one of the treatments or with c failures on both. The sampling rule and the stopping rule of $R_{I,\mathrm{PW}}$ are the same as those of the procedure $R''_{I,\mathrm{PW}}$ of Berry and Sobel (1973) discussed in §4.3.2.

For both $R_{I,\mathrm{VT}}$ and $R_{I,\mathrm{PW}}$, the LFC is given by $p_{[1]}=p_{[2]}$. Barron and Mignona have obtained exact and approximate formulas for the PCS, the expected sample size, and the expected subset size. They have also shown that, asymptotically, it is possible to choose the constants $r=c$, $d_1=d_2=xr^{1/2}$, where $\Phi(x)=P^*$, so that $R_{I,\mathrm{VT}}$ satisfies the P^*-condition and $E(S|R_{I,\mathrm{VT}})<1+\epsilon$ whenever $p_{[2]}-p_{[1]} \geqslant \Delta^*$ where Δ^* and ϵ are specified constants. Similar results for $R_{I,\mathrm{PW}}$ are discussed by Mignogna (1975).

13.2.4 Selection in Terms of Entropy Functions

A *finite scheme A* consists of a set of mutually exclusive and totally exhaustive events $A_1,\ldots,A_m$ along with their probabilities $p_1,\ldots,p_m$. We denote the scheme by

$$A = \begin{pmatrix} A_1 & A_2 & \cdots & A_m \\ p_1 & p_2 & \cdots & p_m \end{pmatrix}$$

where $\Sigma p_i = 1$. Every finite scheme describes a state of uncertainty or randomness. In the preceding scheme A, we have an associated experiment the outcome of each trial of which is one of the events $A_1, \ldots, A_m$ and we know the probabilities of their occurrences. The amount of uncertainty may, of course, vary from scheme to scheme. Consider the two schemes A and B where

$$A = \begin{pmatrix} A_1 & A_2 \\ 0.5 & 0.5 \end{pmatrix} \quad \text{and} \quad B = \begin{pmatrix} B_1 & B_2 \\ 0.99 & 0.01 \end{pmatrix}.$$

The amount of uncertainty in A is obviously much more than that in B. The outcome of an experiment in the second case is "almost surely" B_1, whereas in the first case we will naturally make no attempt to predict the outcome. The *entropy function*

$$H(p_1, p_2, \ldots, p_m) = -\sum_{i=1}^{m} p_i \log p_i \tag{13.28}$$

is a very suitable measure of the uncertainty of a finite scheme introduced by Shannon [1948] in his now classical paper out of which grew the present information theory.

Our present interest is in comparing k finite schemes, $\pi_1, \ldots, \pi_k$, each with $m = 2$ states. It is assumed that the state-probabilities p_{i1}, p_{i2}; $i = 1, \ldots, k$, are unknown. (Of course, we know that $p_{i1} + p_{i2} = 1$ for all i). Let λ_i denote the entropy function of the scheme π_i. Our goal is to select the scheme associated with $\lambda_{[k]}$.

In other words, using the earlier notations of this chapter, we have k binomial populations, π_i having the success probability θ_i. The ranking parameter is

$$\lambda_i = \psi(\theta_i) = -\theta_i \log \theta_i - (1-\theta_i)\log(1-\theta_i). \tag{13.29}$$

Gupta and D. Y. Huang (1976b) have proposed and investigated the rule R_6 which selects π_i if and only if

$$\psi\left(\frac{X_i}{n}\right) \geqslant \max_{1 \leqslant j \leqslant k} \psi\left(\frac{X_j}{n}\right) - d, \tag{13.30}$$

where $d = d(k, n, P^*)$ is a nonnegative constant such that $0 < d \leqslant \psi([n/2]/n)$, where $[x]$ denotes the largest integer less than or equal to x. For this rule, Gupta and Huang have shown that $\inf_{\Omega} P(CS|R_6) = \inf_{\Omega_0} P(CS|R_6)$, where $\Omega_0 = \{\boldsymbol{\theta} | \theta_1 = \cdots = \theta_k = \theta\}$. However, as in the case of R_1, the common value of θ for which the infimum is attained is not known and so a conservative value of d is obtained.

Let $M(k,d(t),t,n)=\Sigma\binom{n}{s_1}\cdots\binom{n}{s_k}$ where the summation is over the set of all nonnegative integers s_i such that $\Sigma_{i=1}^{k}s_i=t$, $0\leqslant s_i\leqslant n$, $i=1,\ldots,n$, $0<t\leqslant kn$, and $\psi(s_k/n)\geqslant\max_{1\leqslant j\leqslant k-1}\psi(s_j/n)-d(t)$ for some constant $d(t)$ depending on t and $0<d(t)\leqslant\psi([n/2]/n)$. The following theorem of Gupta and D. Y. Huang (1976b) gives the conservative value d to be used in R_6 to satisfy the P^*-condition.

Theorem 13.3. For given P^*, and for each t, $0\leqslant t\leqslant kn$, let $d(t)$ be the smallest constant, $0\leqslant d(t)\leqslant\psi([n/2]/n)$, such that

$$M(k;d(t),t,n)\geqslant\binom{kn}{t}P^*. \tag{13.31}$$

If $d=\max\{d(t)\colon 0\leqslant t\leqslant kn\}$, then

$$\inf_{\Omega}P(CS|R_6)\geqslant P^*. \tag{13.32}$$

The d-values required for the procedure R_6 are tabulated by Gupta and Huang (1976b) for $k=2(1)7$, $n=4(1)15$, and $P^*=0.75(0.05)0.95$.

Gupta and Huang have also derived upper bounds for $E(S|R_6)$ when $\boldsymbol{\theta}$ is restricted to Ω_δ given by

$$\Omega_\delta=\left\{\boldsymbol{\lambda}\colon 0<\psi(\delta)\leqslant\lambda_i\leqslant\log 2-\psi(\delta),\ 0<\delta\leqslant\tfrac{1}{2}\right\}. \tag{13.33}$$

Let us define

$$\varphi(y)=y\log 2y+(1-y)\log\left[2(1-y)\right],\ 0\leqslant y\leqslant 1. \tag{13.34}$$

Then the rule R_6 can equivalently be stated as:

Select π_i if and only if

$$\varphi\left(\frac{X_i}{n}\right)\leqslant\min_{1\leqslant j\leqslant k}\varphi\left(\frac{X_j}{n}\right)+d. \tag{13.35}$$

It is known (see Kullback [1959]) that, for $0<\theta_i<1$, $2n\varphi(X_i/n)$ is asymptotically distributed as $Y+2n\{\theta_i\log(\theta_i/(1-\theta_i))+\log 2(1-\theta_i)\}$, where Y has a chi-square distribution with 1 degree of freedom. In this case,

$$\inf_{\Omega_0}P(CS|R_6)=\int_0^\infty\left[1-G_1(y-2nd)\right]^{k-1}dG_1(y), \tag{13.36}$$

where $G_1(\cdot)$ is the cdf of Y.

Gupta and D. Y. Huang (1976b) have also shown that the procedure R_6 is strongly monotone and unbiased. Further the sequence of rules $\{R_6(n)\}$ is consistent with respect Ω [see the definitions in §11.3.4].

The case of general m is considered in §13.6.2 under the discussion of selection from multinomial populations.

Selection in Terms of a Generalized Entropy Function. Let $f(u)$ be a real-valued nonnegative convex function defined on $(0,1]$ with a continuous derivative on $(0,1]$ and $f(1)=0$. Define

$$\psi_f(\theta) = \inf_{z>0} \{\theta f(z) + (1-\theta) f(1-z)\}. \tag{13.37}$$

The function $\psi_f(\theta)$ is a *generalized entropy function*. Some useful properties of the function have been proved by Arimoto [1971]. The modified procedure R_6^f of Gupta and D. Y. Huang (1976b) selects π_i if and only if

$$\psi_f\left(\frac{X_i}{n}\right) \geqslant \max_{1 \leqslant j \leqslant k} \psi\left(\frac{X_j}{n}\right) - d', \tag{13.38}$$

where $d' = d'(k,n,P^*)$ is a nonnegative constant such that $0 \leqslant d' \leqslant \psi_f([n/2]/n)$. Letting $\lambda_i^f = \psi_f(\theta_i)$, Gupta and Huang have shown that $\inf_\Omega(CS|R_6^f)$ is attained when $\lambda_1^f = \cdots = \lambda_k^f$. A conservative constant d' can be obtained as in the case of R_6.

13.2.5 Elimination of Inferior Populations

Gupta, Huang, and Huang (1976) have also considered the selection problem under the formulation of Desu (1970) discussed in §11.4. Any population π_i is called inferior if $\theta_i < \theta_{[k]} - \delta$, where δ is a given positive constant. The goal here is to exclude all the inferior populations. The procedure R_7 of Gupta, Huang, and Huang (1976) includes π_i in the selected subset if and only if

$$X_i > X_{[k]} - (\delta - c)n, \tag{13.39}$$

where $0 < c = c(k,n,P^*) < \delta \leqslant 1$ must be chosen subject to the P^*-condition. A conservative constant c can be obtained by having

$$\inf_{\Omega(\delta)} P\left(\max_{1 \leqslant \alpha \leqslant k-1} X_{(\alpha)} < X_{(k)} - (\delta - c)n \right) \geqslant P^* \tag{13.40}$$

where $X_{(i)}$ is the number of successes from the population having $\theta_{[i]}$ as the success probability and $\Omega(\delta) = \{\boldsymbol{\theta}\colon\ \theta_{[1]} = \cdots = \theta_{[k-1]} = \theta - \delta,\ \theta_{[k]} = \theta,\ \delta < \theta < 1\}$.

For large n, we can evaluate the probability in (13.40) for $\boldsymbol{\theta}\in\Omega(\delta)$ using the normal approximation. Gupta and Huang have shown that $\Phi^{k-1}\{c(2n/(1-\delta^2))^{1/2}\}$ is a lower bound for the infimum in (13.40).

13.3 SELECTION FROM POISSON POPULATIONS

Let π_i be a Poisson process with parameter λ_i. Our observation from π_i is the number of occurrences in a specified interval of time with mean rate of occurrence being λ_i. We first consider the usual location- and scale-type procedures for selecting the Poisson population associated with $\lambda_{[k]}$ for which the P^*-condition cannot be met for all permissible values of P^*.

13.3.1 Usual Location- and Scale-Type Procedures

Let $X_1, X_2, \ldots, X_k$ denote the numbers of occurrences from the k populations. One can define the following procedures R_1 and R_2, which are of the location and scale types, respectively.

$\boldsymbol{R_1}$: Select π_i if and only if

$$X_i \geqslant X_{[k]} - d. \tag{13.41}$$

$\boldsymbol{R_2}$: Select π_i if and only if

$$X_i \geqslant cX_{[k]}. \tag{13.42}$$

The constants $d = d(k, P^*) \geqslant 0$ and $0 < c = c(k, P^*) \leqslant 1$ must be chosen so that the P^*-condition is met. It has been shown by Goel (1972) that

$$\inf_{\Omega} P(CS|R_i) \leqslant \begin{cases} 0.5, & i=1, \\ 0.75, & i=2. \end{cases} \tag{13.43}$$

In the case of $k=2$, we have equality in (13.43). Hence the rules R_1 and R_2 cannot satisfy the P^*-conditions for all values of P^* such that $k^{-1} < P^* < 1$. Gupta and D. Y. Huang (1975a) have proposed a modified rule which is discussed in the next subsection for which the P^*-requirement is met for all permissible values of P^*.

13.3.2 A Modified Procedure

Gupta and D. Y. Huang (1975a) investigated the procedure R_3 which selects π_i if and only if

$$X_i + 1 \geqslant c_1 X_{[k]} \tag{13.44}$$

where $0 < c_1 = c_1(k, P^*) \leqslant 1$ is to be chosen subject to the P^*-condition. The motivation behind this procedure is a result of Chapman [1952] which states that there is no unbiased estimator of the ratio λ_1/λ_2 but the estimator $X_1/(X_2+1)$ is "almost unbiased."

For the procedure R_3 it has been shown that $\inf_\Omega P(CS|R_3) = \inf_{\Omega_0} P(CS|R_3)$, where $\Omega_0 = \{\boldsymbol{\lambda}|\lambda_1 = \cdots = \lambda_k\}$. The values of PCS in Ω_0 depends on the equal value of the parameters. As in the case of the procedure R_1 of §13.2 defined by (13.1), Gupta and D. Y. Huang (1975a) have obtained a conservative value of c_1 given by the following theorem.

Theorem 13.4. For given P^*, let $P_1^* = (P^*)^{1/(k-1)}$. For any nonnegative integer r, let $c_1(r)$ be the largest value such that

$$\Pr\left[X_1(1+c_1(r)) \leqslant r+1 | X_1 + X_2 = r\right] \geqslant P_1^* \qquad \text{for } \boldsymbol{\lambda} \in \Omega_0. \tag{13.45}$$

If $c_1 = \inf(c_1(r)\colon r \geqslant 0)$ then $\inf_\Omega P(CS|R_3) \geqslant P^*$.

An upper bound for $E(S)$ has been obtained by Gupta and Huang for the case where $\lambda_{[1]} = \cdots = \lambda_{[k-1]} = \lambda$; $\lambda_{[k]} = \delta\lambda$, $\delta > 1$.

13.3.3 Conditional Procedures

As in the case of the binomial populations, Gupta and D. Y. Huang (1975a) defined a conditional procedure R_4 which selects π_i if and only if

$$X_i + 1 \geqslant C_2(t) X_{[k]}, \text{ given } T = \sum_{i=1}^{k} X_i = t, \tag{13.46}$$

where $0 \leqslant t$ and $0 < C_2(t) \leqslant 1$ is chosen to satisfy the P^*-condition. Their result is summarized in the following theorem.

Theorem 13.5. For a given $P^*(k^{-1} < P^* < 1)$, and any $r(0 \leqslant r \leqslant t)$, let $c_2(r)$ be the largest value such that, for $\boldsymbol{\lambda} \in \Omega_0$,

$$\Pr(X_2 + 1 \geqslant X_1 c_2(r) | X_1 + X_2 = t) \geqslant \begin{cases} P^* & \text{for } k=2, \\ P_2^* & \text{for } k \geqslant 3, \end{cases} \tag{13.47}$$

where $P_2^* = 1 - (1 - P^*)(k-1)^{-1}$. Define

$$C_2(t) = \begin{cases} c_2(t), & k=2, \\ \min(c_2(r)\colon 0 \leqslant r \leqslant t), & k \geqslant 3. \end{cases} \tag{13.48}$$

Then $\inf_\Omega P(CS|R_4) \geqslant P^*$.

13.3.4 Selecting the Population Associated with $\lambda_{[1]}$

For this problem, an analog of the procedure R_3 of Gupta and D. Y. Huang (1975a) does not work. Gupta, Leong, and Wong (1978) have proposed a modified procedure R_3' which selects π_i if and only if

$$X_i \leqslant c_1' \min_{1 \leqslant j \leqslant k} X_j + c_1' \tag{13.49}$$

where $c_1' = c_1'(k, P^*) \geqslant 1$ is the smallest number to be chosen to satisfy the probability requirement.

Let

$$A(k,t,c_1'(t)) = \sum \frac{t!}{x_1! \cdots x_k!} \left(\frac{1}{k}\right)^t \tag{13.50}$$

where the summation is over all k-tuples of nonnegative integers $(x_1, \ldots, x_k)$ such that $x_1 \leqslant c_1'(t) \min_{2 \leqslant j \leqslant k} x_j + c_1'(t)$ and $x_1 + \cdots + x_k = t$.

To implement the procedure R_3', the following theorem of Gupta, Leong, and Wong (1978) gives a conservative value of c_1'.

Theorem 13.6. For given P^* and any $t \geqslant 0$, let $c_1'(t)$ be the smallest number such that $A(k, t, c_1'(t)) \geqslant P^*$. If $c_1' = \sup_{t \geqslant 0}\{c_1'(t)\}$, then $\inf_\Omega P(CS | R_3') \geqslant P^*$.

Analogous to R_4, one can also define a conditional procedure, namely, R_4' which selects π_i if and only if

$$X_i \leqslant C_2'(t) \min_{1 \leqslant j \leqslant k} X_j + C_2'(t), \text{ given } \sum_{i=1}^{k} X_i = t, \tag{13.51}$$

where $C_2'(t) \geqslant 1$ is the smallest value chosen so that the probability requirement is met. In this case, the following theorem which gives a result similar to Theorem 13.5 has been established by Gupta, Leong, and Wong (1978).

Theorem 13.7. For a given $P^*(k^{-1} < P^* < 1)$, and any $r(0 \leqslant r \leqslant t)$, let $c_2'(r)$ be the smallest value such that for $\boldsymbol{\lambda} \in \Omega_0$

$$\Pr\left(X_1 \leqslant \frac{c_2'(r)(1+t)}{1+c_2'(r)} | X_1 + X_2 = t \right) \geqslant \begin{cases} P^* & \text{for } k=2, \\ P_2^* & \text{for } k \geqslant 3, \end{cases} \tag{13.52}$$

where $P_2^* = 1 - (1 - P^*)(k-1)^{-1}$. Define

$$C_2'(t) = \begin{cases} c_2'(t), & k=2, \\ \max\{c_2'(r) \colon 0 \leqslant r \leqslant t\}, & k \geqslant 3. \end{cases} \tag{13.53}$$

Then $\inf_\Omega P(CS | R_4') \geqslant P^*$.

Gupta, Leong, and Wong (1978) have also investigated a procedure R_3'' which selects π_i if and only if

$$X_i \leqslant c_3' + \frac{c_3'}{k-1} \sum_{j \neq i} X_j \qquad (13.54)$$

where $c_3' \geqslant 1$ is the smallest constant to be chosen subject to the probability requirement. The LFC is where all the λ_i are equal. As in other cases, here also the PCS depends on the equal value of the λ_i and one can obtain a conservative value of the constant c_3'.

Gupta, Leong, and Wong have tabulated approximate values (determined by numerical methods) of the infimum of PCS for the rules R_3' and R_3'' for $k=2(1)10$ and $c=1.6, 1.8, 2.0, 2.4, 2.8, 3.0, 3.5(0.5)5.0$ where c denotes c_1' in the case of R_3' and c_3' in the case of R_3''. They have also tabulated (a) the PCS, (b) the probability of selecting any nonbest population, and (c) the expected proportion of the selected populations, under the slippage configuration $(\delta\lambda, \lambda, \ldots, \lambda)$ for $k=3,5$, and $\delta=0.3, 0.5$ using both R_3' and R_3''.

A rule similar to R_5 of §13.2.2. has been constructed by Gupta and Nagel (1971) in the Poisson case. We denote this rule also by R_5. It has the same form as (13.22). The constants c_T and ρ in this case are determined as before except that the conditional distribution of X_k given $T=t$ is given by

$$\Pr(X_k = x | T=t) = \binom{t}{x} \frac{(k-1)^{t-x}}{k^t}, \qquad x=0,\ldots,t. \qquad (13.55)$$

Gupta and Nagel (1971) have tabulated the values of c_T and ρ for $k=2,3,5$; $t=0(1)50$; and $P^*=0.75, 0.90, 0.95, 0.99$.

13.3.5 Procedure Using Inverse Sampling

Instead of observing the number of Poisson events during an interval of fixed length, one can observe all the k Poisson processes at discrete time points $t=0,1,2,\ldots$. We fix a positive integer M and observe the random variables $T_{i,M}(i=1,\ldots,k)$ where $T_{i,M}$ is the first time point such that the number of events, for the ith population, occurring during $(0, T_{i,M}]$ is at least M. Goel (1972) considered this inverse sampling rule to define a selection procedure for selecting a subset containing the population associated with $\lambda_{[k]}$.

Let $X_{i,t}$ denote the number of events up to time t in the ith population. Define

$$V_i = T_{i,M} - 1 + U(m_i, n_i), \qquad (13.56)$$

where $n_i = X_{i,T_{i,M}} - X_{i,T_{i,M}-1}$, $m_i = M - X_{i,T_{i,M}-1}$, and $U(m,n)$ is the mth order statistic in a random sample of size n from a uniform distribution on the interval $(0,1)$. If the random samples from the uniform distribution are selected independently for each population, then V_i has a gamma distribution with scale parameter $\theta_i = \lambda_i^{-1}$ and M degrees of freedom. Thus, one can use the procedure of Gupta (1963). So we have the procedure R_6 which selects π_i if and only if

$$V_i \leqslant c^{-1} V_{[1]}, \tag{13.57}$$

where $0 < c = c(k, M, P^*) < 1$ is to be chosen subject to the P^*-condition. The constant c is tabulated by Gupta and Sobel (1962b) discussed in §12.4.1.

Since $V_i > T_{i,M}^{-1}$ for all i, it is clear from the rule R_6 that we do not have to observe $T_{i,M}$ for those populations for which $T_{i,M} > c^{-1}V_{[1]} + 1$. Therefore, by putting a bound on $c^{-1}V_{[1]}$, we will in effect put a bound on the sampling costs. Goel (1972) proposed the following criterion for determining M.

Writing c as c_M, choose the largest M such that

$$\sup_{\Omega} \Pr\left[c_M^{-1} V_{[1]} > t\lambda_{[1]}^{-1} \right] < \beta^* \tag{13.58}$$

for given values of t and β^*. The supremum in (13.58) occurs when $\lambda_{[1]} = \cdots = \lambda_{[k]}$. Thus (13.58) reduces to

$$G_{2M}(2c_M t) \geqslant 1 - (\beta^*)^{1/k}, \tag{13.59}$$

where $G_\nu(x)$ is the cdf of a χ^2-variable with ν degrees of freedom. For given values of P^*, β^*, and t one can easily obtain M with the help of the table of c_M-values and tables of the incomplete gamma function ratio. Extensive tables for $G_{2M}(x)$ by Khamis [1965] could be used for this purpose. Goel (1972) has tabulated M values for $P^* = 0.75$; $k = 2(1)5$; $\beta = .01, .05, .10, .25$; and $t = 3(1)10, 15, 20, 25$.

Gupta and Wong (1977) have also discussed selection procedures for selecting the Poisson process associated with the largest mean rate of occurrence λ_i. One of their procedures uses continuous inverse sampling. Each process is observed continuously in time. As before, $X_{i,t}$ denotes the number of events up to time t in the process π_i. Sampling is terminated when $X_{i,t}$ is equal to a specified positive integer N for some i. Then the rule R_7 of Gupta and Wong (1977) selects the process π_i if and only if

$$X_{i,t} \geqslant N - b_1 \tag{13.60}$$

where $b_1 = b_1(k, P^*, N)$ is the smallest nonnegative integer subject to the probability requirement. For this procedure, the infimum of the PCS is attained when $\lambda_1 = \cdots = \lambda_k$ and is independent of the common value. The constant b_1 is given by

$$\int_0^{\infty} [1 - G_N(y)]^{k-1} dG_{N-b_1}(y) \tag{13.61}$$

where

$$G_r(y) = \int_0^y \frac{1}{\Gamma(r)} x^{r-1} e^{-x} dx. \tag{13.62}$$

Gupta and Wong (1977) have tabulated the smallest integer b_1 necessary to meet the P^*-requirement and the actual minimum PCS attained for $k = 2(1)6$, $N = 3(1)10$, and $P^* = 0.75, 0.80, 0.90$, 0.95. They have also tabulated (a) the actual PCS, (b) the probability of selecting a nonbest population, and (c) the expected proportion of populations selected in the subset for $k = 2(1)6$, $N = 4(1)10$, and $P^* = 0.75, 0.80, 0.90$.

Besides R_7, Gupta and Wong (1977) have discussed three different procedures for selecting a subset containing the process with the largest mean rate of occurrence. One of them, say R_8, uses the sampling scheme similar to that of Alam (1971b). Another procedure is R_9 which selects π_i if and only if $X_{i,t_0}/t_0 + 1 \geqslant b_3 \max_{1 \leqslant j \leqslant k} X_{j,t_0}/t_0$ where $b_3 = b_3(k, P^*, t_0)$ is the largest nonnegative number subject to the P^*-condition. The third procedure R_{10} is the conditional analog of R_3. The investigations of R_9 and R_{10} are similar to those of Gupta and D. Y. Huang (1975a) discussed in §13.3.2.

13.4 SELECTION FROM NEGATIVE BINOMIAL AND FISHER'S LOGARITHMIC DISTRIBUTIONS

When we use inverse sampling rule and terminate Bernoulli trials on obtaining a given number of successes, the total number of trials has a negative binomial distribution. Let θ_i be the probability of a success on a single trial associated with π_i and let the sampling be terminated at the rth success. Then $X =$ the total number of trials until the rth success has a negative binomial distribution given by

$$\Pr\{X = x\} = \binom{x-1}{r-1} \theta_i^r (1-\theta_i)^{x-r}, \qquad x = r, r+1, \ldots \tag{13.63}$$

Gupta and Nagel (1971) have considered a rule R similar to their rule R_5 in the case of binomial populations defined by (13.22). The rule R is defined by

$$p_k(\mathbf{x}) = \begin{cases} 1 & \text{if } x_k < c_T, \\ \rho & \text{if } x_k = c_T, \\ 0 & \text{if } x_k > c_T, \end{cases} \tag{13.64}$$

where $\rho = \rho(T, P^*, k)$ and $c_T = c_T(P^*, k)$ satisfying (13.23) are tabulated by Gupta and Nagel for $k = 2, 3, 5$; $r = 5, 10$; $P^* = 0.75, 0.90, 0.95, 0.99$; and the values of T ranging from rk to $rk + 50$.

A similar rule has been discussed by Nagel (1970) for Fisher's logarithmic distributions. The probability distribution for π_i is given by

$$\Pr(X_i = x) = \frac{-1}{\log(1-\theta_i)} \frac{\theta_i^x}{x}, \qquad x = 1, 2, \ldots, \qquad 0 < \theta_i < 1. \tag{13.65}$$

We can define a rule R' similar to R_5 in the Poisson case. However, R' does not work satisfactorily in this case because, unless θ_i is extremely close to 1, there is a very high probability concentrated at $x = 1$. For example, $\Pr(X = 1)$ takes the values 0.72, 0.39, and 0.21 corresponding to $\theta = 0.5$, 0.9, and 0.99. Hence, even for large θ, small values of X are very likely to be observed. This leads to small values of c_T and hence the rule includes with a very high probability all populations in the selected subset. This situation can be improved by taking repeated observations $X_{i1}, \ldots, X_{in}$, $i = 1, \ldots, k$. Then $T_i = \Sigma_{j=1}^{n} X_{ij}$ is sufficient for θ_i. But unlike the Poisson distribution, the logarithmic distribution is not reproductive, that is, T_i does not have a logarithmic distribution. Thus for each value of n we must use a different table. Nagel (1970) has constructed a table of c_T- and ρ-values for $k = 2, 3, 5, 10$; $n = 5$; $P^* = 0.75, 0.90, 0.95, 0.99$, and for 50 values of T starting from $T = kn$ in steps of 2 in the case of $k = 2$ and 3, and in steps of 5 in the case of $k = 5$ and 10.

13.5 SELECTION FROM HYPERGEOMETRIC DISTRIBUTIONS

Let π_i be a finite population of size N and θ_i be the proportion of elements having a desirable attribute. These elements could be called good items or nondefectives. To select a subset containing the population associated with $\theta_{[k]}$, Bartlett and Govindarajulu (1970) considered the procedure R which selects π_i if and only if

$$X_i \geqslant X_{[k]} - c \tag{13.66}$$

where X_i is the number of good items in a random sample of size n drawn from π_i without replacement and $c=c(k,N,n,P^*)>0$ is chosen subject to the P^*-condition. The probability of a correct selection using R is minimized over Ω for a configuration such that $\theta_{[1]}=\cdots=\theta_{[k]}=\theta$, that is, for $\theta\in\Omega_0$. For $k=2$, $\inf_{\Omega_0}P(CS|R)$ is attained when $\theta=\frac{1}{2}$. For $k>2$, the infimum occurs near $\theta=\frac{1}{2}$, but not always at the integers nearest to $N/2$. The c-values needed to guarantee the P^*-condition are tabulated by Bartlett and Govindarajulu for $k=2,3$; $N=5(5)50,\infty$; $n=1(1)\min(15,[N/2])$; and $P^*=0.75,0.90,0.95,0.99$.

If we now assume that the population is large, we can consider the sampling with replacement as an approximation. In other words, X_i has the binomial distribution $b(n,\theta_i)$.

13.6 SELECTION FROM MULTINOMIAL DISTRIBUTIONS

In this section we discuss two different problems. The first is a subset selection procedure for selecting the most or the least probable cell of a single multinomial distribution. The other problem is a generalization of the selection problem for k finite schemes each with two states discussed in §13.2.4 to finite schemes with m (arbitrary) states.

13.6.1 Selection of the Best Multinomial Cell

Let $p_1,\ldots,p_k$ be the unknown cell-probabilities in a multinomial distribution with k cells. Let us first consider the case where the cell associated with $p_{[k]}$ is defined as the best cell. Gupta and Nagel (1967) defined the following procedure R_1 based on a fixed sample of size N. Let X_i be the observed frequency of π_i, the ith cell. The procedure R_1 selects π_i if and only if

$$X_i\geqslant X_{[k]}-D \tag{13.67}$$

where $D=D(k,N,P^*)$ is a nonnegative integer. The probability of a correct selection is given by

$$P(CS|R_1)=\sum\frac{N!}{\nu_1!\cdots\nu_k!}p_{[1]}^{\nu_1}\cdots p_{[k]}^{\nu_k} \tag{13.68}$$

where the summation is over nonnegative integers $\nu_1,\ldots,\nu_k$ such that $\Sigma\nu_i=N$ and $\nu_i\leqslant\nu_k+D$, $i=1,\ldots,k-1$. Gupta and Nagel (1967) have proved the following theorem regarding the least favorable configuration.

Theorem 13.8. Let μ be the smallest integer such that $p_{[\mu]}>0$ and let ν be the largest integer such that $p_{[\nu]}<p_{[k]}$. For a configuration minimizing

$P(CS|R_1)$ we must have

$$\mu \geqslant \nu. \tag{13.69}$$

Further, the inequality in (13.69) is strict if $\mu = k-1$.

According to this theorem the LFC is of the type

$$(0,\ldots,0,s,p,\ldots,p),\ s \leqslant p. \tag{13.70}$$

Let $F(k,N,D;\ \mathbf{p})$ denote the $P(CS|R_1)$ for any $\mathbf{p}\in\Omega$. If r is the number of positive p_i, then

$$\min_{\Omega} F(k,N,D;\ \mathbf{p}) = \min_{r=2,\ldots,k}\left\{ \min_{\frac{1}{r}<p<\frac{1}{r-1}} F(k,N,D;\ (0,\ldots,0,s,p,\ldots,p))\right\}$$

$$= \min_{r=2,\ldots,k}\ \min_{\frac{1}{r}<p<\frac{1}{r-1}} F(r,N,D;\ (s,p,\ldots,p)). \tag{13.71}$$

Thus it is not necessary to consider the cases where r is less than k, when the problem is already solved for all smaller values of k with same N and D. Hence one need consider only vectors of the type $(s,p,\ldots,p)$, $s=1-(k-1)p$. The numerical computations of Gupta and Nagel (1967) showed that the minimum of $F(k,N,D;\ (s,p,\ldots,p))$ over $p\in[1/k,1/(k-1)]$ usually took place at either end of the interval. In PCS evaluations for $D=0(1)4$, $k=2(1)10$, and $N=2(1)15$, the minimum was attained at an interior point of the interval in only one case, namely, $k=3$, $N=6$, and $D=4$. Gupta and Nagel have constructed tables giving (i) probability of a correct selection, (ii) probability of selecting a fixed nonbest population, and (iii) expected proportion of the populations selected for the configuration $(p,\ldots,p,Ap)$, $A\geqslant 1$, for $A=1,3,5$; $D=0,1,2$; $k=2(1)10$; and $N=2(1)15$. For fixed N, k, and P^*, the minimum D-value needed to guarantee the P^*-condition can be obtained from the preceding tables by a simple search procedure. These D-values have been tabulated by Gupta and Nagel for $k=2(1)10$; $N=2(1)15$; and $P^*=0.75$, 0.90.

For selecting a subset containing the most probable multinomial cell, Panchapakesan (1971) has investigated a procedure R_2 based on inverse sampling. Observations are taken one at a time until the count in any one of the cells reaches a predetermined integer M. Let X_i be the count in cell π_i at termination of sampling. Then the procedure R_2 selects π_i if and only if

$$X_i \geqslant M-D \tag{13.72}$$

where D is an integer such that $0 \leqslant D \leqslant M$ and the constants are to be chosen subject to the P^*-condition. It has been shown by Panchapakesan (1971) that the least favorable configuration for $P(CS|R_2)$ is of the type $(0,\ldots,0,r^{-1},\ldots,r^{-1})$, where $r(1 \leqslant r \leqslant k)$ is the number of cells with positive probabilities. Further the configuration $(k^{-1},\ldots,k^{-1})$ has been proved to be asymptotically the least favorable in Ω when $M \to \infty$ and $(M-D)/M \to \lambda$, $0<\lambda<1$. This configuration is, in fact, the LFC for finite M [see Panchapakesan (1973)]. In the case of $k=2$, $\sup_\Omega E(S|R_2)$ is attained when the cell-probabilities are equal.

Nagel (1970) proposed a randomized rule R_3 defined by

$$p_k(\mathbf{x}) = \begin{cases} 1 & \text{if } x_k > d, \\ \rho & \text{if } x_k = d, \\ 0 & \text{if } x_k < d, \end{cases} \tag{13.73}$$

where the $p_i(\mathbf{x})$ are the individual selection probabilities (see (11.27) of §11.2.5) and the integer d is determined such that

$$\begin{aligned} k^{-N} \sum_{i=d+1}^{N} \binom{N}{i} (k-1)^{N-i} &< P^*, \\ k^{-N} \sum_{i=d}^{N} \binom{N}{i} (k-1)^{N-i} &\geqslant P^*. \end{aligned} \tag{13.74}$$

It follows that

$$\rho = \frac{P^* k^N - \sum_{i=d+1}^{N} \binom{N}{i} (k-1)^{N-i}}{(k-1)^{N-d}}. \tag{13.75}$$

In §13.3.3 we discussed conditional procedures for selection from Poisson distributions. If $Y_1,\ldots,Y_k$ are independent Poisson variables with parameters $\lambda_1,\ldots,\lambda_k$, respectively, then the conditional joint distribution of $Y_1,\ldots,Y_k$ given $Y_1 + \cdots + Y_k = N$ is multinomial with cell-probabilities $p_i = \lambda_i/\Sigma\lambda_i$. Using this idea, Gupta and D. Y. Huang (1975a) have proposed the following procedure R_4 for selecting a subset containing the most probable multinomial cell based on a fixed sample size.

***R*$_4$:** Select the cell π_i if and only if

$$X_i + 1 \geqslant c X_{[k]}, \tag{13.76}$$

where $0<c=c(k,N,P^*)\leqslant 1$ is to be chosen so as to satisfy the P^*-condition. This rule R_4 is equivalent to the procedure R_4 of §13.3.3 where p_i is the parameter associated with π_i. Thus the constant c can be obtained from Theorem 13.5.

Suppose the least probable cell is defined as the best. For this problem, Gupta and Nagel (1967) have investigated the fixed-sample rule R' which selects π_i if and only if

$$X_i \leqslant X_{[1]} + C \tag{13.77}$$

where C is a nonnegative integer. In this case $P(CS|R')$ is shown to be minimized for a configuration of the type

$$p_{[1]} = \cdots = p_{[k-1]} = p; \qquad p_{[k]} = 1-(k-1)p. \tag{13.78}$$

In the preceding GLFC, $p \leqslant k^{-1}$. Numerical evaluations of Gupta and Nagel (1967) carried out for $k=2(1)10$; $N=2(1)15$; and $C=0(1)4$ showed that the overall minimum of $P(CS|R')$ is attained for $p=1/k$. As in the case of the most probable cell, tables have been constructed providing (i) the probability of a correct selection, (ii) the probability of selecting a fixed nonbest cell, and (iii) the expected proportion of populations included in the subset corresponding to the configuration $(p/A,p,\ldots,p)$ for $A=1,3,5$; $C=0,1,2$; $k=2(1)10$; and $N=2(1)15$. Using these tables, the minimum C-values satisfying the P^*-condition have been tabulated for $k=2(1)10$; $N=2(1)15$; and $P^*=0.75, 0.90$.

13.6.2 Selection from Finite Schemes

In §13.2.4 we considered finite schemes each consisting of m states where the entropy function $\lambda_i \equiv H(p_{i1},\ldots,p_{im}) = -\Sigma_{j=1}^m p_{ij}\log p_{ij}$ represents the amount of uncertainty and discussed selection from k such schemes in the special case of $m=2$. We now consider the general case $m \geqslant 2$ assuming a partial order restriction among the finite schemes.

For any vector $\mathbf{a}=(a_1,\ldots,a_m)$, let $a_{[1]}\leqslant a_{[2]}\leqslant \cdots \leqslant a_{[m]}$ denote the ordered components and $A_r = \Sigma_{i=r}^m a_{[i]}$.

Definition 13.1. A vector $\mathbf{a}=(a_1,\ldots,a_m)$ is said to *majorize* another vector $\mathbf{b}=(b_1,\ldots,b_m)$ of same dimension (written $\mathbf{a}\succ\mathbf{b}$ or $\mathbf{b}\prec\mathbf{a}$) if $A_r \geqslant B_r$ for $r=2,\ldots,m$, and $A_1 = B_1$.

Note that if $\mathbf{x}=(x_1,\ldots,\mathbf{x}_m)$ and $\mathbf{x}'=(x_{i_1},\ldots,x_{i_m})$ where $\{i_1,\ldots,i_m\}$ is a permutation of $\{1,\ldots,m\}$ then $\mathbf{x}\succ\mathbf{x}'$ and $\mathbf{x}'\succ\mathbf{x}$.

Definition 13.2. If a function f satisfies the property that $f(\mathbf{x}) \geqslant (\leqslant) f(\mathbf{x}')$ whenever $\mathbf{x}\succ\mathbf{x}'$, then f is called a *Schur-convex* (*Schur-concave*) function.

Functions that are either Schur-convex or Schur-concave are called *Schur functions* (or alternatively, are said to possess the *Schur property*). It should be noted that a Schur function is necessarily permutation invariant; that is, $f(\mathbf{x})=f(\mathbf{x}')$ whenever $\mathbf{x}'$ can be obtained by permuting the coordinates of $\mathbf{x}$. A theorem of Ostrowski [1952] is very useful in verifying if a function f has the Schur property. This theorem states that a differentiable, permutation invariant function f on R^m (Euclidean m-space) is Schur-convex (Schur-concave) if and only if

$$(x_1-x_2)\left(\frac{\partial f}{\partial x_1}-\frac{\partial f}{\partial x_2}\right)\geqslant(\leqslant)0 \qquad (13.79)$$

for all $\mathbf{x}$.

EXAMPLES 13.1. **1.** $f_1(\mathbf{x})=\Sigma_{i=1}^m g_i(x_i)$ is Schur-convex (Schur-concave) if all g_i are convex (concave) functions. In particular, the entropy function $H(p_1,\ldots,p_m)=-\Sigma_{i=1}^m p_i \log p_i$ is Schur-concave.

2. $f_2(\mathbf{x})=\Sigma_{i=1}^m x_i$ is both Schur-convex and Schur-concave.

Now, let $\pi_1,\ldots,\pi_k$ be k independent finite schemes each with m states where the state-probabilities of π_i are p_{ij}, $j=1,\ldots,m$. The scheme π_i is defined to be better than the scheme π_j if $\mathbf{p}_i=(p_{i1},\ldots,p_{im})\prec\mathbf{p}_j=(p_{j1},\ldots,p_{jm})$. Since the entropy function is Schur-concave, $\mathbf{p}_i\prec\mathbf{p}_j$ implies that $H(\mathbf{p}_i)\geqslant H(\mathbf{p}_j)$.

It should be noted that the same entropy may arise from two different noncomparable vectors. So we will assume that there exists a scheme that is better than the others. This scheme will be called the best scheme. In case of ties we assume one of them to have been tagged as the best.

For selecting a subset containing the best, Gupta and Wong (1976a) have considered a class of procedures R_φ based on n independent observations from each scheme and a Schur-concave function φ. Let X_{ij} be the number of occurrences in the jth state of the scheme π_i.

$\boldsymbol{R_\varphi}$: Select the scheme π_i if and only if

$$\varphi\left(\frac{X_{i1}}{n},\ldots,\frac{X_{im}}{n}\right)\geqslant \max_{1\leqslant j\leqslant k}\varphi\left(\frac{X_{j1}}{n},\ldots,\frac{X_{jm}}{n}\right)-d \qquad (13.80)$$

where $d=d(k,m,n,P^*)>0$ is to be determined subject to the P^*-condition.

Gupta and Wong (1976a) have shown that $\inf_\Omega P(CS|R_\varphi)$ is attained when all the parameter vectors are equal. A conservative value of the constant d can be obtained in a manner similar to that of Gupta and D. Y. Huang (1976b) for the special case of $m=2$ taking φ to be the entropy function. Let $\mathbf{t}=(t_1,\ldots,t_m)$ where $0\leqslant t_i\leqslant kn$, $i=1,\ldots,k$, and $\Sigma_{i=1}^m t_i=kn$.

Let

$$M(k,d(\mathbf{t}),\mathbf{t},m,n)=\sum \prod_{i=1}^{k} \binom{n}{s_{i1}\ldots s_{im}} \tag{13.81}$$

where the summation is over the set of all m-tuples $(s_{i1},\ldots,s_{im})$ such that $\sum_{l=1}^{k} s_{il}=t_i$, $0\leqslant s_{ij}\leqslant n$, $i=1,\ldots,k$; $j=1,\ldots,m$; and

$$\varphi\left(\frac{s_{k1}}{n},\ldots,\frac{s_{km}}{n}\right)\geqslant \max_{1\leqslant j\leqslant k-1} \varphi\left(\frac{s_{j1}}{n},\ldots,\frac{s_{jm}}{n}\right)-d(\mathbf{t})$$

for some constant $d(\mathbf{t})$ depending on $\mathbf{t}$. The following theorem of Gupta and Wong (1976a) gives the conservative value of the constant to be used in the procedure R_φ.

Theorem 13.9. For given P^* and for each $t=(t_1,\ldots,t_m)$, $0\leqslant t_i\leqslant kn$, $i=1,\ldots,m$, $\sum_{i=1}^{m} t_i=kn$, let $d(\mathbf{t})$ be the smallest constant such that

$$M(k,d(\mathbf{t}),\mathbf{t},n,m)\geqslant \binom{kn}{t_1\ldots t_m}P^*.$$

If $d=\max_{\mathbf{t}}\{d(\mathbf{t})\}$, then

$$\inf_{\Omega} P(CS|R_\varphi)\geqslant P^*.$$

13.7 NOTES AND REMARKS

It can be seen from the investigations of the procedures discussed in this chapter that there is potential and need for further investigations in the area of selection procedures for discrete populations. There are generally difficulties in evaluating the exact infimum of the probability of a correct selection. Consequently lower bound to this infimum is obtained in many cases to evaluate conservative constants for the procedures. The bounds on the supremum of the expected subset size have been obtained only under slippage configurations. There is the problem of comparing several alternative procedures and also that of comparing several sampling schemes.

Gupta, Huang, and Huang (1976) have also discussed the related problem of tests of homogeneity for binomial populations. The mean and the variance of the largest and the smallest order statistic in a sample of size N from a k-cell multinomial distribution have been tabulated by Gupta and Nagel (1967) for the configuration $(p,\ldots,p,Ap)$, $A\geqslant 1$, in the case of the smallest order statistic, and for the configuration $(p/A,p,\ldots,p)$, $A\geqslant 1$, in the case of the largest order statistic. The ranges covered by their tables are: $k=2(1)10$, $N=2(1)15$, and $A=1,2,3,5$.

The concepts of majorization and Schur function discussed in §13.6.2 have been used by many authors to develop useful inequalities in many branches of mathematics. Many well-known inequalities arising in probability and statistics are equivalent to saying that certain functions are Schur functions. These techniques as applied to probability and statistics have become increasingly useful and understandably popular in recent years. One of the early papers in this area of applications was by Cacoullos and Olkin [1965]. Some of the recent contributors are Marshall and Proschan [1965], Marshall, Olkin and Proschan [1967], Marshall and Olkin [1974], Olkin [1972], Proschan and Sethuraman [1977], and Nevius, Proschan, and Sethuraman [1977a, b]. The reader is also referred to the book by Marshall and Olkin [1979].

In this chapter, we discussed some conditional procedures studied by Gupta and Nagel (1971), Gupta and D. Y. Huang (1975a), and Gupta, Huang and Huang (1976). Recently, Kiefer (1977) has discussed conditional confidence procedures and estimated confidence coefficients, and has briefly indicated their applications to selection and ranking problems both under indifference zone and subset selection approaches.

CHAPTER 14

Selection from Multivariate Normal Populations

14.1 INTRODUCTION

In this chapter we discuss subset selection procedures relating to a single multivariate normal population as well as several multivariate normal distributions. Of course, the former problems can be considered as selection from a set of correlated univariate normal populations. As we saw in Chapter 7, the first problem in ranking multivariate populations is to choose an appropriate measure for comparing two populations. Investigations in this area have been generally carried out by choosing a scalar measure which allows a complete order of the populations. The choice of the ranking measure thus depends on the specific situations. We discuss procedures for selection in terms of distance functions, generalized variance, multiple correlation coefficients, and coefficients of alienation.

14.2 SELECTION OF THE BEST COMPONENT OF A MULTIVARIATE NORMAL POPULATION

In this section we consider ranking of the components in terms of their (a) means, (b) variances, and (c) generalized variances. Let the normal distribution have the mean vector $\boldsymbol{\mu}'=(\mu_1,\ldots,\mu_p)$ and covariance matrix $\Sigma=((\sigma_{ij}))$, which is assumed to be positive definite.

14.2.1 Selection in Terms of the Means

We want to select a subset of the components containing the one associated with $\mu_{[p]}$. It is assumed that Σ is positive definite. We first discuss the case in which Σ is known assuming without loss of generality that $\sigma_{ii}=1$ for $i=1,\ldots,k$.

Let $\mathbf{X}'=(X_1,\ldots,X_p)$ be the vector of sample means based on n observations. Gnanadesikan (1966) considered the procedure

$\boldsymbol{R}_1$: Select the ith component if and only if

$$X_i \geqslant X_{[p]} - \frac{d}{\sqrt{n}}, \tag{14.1}$$

where $d=d(n,p,\Sigma)>0$ is to be determined so as to satisfy the P^*-condition. It is easy to show that

$$\inf_{\Omega} P(CS|R_1)=\Pr\{ Y_p \geqslant Y_j - d,\ j=1,\ldots,p-1\}, \tag{14.2}$$

where $Y_i=\sqrt{n}\,(X_{(i)}-\mu_{[i]})$ and $X_{(i)}$ is the sample mean associated with $\mu_{[i]}$. Let $A=((a_{ij}))$ denote the dispersion of $\mathbf{Y}$. Then (14.2) can be rewritten as

$$\inf_{\Omega} P(CS|R_1)=\Pr\left\{ Z_j \leqslant d/\sqrt{2(1-a_{jp})}\ ,\ j=1,\ldots,p-1\right\}, \tag{14.3}$$

where the random vector $\mathbf{Z}'=(Z_1,\ldots,Z_{p-1})$ is given by $\mathbf{Z}=B\mathbf{Y}$ where $B=((b_{ij}))$ is the $(p-1)\times p$ matrix with

$$b_{ij}=\begin{cases} \{2(1-a_{ip})\}^{-\frac{1}{2}}, & i=j=1,\ldots,p-1,\\ -\{2(1-a_{ip})\}^{-\frac{1}{2}}, & i=1,\ldots,p-1,\ j=p,\\ 0, & i,j=1,\ldots,p-1, i\neq j. \end{cases} \tag{14.4}$$

For $p=2$, the right-hand side of (14.3) is equal to $\Phi\{d/\sqrt{2(1-\sigma_{12})}\}$ and so the constant d needed for the procedure is given by

$$d=\sqrt{2(1-\sigma_{12})}\ \Phi^{-1}(P^*). \tag{14.5}$$

However, when $p>2$, the dispersion matrix $\Sigma_0=BAB'$ of $\mathbf{Z}$ is not known. To obtain a conservative value of d, one would obtain a lower bound of the probability in (14.3). Let $d_0=\min\{d/\sqrt{2(1-a_{pj})}\ , j=1,\ldots,p-1\}$. Then

$$\inf_{\Omega} P(CS|R_1)\geqslant \Pr\{Z_j\leqslant d_0, \quad j=1,\ldots,p-1\}. \tag{14.6}$$

Gnanadesikan (1966) considered two different lower bounds for the right-hand side of (14.6). The first bound is $\Phi^{p-1}(d_0)$ which is obtained by using an inequality due to Slepian [1962] under the condition that $a_{ij.l}\geqslant 0$ for $1\leqslant i,j,l\leqslant p(i,j,l$ all different), where $a_{ij.l}$ is the partial correlation coefficient between Y_i and Y_j given Y_l. The other bound is $(2-p)+(p-1)\Phi(d_0)$ which is obtained by using a Bonferroni inequality. Let the approximate values of d using these two bounds be denoted by $d_0^{(1)}$ and $d_0^{(2)}$, respectively. For $p=2$, both are equal to the exact value. The computations of Gnanadesikan (1966) for $p=2(1)10$, and $P^*=0.75$, 0.90, 0.95, show that $d_0^{(1)}$ is closer to the exact value than $d_0^{(2)}$ and that the difference between the two approximate values decreases as P^* increases and is very small when $P^*\geqslant .90$.

One can obtain a better value of d_0 for special cases of Σ by equating the right-hand side of (14.6) to P^*. This involves the determination of the equicoordinate percentage points of a multivariate normal distribution. For various methods of evaluating these percentage points the reader is referred to Gupta [1963a]. The determination of the constant d is considerably simplified when $\sigma_{ij}=\rho>0, i\neq j$. In this case we have

$$\inf_{\Omega} P(CS|R_1)=\int_{-\infty}^{\infty}\Phi^{p-1}\{x+c(1-\rho)^{-\frac{1}{2}}\}\ d\Phi(x) \tag{14.7}$$

and the values of $H=c\{2/(1-\rho)\}^{\frac{1}{2}}$ are tabulated by Gupta [1963a] and also by Gupta, Nagel, and Panchapakesan [1973] who have also considered the selection problem for this special case.

Suppose the covariance matrix Σ is not known. However, we make an assumption that $\sigma_{ii}=\sigma, i=1,\ldots,p$, and let s_ν^2 denote an estimator of σ^2 on ν degrees of freedom independent of the X_i. In this case, the procedure proposed by Gnanadesikan (1966) is

R_1': Select the ith component if and only if

$$X_i \geqslant X_{[p]} - d's_\nu/\sqrt{n} \tag{14.8}$$

where $d' = d'(\nu, p, P^*) > 0$ is to be determined subject to the P^*-condition. Then

$$\begin{aligned}\inf_{\Omega} P(CS|R_1') &\geqslant \Pr\{t_i \leqslant d_0, i=1,\ldots,p-1\}\\ &\geqslant 1 - \sum_{i=1}^{p-1} \Pr\{t_i \geqslant d_0\}\end{aligned} \tag{14.9}$$

where $t_i = Z_i/s_\nu, \mathbf{Z} \sim N_p(\mathbf{0}, \Sigma_0)$, and $\nu s_\nu^2 \sim \chi_\nu^2 \sigma^2$. The last member in the inequalities in (14.9) is equal to $(2-p)+(p-1)G_\nu(d_0)$, where $G_\nu(\cdot)$ is the cdf of a Student's t-variable with ν degrees of freedom. We can get an approximate value of d_0 by equating this to P^*. A better approximation obviously involves the evaluation of equicoordinate percentage points of a multivariate t-distribution. As we pointed out in Chapter 2, for the methods of evaluation of these percentage points, see Dunnett and Sobel [1954], Dunnett (1955), Dunnett and Sobel [1955], Gupta [1963a], Gupta and Sobel (1957), John [1964], [1966], and Krishnaiah and Armitage [1966].

14.2.2 Selection in Terms of the Variances

We now define the best component as the one associated with $\sigma_{[1]}^2$, where $\sigma_i^2 \equiv \sigma_{ii}$. The natural procedure on the analogy of Gupta and Sobel (1962a) is

R_2: Select the ith component if and only if

$$s_i^2 \leqslant c^{-1} s_{[1]}^2, \tag{14.10}$$

where $0 < c = c(p, n, P^*) \leqslant 1$ and $s_i^2 \equiv s_{ii}$, the ith diagonal element of the sample covariance matrix S based on n independent observations on the random vector. Frischtak (1973) has considered this procedure. For $p=2$, he has shown that the infimum of $P(CS|R_2)$ is attained when $\sigma_{[1]}^2 = \sigma_{[2]}^2$ and $\sigma_{12} = 0$ and that the constant c satisfying the P^*-condition is given by

$$\int_c^\infty \frac{\Gamma(n-1)}{\left[\Gamma\left(\frac{n-1}{2}\right)\right]^2} z^{(n-3)/2}(1+z)^{-n+1}\,dz = P^*. \tag{14.11}$$

For $p \geq 3$, we can obtain an asymptotic solution using the asymptotic normality of

$$Y_j = \sqrt{(n-1)}\left\{\frac{\log\left(s_{(1)}^2/s_{(j)}^2\right) - \log\left(\sigma_{[1]}^2/\sigma_{[j]}^2\right)}{2\left(1-\rho_{1j}^2\right)}\right\}, \qquad j \neq 1, \quad (14.12)$$

where $s_{(i)}^2$ is associated with $\sigma_{[i]}^2$ and ρ_{1j}^2 is the correlation between the components associated with $\sigma_{[1]}^2$ and $\sigma_{[j]}^2$. The asymptotic ($n \to \infty$) least favorable configuration is given by

$$\sigma_1^2 = \cdots = \sigma_k^2, \qquad \rho_{ij} = 0 (i \neq j).$$

Thus

$$\inf P(CS|R_2) \approx \Pr\left(Y_j \leq (n-1)^{\frac{1}{2}} 2^{-1} \log c, \, j \neq 1\right) \quad (14.13)$$

where the Y_j are standard normal variables with equal correlation $\frac{1}{2}$.

14.2.3 Selection of the Best Subclass of Components in Terms of the Generalized Variances

Consider a kp-variate normal population with unknown mean vector $\boldsymbol{\mu}' = (\boldsymbol{\mu}_1', \ldots, \boldsymbol{\mu}_k')$, where $\boldsymbol{\mu}_i' = (\mu_{i1}, \ldots, \mu_{ip}), i = 1, \ldots, k$, and unknown covariance matrix

$$\Sigma = \begin{bmatrix} \Sigma_{11} & \Sigma_{12} & \cdots & \Sigma_{1k} \\ \Sigma_{21} & \Sigma_{22} & \cdots & \Sigma_{2k} \\ \vdots & \vdots & \vdots & \vdots \\ \Sigma_{k1} & \Sigma_{k2} & \cdots & \Sigma_{kk} \end{bmatrix}$$

where $\Sigma_{ii} (1 \leq i \leq k)$ are $p \times p$ symmetric positive definite matrices. The kp variates are divided into k subclasses each containing p variates. Let $\lambda_i = |\Sigma_{ii}|, i = 1, \ldots, k$. Then λ_i is the generalized variance of the ith subclass. Our goal is to select a subset of the k subclasses to contain the subclass associated with $\lambda_{[1]}$. Let S denote the sample covariance matrix based on a sample of size n. Let S_i denote the sample covariance matrix for the ith subclass and $Z_i = |S_i|, i = 1, \ldots, k$.

Frischtak (1973) considered the procedure R_3 which selects the ith subclass if and only if

$$cZ_i \leq Z_{[1]} \quad (14.14)$$

where $c' = c'(k, n, P^*) \in (0, 1]$. He obtained only an asymptotic ($n \to \infty$) solution by using the fact that $\sqrt{n-1}\,(|S_1| - |\Sigma_1|, \ldots, |S_k| - |\Sigma_k|)$ is asymptotically normal, with zero means, variances equal to $2p|\Sigma_i|^2$, and covariances equal to $2|\Sigma_i||\Sigma_j|\mathrm{tr}\Sigma_i^{-1}\Sigma_{ij}\Sigma_j^{-1}\Sigma_{ji}$. The LFC is given by $\Sigma_1 = \cdots = \Sigma_k$, and $\Sigma_{ij} = 0 (i \neq j)$. Thus the constant c' is given by

$$\Pr\left\{ U_j \leqslant -\sqrt{(n-1)p}\; 2^{-1}\log c',\ j = 1, \ldots, k-1 \right\} = P^* \qquad (14.15)$$

where the U_j are equally correlated $N(0, 1)$ variables with correlation equal to $\frac{1}{2}$.

14.3 SELECTION FROM SEVERAL MULTIVARIATE NORMAL POPULATIONS

In this section, we consider selection from several multivariate normal populations using different ranking measures such as the Mahalanobis distance function, the generalized variance, the multiple correlation coefficient, and the coefficient of alienation.

Let π_i be a p-variate normal population with mean vector $\boldsymbol{\mu}_i$ and covariance matrix Σ_i which is assumed to be positive definite. Let $\mathbf{X}_{ij}$ denote the jth observation on the p-dimensional random vector associated with π_i.

14.3.1 Selection in Terms of Mahalanobis Distance Function

Let $\lambda_i = \boldsymbol{\mu}_i'\Sigma_i^{-1}\boldsymbol{\mu}_i$, the Mahalanobis distance of π_i from the origin. Define $Y_{ij} = \mathbf{X}_{ij}'\Sigma_i^{-1}\mathbf{X}_{ij} (i = 1, \ldots, k : j = 1, \ldots, n)$ and let $Y_i = \Sigma_{j=1}^n Y_{ij}$. For selecting the subset containing the population associated with $\lambda_{[k]}$, Gupta (1966b) proposed the procedure R_1 which selects π_i if and only if

$$Y_i \geqslant c_1 Y_{[k]} \qquad (14.16)$$

where $0 < c_1 = c_1(k, p, n, P^*) < 1$ is to be suitably determined. It is known that Y_i has a noncentral χ^2 distribution with noncentrality parameter $\lambda_i' = n\lambda_i$ and $p' = np$ degrees of freedom. It is easy to see that

$$\inf_{\Omega} P(CS|R_1) = \inf_{\lambda' \geqslant 0} \int_0^\infty F_{\lambda'}^{k-1}\left(\frac{x}{c_1}\right) dF_{\lambda'}(x), \qquad (14.17)$$

where $F_{\lambda'}(\cdot)$ is the cdf of a noncentral χ^2 variable with noncentrality parameter λ' and degrees of freedom p'. Gupta (1966b) showed that the integral on the right-hand side of (14.17) is nondecreasing in λ' in the case of $k = 2$ populations. Later Gupta and Studden (1970) proved the result for

$k \geqslant 2$. They provided a sufficient condition for the integral in (14.17) to be nondecreasing in λ'. Actually, they proved a general result stated as follows.

Theorem 14.1. Let $g_j(x), j=0,1,2,\ldots$ be a sequence of density functions on the interval $[0,\infty)$ and define

$$f_\lambda(x)=\sum_{j=0}^{\infty}\frac{e^{-\lambda}\lambda^j}{j!}g_j(x), \qquad x\geqslant 0. \tag{14.18}$$

For a fixed integer $k\geqslant 2$ and $c>1$, let

$$I(\lambda)=\int_0^{\infty}\left[F_\lambda(cx)\right]^{k-1}f_\lambda(x)\,dx \tag{14.19}$$

and

$$J(\lambda)=\int_0^{\infty}\left[1-F_\lambda\left(\frac{x}{c}\right)\right]^{k-1}f_\lambda(x)\,dx \tag{14.20}$$

where $F_\lambda(x)$ is the cdf associated with $f_\lambda(x)$. Then $I(\lambda)$ and $J(\lambda)$ are nondecreasing in λ provided that

$$\sum_{i=0}^{l}\binom{l}{i}\{\Delta G_i(cx)g_{l-i}(x)-cg_i(x)\Delta G_{l-i}(x)\}\geqslant 0 \tag{14.21}$$

for every $l\geqslant 0$, where $\Delta G_i(x)=G_{i+1}(x)-G_i(x)$. The monotonicity is strict if we have a strict inequality in (14.21) for some integer l.

The noncentral χ^2 density with noncentrality parameter λ is of the form (14.18). It should be recalled that we have discussed densities of this form in §11.2.4 and the condition (14.21) is a special case of (11.23) with $b=1$, $l=0$, $m=1$, and $a_0=1$. The monotonicity of $J(\lambda)$ in (14.20) is required when we use an analogous procedure for selecting a subset containing the population associated with $\lambda_{[1]}$.

The constant c_1 for the procedure R_1 is given by

$$\int_0^{\infty}G_\nu^{k-1}\left(\frac{x}{c}\right)dG_\nu(x)=P^* \tag{14.22}$$

where $G_\nu(x)$ is the cdf of a standardized gamma variable with degrees of freedom $\nu=np/2$. The values of the constant c are tabulated by Gupta (1963) and Armitage and Krishnaiah [1964]. The tables for the selection procedure for the population with the smallest parameter are given by

Gupta and Sobel (1962b), and Krishnaiah and Armitage [1964]. We have referred to these tables in Chapter 12.

Now, for the procedure R_1, we can obtain large sample solution by using normal approximation. Gupta (1966b) used the approximation based on the Wilson–Hilferty cube root transformation. This and other approximations have been discussed by Abdel-Aty [1954]. The approximate values of c have been tabulated by Gupta (1966b) for $k=2(1)10, 2\nu=5(1)10(5)50$, and $P^*=0.75$, 0.90, 0.95, 0.99.

REMARK 14.1. It should be noted that the procedure R_1 is based on the statistics $Y_i=\sum_{j=1}^{n}\mathbf{X}_{ij}'\Sigma_i^{-1}\mathbf{X}_{ij}$. Let $\overline{\mathbf{X}}_i$ denote the sample mean vector for π_i and let $Z_i=\overline{\mathbf{X}}_i'\Sigma_i^{-1}\overline{\mathbf{X}}_i$. One may want to use a procedure that selects π_i if and only if

$$Z_i \geqslant c' Z_{[k]} \tag{14.23}$$

where $0<c'<1$. This procedure was in fact used by Alam and Rizvi (1966). However, this procedure is unsatisfactory in the sense that the constant c' does not depend on n. We may alternately try a location type procedure based on the Z_i. Such a procedure and its performance relative to R_1 have not been investigated. □

Let us now consider the case of Σ_i unknown. In this case, Gupta and Studden (1970) proposed the procedure R_2 which selects π_i if and only if

$$c_2 T_i \geqslant T_{[k]} \tag{14.24}$$

where $T_i=\overline{\mathbf{X}}_i'S_i^{-1}\overline{\mathbf{X}}_i$, S_i is the usual sample covariance matrix with $(n-1)$ as the divisor, and $c_2=c_2(k,n,p,P^*)>1$. It is known that $(n-p)p^{-1}T_i$ is distributed as a noncentral F variable whose density is of the form (14.18). It has been shown by Gupta and Studden that

$$\inf_{\Omega} P(CS|R_2)=\int_0^{\infty} F_{p,n-p}^{k-1}(c_2x)\,dF_{p,n-p}(x), \tag{14.25}$$

where $F_{p,n-p}(x)$ is the cdf of a central F-variable with p and $n-p$ degrees of freedom. The values of c_2^{-1} have been tabulated by Gupta and Panchapakesan (1969a) for $k=2(1)5$; $P^*=0.75$, 0.90, 0.95, 0.99, and the following combinations of $q=p/2$ and $m=(n-p)/2: m=1(1)8$ for $q=1(1)4$, and $m=1(1)4$ for $q=5(1)8$.

Alam and Rizvi (1966) have also considered the procedure R_2 which was originally studied by Gupta and Studden in a 1965 technical report. The monotonicity of the integral involved was established by Alam and Rizvi by direct differentiation in the specific case.

14.3.2 Selection in Terms of Generalized Variances

It is meaningful to rank multivariate normal populations according to the amounts of dispersion. A natural measure of the dispersion is $|\Sigma|$, the determinant of Σ, called the generalized variance. Let $\mathbf{X}'=(X_1,\dots,X_p)$ be distributed as $N_p(\boldsymbol{\mu},\Sigma)$ and S denote the sample covariance matrix based on a sample of size n. Then $|S|$ is an estimator of $|\Sigma|$, and $|S|$ is proportional to the sum of squares of the volumes of all parallelotopes formed by using as principal edges, p vectors with p of $\mathbf{X}_1,\dots,\mathbf{X}_n$ as one set of end points and $\overline{\mathbf{X}}$ (the mean of the $\mathbf{X}_i$) as the other, and the factor of proportionality is $(n-1)^{-p}$ (see Anderson [1958]).

In this section we discuss a procedure of Gnanadesikan and Gupta (1970) for selecting a subset containing the population associated with the smallest $|\Sigma_i|$. Let $\lambda_i=|\Sigma_i|$ and $Z_i=|S_i|, i=1,\dots,k$. This problem is a special case of the problem discussed in §14.2.3 with $\Sigma_{ij}=0, i\neq j$. The procedure of Gnanadesikan and Gupta is

$\boldsymbol{R_3}$: Select π_i if and only if

$$c_3 Z_i \leqslant Z_{[1]} \tag{14.26}$$

where $c_3=c_3(k,n,p,P^*)\in(0,1]$. It is known that Z_i is distributed as $\lambda_i(n-1)^{-p}$ times the product of p independent factors, the rth factor being distributed as a χ^2 variable with $(n-r)$ degrees of freedom. Using this fact we get

$$\inf_{\Omega} P(CS|R_3)=\Pr\{\eta_i \geqslant c_3\eta_1, i=2,\dots,k\}, \tag{14.27}$$

where $\eta_i, i=1,\dots,k$, are i.i.d. random variables, each being the product of p independent factors where the rth factor has a chi-square distribution with $(n-r)$ degrees of freedom. Thus the constant c_3 is the $100(1-P^*)$ percentage point of $\eta'=\min\{\eta_i/\eta_1, i=2,\dots,k\}$. The distribution of η_i is known only when $p=2$. In this case, $2(n-1)^{p/2}(Z_i/\lambda_i)^{\frac{1}{2}}$ is distributed as a χ^2 variable with $2(n-2)$ degrees of freedom and the constant c_3 can be obtained from the tables of Gupta and Sobel (1962b) and Krishnaiah and Armitage [1964]. When $k=2, c^{\frac{1}{2}}$ is the $100(1-P^*)$ percentage point of the F-distribution with $(2n-4, 2n-4)$ degrees of freedom. In the particular case when $p=2$, and the variances of both the components of the k populations are equal, the problem is equivalent to the selection of a subset containing the population with the maximum absolute value of the correlation coefficient between the two components.

When $p>2$, Gnanadesikan and Gupta (1970) have considered the relative merits of different approximations to the distribution of $\eta_i = U_{i1}U_{i2}\dots U_{ip}$ where the U_{ij} are independent and the distribution of U_{ij} is χ^2 with $(n-j)$ degrees of freedom. Hoel [1937] suggested approximating the distribution of $\eta_i^{1/p}$ by the gamma distribution with density

$$g(y) = [\Gamma(m)]^{-1}\theta^m e^{-\theta y} y^{m-1}, \tag{14.28}$$

where $m = 2^{-1}p(n-p)$ and $2\theta = p[1-(2n)^{-1}(p-1)(p-2)]^{1/p}$. Another approximation is that of $p^{-1}\log\eta_i$ using the normal approximation of $\log\chi^2$ suggested by Bartlett and Kendall [1946]. The investigations of Gnanadesikan and Gupta (1970) show that (1) Hoel's approximation of $\eta_i^{1/p}$ decreases in accuracy as p increases, (2) the approximation of the distribution of $\log\chi^2$ by the normal distribution improves with the degrees of freedom of the χ^2 variable, and (3) the normal approximation to the distribution of the logarithm of the generalized variance improves with both p and n.

Approximating the distribution of $p^{-1}\log\eta$ by the normal distribution, we get

$$\inf_{\Omega} P(CS|R_3) \simeq \int_{-\infty}^{\infty} \Phi^{k-1}(y-c_0)\,d\Phi(y), \tag{14.29}$$

where

$$c_0 = \left\{ \sum_{i=1}^{p} V(\log\chi^2_{n-i}) \right\}^{\frac{1}{2}} \log c_3.$$

As we have seen earlier, the values of c_0 have been tabulated by Gupta [1963a] and Gupta, Nagel, and Panchapakesan [1973] for various values of k and P^*.

Regier (1976) proposed two procedures R_3' and R_3'' as alternatives to R_3. These procedures are based on the geometric mean and the arithmetic mean, respectively, of the sample generalized variances. The procedure R_3' selects π_i if and only if $Z_i \leqslant a(\Pi_{j=1}^k Z_j)^{1/k}$ whereas R_3'' includes π_i in the selected subset if and only if $Z_i \leqslant b\Sigma_{j=1}^k Z_j/k$, where the constants a and b should be determined subject to the P^*-condition. The determination of an approximate value of a is based on the normal approximation to $\log\chi^2$ given by Bartlett and Kendall [1946]. For determining b, we use the asymptotic distribution of the sample variance. Some numerical comparisons of the three procedures are given by Regier (1976).

14.3.3 Selection in Terms of Multiple Correlation Coefficients

In some situations we may be interested in comparing the populations in terms of the association between a particular component and the rest. This component is assumed to be the first one. A measure of this association in the population π_i is $\rho_i \equiv \rho_{1.2\ldots p}^{(i)}$, the multiple correlation coefficient between X_{i1} and $\{X_{i2},\ldots,X_{ip}\}$, which is defined by

$$1-\rho_i^2 = \frac{|\Sigma_i|}{\sigma_{11}^{(i)}|\Sigma_{i(11)}|}, \tag{14.30}$$

where $\sigma_{11}^{(i)}$ is the leading element of Σ_i and $\Sigma_{i(11)}$ is the matrix obtained from Σ_i by deleting the first row and the first column. The multiple correlation ρ_i is the positive square root of ρ_i^2 and it is the maximum of the correlation between X_{i1} and a linear combination of $X_{i2},\ldots,X_{ip}$ over all possible linear combinations. We discuss in the following paragraphs procedures studied by Gupta and Panchapakesan (1969a) for selecting a subset containing the population associated with $\rho_{[k]}$ (or $\rho_{[1]}$).

Let $R_i \equiv R_{1.2\ldots p}^{(i)}$ be the sample multiple correlation coefficient between X_{i1} and $X_{i2},\ldots,X_{ip}$, defined analogous to ρ_i by replacing Σ_i by the sample covariance matrix S_i in the definition of ρ_i. Two cases arise: (i) the case in which $X_{i2},\ldots,X_{ip}$ are fixed, called the conditional case; (ii) the case in which $X_{i2},\ldots,X_{ip}$ are random, called the unconditional case. The density of R_i^2 is given by

$$f_{\lambda_i}(x) = \sum_{j=0}^{\infty} w(q,m;\lambda_i,j)\,b(x;q+j,m), \tag{14.31}$$

where

$$\lambda_i = \rho_i^2, \quad q = \frac{(p-1)}{2}, \quad m = \frac{(n-p)}{2},$$

$$b(x;q+j,m) = \frac{\Gamma(q+j+m)}{\Gamma(q+j)\Gamma(m)} x^{q+j-1}(1-x)^{m-1}, \qquad 0 \leqslant x \leqslant 1, \tag{14.32}$$

and the weight function $w(q,m;\lambda_i,j)$ is given by

$$w(q,m;\lambda_i,j) = \begin{cases} \dfrac{\Gamma(q+m+j)}{\Gamma(q+m)} \dfrac{\lambda_i^j}{j!}(1-\lambda_i)^{q+m} & \text{(unconditional case)}, \\ e^{-m\lambda_i}(m\lambda_i)^j/j! & \text{(conditional case)}. \end{cases} \tag{14.33}$$

In other words, the density of R^2 is a mixture of beta densities with negative binomial weights in the unconditional case and Poisson weights in the conditional case. In either case the procedure proposed by Gupta and Panchapakesan (1969a) for selecting the population with the largest ρ_i is

***D*:** Select π_i if and only if

$$R_i^{*2} \geqslant c_4 R^{*2}_{[k]} \tag{14.34}$$

where $R_i^{*2} = R_i^2(1-R_i^2)^{-1}$ and $0 < c_4 = c_4(k, P^*, q, m) < 1$ is chosen subject to the P^*-condition. Using the stochastic ordering of the distributions, we have

$$\inf_{\Omega} P(CS|D) = \inf_{\lambda} \int_0^{\infty} H_{\lambda}^{k-1}\left(\frac{x}{c_4}\right) dH_{\lambda}(x), \tag{14.35}$$

where $H_\lambda(x)$ is the cdf of R^{*2}. Gupta and Panchapakesan (1969a) have shown that the integral in (14.35) is nondecreasing in λ using the result of Gupta and Studden stated in Theorem 14.1 in the conditional case, and using the following theorem obtained by them in the unconditional case.

Theorem 14.2. Let $g_j(x)$, $j = 0, 1, \ldots$ be a sequence of density functions on the interval $[0, \infty)$ and define

$$f_\lambda(x) = \sum_{j=0}^{\infty} \frac{\Gamma(q_1 + j)}{\Gamma(q_1)} \frac{\lambda^j}{j!} (1-\lambda)^{q_1} g_j(x), x \geqslant 0, 0 \leqslant \lambda < 1. \tag{14.36}$$

For a fixed integer $k \geqslant 2$ and $0 < c < 1$, let

$$A(\lambda) = \int_0^{\infty} \left[F_\lambda\left(\frac{x}{c}\right)\right]^{k-1} f_\lambda(x)\, dx \tag{14.37}$$

and

$$B(\lambda) = \int_0^{\infty} [1 - F_\lambda(cx)]^{k-1} f_\lambda(x)\, dx. \tag{14.38}$$

Then $A(\lambda)$ and $B(\lambda)$ are nondecreasing in λ provided that, for each integer $l \geqslant 0$,

$$\sum_{i=0}^{l} \binom{l}{i} (q_1)_i (q_1)_{l-i} \left\{ (q_1 + i) \Delta G_i\left(\frac{x}{c}\right) g_{l-i}(x) \right.$$
$$\left. - c^{-1}(q_1 + l - i) g_i\left(\frac{x}{c}\right) \Delta G_{l-i+1}(x) \right\} \geqslant 0, \tag{14.39}$$

where $(q_1)_s = q_1(q_1+1)\ldots(q_1+s-1), \Delta G_i(x) = G_{i+1}(x) - G_i(x)$, and $G_j(x)$ is the cdf corresponding to $g_j(x)$. The monotonicity is strict if we have a strict inequality in (14.39) for some integer l.

The density $h_\lambda(x)$ of R^{*2} is of the form (14.18) or (14.36) depending on the case, and the $g_j(x)$ are central F-densities with degrees of freedom $2(q+j)$ and m. It should be noted that the density of R^{*2} is a special case of the convex mixtures discussed in §11.2.4 with $m=1, l=0$, and $a_0=1$ in the conditional case, and with $l=1$ and $a_0=1$ in the unconditional case. Thus the condition (14.39) is a special case of (11.23). In fact, the more stringent condition (11.24) is satisfied in this case. Thus the infimum on the right-hand side of (14.35) is attained when $\lambda=0$, and the constant c_4 (in either case) is given by

$$\int_0^\infty F_{2q,2m}^{k-1}\left(\frac{x}{c_4}\right) dF_{2q,2m}(x) = P^*, \tag{14.40}$$

where $F_{2q,2m}(x)$ is the cdf of a central F variable with $(2q, 2m)$ degrees of freedom.

If we assume that q and m are integers, that is, p and n are odd, we can write (14.40) after a change of the variable of integration and some simplifications in the form

$$P^* = \int_0^1 \frac{\Gamma(q+m)t^{m-1}(1-t)^{qk-1}}{c_4^m\Gamma(q)\Gamma(m)\{1+c_4^{-1}(1-c_4)t\}^{m+q}} \times \left\{1 + \frac{(q)_1}{1!}t + \cdots + \frac{(q)_{m'}t^{m'}}{m'!}\right\}^{k-1} dt \tag{14.41}$$

where $m' = m-1$. The values of the constant c_4 are tabulated by Gupta and Panchapakesan (1969a) for $k=2(1)4; P^*=0.75$, 0.90, 0.95, 0.99, and for $m=1(1)8$ when $q=1(1)4$ and $m=1(1)4$ when $q=5(1)8$. As for the properties of the procedure D, it has been shown that it is monotone and the supremum of $E(S|D)$ is attained when $\lambda_1 = \cdots = \lambda_k$.

14.3.4 Selection in Terms of Measures of Association between Two Subclasses of Variates

In comparing k p-variate normal distributions, we consider the p-variates to be made up of two subsets of q_1 and $q_2 (q_1+q_2=p)$ variates and rank the populations according to the values in these populations of a suitable measure of association between the two sets of variates. There are several such possible measures.

Let $(\mathbf{X},\mathbf{Y})$ be a (q_1+q_2)-dimensional random vector with covariance matrix

$$\Sigma=\begin{pmatrix}\Sigma_{xx} & \Sigma_{xy}\\ \Sigma_{yx} & \Sigma_{yy}\end{pmatrix}.$$

We assume that $q_1 \leqslant q_2$ and let $\nu_1^2,\dots,\nu_{q_1}^2$ be the canonical correlations associated with $\mathbf{X}$ and $\mathbf{Y}$. The *conditional generalized variance* of $\mathbf{Y}$ given $\mathbf{X}$ is

$$|\Sigma_{y.x}|=\frac{|\Sigma|}{|\Sigma_{xx}|}=|\Sigma_{yy}-\Sigma_{yx}\Sigma_{xx}^{-1}\Sigma_{xy}|. \tag{14.42}$$

Selection in terms of the conditional generalized variance of the q_2-set given the q_1-set has been considered by Gupta and Panchapakesan (1969a).

The *coefficient of alienation* between $\mathbf{X}$ and $\mathbf{Y}$ is γ_{xy}, where

$$\gamma_{xy}^2=\frac{|\Sigma_{y.x}|}{|\Sigma_{xx}|}=\frac{|\Sigma|}{|\Sigma_{xx}||\Sigma_{yy}|}. \tag{14.43}$$

Selection in terms of the vector coefficient of alienation was considered by Frischtak (1973). We briefly discuss these two problems.

Let $\pi_i(i=1,\dots,k)$ be a p-variate normal distribution with mean vector $\boldsymbol{\mu}_i$ and covariance matrix Σ_i. The p-variates are partitioned into two sets of q_1 and q_2 components. The corresponding partition of Σ_i is denoted by

$$\Sigma_i=\begin{pmatrix}\Sigma_{11}^{(i)} & \Sigma_{12}^{(i)}\\ \Sigma_{21}^{(i)} & \Sigma_{22}^{(i)}\end{pmatrix}.$$

Let $\theta_i=|\Sigma_i|/|\Sigma_{11}^{(i)}|$, the conditional generalized variance of the q_2-set given the q_1-set. Gupta and Panchapakesan (1969a) considered selection of a subset containing the population with the smallest θ_i.

If the observations are taken on the variables of the q_2 set, holding the variables of the q_1 set fixed, then the problem reduces to selecting in terms of the generalized variances for the conditional normal distributions with dimensionality q_2. This has been solved by Gnanadesikan and Gupta (1970) and discussed in §14.3.2. Suppose we consider the unconditional case and still rank the populations according to the θ-values. Let S_i be the sample covariance matrix from π_i based on $n(>p)$ observations and let the partition of S_i be denoted by

$$S_i = \begin{pmatrix} S_{11}^{(i)} & S_{12}^{(i)} \\ S_{21}^{(i)} & S_{22}^{(i)} \end{pmatrix}.$$

Define $s_i = |S_i|/|S_{11}^{(i)}|$. Gupta and Panchapakesan (1969a) defined the procedure

R: Select π_i if and only if

$$s_i \leqslant D^{-1} s_{[1]}, \tag{14.44}$$

where $0 < D = D(k, P^*, n, q_1, q_2) \leqslant 1$ is to be chosen subject to the P^*-condition.

In this case it has been shown by them that

$$\inf_{\Omega} P(CS|R) = \int_0^{\infty} [1 - G(Dx)]^{k-1} dG(x), \tag{14.45}$$

where $G(x)$ is the cdf of a random variable that is distributed as the product of q_2 independent χ^2 variables with degrees of freedom $n - q_1 - 1, n - q_1 - 2, \ldots, n - q_1 - q_2$, respectively. The problem of evaluation of D is similar to that of c_3 in the procedure R_3 of Gnanadesikan and Gupta (1970) defined by (14.26).

For selection in terms of the coefficient of alienation, let $\gamma_i^2 = |\Sigma_i|/|\Sigma_{11}^{(i)}||\Sigma_{22}^{(i)}|$ and $G_i^2 = |S_i|/|S_{11}^{(i)}||S_{22}^{(i)}|$. Then the procedure of Frischtak (1973) is

R': Select π_i if and only if

$$G_i^2 \leqslant c^{-1} G_{[1]}^2, \tag{14.46}$$

where $0 < c = c(k, P^*, n, q_1, q_2) \leqslant 1$ is a constant to be determined subject to the P^*-condition. Frischtak has only obtained an asymptotic $(n \to \infty)$ solution. The value of c is given by

$$\Pr\left\{ Y_j \leqslant \frac{-\sqrt{n-1}\ \log c}{2(2q)^{\frac{1}{2}}}, \qquad j = 1, \ldots, k-1 \right\} = P^*, \tag{14.47}$$

where the Y_j are equally correlated $N(0,1)$ variables with correlation equal to $1/2$.

Finally we note that when $q_1 = 1$, we have $1 - \gamma^2 =$ the square of the multiple correlation coefficient between the first variable and the rest.

14.3.5 Other Ranking Measures

There are several other measures that could be used for ranking multivariate normal populations. Suppose π_i is $N_p(\boldsymbol{\mu}_i, \Sigma_i), i=1,\ldots,k$. If $\Sigma_i = \Sigma$ for all i, one may want to order the populations on the basis of the mean vectors. For example, π_i can be defined to better than π_j if $\mu_{i\alpha} \geqslant \mu_{j\alpha}$, for $\alpha = 1,\ldots,p$. This leads to only a quasi-ordering because any two populations may not be comparable. For a population to be best, it must be so componentwise. Such a population can be called "uniformly best." However, we do not know if there is one such among a given set of populations. Gnanadesikan (1966) has done some investigation of this problem assuming that (i) there does exist a uniformly best population, and (ii) a uniformly best population exists with a specified probability.

As a measure of association between the sets **X** and **Y**, we defined the coefficient of alienation, γ_{xy}, in (14.43). Two other possible measures mentioned in this context are the *vector multiple correlation coefficient*, R_{xy}, and the *trace correlation coefficient*, ρ_{xy}, defined respectively by

$$R_{xy}^2 = \frac{|\Sigma_{xy}\Sigma_{yy}^{-1}\Sigma_{yx}|}{|\Sigma_{yy}|} \tag{14.48}$$

and

$$\rho_{xy}^2 = q_1^{-1}\operatorname{tr}\Sigma_{xy}\Sigma_{yy}^{-1}\Sigma_{yx}\Sigma_{xx}^{-1}. \tag{14.49}$$

We assume that $q_1 \leqslant q_2$. Let the canonical correlations associated with the two sets of variates be $\nu_1^2,\ldots,\nu_{q_1}^2$. Then we have

$$\begin{aligned}\gamma_{xy}^2 &= (1-\nu_1^2)\cdots(1-\nu_{q_1}^2),\\ R_{xy}^2 &= \nu_1^2\cdots\nu_{q_1}^2,\\ \rho_{xy}^2 &= q_1^{-1}(\nu_1^2+\cdots+\nu_{q_1}^2).\end{aligned} \tag{14.50}$$

Another possible measure is an analog of the product moment correlation, namely, $|\Sigma_{xy}|/|\Sigma_{xx}||\Sigma_{yy}|$.

14.4 NOTES AND REMARKS

As we have seen in this chapter as well as in Chapter 7, most of the investigations involve a scalar measure for comparing two populations. As explained in §7.4, it is interesting and useful to investigate when comparisons are based on vector-valued measures which induce a partial ordering.

Another problem of practical interest is to select the best subset of regression variables where bestness can be defined in different ways. Some recent work in this area has been done by McCabe and Arvesen (1974), Arvesen and McCabe (1975), and Gupta and D. Y. Huang (1977e). These problems are discussed in greater detail in §22.3.

CHAPTER 15

Nonparametric Procedures

15.1 INTRODUCTION

In this chapter we assume that the functional forms of the underlying distributions are not known. However, we do assume some general property of the class to which they belong. In the problem of selecting the population with the largest α-quantile we only assume that the underlying distributions are absolutely continuous. In some other problems, we assume that the distributions differ in a location or a scale parameter. There are situations in which we can assume that the distributions belong to a family which can be described by a partial order relation with respect to a known distribution. For example, the family of distributions with increasing failure rates is of this type. Such distributions are convex ordered with respect to an exponential distribution. These families of distributions have been called restricted families in the literature. We briefly discussed such families in §9.2.4. The selection procedures concerning these families could be called nonparametric in a sense; however, these are discussed separately in the next chapter in view of their special importance in reliability problems.

15.2 SELECTION IN TERMS OF QUANTILES

It is assumed that the population π_i has associated with it a continuous cdf $F_i(x)$, $i=1,\ldots,k$, that a sample of size n is taken from each population and that observations from the same or different populations are all independent. We follow the notations of §8.2 wherein we discussed the procedure of Sobel (1967) for selecting the populations with the t-largest α-quantiles under the indifference zone formulation. A procedure of Desu and Sobel (1971) under the fixed-size subset selection approach was discussed in §9.2.2. We now discuss a selection procedure of Rizvi and Sobel (1967) for selecting a subset containing the population with the largest α-quantile. It is assumed that n is sufficiently large so that

$$1 \leqslant (n+1)\alpha \leqslant n \tag{15.1}$$

and a positive integer r is defined by the inequalities

$$r \leqslant (n+1)\alpha < r+1 \tag{15.2}$$

It follows that $1 \leqslant r \leqslant n$. The procedure $R_1 = R_1(c)$ of Rizvi and Sobel (1967) given below is defined in terms of a positive integer $c(1 \leqslant c \leqslant r-1)$ and the order statistics $Y_{j,i}$, where $Y_{j,i}$ denotes the jth order statistic from π_i.

***R*$_1$**: Select π_i if and only if

$$Y_{r,i} \geqslant \max_{1 \leqslant j \leqslant k} Y_{r-c,j} \tag{15.3}$$

where c is the smallest integer with $1 \leqslant c \leqslant r-1$ for which the P^*-condition is satisfied.

For this procedure, we have

$$P(CS|R_1) = \int_{-\infty}^{\infty} \prod_{i=1}^{k-1} H_{r-c,i}(y)\, dH_{r,k}(y) \tag{15.4}$$

where

$$H_{r,i}(y) = \sum_{j=r}^{n} \binom{n}{j} F_{[i]}^{j}(y)\left[1 - F_{[i]}(y)\right]^{n-j}. \tag{15.5}$$

The infimum of $P(CS|R_1)$ over Ω, the space of all k-tuples $(F_1,\dots,F_k)$, is attained when the distributions are identical and after a simple transformation we get

$$\inf_{\Omega} P(CS|R_1) = \int_0^1 G_{r-c}^{k-1}(u)\,dG_r(u), \tag{15.6}$$

where $G_r(u) = I_u(r, n-r+1)$, the standard incomplete beta function. For $c=0$, the right-hand side of (15.6) is clearly $1/k$. It can be easily shown that $G_{r-c}(u)$ is an increasing function of c for fixed r and u, or that $G_r(u)$ is decreasing in r for fixed u. Thus the maximum permissible value of P^* such that a c-value satisfying the P^*-condition exists is obtained by setting $c=r-1$ in (15.6) and this maximum value is given by

$$P_1 = P_1(n,\alpha,k) = \binom{n}{r} \sum_{i=0}^{k-1} \frac{(-1)^i \binom{k-1}{i}}{\binom{n(i+1)}{r}}. \tag{15.7}$$

A short table of P_1-values is given by Rizvi and Sobel for $\alpha = \frac{1}{2}$ and $k = 2(1)10$. The n-value increases from 1 in steps of 2 until a value (depending on k) for which P_1 is very close to 1. Another table gives the largest value of $r-c$ for $\alpha = \frac{1}{2}$ (and so $r = (n+1)/2$), $k = 2(1)10$, $n = 5(10)95(50)495$, and $P^* = 0.75, 0.90, 0.95, 0.975, 0.99$.

Rizvi and Sobel (1967) have shown that the procedure R_1 is monotone and that $E(S|R_1)$ is maximized when the distributions are identical. They have compared their procedure asymptotically with the procedure R_2 based on sample means. To make this comparison it is assumed that the F_i have a finite, known and (for simplicity) common variance σ^2, and that the sample mean $\bar{X}_i$ from the distribution $F_{[i]}$ has an asymptotic normal distribution with mean $\theta_{[i]}$. The procedure R_2 of Gupta (1965) selects π_i if and only if

$$\bar{X}_i > \bar{X}_{[k]} - \frac{d\sigma}{\sqrt{n}}, \tag{15.8}$$

where $d = d(n, P^*) > 0$ is given by (see §12.2)

$$\int_{-\infty}^{\infty} \Phi^{k-1}(y+d)\,d\Phi(y) = P^*.$$

The asymptotic relative efficiency (ARE) of R_1 relative to R_2 for translation alternatives (TA) is the limit as $\epsilon \to 0$ of the ratio of $n_2(\epsilon)/n_1(\epsilon)$, where $n_i(\epsilon)$, $i = 1, 2$, is an approximation to the sample size needed to satisfy

$$E(S|R_i, \Omega_\Delta) \leqslant 1 + \epsilon \tag{15.9}$$

and $\Omega_\Delta = \{\boldsymbol{\theta}: \theta_{[1]} = \cdots = \theta_{[k-1]} = \theta_{[k]} - \Delta, \Delta > 0\}$. It has been shown that

$$\text{ARE}(R_1, R_2; \text{TA}) = \frac{\sigma^2 f^2(\lambda)}{\alpha\bar{\alpha}}, \tag{15.10}$$

where $F(\lambda) = \alpha$.

The asymptotic efficiency of R_1 has been discussed by Rizvi and Sobel (1967) relative to procedures of Bartlett and Govindarajulu (1968) to be defined in the next section and by Barlow and Gupta (1969) relative to their quantile selection procedure for a restricted family which is discussed in the next chapter.

15.3 PROCEDURES BASED ON RANKS

Let π_i have an associated continuous distribution function $F(x, \theta_i)$. Let X_{ij}, $j = 1, \ldots, n_i$, be independent observations from π_i, and R_{ij} be the rank of X_{ij} among all the $N = \sum_{i=1}^k n_i$ observations in the pooled sample. Let $Z(1) < Z(2) < \cdots < Z(N)$ be the ordered observations based on N independent observations from a continuous distribution G with finite expectations. Define

$$a(r) = E[Z(r)|G], \qquad r = 1, \ldots, N,$$

$$H_i = n_i^{-1} \sum_{j=1}^{n_i} a(R_{ij}), \qquad i = 1, \ldots, k, \tag{15.11}$$

$$H_i' = n_i^{-1} \sum_{j=1}^{n_i} Z(R_{ij}), \qquad i = 1, \ldots, k.$$

When $F(x, \theta_i) = F(x - \theta_i)$, Gupta and McDonald (1970) proposed the rule $R_1(G)$: Select π_i if and only if

$$H_i \geqslant H_{[k]} - d \tag{15.12}$$

where $0 < d = d(k, n_1, \ldots, n_k, P^*)$ is to be chosen subject to the P^*-condition. It has been shown by Gupta and McDonald that the infimum of $P(CS|R_1(G))$ over Ω is attained for $\boldsymbol{\theta} \in \Omega_k = \{\boldsymbol{\theta}: \theta_{[k-1]} = \theta_{[k]}\}$. This shows that for $k = 2$ the infimum occurs at an equi-parameter configuration. However, for $k \geqslant 3$ the LFC is not given by the equi-parameter configuration as claimed earlier by Bartlett and Govindarajulu (1968). The difficulty in establishing the LFC can be seen from the counterexamples of Rizvi and Woodworth (1970).

Suppose that $n_i = n$. It is easy to see that

$$(k-1)^{-1}\sum_{j=1}^{k-1} H_{(j)} \leqslant \max_{1 \leqslant j \leqslant k-1} H_{(j)} \leqslant n^{-1} \sum_{r=N-n+1}^{N} a(r), \tag{15.13}$$

where $H_{(j)}$ is the H_i corresponding to the population associated with $\theta_{[j]}$. Using (15.13) and the result that $\Pr(H_{(k)} \geqslant l)$ is minimized for $\theta_1 = \cdots = \theta_k$, it has been shown by Gupta and McDonald (1970) that

$$P_\theta(H_{(k)} \geqslant v) \leqslant \inf_{\Omega} P(CS|R_1(G)) \leqslant P_\theta(H_{(k)} \geqslant u) \tag{15.14}$$

where the bounds are for the configuration $\theta_1 = \cdots = \theta_k = \theta$, and

$$v = v(d,k,n) = n^{-1} \sum_{r=N-n+1}^{N} a(r) - d,$$

$$u = u(d,k,n) = (nk)^{-1}\left[\sum_{r=1}^{N} a(r) - nd(k-1)\right]. \tag{15.15}$$

In the special case of $a(r) = r$, a more useful form of the lower bound appearing in (15.14) is given by $\Pr(U \leqslant nd)$, where U is the Mann–Whitney statistic associated with samples of size n and $(k-1)n$ taken from identically distributed populations. When $a(r) = r$ and $n_i = n$, let $T_i = n_i H_i$. Then T_i is the rank-sum statistic. The distribution of $\mathbf{T}' = (T_1, \ldots, T_k)$ is asymptotically normal with mean vector $\boldsymbol{\mu}'_T = (\mu_1, \ldots, \mu_k)$ and covariance matrix $\Sigma_T = ((\sigma_{ij}))$, where $\mu_i = E(T_i)$, $\sigma_{ii} = V(T_i)$, and $\sigma_{ij} = \mathrm{cov}(T_i, T_j)$. Let

$$\mathbf{W} = A\mathbf{T}, \tag{15.16}$$

where A is a $(k-1) \times k$ matrix given by

$$A = \begin{bmatrix} 1 & 0 & \cdots & 0 & -1 \\ 0 & 1 & \cdots & 0 & -1 \\ \vdots & \vdots & \vdots & \vdots & \vdots \\ 0 & 0 & \cdots & 1 & -1 \end{bmatrix}. \tag{15.17}$$

Thus $W_i = T_i - T_k$, $i = 1, \ldots, k-1$, and for large n, $\mathbf{W}' = (W_1, \ldots, W_{k-1})$ is approximately distributed normally with mean vector $\boldsymbol{\mu}_W = A\boldsymbol{\mu}_T$ and covariance matrix $\Sigma_W = A\Sigma_T A'$. Now define

$$\mathbf{W}_\nu = A_\nu \mathbf{T},$$

where A_ν is the matrix obtained from A in (15.17) by moving column j to column $j+1$ for $j=\nu,\nu+1,\ldots,k$, and replacing column ν by column k. Let $\Sigma_\nu = A_\nu \Sigma_T A_\nu'$ and assume that Σ_ν is positive definite. Then the asymptotic expression for $E(S|R_1)$ ($R_i(G)$ is called R_i in the special case) is given by

$$E(S|R_1) \approx \sum_{\nu=1}^{k} \left[(2\pi)^{k-1}|\Sigma_\nu|\right]^{-1/2}$$

$$\times \int_{-\infty}^{d} \cdots \int_{-\infty}^{d} \exp\left[-\frac{1}{2}(\mathbf{w}-\boldsymbol{\mu}_\nu)'\Sigma_\nu^{-1}(\mathbf{w}-\boldsymbol{\mu}_\nu)\right] d\mathbf{w} \quad (15.18)$$

We saw that $\inf P(CS|R_1)$ over $\Omega' = \{\boldsymbol{\theta}: \theta_{[1]} = \cdots = \theta_{[k-1]} \leqslant \theta_{[k]}\}$ is attained when $\theta_1 = \cdots = \theta_k$. Suppose we want to evaluate d for which this infimum is at least P^*. Using the rank-sum statistics, this means that we want the smallest integer $m = nd$ such that

$$\Pr(T_k \geqslant T_{[k]} - m) \geqslant P^* \quad (15.19)$$

where the T_i are i.i.d. random variables. McDonald (1971) has discussed two methods of approximating the solution. The first method uses the asymptotic ($n \to \infty$) expression for the probability given by

$$\Pr(T_k \geqslant T_{[k]} - m) \approx \int_{-\infty}^{\infty} \Phi^{k-1}(x + mz^{-1})\, d\Phi(x), \quad (15.20)$$

where $z = z(n,k) = n[k(nk+1)/12]^{1/2}$. The approximate value can be calculated using the tables of Gupta [1963a]. The second approximation is for large P^* (near 1). Suppose $Z_1,\ldots,Z_k$ are $N(0,1)$ random variables with the correlation matrix Σ. Let

$$\Pr\{Z_{[1]} \geqslant -\delta\} = P^*. \quad (15.21)$$

Dudewicz (1969a) has shown that, for large P^* (near 1), an approximation to δ is given by

$$\delta^2 \approx -2[\log(1-P^*)] \quad (15.22)$$

in the sense that the ratio tends to 1 as $P^* \to 1$. Using this approximation and the joint asymptotic normal distribution of $[n^2k(nk+1)/6]^{-1/2}(T_k - T_i)$, $i = 1,\ldots,k-1$, we obtain the approximation

$$m^2 \approx -\left[n^2k\frac{(nk+1)}{3}\right]\log(1-P^*). \quad (15.23)$$

One can also obtain this approximation from (15.20) by noting that $mz^{-1} \approx \sqrt{2}\,\Phi^{-1}(P^*)$ as $P^* \to 1$, a result of Rizvi and Woodworth (1970), and using the well-known fact that

$$\Phi^{-1}(P^*) \approx \left[-2\log(1-P^*)\right]^{1/2} \quad \text{as} \quad P^* \to 1. \tag{15.24}$$

The two approximations have been compared by McDonald (1971) in the case of $P^*=0.99$ for $k=2(1)5$, $n=5(5)25$.

Let $\hat{m}_1$ and $\hat{m}_2$ denote the approximate values of m given by (15.20) and (15.23), respectively. The numerical evaluations of $\hat{m}_1$ and $\hat{m}_2$ show that (a) $\hat{m}_2 - \hat{m}_1$ increases in n for fixed k, and decreases in k for fixed n, (b) $\hat{m}_1/\hat{m}_2$ increases in k for fixed n, and is constant for fixed k over various values of n, and (c) both approximations are conservative, $\hat{m}_2$ being more so than $\hat{m}_1$. For $k=2$, McDonald (1971) has analytically shown that $\hat{m}_2 - \hat{m}_1$ is positive and increasing in n, and that $\hat{m}_2/\hat{m}_1$ is independent of n.

If G is the distribution of a nonnegative random variable, then $H_i \geqslant 0$ for all i. In this case, Gupta and McDonald (1971) defined the class of procedures $R_2(G)$: Select π_i if and only if

$$H_i \geqslant c^{-1} H_{[k]} \tag{15.25}$$

where $c = c(k, P^*, n_1, \ldots, n_k)$ is a constant $\geqslant 1$. As in the case of $R_1(G)$, the infimum of $P(CS|R_2(G))$ over Ω is attained at an equi-parameter configuration only for $k=2$. Corresponding to (15.14), Gupta and McDonald have shown that

$$P_\theta(H_{(k)} \geqslant v') \leqslant \inf_{\Omega} P(CS|R_2(G)) \leqslant P_\theta(H_{(k)} \geqslant u') \tag{15.26}$$

where the bounds are for the configuration $\theta_1 = \cdots = \theta_k = \theta$, and

$$v' = v'(d,k,n) = (nc)^{-1} \sum_{r=N-n+1}^{N} a(r),$$

$$u' = u'(d,k,n) = \left[n\{1 + c(k-1)\}\right]^{-1} \sum_{r=1}^{N} a(r). \tag{15.27}$$

A third class of rules proposed by Gupta and McDonald (1971) is $R_3(G)$: Select π_i if and only if

$$H_i \geqslant D, \tag{15.28}$$

where $D \in (-\infty, \infty)$. $R_3(G)$ differs from the other two classes of procedures in the sense that the subset chosen could be empty. However, $\inf_{\Omega} P(CS|R_3(G))$ is attained when $\theta_1 = \cdots = \theta_k$.

It should be noted that for $k=2$ and $n_1 = n_2$, the rules $R_i(G)$, $i=1,2,3$, are equivalent. For the special case of rank-sum statistics based on equal sample sizes, Gupta and McDonald (1971) have studied the asymptotic efficiency of R_1 relative to the means procedure of Gupta (1956) for normal populations (procedure R_1 of §12.2.1) and the efficiency of R_2 relative to the procedure of Gupta (1963) for gamma populations (procedure R_{16} of §12.6) both in the case of $k=2$ populations.

Let π_1 and π_2 be independent normal populations with means θ_0 and $\theta_0 + \theta (\theta \geq 0)$ and common unit variance. Let R denote Gupta's means procedure. For both R_1 and R satisfying the P^*-condition, the asymptotic efficiency of R_1 relative to R is $\text{ARE}(R_1, R; \theta) = \lim_{\epsilon \to 0} n_R(\epsilon)/n_{R_1}(\epsilon)$, where $n_R(\epsilon)$ and $n_{R_1}(\epsilon)$ are the samples sizes for which $E(S) - \text{PCS} = \epsilon$ for R and R_1, respectively. It is shown by Gupta and McDonald that

$$\text{ARE}(R_1, R; \theta) = \left\{ \frac{2\Phi(\theta/\sqrt{2}\,) - 1}{2\theta B(\theta)} \right\}^2, \tag{15.29}$$

where

$$B^2(\theta) = \int_{-\infty}^{\infty} \Phi^2(x+\theta)\varphi(x)\,dx - \Phi^2\left(\frac{\theta}{\sqrt{2}}\right).$$

As θ decreases to zero, we see that $\text{ARE}(R_1, R; \theta) \to 3/\pi = 0.9549$.

Let R' denote Gupta's procedure for gamma populations. Let the scale parameters of π_1 and π_2 be θ_0 and $\theta_0\theta$, $\theta > 1$. In this case

$$\text{ARE}(R_2, R'; \theta) = \left[\frac{(\theta - 1)}{4(\theta+1)B(\theta)\log\theta} \right]^2 \tag{15.30}$$

and $\lim_{\theta \downarrow 1} \text{ARE}(R_2, R'; \theta) = \frac{3}{4}$.

In another paper (1972), Gupta and McDonald have compared the procedures R_1, R_2, and R_3 based on rank-sum statistics with a procedure R_m which they proposed for selection from gamma populations in terms of the guaranteed life. Let π_i have the associated density function

$$f(x-\theta_i) = \begin{cases} [\lambda\Gamma(r)]^{-1}[(x-\theta_i)/\lambda]^{r-1} e^{-(x-\theta_i)/\lambda}, & x \geq \theta_i \\ 0 & \text{elsewhere} \end{cases}$$

where $r(>0)$ and $\lambda(>0)$ are known common parameters. In life-testing problems, the parameter θ is called the guaranteed life time. We can assume with no loss of generality that $\lambda=1$. Let $Y_i=X_{i[1]}$, the smallest order statistic based on n independent observations from π_i. It is known that Y_i is a complete sufficient statistic for θ_i. The procedure R_m of Gupta and McDonald (1972) for selecting a subset containing the population with the largest guaranteed life is

R_m: Select π_i if and only if

$$Y_i \geqslant Y_{[k]} - b, \tag{15.31}$$

where $b=b(k,n,P^*)>0$ is chosen to satisfy the P^*-requirement. They have shown that

$$\inf_{\Omega} P(CS|R_m) = \int_0^{\infty} H^{k-1}(x+b)\,dH(x), \tag{15.32}$$

where $H(x)$ is the cdf Y_i when $\theta_i=0$.

In the special case of $r=1$ (the exponential case) (15.32) reduces to

$$\inf_{\Omega} P(CS|R_m) = k^{-1}(1-w)^{-1}(1-w^k), \tag{15.33}$$

where $w=1-e^{-nb}$. For this special case, Gupta and McDonald (1972) have tabulated the values of b for $k=5,10$; $n=2(1)25$; and $P^*=0.75, 0.90, 0.95, 0.975, 0.99$.

Consider three exponential populations with location parameters $\theta_1=0$, $\theta_2=\theta_3=\theta>0$. In this case, Gupta and McDonald (1972) have compared the expected subset sizes for the procedures R_1, R_2, R_3, and R_m for $\theta^*=0(0.1)1.5$ and $P^*=0.6, 14/15$. The computations indicate that (1) R_1 and R_2 perform equally well for $P^*=0.6$, (2) R_2 and R_3 perform equally well for $P^*=14/15$, (3) $E(S|R_2)=E(S|R_3)\leqslant E(S|R_1)$ for all θ, equality holding for $\theta=0$, (4) R_m performs better than all the distributionfree procedures for the smaller value of P^*, (5) for the larger P^*, the distributionfree procedures are better than R_m for $\theta \leqslant 0.5$, and (6) for larger values of θ ($\theta \geqslant 1.1$) R_m is the best among the four rules.

More details on the preceding distributionfree procedures are given by McDonald (1969a). In the location parameter case, Bartlett and Govindarajulu (1968) have discussed the randomized version of R_1 which uses H_i' defined in (15.11) in the place of H_i. The randomized and the nonrandomized rules are asymptotically equally efficient.

In the preceding procedures R_1, R_2, and R_3, the statistics H_i are defined using the ranks of the observations in the pooled sample. When we have equal samples, we can use vector-at-a-time sampling. Let $(X_{1j},X_{2j},\ldots,X_{kj})$ be the jth vector and R_{ij} be the rank of X_{ij} among the k observations of the vector. Let $Z(1)\leqslant Z(2)\leqslant\cdots\leqslant Z(k)$ denote an ordered sample of size k from a continuous distribution G. Define

$$b(r)=E[Z(r)|G], \qquad r=1,\ldots,k,$$

$$J_i=n^{-1}\sum_{j=1}^{n} b(R_{ij}), \qquad i=1,\ldots,k. \tag{15.34}$$

McDonald (1972) investigated the classes of procedures $R_1'(h;G)$ and $R_2'(g;G)$ which are defined using the two classes of functions $\{h(x)\}$ and $\{g(x)\}$, where h and g are nondecreasing real-valued functions defined on the interval $I=[b(1),b(k)]$ and h satisfies the additional property that $h(x)\geqslant x$ for all $x\in I$. The two classes of procedures are

$\boldsymbol{R_1'(h;G)}$: Select π_i if and only if

$$h(J_i)\geqslant J_{[k]}, \tag{15.35}$$

and

$\boldsymbol{R_2'(g;G)}$: Select π_i if and only if

$$g(J_i)\geqslant 0 \tag{15.36}$$

Of course, $R_2'(g;G)$ can select an empty set. Particular members of these classes that are of special interest are $R_1'(G)$ with $h(x)=x+b$, $b>0$, and $R_2'(G)$ with $g(x)=x-d$, d real. The treatment of $R_1'(G)$ and $R_2'(G)$ is similar to that of $R_1(G)$ and $R_2(G)$ described earlier in this section. The infimum of PCS over Ω is attained at a point in Ω_k in the case of $R_1'(h;G)$. However as in the case of $R_1(G)$, it is not generally true that the infimum is attained at an equi-parameter configuration. But the statement is true in the case of $R_2'(g;G)$.

When $b(r)=r$, $nH_i=T_i$, the rank-sum statistic associated with π_i. McDonald (1973) has discussed the related distribution of $U=\max_{1\leqslant j\leqslant k}T_j-T_1$, where the distributions F_i are identical. He has tabulated $\Pr(U\leqslant b)$ for $k=2$, $n=2(1)20$; $k=3$, $n=2(1)8$; $k=4$, $n=2(1)5$; and $k=5$, $n=2,3$. For $\Pr(U\leqslant b)=P^*=0.75$, 0.90, 0.95, 0.975, and 0.99, he has tabulated the asymptotic value of b for $k=2$, $n=10(5)20$; $k=3$, $n=6(1)8$; $k=4$, $n=3(1)5$; and $k=5$, $n=3$.

The investigations of McDonald (1973) with respect to slippage configuration based on simulations show that R_1' and R_2' (which are $R_1'(G)$ and $R_2'(G)$, respectively, in the special case with $b(r)=r$) are roughly equivalent when the underlying distribution has a long tail and the slippage is small, and that R_1 is better otherwise.

In another paper, McDonald (1975) considered the case of three exponential distributions with parameters (guaranteed lives) $0=\theta_1=\theta_2\leqslant\theta_3=\theta$ with samples of size two. For the rules R_1, R_2, and R_3 (using rank-sum statistics T_i) the infimum of PCS takes place when $\theta_1=\theta_2=\theta_3$. However, it is shown that the expected subset size is not bounded above by kP^*, a property enjoyed by many parametric procedures [see Gupta (1965)] for the location parameter case under monotone likelihood ratio conditions.

Blumenthal and Patterson (1969) have considered selecting a subset containing the population with the smallest scale parameter from a set of k populations with associated distribution functions $F(x/\sigma_i)$, $i=1,\ldots,k$, which are assumed to be continuous with mean zero but otherwise unknown. They would like to define a procedure that satisfies the P^*-condition and does not select a large subset when $\sigma_{[1]}$ and $\sigma_{[2]}$ are sufficiently apart. In absence of any general result relating to the configurations that will enable one to achieve this, they consider the sample size required to guarantee that $E(S)=1+\epsilon^*$ when $\theta^*\sigma_{[1]}=\sigma_{[2]}=\cdots=\sigma_{[k]}$, and $\mathrm{PCS}=P^*$ when $\sigma_{[1]}=\cdots=\sigma_{[k]}$, where ϵ^* and θ^* are given.

Let H_i and H_i' be defined as in (15.11) with $n_i=n$ for all i except for the fact that R_{ij} is now the rank of $|X_{ij}|$ among the absolute values of all the observations in the pooled sample. Blumenthal and Patterson (1969) considered a randomized procedure $R_3(G)$ which selects π_i if and only if

$$H_i'^2\leqslant b_3^{-1}H_{[1]}'^2, \tag{15.37}$$

and a rank scores procedure $R_4(G)$ which selects π_i if and only if

$$H_i^2\leqslant b_4^{-1}H_{[1]}^2, \tag{15.38}$$

where $0<b_3,\ b_4\leqslant 1$. The two procedures are asymptotically equivalent and the large sample solution for b_3 (and hence for b_4) is obtained using the techniques of Lehmann (1963) and Puri and Puri (1968). For large n, the region of possible bias—where PCS may fall below P^*—is claimed to be confined at most to a circle of radius $\sigma/\sqrt{n}$ about the "origin" ($\sigma_{[1]}=\cdots=\sigma_{[k]}$).

The procedures that we have discussed so far in this chapter and the nonparametric procedures under the indifference zone approach have been reviewed by Lee and Dudewicz (1974) who have considered among other things procedures based on vector ranks under the indifference zone formulation.

15.4 PROCEDURES BASED ON ESTIMATORS

In §8.3.3 we discussed selection procedures based on one-sample and two-sample Hodges–Lehmann estimators. One could use these estimators to define subset selection procedures. The advantage of using these estimators is in establishing the least favorable configuration for the PCS. However, it should be noted that it is assumed that the populations differ by a location parameter. In this section we deal with subset selection procedures based on one-sample Hodges–Lehmann estimators. These procedures were investigated by Gupta and D. Y. Huang (1974) under the goal of eliminating all strictly non-t-best populations discussed in §11.4 and under the generalized subset selection goal discussed in §11.2.7.

Let X_{ij}, $j=1,\ldots,n$, be independent observations from π_i which has the associated continuous distribution function $F(x-\theta_i)$, $i=1,\ldots,k$. It is assumed that the density $f(\cdot)$ is symmetric about the origin and that the populations have a common known variance taken to be unity.

15.4.1. Elimination of Strictly Non-t-Best Populations

Any population π_i is called "strictly non-t-best" if $\theta_{[k-t+1]}-\theta_i>\Delta$, where Δ is a given positive constant. The goal as we pointed out earlier is to select a nonempty subset with a minimum guaranteed probability P^* of eliminating all the strictly non-t-best populations. The procedure R of Gupta and D. Y. Huang (1974) rejects a population π_i if and only if

$$\hat{\theta}_i<\hat{\theta}_{[k-t+1]}-(\Delta-d_1), \tag{15.39}$$

where d_1 is the smallest number such that $0<d_1<\Delta$ for which $\text{PCS}\geqslant P^*$ for all $\boldsymbol{\theta}\in\Omega$, and $\hat{\theta}_i$ is given by (8.55).

Let $S'_{s,\beta}=\{\hat{\theta}_{\beta_1},\ldots,\hat{\theta}_{\beta_s}\}$ be any subset of $\{\hat{\theta}_1,\ldots,\hat{\theta}_k\}$ and $S_{s,\beta}=\{\alpha|\hat{\theta}_\alpha\in S'_{s,\beta}\}$. It is easy to see that (15.39) is equivalent to

$$\hat{\theta}_i<\max_{(\beta_1,\ldots,\beta_t)}\ \min_{\alpha\in S_{t,\beta}}\hat{\theta}_\alpha-(\Delta-d_1). \tag{15.40}$$

It has been shown by Gupta and Huang (1974) that the infimum of $P(CS|R)$ over Ω is attained at a configuration of the type

$$\theta_{[1]}=\theta_{[k-t]}=\theta-\Delta;\qquad \theta_{[k-t+1]}=\theta_{[k]}=\theta. \tag{15.41}$$

Define

$$A^2=4^{-1}\int_0^1 J^2(u)\,du,$$

$$B(F)=\int_0^\infty \frac{d}{dx}J(2F(x)-1)\,dF(x), \tag{15.42}$$

where $J(u)=\psi^{-1}\{(1+u)/2\}$ and $\psi(\cdot)$ is a symmetric distribution function. Now, let $\hat{\theta}_{(i)}$ denote the estimator associated with $\theta_{[i]}$ and $S_{s,(\beta)}=\{\alpha|\theta_{(\alpha)}\in S'_{s,(\beta)}\}$, where $S'_{s,(\beta)}=\{\hat{\theta}_{(\beta_1)},\ldots,\hat{\theta}_{(\beta_s)}\}$. It can be seen that

$$\inf_{\Omega} P(CS|R)=\inf_{\Omega_0^t}\Pr\left\{\frac{\max\limits_{1\leqslant i\leqslant k-t}\sqrt{n}\left[\hat{\theta}_{(i)}-(\theta-\Delta)\right]B(F)}{A}\right.$$

$$<\frac{\max\limits_{(\beta_1,\ldots,\beta_t)}\ \min\limits_{\alpha\in S_{t,(\beta)}}\sqrt{n}\left[\hat{\theta}_{(\alpha)}-\theta\right]B(F)}{A}$$

$$\left.+B(F)d_1\sqrt{n}\,/A\right\}, \tag{15.43}$$

where Ω_0^t is the space of $\boldsymbol{\theta}$ satisfying (15.41). The reason for rewriting the PCS as in (15.43) is to obtain the asymptotic value using the result (see Hodges and Lehmann [1963]) that, under some regularity conditions, $\sqrt{n}\,(\hat{\theta}_{(i)}-\theta_{[i]})B(F)/A$ is asymptotically $(n\rightarrow\infty)$ distributed as $N(0,1)$, where A and $B(F)$ are defined in (15.42).

We are interested in two special cases. For $J(u)=u$ (Wilcoxon case), $A^2=\frac{1}{12}$ and $B(F)=\int_0^\infty 2f^2(x)dx$. Hodges and Lehmann [1956] showed that for all F, $12[\int_{-\infty}^{\infty}f^2(x)dx]^2\geqslant 0.864$. Thus we have $B(F)/A\geqslant\sqrt{0.864}$. The lower bound is attained for some distribution with parabolic density. If $J(u)=\chi_1^{-1}(u)$ (normal score case), then $B(F)/A\geqslant 1$ for all F (see Puri and Sen [1971], pp. 117–118). The lower bound is attained if $F(x)=\Phi(x)$.

Thus we have

$$\inf_{\Omega} P(CS|R)\geqslant\inf_{\theta}\Pr\left\{\max_{1\leqslant i\leqslant k-t}\frac{\sqrt{n}\left[\hat{\theta}_{(i)}-(\theta-\Delta)\right]B(F)}{A}\right.$$

$$\left.<\max_{(\beta_1,\ldots,\beta_t)}\ \min_{\alpha\in S_{t,\beta}}\frac{\sqrt{n}\left[\hat{\theta}_{(\alpha)}-\theta\right]B(F)}{A}+d\sqrt{n}\right\}, \tag{15.44}$$

where $d=\sqrt{0.864}\ d_1$ in the Wilcoxon case, and $d=d_1$ in the normal score case. Now we can use the asymptotic theory and obtain the following theorem of Gupta and D. Y. Huang (1974).

Theorem 15.1. For large n,

$$\inf_{\Omega} P(CS|R) \geqslant (t-r)\binom{t}{r}\int_{-\infty}^{\infty} \Phi^{k-t}(x+d\sqrt{n}\,)\Phi^{r}(x)[1-\Phi(x)]^{t-r-1}d\Phi(x), \tag{15.45}$$

where $r=\min(t,k-t)-1$.

For the normal score case with $F(x)=\Phi(x)$ we have equality in (15.45) and the right-hand side of (15.45) is same as the result of Carroll, Gupta, and Huang (1975) discussed in §11.4.3.

The supremum of the expected number of strictly non-t-best populations included in the selected subset is shown to tend to zero as $n\rightarrow\infty$ in the Wilcoxon case for some distribution with parabolic density and in the normal score case with $F(x)=\Phi(x)$.

If we consider the sequence of rules $\{R_1(n)\}$, it is consistent with respect to Ω in the sense that $\lim_{n\rightarrow\infty}\inf_{\boldsymbol{\theta}\in\Omega}P(CS|R_1(n))=1$.

15.4.2 Generalized Goal of Selecting a Subcollection of Subsets of Fixed Size

In §11.2.7, we discussed a generalized goal of selecting a random subcollection from the collection of all the $\binom{k}{s}$ subsets of size s so that there is at least one s-subset having at least c of the t best populations. Of course, we recall that k, t, s, and c are known integers such that $\max[1,s+t+1-k]\leqslant c\leqslant\min[s,t]$. Parametric procedures for this goal have been investigated by Deverman (1969), and Deverman and Gupta (1969). Gupta and D. Y. Huang (1974) have discussed a nonparametric procedure based on Hodges–Lehmann estimators. Their procedure R_s selects the s-subset $\{\pi_{\beta_1},\ldots,\pi_{\beta_s}\}$ if and only if

$$\min\{\hat{\theta}_{\beta_1},\ldots,\hat{\theta}_{\beta_s}\}\geqslant\max\{\hat{\theta}_{\beta_{s+1}},\ldots,\hat{\theta}_{\beta_k}\}-d. \tag{15.46}$$

The least favorable configuration can be shown to be $\theta_1=\cdots=\theta_k$ and one can obtain lower bound for the infimum of the PCS by using the same arguments as in the case of the procedure R in §15.4.1. For large n, conservative constant d can be computed using the results of Deverman (1969).

15.5 SELECTION OF REGRESSION EQUATION WITH THE LARGEST SLOPE

Let Y_{ij}, $j=1,\ldots,n$, be independent observations from population π_i which has the associated distribution function $F(y;\alpha_i,\beta_i)=F(y-\alpha_i-\beta_i x_i)$. The populations are assumed to have a common known variance σ^2 (taken to be unity) and the x_i are known constants such that

$$\lim\left[\max_j \frac{(x_j-\bar{x}_n)^2}{\left\{\sum_j (x_j-\bar{x}_n)^2\right\}}\right]=0,$$

$$0<\lim\left[n^{-1}\sum_j (x_j-\bar{x}_n)^2\right]<\infty, \qquad (15.47)$$

$$|\lim \bar{x}_n|<\infty,$$

where $\bar{x}_n = n^{-1}\Sigma_j x_j$, the summation goes from 1 to n, and the limits are taken as $n\rightarrow\infty$. We also assume that the underlying distribution function F belongs to a class $\mathfrak{F}$ of absolutely continuous symmetric distributions functions whose densities are also absolutely continuous and square integrable. With these regularity conditions, Adichie [1967b] defined an estimate of β that is similar to the Hodges–Lehmann estimate used in the last section.

Let $T_i(\mathbf{Y}_i)=T_i(Y_{i1},\ldots,Y_{in})$ be a statistic that satisfies the following two conditions:

A. For fixed a, $T(\mathbf{y}+a+b\mathbf{x})$ is nondecreasing in b for each $\mathbf{y}$ and $\mathbf{x}$. Here $\mathbf{y}+a+b\mathbf{x}$ stands for $(y_1+a+bx_1,\ldots,y_n+a+bx_n)$.

B. When $\alpha=\beta=0$, the distribution of $T(\mathbf{Y})$ is symmetric about a fixed point μ, independent of $F\in\mathfrak{F}$.

Let

$$\beta_i^* = \sup\{b:\ T_i(\mathbf{y}-a-b\mathbf{x})>\mu \quad \text{for all} \quad a\},$$

$$\beta_i^{**} = \inf\{b:\ T_i(\mathbf{y}-a-b\mathbf{x})<\mu \quad \text{for all} \quad a\}. \qquad (15.48)$$

For a suitable function T_i, Adichie [1967b] proposed

$$\hat{\beta}_i = \frac{(\beta_i^* + \beta_i^{**})}{2} \qquad (15.49)$$

as an estimate of β. Many existing estimates of β belong to this class. In particular, the least-squares estimate of β can be obtained as a special case by taking $T_i(\mathbf{Y}_i)=\Sigma_{j=1}^n(x_j-\bar{x}_n)(Y_{ij}-\bar{Y}_i)$, where $\bar{Y}_i=n^{-1}\Sigma_{j=1}^n Y_{ij}$.

For any distribution $G \in \mathcal{F}$ and $0<u<1$, define

$$\psi(u) = -\frac{g'\left(G^{-1}\left(\frac{u+1}{2}\right)\right)}{g\left(G^{-1}\left(\frac{u+1}{2}\right)\right)},$$

$$\psi_0(u) = -\frac{g'(G^{-1}(u))}{g(G^{-1}(u))}, \tag{15.50}$$

where G^{-1} is the inverse of G. Consider the statistics

$$T_i(\mathbf{Y}_i) = n^{-1}\sum_{j=1}^{n}(x_j - \bar{x}_n)\psi_{on}\left(\frac{R_{ij}}{n+1}\right), \qquad i=1,\dots,k, \tag{15.51}$$

where R_{ij} is the rank of Y_{ij} in the ordered sample from π_i, $\bar{x}_n = n^{-1}\Sigma_{j=1}^{n}x_j$, and

$$\psi_{on}(u) = \psi_0\left(\frac{j}{n+1}\right) \quad \text{for} \quad \frac{j-1}{n} < u \leqslant \frac{j}{n}. \tag{15.52}$$

The statistic $T_i(\mathbf{Y}_i)$ in (15.51) has been studied by Hájek [1962] and it satisfies the conditions (A) and (B) mentioned earlier.

Gupta and D. Y. Huang (1977d) have studied the problem of selecting a subset containing the population associated with $\beta_{[k]}$. Their rule R selects π_i if and only if

$$\hat{\beta}_i \geqslant \hat{\beta}_{[k]} - \frac{d}{\sqrt{n}}, \tag{15.53}$$

where $d = d(k, P^*, n) > 0$ is chosen to satisfy the P^*-condition. Using the stochastic ordering of the distributions of $\hat{\beta}_i$, it is shown that the infimum of the PCS occurs when $\beta_1 = \cdots = \beta_k = \beta$.

Under certain regularity conditions, Adichie [1967b] has shown that $\sqrt{n}\,(\hat{\beta}_{(i)} - \beta_{[i]})c/h(\psi)$ is asymptotically $(n\to\infty)$ distributed as $N(0,1)$ where

$$c^2 = \lim n^{-1}\sum_{j=1}^{n}(x_j - \bar{x}_n)^2,$$

$$h^2(\psi) = \frac{\int_0^1 \psi^2(u)\,du}{\left(\int_0^1 \psi(u)\varphi(u)\,du\right)^2}, \tag{15.54}$$

where $\psi(u)$ is defined in (15.50), and $\varphi(u)$ is defined similar to $\psi(u)$ with F in the place of G. Using this result, we can get the asymptotic expression for $P(CS|R)$ when $\beta_1=\cdots=\beta_k$ as $\Pr\{Z_k \geqslant \max_{i\leqslant j\leqslant k-1} Z_j - cd/h(\psi)\}$, where the Z_i are independent $N(0,1)$ variables.

If we choose G to be the logistic distribution function, then $\psi(u)=u$ and $T_i(\mathbf{Y}_i)$ coincides with the Wilcoxon two-sample statistic. In this case, it is known (see Hodges and Lehmann [1956]) that $[h(\psi)]^{-1}$ is bounded below by $\sqrt{0.864}$, a bound that is attained for some distribution with a parabolic density. If G is chosen to be the normal distribution function Φ, then $\psi(u)=\Phi^{-1}((u+1)/2)$ and T_i is of Van der Waerden type. In this case, the lower bound of $[h(\psi)]^{-1}$ is known (see Puri and Sen [1971]) to be unity.

Using these results, we can obtain a lower bound for the infimum of the PCS for large n. Specifically,

$$\inf P(CS|R) \geqslant \int_{-\infty}^{\infty} \Phi^{k-1}(x+cd_1)\,d\Phi(x), \tag{15.55}$$

where $d_1=\sqrt{0.864}\,d$ or d according as it is the Wilcoxon case or the normal score case.

15.6 PROCEDURES BASED ON PAIRED COMPARISONS

In §8.5 we discussed a selection procedure of Trawinski and David (1963) under the indifference zone formulation for selecting the best treatment. In this section we discuss their procedure for selecting a subset containing the best. The setup and the notations are carried over from the earlier discussion.

The procedure R of Trawinski and David (1963) selects π_i if and only if

$$a_i \geqslant a_{[k]} - \nu, \tag{15.56}$$

where $\nu=\nu(k,n,P^*)$ is a nonnegative integer. It has been shown that the least favorable configuration is given by $\pi_{ij}=\frac{1}{2}$; $i,j=1,\ldots,k$; $i\neq j$, provided that any configuration of the preference probabilities satisfies a linear model. We can evaluate the exact expression of the PCS when all the treatments are equally good with the help of tables of partition functions $G(\boldsymbol{\alpha},n)$ giving the number of permissible partitions of $\frac{1}{2}\,nk(k-1)$ into k scores $a_1,\ldots,a_k$ irrespective of order. The asymptotic $(n\to\infty)$ expression for the infimum of the PCS is given by

$$\inf P(CS|R) \approx \int_{-\infty}^{\infty} \Phi^{k-1}(z+w)\,d\Phi(z), \tag{15.57}$$

where

$$w=(2\nu+1)(nk)^{-1/2}. \tag{15.58}$$

Trawinski and David (1963) have tabulated the values of ν for $k=2(1)20$, $n=1(1)20(5)50(10)100$, and $P^*=0.75$, 0.90, 0.95, 0.975, 0.99. These values are based on asymptotic theory except for the following cases where the exact theory is used: $k=2$, all n; $k=3$, $n=1(1)20$; $k=4$, $n=1(1)8$; $k=5$, $n=1(1)3$; and $k=6(1)8$, $n=1$.

Trawinski (1969) considered the expected subset size corresponding to the slippage configuration given by $\pi_{k,i}=p(>\frac{1}{2})$, $i=1,\ldots,k-1$, and $\pi_{ij}=\frac{1}{2}$, $i,j=1,\ldots,k-1$, $i\neq j$. We denote this configuration by $C(\frac{1}{2},p)$. Let

$$\begin{aligned}
\delta &= nk\left(p-\tfrac{1}{2}\right),\\
\sigma^2 &= n\left\{2p(1-p)+\tfrac{1}{2}(k-1)\right\},\\
\sigma_k^2 &= n\left\{(k+2)p(1-p)+\tfrac{1}{4}(k-2)\right\},\\
\Delta &= \frac{\left(\nu+\frac{1}{2}\right)}{\sigma},\\
\Delta_{k-1} &= \frac{\left(\nu+\frac{1}{2}-\delta\right)}{\sigma_k},\\
\Delta_k &= \frac{\left(\nu+\frac{1}{2}+\delta\right)}{\sigma_k},\\
c_k &= n\left\{(k+1)p(1-p)-\tfrac{1}{4}\right\},\\
\rho &= \frac{\sigma}{2\sigma_k},\\
\rho_k &= \frac{c_k}{\sigma_k^2},\\
z_1 &= z+\sqrt{2}\,\Delta,\\
z_2 &= \sqrt{2}\left(\rho/\sqrt{\rho_k}\,\right)+\frac{\Delta_{k-1}}{\sqrt{\rho_k}},\\
z_3 &= \frac{\left\{\sqrt{\rho_k}\,z+\Delta_k\right\}}{\sqrt{1-\rho_k}}.
\end{aligned}\tag{15.59}$$

Trawinski (1969) has shown that the asymptotic expression for the expected subset size corresponding to the slippage configuration is given by

$$E\{S|R,c(\tfrac{1}{2},p)\}\approx\int_{-\infty}^{\infty}[(k-1)\Phi^{k-2}(z_1)\Phi(z_2)+\Phi^{k-1}(z_3)]\,d\Phi(z). \tag{15.60}$$

His evaluation of the expression assumes c_k to be positive. This is equivalent to the restriction that $\rho_k>0$. Hence the asymptotic approximation (15.60) holds whenever $p<\frac{1}{2}+\frac{1}{2}\{k/(k+1)\}^{1/2}$. For $p=1$,

$$E\{S|R,c(\tfrac{1}{2},1)\}\approx(k-1)\Phi(u)+1, \tag{15.61}$$

where $u=(2\nu+1-nk)/\{n(k-2)\}^{1/2}$.

Expected proportion of the populations included, that is, $k^{-1}E(S)$, has been tabulated by Trawinski (1969) for $P^*=0.90$, $k=3(1)10(2)20$, $n=2(1)6$, 8, 10(5)25, and $p=0.6(0.1)1.0$. Extensive tables have been prepared (but not published) by Trawinski for $k=3(1)20$, $n=1(1)20(5)50(10)100$, $p=0.6(0.1)1.0$, and $P^*=0.75$, 0.90, 0.95, 0.975, 0.99.

15.7 SELECTION FROM MULTIVARIATE POPULATIONS IN TERMS OF RANK CORRELATION

In §8.7, we discussed selection procedures under the indifference zone formulation investigated by Govindarajulu and Gore (1971) for selecting the best of p-variate populations which are ranked according to the usual rank correlation coefficient when $p=2$ and according to a p-variate analog of rank correlation coefficient when $p>2$. In this section we describe their results for subset selection procedures in the preceding cases. The notations of §8.7 are carried over here.

To select a subset containing the bivariate population associated with $\tau_{[k]}$ Govindarajulu and Gore (1971) defined the procedure R_1 which selects π_i if and only if

$$T_i \geqslant T_{[k]}-h \tag{15.62}$$

where T_i is given by (8.101) and $h=h(k,p,n,P^*)>0$ is to be determined subject to the P^*-condition. As in §8.7, a lower bound for $P(CS|R_1)$ has been obtained when n is sufficiently large and the Δ_i (see (8.103)) are small.

When $p>2$, the populations are ranked according to the measure θ_i defined in (8.104). The procedure of Govindarajulu and Gore (1971) is R_1 defined by (15.62) except that T_i is now given by (8.106). They have also discussed other measures for ranking the populations, namely, the sum of bivariate product–moment correlations and the sum of bivariate rank correlations.

15.8 NOTES AND REMARKS

Some interesting applications of nonparametric procedures are given by Lee (1977b), McDonald (1977b), and Sobel (1977b). Hsu (1978b) has studied the construction of asymptotically distributionfree procedures under the location model by using location estimators. He has also studied procedures based on pairwise ranking and their efficiencies. In §8.2, we mentioned a procedure of Sobel (1977) for selection in terms of inter $(\alpha-\beta)$-range. He has also briefly discussed the corresponding subset selection problem. Gupta, Huang, and Nagel (1979) have discussed locally optimal subset selection procedures based on ranks. In this chapter, we have not discussed sequential procedures; they are presented in Chapter 17.

CHAPTER 16

Selection from Restricted Families of Probability Distributions

16.1 INTRODUCTION

In the last chapter we discussed procedures for selecting a subset from a given set of populations whose distribution functions are not known. In some cases we only assumed that they differ by a location or scale parameters. In this chapter we assume that these populations belong to a family of distributions which is characterized by a partial order relation with respect to a known distribution. As we pointed out earlier, the connotation of "nonparametric procedures" is vague and in some sense the procedures for these restricted families of distributions can be said to be nonparametric. However, these partial order relations imply special structural properties which are important in reliability problems. As such these procedures are discussed now rather than in the last chapter.

16.1.1 Partial Ordering in the Space of Probability Distributions

A binary ordering relation ($\precsim$)is called a *partial ordering* in the space of probability distributions if

a. $F \precsim F$ for all distributions F, and
b. $F \precsim G, G \precsim H$ imply $F \precsim H$.

In other words, a partial order relation is reflexive and transitive. It should be noted that $F \precsim G$ and $G \precsim F$ do *not* necessarily imply that $F \equiv G$; that is, the order relation is not antisymmetric.

The preceding definition has been used by early authors and followed by Barlow and Gupta (1969) who were first to investigate selection procedures for restricted families. On the other hand, Barlow, Bartholomew, Bremner, and Brunk [1972, p. 24] have defined a partial order relation by the properties of reflexivity, transitivity, and antisymmetry. According to them the relation without the antisymmetry property is a *quasi-ordering*.

Assuming that all our distributions are absolutely continuous, we now define some of the special order relations of interest to us. In these definitions F and G denote distribution functions. Further the terms "increasing" and "decreasing" are used in a loose sense to mean nondecreasing and nonincreasing, respectively.

1. F is said to be *convex* with respect to (w.r.t.) G (written $F \prec_c G$) if and only if $G^{-1}F(x)$ is convex on the support of F.
2. F is said to be *star-shaped* w.r.t. G (written $F \prec_* G$) if and only if $F(0) = G(0) = 0$, and $G^{-1}F(x)/x$ is increasing in $x \geqslant 0$ on the support of F.
3. F is said to be *r-ordered* w.r.t. $G(F \prec_r G)$ if and only if $F(0) = G(0) = \frac{1}{2}$ and $G^{-1}F(x)/x$ is increasing (decreasing) for x positive (negative).
4. F is said to be *tail-ordered* w.r.t. $G(F \prec_t G)$ if and only if $F(0) = G(0) = \frac{1}{2}$ and $G^{-1}F(x) - x$ is increasing on the support of F.

Let $G(x) = 1 - e^{-x} (x \geqslant 0)$. Then $F \prec_c G$ is equivalent to saying that $-\log(1 - F(x))$ is convex. This implies that $r(x) = f(x)/[1 - F(x)]$ is increasing in x on the support of F. The function $r(x)$ is called the failure rate or the hazard rate or the mortality rate. If F is a life-length distribution of an item (or individual), then $r(x)\,dx$ is the probability that the item fails between x and $x + dx$ given that it survives x units of time. When $r(x)$

is increasing, F is naturally called an *increasing failure rate* (IFR) distribution. Several properties of the class of IFR distributions have been studied by Barlow, Marshall, and Proschan [1963].

When $G(x)$ is exponential, $F \prec_* G$ means that $F(0)=0$ and $R(x)=-x^{-1}\log[1-F(x)]$ increases on the support of F. Since $R(x)=x^{-1}\int_0^x r(t)\,dt$, we see that F has an *increasing failure rate average* (IFRA). The class of IFRA distributions has been studied by Barlow, Esary, and Marshall [1966].

Van Zwet [1964] has studied convex-ordered distributions. Properties of linear combinations of order statistics from star-ordered distributions have been investigated by Barlow and Proschan [1966]. It is easy to see that convex ordering implies star ordering. In particular, F is IFR implies F is IFRA. The implication does not hold the other way.

The r-ordering has been defined and investigated by Lawrence [1975]. The tail ordering was defined by Doksum [1969].

16.1.2 The Setup of the Problem

We assume that π_i has the associated absolutely continuous distribution F_i. The best population is defined in terms of a characteristic such as the mean or quantile of a given order. Let $F_{[k]}$ denote the cdf of the best population. We assume that

$$F_{[i]}(x) \geqslant F_{[k]}(x) \quad \text{for all} \quad x, \qquad i=1,2,\ldots,k; \tag{16.1a}$$

there exists an absolutely continuous distribution G such that $F_{[i]} \precsim G, \qquad i=1,2,\ldots,k,$ (16.1b)

where $\precsim$ denotes a partial ordering relation in the space of probability distributions. The space of the k-tuples $(F_1,\ldots,F_k)$ is denoted by Ω. The procedure is based on suitable statistics $T_i(X_{i1},\ldots,X_{in}), i=1,\ldots,k$, where $X_{ij}, j=1,\ldots,n$, are independent observations from π_i. The statistics chosen are such that the properties in (16.1) are preserved. The property (a) ensures that the infimum of the PCS takes place when $F_1=\cdots=F_k$, and the property (b) enables us to obtain a lower bound for this infimum which can be evaluated with the knowledge of G.

16.2 SELECTION IN TERMS OF QUANTILES FROM DISTRIBUTIONS STAR-SHAPED W.R.T. G

Let $\xi_{\alpha i}(i=1,\ldots,k)$ be the α-quantile (assumed to be unique) of F_i and let $F_{[i]}$ denote the cdf of the population with the ith smallest α-quantile. We make the assumptions (a) and (b) of (16.1) with star ordering as the partial ordering. For selecting a subset containing the population with the largest α-quantile Barlow and Gupta (1969) proposed the rule R_1 which selects π_i if and only if

$$T_{j,i} \geqslant c_1 \max_{1 \leqslant r \leqslant k} T_{j,r}, \tag{16.2}$$

where $T_{j,i}$ denotes the jth order statistic based on n independent observations from π_i, $j \leqslant (n+1)\alpha < j+1$, and $0 < c_1 = c_1(k, P^*, n, j) < 1$ is determined so as to satisfy the P^*-condition.

Let $H_{j,i}$ denote the cdf of $T_{j,i}$ and G_j denote the cdf of the jth order statistic based on n independent observations from G. Since $G^{-1}H_{j,i}(x) = G^{-1}F_i(x)$, the properties (a) and (b) of (16.1) are preserved by the order statistics, that is,

(a′) $H_{j,[i]}(x) \geqslant H_{j,[k]}(x)$ for all x and $i = 1, 2, \ldots, k$;

(b′) $H_{j,i} \prec_* G_j$, $i = 1, 2, \ldots, k$.

The following theorem of Barlow and Gupta (1969) enables us to evaluate the constant c_1.

Theorem 16.1. If $F_{[i]}(0) = G(0) = 0$, $F_{[i]}(x) \geqslant F_{[k]}(x)$, $x \geqslant 0, i = 1, 2, \ldots, k$, and $F_{[k]} \prec_* G$, then

$$\inf_{\Omega} P(CS|R_1) = \int_0^\infty \left[G_j\left(\frac{x}{c_1}\right)\right]^{k-1} dG_j(x). \tag{16.3}$$

Thus the constant c_1 which defines the procedure R_1 is determined by equating the right-hand side of (16.3) to P^*. As we have noted earlier, when $G(x) = 1 - e^{-x} (x \geqslant 0)$, the populations F_i have increasing failure rate averages. For this case, the values of c_1 have been tabulated by Barlow, Gupta, and Panchapakesan (1969) for $P^* = 0.75$, 0.90, 0.95, 0.99 and the following values of k, n, and j corresponding to each selected value of P^*:

(i) $j = 1$, $k = 2(1)11$ (in this case c_1 is independent of n);
(ii) $k = 2(1)4$, $n = 5(1)15$, $j = 2(1)n$;
(iii) $k = 5$, $n = 5(1)12$, $j = 2(1)n$;
(iv) $k = 6$, $n = 5(1)10$, $j = 2(1)n$.

These tables can be used also when the F_i are star-shaped w.r.t. the Weibull distribution $G_\lambda(x) = 1 - \exp(-x^\lambda/\theta)$ for $x \geqslant 0$ and $\theta, \lambda > 0$. The class of such distributions is the smallest class of continuous distributions containing the Weibull class of distributions with shape parameter λ which is closed under the formation of coherent structures and limits in distribution. For the selection problem in this case, we use the rule R_1 with the constant $c_1' = c_1^{1/\lambda}$, where c_1 is obtained from the table of Barlow, Gupta, and Panchapakesan (1969).

Assuming that $F_{[k]}$ (G) has a differentiable density $f_{[k]}$ (g) in a neighborhood of the α-quantile ξ_α (η_α) and that $f_{[k]}(\xi_\alpha) \neq 0$ $(g(\eta_\alpha) \neq 0)$, we can obtain the asymptotic $(n \to \infty)$ expression for the infimum of the PCS using the asymptotic normality of the sample quantile. Barlow and Gupta (1969) have shown that

$$\begin{aligned}\lim_{n\to\infty} \inf_{\Omega} P(CS|R_1) &= \int_{-\infty}^{\infty} \Phi^{k-1}\left\{ c_1^{-1}x + (1-c_1)\xi_\alpha f_{[k]}(\xi_\alpha) c_1^{-1}\left(\frac{n}{\alpha\bar{\alpha}}\right)^{\frac{1}{2}} \right\} d\Phi(x) \\ &\geqslant \int_{-\infty}^{\infty} \Phi^{k-1}\left\{ c_1^{-1}x + (1-c_1)\eta_\alpha g(\eta_\alpha) c_1^{-1}\left(\frac{n}{\alpha\bar{\alpha}}\right)^{\frac{1}{2}} \right\} d\Phi(x),\end{aligned} \tag{16.4}$$

where $j/n \to \alpha$ as $n \to \infty$, and $\bar{\alpha} = 1 - \alpha$. Setting the right-hand side of the inequality in (16.4) to P^*, we see that

$$c_1(k, P^*, n, j) \approx 1 - \omega(k, P^*, \alpha) n^{-\frac{1}{2}} \quad \text{as } n \to \infty, \tag{16.5}$$

where $\omega(k, P^*, \alpha)$ is some constant independent of n. For $k=2$, and $g(x) = e^{-x}$, we have

$$c_1(2, P^*, n, j) = 1 - \frac{2^{\frac{1}{2}}C}{n^{\frac{1}{2}}} + \frac{C}{n} - \frac{3C^3}{2^{\frac{1}{2}}n^{\frac{3}{2}}} + 0(n^{-2}), \tag{16.6}$$

where $C = \Phi^{-1}(P^*)(\alpha\bar{\alpha})^{\frac{1}{2}}/(1-\alpha)[-\log(1-\alpha)]$.

For the quantile selection when there is no partial order restriction, we have discussed in §15.2 a procedure of Rizvi and Sobel (1967) defined by (15.3). Barlow and Gupta (1969) have studied in the case of $k=2$ the asymptotic efficiency of R_1 relative to the Rizvi–Sobel procedure (denoted here by R_S) and the means procedure (denoted here by R_G) of Gupta (1963) for selection from gamma population discussed in §12.6.

When $k=2$, let $F_{[1]}(x) = F(x/\delta)$ and $F_{[2]}(x) = F(x), 0 < \delta < 1$. It has been shown that

$$E(S|R_1) - P(CS|R_1) \approx \Phi\left\{ -f(\xi_\alpha)\xi_\alpha\left(\frac{1-\delta}{c_1}\right) n^{\frac{1}{2}}(\alpha\bar{\alpha})^{-\frac{1}{2}}\left(1 + \left(\frac{\delta}{c_1}\right)^2\right)^{-\frac{1}{2}} \right\}. \tag{16.7}$$

Setting the right-hand side of (16.7) equal to ϵ and using $c_1 \approx 1 - C(2/n)^{\frac{1}{2}}$ (from 16.6)) we obtain

$$n_{R_1}(\epsilon) \approx \left[-(\alpha\bar{\alpha})^{\frac{1}{2}}\Phi^{-1}(\epsilon)(1+\delta^2)^{\frac{1}{2}} + 2^{\frac{1}{2}}C\delta\xi_\alpha f(\xi_\alpha)\right]^2 \left[\xi_\alpha^2 f^2(\xi_\alpha)(1-\delta)^2\right]^{-1}. \tag{16.8}$$

Similarly, we get

$$n_{R_S}(\epsilon) \approx \left[(\alpha\bar{\alpha})^{\frac{1}{2}}(1+\delta^2)^{\frac{1}{2}}\Phi^{-1}(\epsilon) - \gamma\right]^2 (1-\delta)^{-2}\left[\xi_\alpha f(\xi_\alpha)\right]^{-2}, \tag{16.9}$$

where $\gamma/\sqrt{n} = a/(n+1)$ and a is the integer (earlier denoted by c) which defines the procedure R_S [see (15.3)].

For the procedure R_G we have θ_1 and θ_2 as scale parameters where $\theta_{[1]} = \delta\theta_{[2]}, 0 < \delta < 1$. It is shown that

$$n_{R_G}(\epsilon) \approx 2\left[\Phi^{-1}(\epsilon) - \Phi^{-1}(P^*)\right]^2 \left[r(\log\delta)^2\right]^{-1}, \tag{16.10}$$

where r is the common shape parameter of the gamma distributions.

It has been shown by Barlow and Gupta (1969) that $ARE(R_1, R_S; \delta) \equiv \lim_{\epsilon\to 0} n_{R_S}(\epsilon)/n_{R_1}(\epsilon) = 1$, and that

$$ARE(R_1, R_G; \delta) \geqslant 2(1-\delta)^2\bar{\alpha}^2\left[\log\bar{\alpha}\right]^2\left[r(\log\delta)^2\alpha\bar{\alpha}(1+\delta^2)\right]^{-1}, r \geqslant 1; \tag{16.11}$$

when $r = 1$ and $\alpha = \frac{1}{2}$, we get $ARE(R_1, R_G; \delta\uparrow 1) \geqslant 0.493$.

The selection of the population with the smallest α-quantile can be handled analogously, Here, of course, we assume that $F_{[i]}(x) \leqslant F_{[1]}(x), i = 1, 2, \ldots, k$ and all $x \geqslant 0$, and $F_{[i]} \prec_* G$. The procedure proposed by Barlow and Gupta (1969) is R_2 which selects π_i if and only if

$$dT_{j,i} \leqslant \min_{1 \leqslant r \leqslant k} T_{j,r}, \qquad j \leqslant (n+1)\alpha < j+1, \tag{16.12}$$

where $d = d(k, P^*, n, j)$ is a constant between 0 and 1, and its value satisfying the P^*-condition is given by

$$\int_0^\infty \left[1 - G_j(xd)\right]^{k-1} dG_j(x) = P^*. \tag{16.13}$$

For $G(x) = 1 - e^{-x}$, the d-values have been tabulated by Barlow, Gupta, and Panchapakesan (1969) for $P^*=0.75$, 0.90, 0.95, 0.99 and the following ranges of k,n, and j corresponding to each P^*-value:

(i) $j=1$, $k=2(1)11$ (d is independent of n);
(ii) $k=2(1)5$, $n=5(1)15$, $j=2(1)n$;
(iii) $k=6$, $n=5(1)12$, $j=2(1)n$.

Another special case of interest is where G is the folded normal distribution obtained by folding $N(0,\sigma^2)$ at the origin, where we assume σ to be known. The cdf $G(y)$ is given by

$$G(y)=\Phi\left(\frac{y}{\sigma}\right)-\Phi\left(\frac{-y}{\sigma}\right), \qquad y \geqslant 0. \tag{16.14}$$

$G(y)$ is also the cdf of a chi random variable with 1 degree of freedom. Of course, it will have a noncentrality parameter $\lambda=|\mu|/\sigma$, if the initial distribution is $N(\mu,\sigma^2)$. The failure rate function for the folded normal $G(y)$ is given by $r(t)=\sigma^{-1}\varphi(t/\sigma)/[1-\Phi(t/\sigma)], t \geqslant 0$. It should be noted that $r(t)$ is the reciprocal of Mill's ratio. It is known (see Gordon [1941]) that $r(t)$ is increasing in $t \geqslant 0$. Thus the folded normal is an IFR and therefore an IFRA distribution. In other words, the folded normal is star-shaped w.r.t. the exponential distribution. Since the star ordering is transitive, the family of distributions which are star-shaped w.r.t. the folded normal have increasing failure rate averages. Thus our class of distributions is a special subclass of IFRA distributions. The selection of the population with the largest α-quantile from populations belonging to this class has been considered by Gupta and Panchapakesan (1975). We still use the procedure R_1. Let us denote the constant by c_2 for the purposes of later comparison. The values of c_2 have been tabulated by Gupta and Panchapakesan for $k=2(1)10$, $n=5(1)10$, $j=1(1)n$, and $P^*=0.75$, 0.90, 0.95, 0.99.

As we pointed out earlier, the class of distributions which are star-shaped w.r.t. the folded normal is a subclass of IFRA distributions. Thus one could use the procedure R_1 with the constant c_1. It can be seen that $c_1(k,P^*,n,j)<c_2(k,P^*,n,j)$. This means that the use of the constant c_1 will result in a larger subset size. For $k=2$, let $F_{[1]}(x)=F(x/\delta)$ and $F_{[2]}(x)=F(x), 0<\delta<1$. Let $n_i(\epsilon)$ be the sample size needed to satisfy $E(S|R_1)-P(CS|R_1)=\epsilon$ when the constant c_i is used. Gupta and Panchapakesan (1975) have evaluated the n_i using asymptotic $(n\to\infty)$ expressions and shown that $\lim_{\epsilon\downarrow 0} n_2(\epsilon)/n_1(\epsilon)=1$.

16.3 SELECTION IN TERMS OF THE MEDIANS FOR DISTRIBUTIONS r-ORDERED W.R.T. G

Here we consider selection of the population with the largest median from a set of distributions that have lighter tails than a specified distribution G. We say that F_i has a lighter tail than G if F_i centered at its median, Δ_i, is r-ordered w.r.t. $G(G(0)=\frac{1}{2})$ and $(d/dx)F_i(x+\Delta_i)|_{x=0} \geqslant (d/dx)G(x)|_{x=0}$. Let $T_{j,i}$ denote the jth order statistic based on n independent observations from F_i. For selecting a subset containing the population associated with $\Delta_{[k]}$, Barlow and Gupta (1969) proposed the rule R_3 which selects π_i if and only if

$$T_{j,i} \geqslant \max_{1 \leqslant r \leqslant k} T_{j,r} - D, \qquad j \leqslant \frac{(n+1)}{2} < j+1, \tag{16.15}$$

where $D = D(k, P^*, n) > 0$ is chosen subject to the P^*-condition. It has been shown by them that

$$\inf_{\Omega} P(CS|R_3) = \int_{-\infty}^{\infty} G_j^{k-1}(t+D)\,dG_j(t), \tag{16.16}$$

where G_j is as defined in the last section.

If F_i has a lighter tail than $G(G(0)=\frac{1}{2})$, then it follows (Doksum [1969]) that F_i, centered at its median Δ_i, is tail-ordered with respect to G. In fact, the procedure R_3 of Barlow and Gupta (1969) can be used for selecting the population with the largest median from a set of populations which, centered at their respective medians, are tail-ordered w.r.t. G. This has been discussed by Gupta and Panchapakesan (1974) who have proved the following theorem.

Theorem 16.2. If $F_{[i]}(x) \geqslant F_{[k]}(x)$ for all $x, i = 1, 2, \ldots, k, G(0) = \frac{1}{2}$ and $G^{-1}F_{[k]}(x+\Delta_{[k]}) - x$ is nondecreasing in x on $\{x : 0 < F_{[k]}(x+\Delta_{[k]}) < 1\}$, then

$$\inf_{\Omega_1} P(CS|R_3) = \int_{-\infty}^{\infty} G_j^{k-1}(t+D)\,dG_j(t) \tag{16.17}$$

where Ω_1 is the space of the k-tuples $(F_1, \ldots, F_k)$ such that $F_i(x+\Delta_i)$ is tail-ordered w.r.t. G.

The values of $D = D(k, P^*, n, j)$ satisfying the equation

$$\int_{-\infty}^{\infty} G_j^{k-1}(t+D)\,dG_j(t) = P^* \tag{16.18}$$

have been tabulated by Gupta and Panchapakesan (1974) for $k=2(1)10, n=5(2)15$, and $P^*=0.75, 0.90, 0.95, 0.99$ when G is chosen to be the logistic distribution $G(x)=[1+e^{-x}]^{-1}$. In this case, the distributions F_i belong to the class for which $f_i(x+\Delta_i) \geqslant F_i(x+\Delta_i)[1-F_i(x+\Delta_i)]$ for all x on the support of F_i.

The definition of tail ordering can be generalized by centering F and G at their respective α-quantiles. In other words, F is *α-quantile tail ordered* w.r.t. $G(F \prec_{t_\alpha} G)$ if $G^{-1}F(x+\xi_\alpha)-x-\eta_\alpha$ is nondecreasing in x on the support of F, where ξ_α and η_α are, respectively, the α-quantiles of F and G. The problem can now be stated in general as one of selecting the distribution with the largest α-quantiles from distributions which are α-quantile tail-ordered w.r.t. G. The rule R_4 in this case is same as R_3 except that $T_{j,i}$ is now the jth order statistic where $j \leqslant (n+1)\alpha < j+1$. For this rule, the constant D is shown [Gupta and Panchapakesan (1974)] to satisfy (16.18).

16.4 SELECTION PROCEDURES BASED ON GENERALIZED TOTAL LIFE STATISTICS

Let $\pi_1, \ldots, \pi_k$ be k independent populations whose associated distribution functions F_i $(1 \leqslant i \leqslant k)$ belong to a restricted family satisfying the properties in (16.1) with $\precsim$ denoting the convex ordering. We assume that $F_i^{-1}(0)=0$ for $i=1,\ldots,k$. Gupta and Lu (1977) have studied procedures based on *generalized total life statistics*. Let $X_{i:j,n}(Y_{j,n})$ denote the jth order statistic in a random sample of n observations from $F_i(G)$. Define

$$T_i = \sum_{j=1}^{r} a_j X_{i:j,n}, \qquad i=1,\ldots,k,$$

$$T = \sum_{j=1}^{r} a_j Y_{j,n}, \tag{16.19}$$

where $r (1 \leqslant r \leqslant n)$ is a fixed integer and

$$a_j = gG^{-1}\left(\frac{j-1}{n}\right) - gG^{-1}\left(\frac{j}{n}\right), \qquad j=1,\ldots,r-1.$$

$$a_r = gG^{-1}\left(\frac{r-1}{n}\right). \tag{16.20}$$

REMARK 16.1. If $G(x)=1-e^{-x}, x \geqslant 0$, then $a_j = 1/n$ for $j=1,\ldots,r-1$, and $a_r = (n-r+1)/n$. Thus $T_i = n^{-1}[X_{i:1,n} + \cdots + X_{i:r-1,n} + (n-r+1)X_{i:r,n}]$, which is the well-known total life statistic until the rth failure from F_i. □

Gupta and Lu (1977) proposed the following procedure R_5 for selecting a subset containing the best population, namely, $F_{[k]}$.

R_5: Select π_i if and only if

$$T_i \geqslant c_5 \max_{1 \leqslant j \leqslant k} T_j \tag{16.21}$$

where $c_5 = c_5(k, P^*, n, r)$ is the largest number between 0 and 1 subject to the P^*-condition. The following theorem of Gupta and Lu (1977) states the result regarding the infimum of the PCS.

Theorem 16.3. If $F_i(x) \geqslant F_{[k]}(x)$ for all x and $i = 1, \ldots, k, F_{[k]} <_c G, a_j \geqslant 0$ for $j = 1, \ldots, r$, $g(0) \leqslant 1$, and $a_r \geqslant c_5$, then

$$\inf_{\Omega} P(CS|R_5) = \int_0^{\infty} G_T^{k-1}\left(\frac{x}{c_5}\right) dG_T(x) \tag{16.22}$$

where $G_T(x)$ is the cdf of T. Thus the constant c_5 is the largest number between 0 and 1 determined by

$$\int_0^{\infty} G_T^{k-1}\left(\frac{x}{c_5}\right) dG_T(x) \geqslant P^* \text{ and}$$

$$gG^{-1}\left(\frac{r-1}{n}\right) \geqslant c_5. \tag{16.23}$$

Let us consider some special choices of G. If $G(x) = 1 - e^{-x}, x \geqslant 0$, then for $n \geqslant \max(r, (r-1)/(1-c_5))$ we have

$$\inf_{\Omega} P(CS|R_5) = \int_0^{\infty} H^{k-1}\left(\frac{x}{c_5}\right) dH(x) \tag{16.24}$$

where $H(x)$ is the cdf of a chi-square random variable with $2r$ degrees of freedom. This result is a slight generalization over that of Patel (1976). If $G(x) = x, 0 < x < 1$, the statistic T_i becomes the rth order statistic based on n observations from π_i and Theorem 16.1 of Barlow and Gupta (1969) applies.

Gupta and Lu (1977) have studied the asymptotic efficiency of their procedure R_5 relative to the procedure R_1 (defined in (16.2)) of Barlow and Gupta (1969) and the procedure of Gupta (1963) for selection from gamma populations (see §12.6). The efficiency comparisons are for $k = 2$ and are similar in approach to the comparisons discussed in §16.2.

Other problems considered by Gupta and Lu (1977) include selection from distributions that are convex ordered with respect to Weibull distribution and selection with respect to the means from gamma populations.

16.5 SELECTION IN TERMS OF THE MEANS FOR THE CLASS OF IFR DISTRIBUTIONS

Let μ_i be the mean of the distribution $F(x;\mu_i), i=1,2,\ldots,k$, and assume that

$$F(x,\mu_{[i]}) \geqslant F(x;\mu_{[k]}) \quad \text{for all } x, i=1,2,\ldots,k; \tag{16.25a}$$

$$F(x;\mu_{[i]}) \prec_c G(x) = 1-e^{-x}, \qquad i=1,2,\ldots,k. \tag{16.25b}$$

For convenience we assume that $F(0;\mu_i)=0$ for all i. For selecting a subset containing the population associated with $\mu_{[k]}$, Barlow and Gupta (1969) proposed the rule

$\boldsymbol{R_6}$: Select π_i if and only if

$$\overline{X}_i \geqslant c'\overline{X}_{[k]}, \tag{16.26}$$

where $\overline{X}_i$ is the sample mean based on a random sample of n observations from π_i. As a consequence of (16.25a), the distribution $U_{[k]}$ is stochastically larger than $U_{[i]}$ $(i=1,\ldots,k-1)$, where $U_{[i]}$ is the cdf of the sample mean from $F(x;\mu_{[i]})$. Since the class of IFR distributions is closed under convolution (see Barlow, Marshall, and Proschan [1963], we have $U_{[i]} \prec_c G$, $i=1,\ldots,k$. Using these properties, Barlow and Gupta (1969) have shown that

$$P(CS|R_6) \geqslant \int_0^\infty G^{k-1}\left(\frac{x}{c'}\right) dG(x). \tag{16.27}$$

An obvious disadvantage of the previous procedure is that the bound in (16.27), and hence the constant c' satisfying the P^*-condition, is independent of n. However, if we restrict the class of distributions to the gamma family, we can obtain a lower bound for the PCS which depends on n by taking the density $f(x;\theta_i) = \theta_i^\alpha x^{\alpha-1}\exp\{-\theta_i x\}/\Gamma(\alpha), x \geqslant 0, \theta_i > 0, i = 1,\ldots,k$, and $\alpha \geqslant 1$. The value of α is unknown. In this case, selecting the population with the largest mean is equivalent to selecting the population with the smallest θ_i. It has been shown by Barlow and Gupta (1969) that

$$P(CS|R_6) \geqslant \int_0^\infty \left[G^{(n)}\left(\frac{x}{c'}\right)\right]^{k-1} dG^{(n)}(x), \tag{16.28}$$

where $G^{(n)}$ denotes the distribution of a gamma variate with parameter $\alpha = n$. Thus the constant $c' = c'(k, P^*, n)$ satisfying the P^*-condition is obtained by equating the right-hand side of (16.28) to P^*. The value of this constant c' is same as the constant b of the procedure R_{16} of Gupta (1963) defined in (12.66) and is tabulated by him for $k = 2(1)4, n = 2(2)50$, and $P^* = 0.75, 0.90, 0.95, 0.99$.

16.6 A GENERAL PARTIAL ORDER RELATION AND A RELATED SELECTION PROBLEM

Gupta and Panchapakesan (1974) have defined a general partial order relation and discussed a related selection problem. Let $\mathcal{H} = \{h(x)\}$ be a class of real-valued functions $h(x)$ defined on the real line. Let F and G be distributions on the real line such that $F(0) = G(0)$. We say that F is $\mathcal{H}$-*ordered* w.r.t. $G(F \prec_{\mathcal{H}} G)$ if $G^{-1}F(h(x)) \geqslant h(G^{-1}F(x))$ for all $h \in \mathcal{H}$ and all x on the support of F. It is easy to verify that the order relation is reflexive and transitive. It is readily seen that (a) the star ordering is obtained by taking $\mathcal{H} = \{ax, a \geqslant 1\}$ and $F(0) = G(0) = 0$, and (b) the tail ordering is the result if $\mathcal{H} = \{x + b, b \geqslant 0\}$ and $F(0) = G(0) = \frac{1}{2}$.

Let $\pi_i (i = 1, \ldots, k)$ have the associated absolutely continuous distribution F_i. We assume that there is one among the k populations, denoted by $F_{[k]}$, which is stochastically larger than any other. We assume that $F_i \prec_{\mathcal{H}} G, i = 1, 2, \ldots, k$, where G is a specified continuous distribution, that the F_i have the same support I and that $\mathcal{H} = \{h\}$ is a class of continuous distributions satisfying the following properties:

(i) $h_0 : x \to x$ is a member of $\mathcal{H}$,

(ii) $h(x) \geqslant x$ for every $h \in \mathcal{H}$ and $x \in I$,

(iii) for every $x \in I$, and $h \in \mathcal{H}$, there exists a sequence $\{h_m\}$ in $\mathcal{H}$ such that $\lim_{m \to \infty} h_m(x) = h(x)$,

(iv) for every $x \in I$, except perhaps on a set of Lebesgue measure zero, there is a sequence $\{h_m\}$ in $\mathcal{H}$ such that $\lim_{m \to \infty} h_m(x) = \infty$. (16.29)

Let $T_i = T(X_{i1}, X_{i2}, \ldots, X_{in}), i = 1, \ldots, k$, and $T = T(Y_1, Y_2, \ldots, Y_n)$ be statistics based on independent observations $X_{i1}, \ldots, X_{in}$ from $F_i, i = 1, 2, \ldots, k$, and $Y_1, Y_2, \ldots, Y_n$ from G. It is assumed that the T_i and T preserve the stochastic and $\mathcal{H}$-ordering of the parent distributions. For selecting a subset containing the distribution $F_{[k]}$, Gupta and Panchapakesan (1974) proposed the rule R_7 which selects π_i if and only if

$$h(T_i) \geqslant T_{[k]}, \tag{16.30}$$

where $h \in \mathcal{H}$ is chosen so that the P^*-condition is satisfied. It has been shown by Gupta and Panchapakesan that

$$\inf_{\Omega} P(CS|R_7) = \int_{-\infty}^{\infty} G_T^{k-1}(h(x))\, dG_T(x), \tag{16.31}$$

where G_T is the cdf of T. The preceding result is proved by using the fact that the T_i associated with $F_{[k]}$ is stochastically larger than the statistic associated with any other distribution and the following lemma of Gupta and Panchapakesan (1974).

Lemma 16.1. If $F \prec_{\mathcal{H}} G$, then for any positive integer m,

$$\int_{-\infty}^{\infty} F^m(h(x))\, dF(x) \geqslant \int_{-\infty}^{\infty} G^m(h(x))\, dG(x) \tag{16.32}$$

for all $h \in \mathcal{H}$.

Gupta (1966b) proved a lemma which is stated as follows essentially in its original form.

Lemma 16.2. Let X be a random variable having the distribution F_λ. Let $h_b(x)$ be a class of functions and suppose there exists a distribution function F such that $h_b(g_\lambda(x)) \geqslant g_\lambda(h_b(x))$ for all λ and all x, where $g_\lambda(x)$ is defined by $F_\lambda(g_\lambda(x)) = F(x)$ for all x. Then, for any $t > 0$,

$$\int_{-\infty}^{\infty} F_\lambda^t(h_b(x))\, dF_\lambda(x) \geqslant \int_{-\infty}^{\infty} F^t(h_b(x))\, dF(x). \tag{16.33}$$

Although the preceding lemma is not stated in terms of a partial order relation, it can be seen that the hypothesis of the lemma amounts to saying that $F_\lambda \prec_{\mathcal{H}} F$, where $\mathcal{H} = \{h_b\}$. It should be pointed out that Lemma 16.1 is, in fact, true for any $m > 0$.

16.7 NOTES AND REMARKS

In §9.2.4, we discussed fixed-size subset selection procedures of Hooper and Santner (1979) for selecting good populations in terms of α-quantiles for star- and tail-ordered distributions. Hooper and Santner have also studied the problem of selecting good populations using the restricted subset size approach of Santner (1975) discussed in §11.3. The procedure and analysis are analogous to those of Santner (1975). There are other important classes of distributions which have not been considered so far. For example, one may consider NBU (new better than used) or NWU (new worse than used) distributions. These are distributions that are super-additive or sub-additive with respect to the exponential distribution. It is known that NBU is a strictly larger class that includes IFRA distributions. One can rank life length distributions of certain components in terms of the optimal replacement period, that is, the period after which it is profitable to use a new item in the place of a used one. Also, little has been done for restricted families of distributions in more than one dimension. Some important sources of information on many of these and other related problems in reliability and related areas are Barlow and Proschan [1965], Barlow, Bartholomew, Bremner, and Brunk [1972], Proschan and Serfling [1974], Barlow and Proschan [1975], and Barlow, Fussell, and Singpurwalla [1975].

CHAPTER 17

Sequential Procedures

17.1 INTRODUCTION

In this chapter we discuss sequential subset selection procedures both parametric and nonparametric. As we have seen in Chapter 8, the sequential procedures are either of eliminating type or of noneliminating type. The procedures of Barron (1968) and Barron and Gupta (1972) are of noneliminating type. Gupta and D. Y. Huang (1975b) have investigated procedures of both types using a generalization of the formulation of Desu (1970). Swanepoel and Geertsema (1973a) considered a sequential procedure for which the expected subset size does not exceed a specified number. Guttman (1963) defined a procedure which is pseudo-sequential. In other words, it is sequential in that after each stage a decision is made as to whether sampling is to be continued or not. It is not sequential in the conventional sense because at each stage for making a decision the observations of the previous stages are not used. We discuss these procedures in the following sections.

17.2 A CLASS OF NONELIMINATING PROCEDURES FOR NORMAL POPULATIONS

Let π_i be a normal population with unknown mean θ_i and known variance σ^2. It is assumed that the successive differences of the ordered θ_i are known. The procedure of Barron and Gupta (1972) uses the subset selection procedure of Gupta (1956) at each stage. Suppose $\bar{X}_i$ is the mean of a sample of n independent observations from π_i. Then the fixed-sample procedure of Gupta (1956), which is denoted by $R(n)$ here, selects π_i if and only if

$$\bar{X}_i \geqslant \bar{X}_{[k]} - \frac{d\sigma}{\sqrt{n}} \tag{17.1}$$

and we saw in §12.2.1 that the constant $d > 0$ satisfying the P^*-condition is given by

$$\int_{-\infty}^{\infty} \Phi^{k-1}(x+d)\,d\Phi(x) = P^*. \tag{17.2}$$

Let p_i denote the probability of selecting $\pi_{(i)}$, the population associated with $\theta_{[i]}$. Then it is known that $p_1 \leqslant p_2 \leqslant \cdots \leqslant p_k$, where

$$p_i = \int_{-\infty}^{\infty} \prod_{\substack{j=1 \\ j \neq i}}^{k} \Phi\left\{x + d + (\theta_{[i]} - \theta_{[j]})\sqrt{n}\,\sigma^{-1}\right\} d\Phi(x). \tag{17.3}$$

Since the successive differences of the ordered θ_i are assumed to be known, the p_i are known. The approach of Barron and Gupta (1972) is to make use of this fact to convert the problem into one involving Bernoulli random variables as follows.

Sampling is done sequentially taking one observation from each population at every stage. Let X_{ij} denote the jth stage observation from π_i. At each stage, we perform $R(1)$ and define

$$Y_{ij} = \begin{cases} 1 & \text{if } \pi_i \text{ is selected by } R(1), \\ 0 & \text{otherwise.} \end{cases} \tag{17.4}$$

At any stage $m (\geqslant 1)$ we define

$$S_{im} = \sum_{j=1}^{m} Y_{ij} \tag{17.5}$$

so that S_{im} has a binomial distribution with parameters m and p_i.

Let us consider now a pair of sequence of real numbers $\eta \equiv \eta_{b,c} = (\{b_m\}, \{c_m\})$ such that for all $m \geqslant 1$,

$$\text{(i)}\quad b_m \leqslant b_{m+1}, \qquad c_m \leqslant c_{m+1},$$

$$\text{(ii)}\quad b_m < c_m,$$

$$\text{(iii)}\quad \lim_{m\to\infty} b_m = \infty,$$

$$\text{(iv)}\quad \Pr\left\{ \bigcap_{m=1}^{\infty} [b_m < S_{im} < c_m] \right\} = 0 \qquad \text{for all} \quad i = 1, \dots, k. \tag{17.6}$$

The existence of such sequences is well known and one such pair is discussed later. We now define the class of sequential procedures $\mathcal{S}(\eta_{b,c})$ of Barron and Gupta (1972).

> **$\mathcal{S}(\eta_{b,c})$:** Tag each unmarked population π_i at stage m ($m \geqslant 1$) for which $S_{im} \notin (b_m, c_m)$; mark it "rejected" if $S_{im} \leqslant b_m$ and "accepted" if $S_{im} \geqslant c_m$. Terminate sampling when all the populations are tagged and include in the selected subset all those populations which are marked "accepted." (17.7)

It should be noted that a population π_i which is tagged "accepted" or "rejected" stays that way irrespective of later changes in S_{im}. To describe the properties of this class of procedures, we need the following notations and definitions.

Let $a_i(m) \equiv a_i(m, \eta_{b,c}) = \Pr\{\text{accepting } \pi_{(i)} \text{ at stage } m | \mathcal{S}(\eta_{b,c})\}$ and $r_i(m) \equiv r_i(m, \eta_{b,c}) = \Pr\{\text{rejecting } \pi_{(i)} \text{ at stage } m | \mathcal{S}(\eta_{b,c})\}$. Further, let $a_i \equiv a_i(\eta_{b,c}) = \sum_{m=1}^{\infty} a_i(m)$ and $r_i \equiv r_i(\eta_{b,c}) = \sum_{m=1}^{\infty} r_i(m)$. Then a_i is the probability that $\pi_{(i)}$ is included in the selected subset and r_i is the probability that it is not included. When there is no ambiguity we use $\mathcal{S}(\eta)$ for $\mathcal{S}(\eta_{b,c})$. It should be noted that $a_k = P(CS | \mathcal{S}(\eta))$.

Definition 17.1. Two sequences $\{b_m\}$ and $\{b'_m\}$ are said to be *pairwise ordered* if and only if $b_m \leqslant b'_m$ for all $m \geqslant 1$. This relation is denoted by $\{b_m\} \prec \{b'_m\}$.

Definition 17.2. Let $\eta = (\{b_m\}, \{c_m\})$ and $\eta' = (\{b'_m\}, \{c'_m\})$ be two pairs of sequences. The pair η is said to be ordered with respect to η' (written $\eta \prec \eta'$) if $\{b_m\} \prec \{b'_m\}$ and $\{c_m\} \prec \{c'_m\}$.

Definition 17.3. Let $\mathcal{C}$ be a class of pairs of sequences η. We say the class $\mathcal{C}$ is *ordered* if for all $\eta, \eta' \in \mathcal{C}$ we have either $\eta \prec \eta'$ or $\eta' \prec \eta$.

Barron and Gupta (1972) have proved the following theorem which gives some general properties of their sequential procedure.

Theorem 17.1. (a) If $\eta' \prec \eta$, then $a_i(\eta') \geqslant a_i(\eta)$ and $r_i(\eta') \leqslant r_i(\eta)$, $i = 1, \ldots, k$. In particular, $P(CS|\mathcal{S}(\eta')) \geqslant P(CS|\mathcal{S}(\eta))$. (b) The procedure $\mathcal{S}(\eta)$ is monotone and unbiased, i.e., $a_k \geqslant a_{k-1} \geqslant \cdots \geqslant a_1$ and $r_k \leqslant r_i$, $i = 1, 2, \ldots, k-1$.

The rest of the investigation of the procedure $\mathcal{S}(\eta)$ relates to the special class $\mathcal{C}_1$ of pairs of sequences given by $b_m = \delta m - \gamma_1$ and $c_m = \delta m + \gamma_2$, where δ is a rational number in (0, 1) and γ_1, γ_2 are positive integers. For fixed γ_1 and γ_2, the class $\mathcal{C}_1$ is ordered in δ. For this class the conditions of (17.6) are satisfied. Let $R_{im} = S_{im} - \delta m$. For any $\eta \in \mathcal{C}_1$, the events $[\delta m - \gamma_1 < S_{im} < \delta m + \gamma_2]$, $[S_{im} \geqslant \delta m + \gamma_2]$, and $[S_{im} \leqslant \delta m - \gamma_1]$ are equivalent to $[-\gamma_1 < R_{im} < \gamma_2]$, $[R_{im} \geqslant \gamma_2]$, and $[R_{im} \leqslant -\gamma_1]$, respectively. By taking $\delta = t/s$ where t and s are relatively prime integers with $t < s$, the problem of evaluating the various probabilities and expectation can be restated in terms of a random walk on the line where the state space is all the points of the form $(Ns - Mt)/s$ for all integers $M > N > 0$. It is now possible to reduce the problem to a random walk on the space of integers. Using this technique, Barron and Gupta have obtained exact expressions for the probability of selecting a specified population and the expected number of stages to reach a decision. However, these expressions are not always easy to compute and hence it is useful to have approximations and bounds. The following two theorems of Barron and Gupta (1972) are in this direction.

Theorem 17.2. For the sequential procedure $\mathcal{S}(\eta)$ where $\eta = (\{\delta m - \gamma\}, \{\delta m + \gamma\})$ and $\delta = t/s > 0$,

$$\lim_{\gamma \to \infty} a_i(\delta, \gamma) = \begin{cases} 0 & \text{if} \quad p_i < t/s, \\ \frac{1}{2} & \text{if} \quad p_i = t/s, \\ 1 & \text{if} \quad p_i > t/s, \end{cases} \tag{17.8}$$

where p_i is given by (17.3) with $n = 1$.

Theorem 17.3. Let m_i be the smallest $m \geqslant 1$ such that $\pi_{(i)}$ is accepted or rejected, and $M_i = E\{m_i|\mathcal{S}(\eta)\}$. Then, for the sequential procedure $\mathcal{S}(\eta)$

specified in Theorem 17.2,

$$M_i \approx \frac{\gamma}{|p_i - t/s|} \tag{17.9}$$

provided that γ is sufficiently large and $p_i \neq t/s$.

Numerical evaluations made for $\delta = 0.75$, $\gamma = 3(1)10$, and $p_i = 0.4, 0.6, 0.8, 0.9$ indicate that the approximations are good. In the case of the probability of selecting a population the approximation becomes better as γ increases.

There still, however, remains the problem of choosing the two constants δ and γ. Theorem 17.2 guarantees that for any choice of $\delta \in (p_{k-1}, p_k)$ there exists a number $\gamma = \gamma(\delta, \epsilon)$ such that for any $\epsilon > 0$

$$\begin{aligned} &\text{(i)} \quad a_k(\delta, \gamma) \geqslant 1 - \epsilon, \\ &\text{(ii)} \quad a_{k-1}(\delta, \gamma) \leqslant \epsilon, \end{aligned} \tag{17.10}$$

regardless of the configuration of $p_1 \leqslant p_2 \leqslant \cdots \leqslant p_k$ and hence that of $\theta_{[1]} \leqslant \theta_{[2]} \leqslant \cdots \leqslant \theta_{[k]}$. Thus, for a sufficiently small ϵ, the P^*-condition can always be satisfied by choosing an appropriate $\eta \in \mathcal{C}_1$. Since $E(S) \leqslant 1 + (k-1)a_{k-1}$, (17.10) can be replaced by

$$\begin{aligned} &\text{(i)} \quad a_k(\delta, \gamma) \geqslant 1 - \epsilon. \\ &\text{(ii)} \quad 1 - \epsilon < E(S) \leqslant 1 + (k-1)\epsilon. \end{aligned} \tag{17.11}$$

For any $\delta \in (p_{k-1}, p_k)$, the experimenter has a countably infinite number of procedures $\mathcal{S}(\eta)$ satisfying (17.11). Given two procedures corresponding to η and η' belonging to $\mathcal{C}_1$ and satisfying (17.11), the procedure with the smaller expected number of stages is preferable in some sense. If $M = \max_{1 \leqslant i \leqslant k} M_i$, then the experimenter will want to choose an η that minimizes M over the subclass $\mathcal{C}_2 \subset \mathcal{C}_1$ of procedures satisfying (17.11). The following theorem of Barron and Gupta has been obtained using approximate value of M.

Theorem 17.4. For $\delta \in (p_{k-1}, p_k)$,

$$\min_{\delta} M = \begin{cases} \displaystyle\min_{\delta^* \leqslant \delta \leqslant \bar{\delta}} \{\gamma_1(\delta)/(\delta - p_{k-1})\} & \text{for } \delta^* \leqslant \bar{\delta}, \\ \displaystyle\min_{\bar{\delta} \leqslant \delta \leqslant \delta^*} \{\gamma_2(\delta)/(p_k - \delta)\} & \text{for } \bar{\delta} \leqslant \delta^*, \end{cases} \tag{17.12}$$

where $\gamma_1(\delta)$ is the first positive integer such that $a_k \geqslant 1-\epsilon$, $\gamma_2(\delta)$ is the first positive integer such that $a_{k-1} \leqslant \epsilon$, δ^* is the value of δ such that $\gamma_1(\delta) = \gamma_2(\delta)$, and $\bar{\delta} = (p_{k-1} + p_k)/2$.

A lemma of Barron and Gupta shows that the approximate unique value δ^* is given by

$$\delta^* = \begin{cases} \dfrac{\log[(1-p_{k-1})/(1-p_k)]}{\log[p_k(1-p_{k-1})/p_{k-1}(1-p_k)]} & \text{if } p_{k-1}+p_k \neq 1, \\ \frac{1}{2} & \text{otherwise.} \end{cases} \qquad (17.13)$$

As we can see, there still remains the question of how to choose a specific δ if $\delta^* \neq \bar{\delta}$. It has been empirically found by Barron (1968) that often $\delta^* \approx \bar{\delta}$ so that the experimenter will not be "far" from the minimum for any choice of δ between $\bar{\delta}$ and δ^*. Numerical evidence indicates that if $\bar{\delta}$ and δ^* are significantly apart, the minimum takes place near δ^*. It seems that an approximate minimax rule which possesses certain desirable properties will be $\mathbb{S}(\eta^*)$ where $\eta^* = (\{\delta^* m - \gamma^*\}, \{\delta^* m + \gamma^*\})$.

Barron and Gupta (1972) have compared the procedure $\mathbb{S}(\eta^*)$ with the fixed-sample procedure $R(n)$ of Gupta (1956) defined by (17.1) in the cases of the slippage configuration given by $\theta_{[1]} = \cdots = \theta_{[k-1]} = \theta, \theta_{[k]} = \theta + \tau, \tau > 0$, and the equally spaced configuration given by $\theta_{[i]} = \theta + (i-1)\tau$, $\tau > 0$, assuming $\sigma = 1$ in both cases. The comparison between $M = \max_{1 \leqslant i \leqslant k} M_i$ for $\mathbb{S}(\eta^*)$ and the sample size n of $R(n)$ was carried out for $k = 2(1)10, 25, 50$; $\tau = 0.05(0.10)0.60, 1.2$; and $P^* = 0.75, 0.90$ in the case of slippage configuration, and for $k = 2(1)5$; $\tau = 0.5, 0.10(0.10)0.60$; and $P^* = 0.75, 0.90$ in the case of equally spaced configuration. The empirical results indicate that $\mathbb{S}(\eta^*)$ does better when the means are close and $R(n)$ is better when any one of the means gets significantly larger than the others.

17.3 PROCEDURES WITH RESTRICTIONS ON THE SUBSET SIZE

Let $\pi_i (i = 1, 2, \ldots, k)$ be a normal population with mean μ_i and variance σ_i^2. To select a subset containing the population $\pi_{(k)}$ associated with $\mu_{[k]}$ but not exceeding in size a given number, Swanepoel and Geertsema (1973a) define what they call a selection sequence. For each $n \geqslant 1$, let B_n denote a subset of the k populations based on n observations from each population. Any sequence $\{B_n\}$ is called a *selection sequence* if

$$\Pr(\pi_{(k)} \in B_n \quad \text{for all} \quad n \geqslant 1) \geqslant P^* \qquad (17.14)$$

for all $(\mu_1,\ldots,\mu_k)\in\Omega$ where $0<P^*<1$ is given. The idea of a selection sequence is analogous to that of a confidence sequence introduced by Robbins [1970].

Let X_{ij} denote the jth observation from π_i and $\bar{X}_i(n)$ the mean of n observations from π_i. Define

$$H^2(i,i',n)=n^{-1}\sum_{j=1}^{n}\left\{(X_{ij}-X_{i'j})-\left(\bar{X}_i(n)-\bar{X}_{i'}(n)\right)\right\}^2,$$

$$s_{ni}=\max_{i'\neq i} H(i,i',n),$$

$$h(t,y)=\left[(ty)^{y^{-1}}-1\right]^{1/2}, \tag{17.15}$$

$$t=\frac{(1+a^2)^2}{2},$$

where a is defined by the equation

$$\frac{1}{2}-\frac{(\arctan a)}{\pi}+\frac{a}{\pi(1+a^2)}=\frac{(1-P^*)}{2(k-1)}. \tag{17.16}$$

Now we define the sequence $\{B_n\}$ where

$$B_1=\{\pi_1,\ldots,\pi_k\},$$

$$B_n=\left\{\pi_r\colon \bar{X}_r(n)\geqslant \max_{1\leqslant i\leqslant k}\bar{X}_i(n)-s_{nr}h(t,n)\right\}. \tag{17.17}$$

Swanepoel and Geertsema (1973a) have shown that the preceding sequence $\{B_n\}$ is a selection sequence. The proof uses a result of Robbins [1970, p. 1402].

A second sequence $\{B_n'\}$ can be defined based on $\{B_n\}$ by taking

$$B_n'=\bigcap_{j=1}^{n}B_j,\qquad n\geqslant 1. \tag{17.18}$$

It follows easily that $\{B_n'\}$ is also a selection sequence. Although B_n' could be empty for some n, the probability of that happening is at most $1-P^*$.

Let $S(n)$ denote the size of the subset B_n and let r be the number of indices j with $\mu_j=\mu_{[k]}$, $j=1,\ldots,k$. Swanepoel and Geertsema have shown that $S(n)\to r$ a.s. as $n\to\infty$, and $B_n=\{\pi_{(k-r+1)},\ldots,\pi_{(k)}\}$ a.s. for all n large.

Now define the stopping variable N to be the first integer $n\geqslant 1$ such that $S(n)\leqslant m$ where m is an integer chosen in advance with $1\leqslant m\leqslant k-1$. If

$N<\infty$, the subset B_N contains at most m populations and includes the best population with a minimum probability P^*. If $r \leqslant m$, then $N<\infty$ a.s. However, for the case $r \geqslant m+1$, it is clear that $N=\infty$ with positive probability.

We can define another stopping variable N' as the first integer $n \geqslant 1$ such that $S'(n) \leqslant m$, where $S'(n)$ is the size of the subset B'_n. The following asymptotic ($P^* \to 1$) results have been obtained by Swanepoel and Geertsema (1973a).

Theorem 17.5. If $r \leqslant m$,

$$\lim_{P^* \to 1} \frac{E(N)}{\{-\log(1-P^*)\}} = \lim_{P^* \to 1} \frac{E(N')}{\{-\log(1-P^*)\}}$$
$$= \frac{4}{\log\left(1+\theta^2_{[m+1]}\right)},$$

where $\theta_{[1]} \leqslant \cdots \leqslant \theta_{[k]}$ are the ordered θ_j given by $\theta_j = (\mu_{[k]} - \mu_{[j]})/\max_{i \neq j}(\sigma^2_{(i)} + \sigma^2_{(j)})^{1/2}$, $j=1,\ldots,k$, and $\sigma^2_{(i)}$ is the variance of $\pi_{(i)}$.

Guttman (1963) has discussed a different procedure for selecting a subset whose size is constrained by cost considerations. Suppose that π_i has the density f_{θ_i}. Let T_i be an appropriate statistic based on a sample of n independent observations in the sense that $E(T_i) = \theta_i$ or a monotonic function of θ_i. Consider the rule R which selects π_i if and only if

$$T_i \in \omega_{n,k}(P^*, \mathbf{T}) \tag{17.19}$$

where $\omega_{n,k}(P^*, \mathbf{T})$ is a random linear set contained in the sample space of T_i and depends on $\mathbf{T} = (T_1, \ldots, T_k)$ and is such that the P^*-condition is satisfied. Since the size of the selected subset is random, a natural question is how to proceed sequentially so that we could select one population as the best or reduce the size of the subset selected subject to certain cost considerations which restrict the number of stages.

Let t denote the stage of the experiment and k_t denote the number of populations retained at the start of the stage. If M units of capital are available to spend on the procedure and at each stage a sample of n_t independent observations are taken from each retained population, let t_0 be the largest integer for which $\sum_{t=1}^{t_0} k_t n_t d \leqslant M$ where d is the cost per observation. The sequential procedure proposed and investigated by Guttman (1963) is defined as follows.

R': At each stage t, use the rule R defined by (17.19) with $P^* = P^*_t$, where $P^*_t = 1 - (1-\beta)2^{-t}$ adopting the following stopping rule.

At the end of stage t,

1. Stop if $t = t_0$.
2. Stop if $t < t_0$ and $k_{t+1} = 1$.
3. Continue if $t < t_0$ and $k_{t+1} > 1$.

It has been shown by Guttman that $P(CS|R') \geqslant \beta$. Suppose that there is infinite capital. The rule R' is said to be in state γ at stage t if $k_t = \gamma$. The initial state is k and the states form a Markov chain with nonstationary transition probabilities

$$p_{\gamma\alpha}(t) = \Pr\{k_{t+1} = \alpha | k_t = \gamma\}, \qquad 1 \leqslant \alpha \leqslant \gamma \leqslant k. \tag{17.20}$$

These are dependent on $\omega_{n_t,\gamma}(P^*, \mathbf{T})$. We note that $p_{\gamma\alpha}(t) = 0$ if $\gamma < \alpha$ and $\Sigma_{\alpha=1}^{\gamma} p_{\gamma\alpha}(t) = 1$. The following theorem has been established by Guttman (1963).

Theorem 17.6. Consider the Markov chain with the preceding structure. Let $p_{\alpha\alpha}(t) = 1 - \delta_n(t)$, $0 < \delta_n(t) < 1$ for $\alpha \neq 1$. Then the Markov chain is absorbed at state 1 (i.e., R' terminates at a finite stage) if and only if $\Sigma_{t=1}^{\infty} \delta_\alpha(t)$ diverges for all $\alpha \neq 1$.

It might be possible to find a "reasonable" value of n_t in some special cases. Suppose that the expected subset size at stage t can be written as a function of n_t, k_t, P_t^* and the differences $\theta_{[j]} - \theta_{[i]}$, $i < j$. Since k_t and P_t^* are known, if we have information about the differences of the ordered θ_i, we can set $E(S) = 1$ and solve for n_t.

As we pointed out in the introduction, the decision at each stage depends only on the observations of that stage and not on those of the previous stages.

17.4 SELECTION OF MILDLY t BEST POPULATIONS

In this section we discuss three sequential procedures of Gupta and D. Y. Huang (1975b) one of which is nonparametric. This nonparametric procedure and one of the parametric procedures are of the nonelimination type.

17.4.1 Parametric Procedures

Let X_{in} denote the nth observation from π_i. The observations $\{X_{in}\}$ are assumed to be independent for all i and n. The distributions associated with the populations π_i are not completely known. The selection procedures depend on the observations through a sequence of statistics $T_{ij}(n) = f_n(X_{i1}, \ldots, X_{in}; X_{j1}, \ldots, X_{jn})$, $i, j = 1, 2, \ldots, k$. In a given problem the function f_n

is chosen to indicate the differences between the populations in a reasonable way. For example, if π_i is normal with unknown mean θ_i and known variance σ^2, a choice of $T_{ij}(n)$ is given by

$$T_{ij}(n) = \sum_{r=1}^{n} (X_{jr} - X_{ir}). \tag{17.21}$$

Now we assume that $\mathbf{T}_{ij}(n) = (T_{ij}(1), \ldots, T_{ij}(n))$ has a joint probability density function $g_n(\mathbf{t}_{ij}(n); \tau_{ij})$ involving the parameter τ_{ij}. Usually the sequence $T_{ij}(1), T_{ij}(2), \ldots$ is chosen so that it is a sufficient and transitive sequence and also invariantly sufficient for τ_{ij}. The parameter τ_{ij} is assumed to be a suitable measure of the distance between π_i and π_j. Next, let $S_{s,\gamma} = \{\gamma_1, \ldots, \gamma_s\}$ be any s-subset of $\{1, 2, \ldots, k\}$ and let $\tau_i = \max_{\{\gamma_1, \ldots, \gamma_t\}} \min_{\alpha \in S_{t,\gamma}} \tau_{i\alpha}$, $i = 1, 2, \ldots, k$.

The t-subset $S_{t,\gamma}$ is called the *t best* if for any $i \in S_{t,\gamma}$, τ_i is one of the t smallest values among $\tau_1, \ldots, \tau_k$. A population π_i is said to be *mildly t best* if $\tau_i \leqslant \Delta$, and *strictly non-t-best* if $\tau_i > \Delta$, where Δ is a specified positive constant.

For the normal means problem with $T_{ij}(n)$ given by (17.21), we have $\tau_{ij} = \theta_j - \theta_i$ and $\tau_i = \theta_{[k-t+1]} - \theta_i$. Thus π_i is mildly t best-if $\theta_i \geqslant \theta_{[k-t+1]} - \Delta$ and is strictly non-t-best if $\theta_i < \theta_{[k-t+1]} - \Delta$. For $t = 1$, mildly t best and strictly non-t-best populations are called superior and inferior populations by Desu (1970). The parameter space Ω can be written as $\Omega = \cup_{m=0}^{k-t} \Omega_m$, where $\Omega_m = \{\boldsymbol{\theta}\colon\ \tau_{[1]} \leqslant \cdots \leqslant \tau_{[k-m]} \leqslant \Delta < \tau_{[k-m+1]} \leqslant \cdots \leqslant \tau_{[k]}\}$. Thus Ω_m is the set of configurations corresponding to exactly m strictly non-t-best populations.

Gupta and Huang (1975b) have studied a nonelimination type procedure R_1 for selecting a subset containing all the superior populations (i.e., mildly t best for $t = 1$) based on log-likelihood ratios $l_n(\mathbf{T}_{ij}(n))$ defined by

$$l_n(\mathbf{T}_{ij}(n)) = \log g_n(\mathbf{T}_{ij}(n); \tau_0) - \log g_n(\mathbf{T}_{ij}(n); \Delta) \tag{17.22}$$

where $\Delta > \tau_{ii}$ and $\tau_0 (> \Delta)$ is a constant chosen preferably close to Δ. Let $\alpha = 2(1 - P^*)/\{(k-1)k(k+1)\}$ and $0 < \beta < 1$ be given. We choose a and $b (b < a)$ such that $a \leqslant \log[(1-\beta)/\alpha]$ and $b \geqslant \log[\beta/(1-\alpha)]$. Then the procedure R_1 of Gupta and Huang is defined as follows.

$\boldsymbol{R_1}$: Begin by taking one observation from each population. Calculate the values of the $k(k-1)$ log-likelihood ratios $l_1(\mathbf{T}_{ij}(1))$, $(i, j = 1, \ldots, k; i \neq j)$. Mark the population π_i "accepted" or "rejected" according as

$$l_1(\mathbf{T}_{ij}(1)) \leqslant b \quad \text{for all } j \neq i, \quad \text{or} \tag{17.23}$$

$$l_1(\mathbf{T}_{ij}(1)) \geqslant a \quad \text{for some } j \neq i. \tag{17.24}$$

If neither of (17.23) and (17.24) is satisfied for any of the populations, then proceed to the next stage and take an additional observation from each population. Repeat the steps of the first with $l_2(\mathbf{T}_{ij}(2))$ in the place of $l_1(\mathbf{T}_{ij}(1))$. Continue in this manner until the stage where every population is marked one way or the other. At this stage terminate sampling and declare the set of populations marked "accepted" as the selected subset. If all populations are marked "rejected" at termination, then include in the selected subset all populations π_i for which $\max_{\substack{1\leqslant j\leqslant k\\ j\neq i}} l_n(\mathbf{T}_{ij}(n))$ is a minimum.

It should be noted that at each stage all the populations are marked afresh. The decision rule is based only on the markings of the terminal stage. The following theorem of Gupta and Huang gives sufficient conditions for meeting the P^*-requirement.

Theorem 17.7. Assume that for all i and j,

$$P_{\tau_{ij}}\left\{b<l_r(\mathbf{T}_{ij}(r))<a \text{ for } r=1,\ldots,n-1 \text{ and } l_n(\mathbf{T}_{ij}(n))\geqslant a \text{ for some } n\right\} \tag{17.25}$$

is a nondecreasing function of τ_{ij}. Then $P(CS|R_1)\geqslant P^*$ for all $\boldsymbol{\theta}\in\Omega$ provided that the procedure terminates with probability one.

The probability in (17.25) is the probability of accepting H_1 by using the sequential probability ratio test (SPRT) for H_0: $\tau=\Delta$ against H_1: $\tau=\tau_0$ with stopping boundaries (b,a). Therefore (17.25) is satisfied if the SPRT has a monotone OC function.

Returning to the normal means problem with $\mathbf{T}_{ij}(n)$ given by (17.21), it can be seen that the inequalities in (17.23) and (17.24) reduce, respectively, to

$$Y_{jn}\leqslant Y_{in}-\frac{n(\tau_0+\Delta)}{2}-\frac{2\sigma^2 b}{(\tau_0-\Delta)} \tag{17.26}$$

and

$$Y_{jn}\geqslant Y_{in}-\frac{n(\tau_0+\Delta)}{2}-\frac{2\sigma^2 a}{(\tau_0-\Delta)}, \tag{17.27}$$

where $Y_{jn}=\Sigma_{r=1}^{n}X_{jr}$. In this case, the procedure R_1 terminates with probability one. Gupta and Huang have obtained an upper bound for the expected subset size.

REMARK 17.1. One can use the procedure R_1 for selecting normal populations with variance $\sigma_i^2\leqslant\Delta\sigma_{[1]}^2$, $\Delta>1$, by taking $T_{ij}(n)=s_{in}^2/s_{jn}^2$, $n=$

$2,3,\ldots$, where s_{in}^2 is the usual unbiased estimator of σ_i^2 based on $(n-1)$ degrees of freedom. □

Gupta and Huang (1975b) have also considered an elimination-type procedure R_2 for selecting the mildly t best populations which is described as follows.

R_2: Begin by taking $n_1(\geqslant 1)$ independent observations from each population. Calculate the values of the $k(k-1)$ log-likelihood ratios $l_{n_1}(\mathbf{T}_{ij}(n_1))$, $(i,j=1,\ldots,k;i\neq j)$. Eliminate π_i from further considerations if $l_{n_1}(\mathbf{T}_{ij}(n_1))\geqslant a$ for all $j\in S_{t,\gamma}, j\neq i$, and any $\{\gamma_1,\ldots,\gamma_t\}$, where

$$a=\left[\binom{k-1}{t}+\binom{k-1}{t-1}(1-\delta_{t1})\right]\log\left[\frac{k(k+1)}{2(1-P^*)}\right]\quad\text{and}$$

$$\delta_{t1}=\begin{cases}1 & \text{if } t=1,\\ 0 & \text{if } t>1.\end{cases}$$

Proceed to the next stage by taking n_2-n_1 independent observations from each of the remaining populations. The log-likelihood ratios for the contending populations are computed again and the same elimination rule is used except that $l_{n_2}(\mathbf{T}_{ij}(n_2))$ replaces $l_{n_1}(\mathbf{T}_{ij}(n_1))$. Continue in this manner (with number of observations equal to n_3-n_2, n_4-n_3, and so on) until the stage where no population is eliminated. At that stage terminate sampling and select all the remaining populations. If after applying the rule at the sth stage (say) the number of remaining populations is less than t, then select the t populations from among those populations π_i sampled at the sth stage such that

$$\max_{\{\gamma_1,\ldots,\gamma_t\}}\ \min_{\substack{\alpha\in S_{t,\gamma}\\ \alpha\neq i}} l_{n_s}(\mathbf{T}_{i\alpha}(n_s))$$

is one of the t smallest values.

Theorem 17.8. $P(CS|R_2)\geqslant P^*$ for all $\boldsymbol{\theta}\in\Omega$ provided that for all i and j

$$P_{\tau_{ij}}\{l_n(\mathbf{T}_{ij}(n))\geqslant c\quad\text{for some}\quad n\}\tag{17.28}$$

is a nondecreasing function of τ_{ij}, where c is a constant.

The condition (17.28) is explained by Hoel (1971a) as a modified SPRT with stopping boundaries $(-\infty,c)$. For the normal means problem, let $\pi_{s_1},\ldots,\pi_{s_r}(r\geqslant t)$ be the remaining populations at the sth stage. Then π_i is eliminated if

$$Y_{in_s}\leqslant Y_{[r-t+1]n_s}-2^{-1}n_s(\tau_0+\Delta)-2\sigma^2c(\tau_0-\Delta)^{-1},\tag{17.29}$$

where $Y_{in_s} = \Sigma_{j=1}^{n_s} X_{ij}$, $i = 1, 2, \ldots, k$, $Y_{[r-t+1]n_s}$ is the tth largest among the Y_{in_s}, and c is chosen such that

$$c = \left[\binom{k-1}{t} + (1 - \delta_{t1}) \binom{k-1}{t-1} \right] \log\left[k(k+1)/2(1-P^*) \right].$$

Gupta and Huang (1975b) have done Monte Carlo study of the expected subset size assuming that $(k-t)$ populations are $N(0,1)$ and t populations are (a) $N(2\Delta, 1)$, (b) $N(4\Delta, 1)$, and (c) $N(6\Delta, 1)$. In each of these cases the values of the other constants chosen are: $k = 2(1)5$, $t = 1(1)k-1$, $\Delta = 0.5(0.5)1.5$, $\tau_0 = 2\Delta$, $n_i = 10i$, and $P^* = 0.90, 0.95$.

17.4.2 A Nonparametric Procedure

Let π_i have the associated absolutely continuous distribution function $F(x - \theta_i)$ where θ_i is unknown and F belongs to the class $\mathcal{F}_0$ of all absolutely continuous distribution functions symmetric about the origin whose densities are Pólya frequency functions of order 2 and are differentiable almost everywhere. Let $\{X_{in}\}$ be a sequence of independent observations from π_i and let $Y_{ij}(n) = X_{jn} - X_{in} - \Delta$. Then $F(y - \tau_{ij})$ is the cdf of Y_{ij} with parameter $\tau_{ij} = \theta_j - \theta_i - \Delta$ [see Randles and Hollander (1971)].

For each pair π_i and π_j, and for every $n \geqslant 1$ and real $d(-\infty < d < \infty)$, let

$$R_{ij}^{n,r}(d) = \tfrac{1}{2} + \sum_{l=1}^{n} c\left(|Y_{ij}(r) - d| - |Y_{ij}(l) - d|\right), \qquad r = 1, \ldots, n, \tag{17.30}$$

where $c(u) = 1, \frac{1}{2}$, or 0 according as u is $>$, $=$, or < 0. Thus $R_{ij}^{n,r}(d)$ is the rank of $|Y_{ij}(r) - d|$ among $|Y_{ij}(l) - d|$, $l = 1, \ldots, n$. Consider a set of scores $J_n(i/n+1) = EJ(U_{ni})$ or $J(i/n+1)$, $i = 1, \ldots, n$, where $U_{n1} < \cdots < U_{nn}$ are the ordered random variables of a sample of size n from the uniform distribution over $(0,1)$, the score function $J(u)$ is defined by

$$J(u) = -g'\left(G^{-1}\left(\frac{1+u}{2}\right)\right) \Big/ g\left(G^{-1}\left(\frac{1+u}{2}\right)\right), \qquad 0 \leqslant u < 1, \tag{17.31}$$

and $G \in \mathcal{F}_0$ is a nondegenerate strongly unimodal distribution function having the density g. As we have seen in Chapter 15, notable examples of G are the normal and the logistic distributions.

Now define

$$Q_{ij}^{n}(d) = \sum_{r=1}^{n} J_n\left((n+1)^{-1} R_{ij}^{n,r}(d)\right) s\left(Y_{ij}(r) - d\right), \tag{17.32}$$

where $s(u)=2c(u)-1$. The $Q_{ij}^n(d)$ are the usual one sample rank order statistics. When $\tau_{ij}=0$, $Q_{ij}^n(0)$ has a known distribution symmetric about the origin. We can select $\{Q_{ij}^{n,\epsilon}\}$ such that

$$\Pr\{|Q_{ij}^m(0)| \leqslant Q_{ij}^{n,\epsilon}\} \to 1-\epsilon \quad \text{as } n\to\infty. \tag{17.33}$$

Let $\theta^{*ij}_{L,n}=\sup\{d: Q_{ij}^n(d)>Q_{ij}^{n,\epsilon}\}$, $\theta^{*i,j}_{U,n}=\inf\{d: Q_{ij}^n(d)<-Q_{ij}^{n,\epsilon}\}$, and $C_n^*(i,j)=2Q_{ij}^{n,\epsilon}/n(\theta^{*ij}_{U,n}-\theta^{*ij}_{L,n})$.

We choose Δ close to zero and let $2^{-1}(k-1)k(k+1)(A-1)(A^2-1)^{-1}=(1-P^*)$ where $A \leqslant (1-\beta)\alpha^{-1}$ and $B \geqslant \beta(1-\alpha)^{-1}$ with $0<B<1<A<\infty$. The following procedure R_3 was proposed by Gupta and Huang (1975b) for selecting a subset containing all superior populations.

$\boldsymbol{R_3}$: Begin by taking a single observation from each population. Calculate the values of the $k(k-1)$ rank order statistics $Q_{ij}^1(\Delta/2)$ and $C_1^*(i,j)$ $(i,j=1,\ldots,k;\ i\neq j)$. A population π_i is marked "rejected" if

$$\Delta C_1^*(i,j)Q_{ij}^1\left(\frac{\Delta}{2}\right) \geqslant av \quad \text{for some } j\neq i, \tag{17.34}$$

and it is marked "accepted" if

$$\Delta C_1^*(i,j)Q_{ij}^1\left(\frac{\Delta}{2}\right) \leqslant -av^2 \quad \text{for all } j\neq i, \tag{17.35}$$

where $\nu^2=\int_0^1 J^2(u)du$. If neither of (17.34) and (17.35) is satisfied for any of the populations, then proceed to the next stage by taking one more observation from each population. Repeat the same rule for marking each population with $Q_{ij}^2(\Delta/2)$ and $C_2^*(i,j)$ replacing $Q_{ij}^1(\Delta/2)$ and $C_1^*(i,j)$, respectively. Continue in this manner until the stage when all the populations are marked one way or the other. At this stage terminate sampling and select all the populations that are marked "accepted" at the terminal stage. If at any stage n (say) all the populations are marked "rejected", then select all the populations π_i for which $\max\limits_{\substack{1\leqslant j\leqslant k\\ j\neq i}} \Delta C_n^*(i,j)Q_{ij}^n(\Delta/2)$ is a minimum.

This procedure is shown by Gupta and Huang (1975b) to be satisfying the P^*-condition.

CHAPTER 18

Bayes, Empirical Bayes, and Γ-Minimax Procedures

18.1 INTRODUCTION

In Chapter 10 we discussed Bayesian selection and ranking procedures for selecting the best population. In this chapter we are concerned with the Bayesian approach to subset selection problems. We also discuss an empirical Bayes approach to the problem. The empirical Bayes approach to statistical decision problem is due to Robbins [1964]. In this approach, we assume the existence of an a priori distribution G on the parameter space but the form of G is unknown. We also discuss some Γ-minimax procedures for which we assume a possible class of prior distributions. Thus Γ-minimax procedures include Bayes procedures as well as the usual minimax procedures.

18.2 BAYES PROCEDURES

Let θ_i denote the unknown characteristic of the population π_i, $i=1,\ldots,k$. Let $X_{i1},\ldots,X_{in}$ be n observations from π_i and $Y_i=T_i(X_{i1},\ldots,X_{in})$ a suitable estimator of θ_i. As we have seen in other chapters, Y_i is usually a sufficient statistic for θ_i. It is assumed that, given $\boldsymbol{\theta}$, the Y_i are independently distributed and their joint distribution $F(\cdot,\boldsymbol{\theta})$ is absolutely continuous with respect to a σ-finite measure μ. Let us denote the pdf of Y_i by $f(y_i|\theta_i)$.

The elements of the action space $\mathcal{Q}$ are all possible subsets (except the null set) of $\pi_1,\ldots,\pi_k$. Let us denote these subsets by S_j, $j=1,2,\ldots,2^k-1$. Since we are interested only in Bayes procedures, we need to consider only nonrandomized decision rules (see DeGroot [1970], §8.5). A selection procedure D is a mapping from $\mathcal{Y}$, the sample space of $\mathbf{Y}$, to the action space $\mathcal{Q}$. Let $\mathbf{y}=(y_1,\ldots,y_k)\in\mathcal{Y}$. Then $D(\mathbf{y})$ is some nonempty subset of the k populations. Let $L(D(\mathbf{y}),\boldsymbol{\theta})$ denote the loss incurred when $\boldsymbol{\theta}$ is the true state of nature and the subset $D(\mathbf{y})$ is selected. We assume that G is the a priori distribution on the parameter space Ω. The Bayes risk of a decision procedure D with respect to the a priori distribution G is given by

$$B(D,G)=\int_{\Omega}\left\{\int_{\mathcal{Y}}L(D(\mathbf{y}),\boldsymbol{\theta})f(\mathbf{y}|\boldsymbol{\theta})\,d\mathbf{y}\right\}dG(\boldsymbol{\theta}). \tag{18.1}$$

18.2.1 Procedures with Linear Loss Functions

Deely and Gupta (1968) considered the loss function $L(S_j,\boldsymbol{\theta})$ corresponding to the choice of S_j given by

$$L(S_j,\boldsymbol{\theta})=\sum_{q\in S_j}\alpha_{jq}(\theta_{[k]}-\theta_q), \tag{18.2}$$

where $\alpha_{jq}\geqslant 0$. This loss function is analogous to the linear loss function considered by Bahadur and Robbins (1950), Dunnett (1960), and Bland (1961). Some examples of such a loss function are:

(i) $L(S_j,\boldsymbol{\theta})=\sum_{q\in S_j}(\theta_{[k]}-\theta_q)$, sum of losses,

(ii) $L(S_j,\boldsymbol{\theta})=\dfrac{1}{|S_j|}\sum_{q\in S_j}(\theta_{[k]}-\theta_q)$, average loss,

(iii) $L(S_j,\boldsymbol{\theta})=(k+1-|S_j|)\sum_{q\in S_j}(\theta_{[k]}-\theta_q)$.

Note that $|S_j|$ denotes the cardinality of the subset S_j. A sufficient condition for the rule D^* to be Bayes is that

$$\psi_G(D^*,\mathbf{y})\leqslant\psi_G(D,\mathbf{y}) \tag{18.3}$$

for any D, where

$$\psi_G(D,\mathbf{y}) = \int_\Omega L(D(\mathbf{y}),\boldsymbol{\theta}) f(\mathbf{y}|\boldsymbol{\theta})\, dG(\boldsymbol{\theta}). \tag{18.4}$$

Let S_j denote the one-element set $\{\pi_j\}$, $j=1,\ldots,k$ and the remaining subsets S_j be associated one-to-one with $j=k+1,\ldots,2^k-1$ arbitrarily. Deely and Gupta (1968) proved the following theorem.

Theorem 18.1. Let the loss function be given by (18.2) in which $\alpha_{jq}=\alpha>0$ for $j=1,2,\ldots,k$ and hence $\Sigma_{q\in S_j}\alpha_{jq}\geqslant\alpha$ for $j=1,2,\ldots,2^k-1$. Then the Bayes procedure with respect to G is given by $D^*=D^*(\mathbf{y})=S_j$ where j is any positive integer $1,2,\ldots,k$ such that $\psi_G(S_j,\mathbf{y})=\min\{\psi_G(S_i,\mathbf{y}): 1\leqslant i\leqslant k\}$.

Several special cases have been considered by Deely (1965).

18.2.2 Nonlinear Loss Functions

Goel and Rubin (1977) have considered the loss function $L(S,\boldsymbol{\theta})$ corresponding to any selected subset S defined by

$$L(S,\boldsymbol{\theta}) = l_1|S| + l_2(\theta_{[k]} - \theta_{[S]}) \tag{18.5}$$

where $\theta_{[S]}=\max_{j\in S}\theta_j$, and l_1 and l_2 are the weights for each component of loss. Without any loss of generality the loss function can be rewritten as

$$L(S,\boldsymbol{\theta}) = c|S| + (\theta_{[k]} - \theta_{[S]}), \tag{18.6}$$

where $c=l_1/l_2>0$.

Before discussing the results of Goel and Rubin we introduce some notations which differ from theirs. Let us recall that $Y_{[1]}<\cdots<Y_{[k]}$ are the ordered observations. The population and the observation associated with $\theta_{[i]}$ are denoted by $\pi_{(i)}$ and $Y_{(i)}$, respectively. The population and the parameter associated with $Y_{[i]}$ are denoted by $\pi_{[i]}$ and $\theta_{(i)}$, respectively. Thus our scheme of notations can be summarized as follows.

$\pi_1,\ldots,\pi_k$	$\pi_{(1)},\ldots,\pi_{(k)}$	$\pi_{[1]},\ldots,\pi_{[k]}$
$\theta_1,\ldots,\theta_k$	$\theta_{[1]}<\cdots<\theta_{[k]}$	$\theta_{(1)},\ldots,\theta_{(k)}$
$Y_1,\ldots,Y_k$	$Y_{(1)},\ldots,Y_{(k)}$	$Y_{[1]}<\cdots<Y_{[k]}$

For $j=1,2,\ldots,k$, let d_j be the rule that chooses the subset $S_j^*=\{\pi_{[k]},\ldots,\pi_{[k-j+1]}\}$ with probability 1. For $j=k+1,\ldots,2^k-1$, the d_j's and the remaining subsets in $\mathscr{Q}$ are associated one-to-one arbitrarily.

Given $\mathbf{y}=(y_1,\ldots,y_k)$, the posterior risk of the decision d, which chooses the subset S with probability 1, is denoted by

$$r(d,\mathbf{y})=c|S|+E\{(\theta_{[k]}-\theta_{[S]})|\mathbf{y}\}, \tag{18.7}$$

where the expectation is taken with respect to the posterior distribution of $\boldsymbol{\theta}$ given $\mathbf{y}$. The following lemma of Goel and Rubin (1977) reduces the number of posterior risks to be compared from 2^k-1 to k.

Lemma 18.1. Assume that the density $f(y,\theta)$ has a monotone likelihood ratio and the prior distribution of $\boldsymbol{\theta}$ is symmetric on Ω (i.e., the random variables $\theta_1,\ldots,\theta_k$ are exchangeable). Then the Bayes rule d^* is given by

$$r(d^*,\mathbf{y})=\min_{j=1,\ldots,k} r(d_j,\mathbf{y}). \tag{18.8}$$

REMARK 18.1. The result of Lemma 18.1 holds for any loss function which satisfies the conditions (3.13) and (3.14) in §3.3. Also it is not necessary that $f(y,\theta)$ possesses a monotone likelihood ratio. Instead, it could be a stochastically increasing family. However, we will need an extra condition on the posterior distribution of $\boldsymbol{\theta}$. If the posterior distribution of $\boldsymbol{\theta}$ has the stochastically increasing property (SIP) defined in §3.3 (see Definition 3.8), then Lemma 18.1 can be proved by using Theorem 3.7 due to Alam (1973). □

Let $\succ$ denote an order relationship whereby $d_i \succ d_j$ means that the subset S_i^* is preferred to S_j^*, that is, the posterior risk of d_i is less than that of d_j. The following theorem of Goel and Rubin (1977) shows that it is sufficient to consider the $(k-1)$ differences $\Delta_m=r(d_{m+1},\mathbf{y})-r(d_m,\mathbf{y})$, $m=1,\ldots,k-1$, for a complete specification of the Bayes rule.

Theorem 18.2. Let the prior distribution of $\boldsymbol{\theta}$ be symmetric on Ω. If the set $A=\{j\colon \Delta_j \geqslant 0\}$ is empty, then the Bayes rule is $d^*=d_k$; otherwise, $d^*=d_m$ where m is the smallest number in the set A.

Goel and Rubin have shown that

$$\Delta_m \geqslant c-E\{\max(0,\theta_{(k-m)}-\theta_{(k)})|\mathbf{y}\}, \qquad m=1,\ldots,k-1. \tag{18.9}$$

This bound can be used to eliminate the computation of Δ_m for some large values of m.

No further simplification of the Bayes rule is possible unless some more assumptions on the prior distribution of $\boldsymbol{\theta}$ are made. Goel and Rubin assume that, given $W=\omega$, $\theta_1,\ldots,\theta_k$ are i.i.d. random variables with density $\xi(\cdot,\omega)$ and the distribution of W is known. This prior is a mixture of i.i.d. random variables.

Now, it follows that, given $\mathbf{Y}=\mathbf{y}$ and $W=\omega$, the random variables $\theta_1,\ldots,\theta_k$ are a posteriori independently distributed. Let $g_i(\theta_{(i)})=g(\theta_{(i)}|y_{[i]},\omega)$ and $G_i(\theta_{(i)})=G(\theta_{(i)}|y_{[i]},\omega)$ denote the density and the cdf of $\theta_{(i)}$ given $y_{[i]}$ and ω, respectively. Furthermore, we let $Q(\omega|\mathbf{y})$ denote the conditional cdf of W given $\mathbf{y}$. With this specification of the prior, define

$$I_m(\omega)=\int_{-\infty}^{\infty}\prod_{i=k-m+1}^{k} G_i(z)\left[1-G_{k-m}(z)\right]dz. \tag{18.10}$$

It is shown by Goel and Rubin that

$$\Delta_m=c-\int_{-\infty}^{\infty} I_m(\omega)\,dQ(\omega|\mathbf{y}), \qquad m=1,\ldots,k-1. \tag{18.11}$$

Let

$$\begin{aligned}
a_m&=E\left[\max(0,\theta_{(k-m)}-\theta_{(k)})\right],\\
v_m&=\int_{-\infty}^{\infty}\int_{-\infty}^{\infty} G_{k-m+1}^m(z)\left[1-G_{k-m}(z)\right]dz\,dQ(\omega|\mathbf{y}),\\
u_m&=\int_{-\infty}^{\infty}\int_{-\infty}^{\infty} G_k^m(z)\left[1-G_{k-m}(z)\right]dz\,dQ(\omega|\mathbf{y}).
\end{aligned} \tag{18.12}$$

Further, let $r_i=\min(a_i,v_i)$, $i=1,\ldots,k-1$, and $r_k=0$.

Theorem 18.3. [Goel and Rubin (1977)]. If the prior distribution of $\boldsymbol{\theta}$ is a mixture of i.i.d. random variables, then the Bayes rule d^* in the selection problem is given by

(i) $c\geqslant r_1 \Rightarrow d^*=d_1$,
(ii) $r_2\leqslant c<r_1\Rightarrow d^*=d_2$, and
(iii) $r_i\leqslant c<r_{i-1}$ and $c\leqslant u_j$ for $i=3,\ldots,k$, and some $j\in\{2,\ldots,i-1\}\Rightarrow d^*=d_m$, where m is the smallest number $\in\{j+1,\ldots,i\}$ which satisfies $\Delta_m\geqslant 0$. (18.13)

For $k=2$, the Bayes rule is given by:

$$d^*=\begin{cases} d_1 & \text{if } c\geqslant a_1,\\ d_2 & \text{otherwise.}\end{cases} \tag{18.14}$$

An "approximate" Bayes rule R, which will select larger subsets than d^*

but is the Bayes rule for $k=2$, is suggested by Theorem 18.3 and is given as follows:

R: If $c \geqslant a_1$, select S_1^*; if $a_m \leqslant c < a_{m-1}$, select S_m^*, $m=2,\ldots,k$.

Let us now assume that the posterior cdf of $\theta_{(i)}$ is of the form

$$G_i(\theta_{(i)}|y_{[i]},\omega)=G\left[\frac{\theta_{(i)}-b_1y_{[i]}-b_0\omega}{b_2}\right], \qquad i=1,\ldots,k, \tag{18.15}$$

for some $b_1,b_2>0$. Then, for $k=2$, the Bayes rule d^* is given by

$$d^*=\begin{cases} d_1 & \text{if } y_{[1]}\leqslant y_{[2]}-b\delta_1, \\ d_2 & \text{otherwise,}\end{cases} \tag{18.16}$$

where $\delta_1=-t_1^{-1}(c/b_2)$ and

$$t_1(z)=\int_{-\infty}^{\infty}G(u+z)(1-G(u))\,du. \tag{18.17}$$

The approximate Bayes rule now reduces to R: Select π_i if $y_i>y_{[k]}-\delta_1 b$, which is the classical subset selection procedure of Gupta (1956). In the classical case, the value of δ_1 is a function of k and P^*, whereas δ_1 here depends only on the cost per population.

Goel and Rubin (1977) have also considered the special case where π_i is normal with mean θ_i and variance σ^2 (known). The prior of $\boldsymbol{\theta}$ is assumed to be exchangeable multivariate normal for all k. It has been shown that $d^*=d_1$ or $d^*\in\{d_1,d_2\}$ according as $c/\gamma\geqslant 1/\sqrt{\pi}$ or $.2821\leqslant c/\gamma<1/\sqrt{\pi}$, where γ is the posterior standard deviation.

Chernoff and Yahav (1977) have also discussed Bayes rules for selecting the best populations. Let π_i be a normal population with mean μ_i and variance σ^2, $i=1,\ldots,k$. It is assumed that the μ_i are unknown and σ^2 is known. The unknown means are assumed to have a joint normal prior distribution. Let $K=\{1,\ldots,k\}$ and let J be any subset of K. Define

$$\bar{\mu}(J)=\frac{\sum_{i\in J}\mu_i}{\sum_{i\in J}1},$$

$$V(\boldsymbol{\mu},J)=\bar{\mu}(J)+r\max_{i\in J}\mu_i, \tag{18.18}$$

where r is a positive constant.

The loss function associated with the subset J is

$$L(\boldsymbol{\mu},J)=(1+r)\mu_{[k]}-V(\boldsymbol{\mu},J). \tag{18.19}$$

This loss function is motivated by the following considerations. We have k varieties of treatments available for use on a land. Suppose we carry out an experiment on each treatment and use the resulting data to select a subset of the treatments. The selected treatments are applied on equal areas of land and the ensuing crop determines the best among the treatments of the selected subset. This best treatment is applied to the land for the next r years. We can see that $\bar{\mu}(J)$ is the average of the mean yields associated with the subset J and represents the expected yield for the first crop. Further $r\max_{i\in J}\mu_i$ stands for the expected yield for the next r years. Thus the loss $L(\boldsymbol{\mu},J)$ is based on the comparison of the value $V(\boldsymbol{\mu},J)$ with the best achieved when a correct decision is made regarding the best treatment.

Chernoff and Yahav (1977) have compared the Bayes procedure with the (random size) subset selection procedure of Gupta (1956) and the fixed-size subset selection procedure of Desu and Sobel (1968). Their results are based on Monte Carlo studies of the integrated risks with respect to exchangeable normal priors.

Bayes procedures typically require numerical integrations to implement them and this makes them sometimes unsuitable for practical use. So if there is an easy-to-implement procedure whose performance is close to that of Bayes procedure, then this procedure ought to be used. This possibility was explored by Hsu (1977) in his thesis in the case of normal populations problem with normal exchangeable priors. For each pair of prior and loss function considered by him, it turns out that there is always a Gupta-type procedure that performs almost as well as the Bayes procedure. However, it is not known how these procedures perform when the prior is not normal.

Bickel and Yahav (1977) have discussed the problem of selecting a set of good populations. Let us again assume that π_i is normal with unknown mean μ_i and variance $\sigma^2=1$, $i=1,\ldots,k$. Let $S=\{\pi_{i_1},\ldots,\pi_{i_r}\}$ be any subset of the k populations. The loss function associated with it is defined by

$$L(\boldsymbol{\mu},S)=\mu_{[k]}-r^{-1}\sum_{j=1}^{r}\mu_{i_j}+KI_{\{\pi(k)\notin S\}}, \tag{18.20}$$

where $\pi_{(k)}$ is the population with the largest mean, K is a positive constant, and I_A is the indicator function of A. Bickel and Yahav (1977) have investigated the optimal solution when k goes to infinity under the assumption that the "empirical distribution" of the means $\mu_i^{(k)}$, $i=1,\ldots,k$, converges in a suitable sense to a smooth limiting probability distribution. Their asymptotic solution is "select the populations that generated the last $100\lambda_0$ percent of the order statistics", where λ_0 depends on the limiting distribution of the $\mu_i^{(k)}$ and on the penalty associated with a wrong decision.

Gupta and Hsu (1978) and Bjørnstad (1978) have also studied comparative performances of subset selection procedures under various loss functions. The Gupta type procedures are referred to as the maximum type procedures as opposed to the average-type procedures of Seal (1955) discussed in §12.2.6. For the purposes of comparison, Gupta and Hsu (1978) have slightly modified the average-type procedure so that it will not select an empty set. For observations $X_1,\ldots,X_k$ from the k populations the modified average-type procedure selects π_i if and only if $X_i > X_{[k-1]}$ and/or

$$X_i > (k-1)^{-1} \sum_{\substack{j=1 \\ j\neq i}}^{k} X_j - d, \text{ where } d = d(k,P^*) > 0.$$

Let us consider the case of location parameters. For any subset S, let $\mathrm{ICS}(\boldsymbol{\theta}, S)$ denote the simple loss due to an incorrect selection and $|S|$ denote, as before, the size of the subset. The comparisons have been made using different combinations of loss components. The four loss combinations that have been used are shown in Figure 18.1. As we have seen before, Goel and Rubin (1977) used linear combinations represented by (2). Bickel and Yahav (1977) used the combination indicated by (3) whereas Chernoff and Yahav (1977) employed the combination denoted by (4). Gupta and Hsu (1978) used linear combinations of $\mathrm{ICS}(\boldsymbol{\theta}, S)$ and $|S|$ in their Monte Carlo studies in the case of normal means with exchangeable normal priors. Their results also indicate that the maximum-type procedures do almost as well as the Bayes procedures.

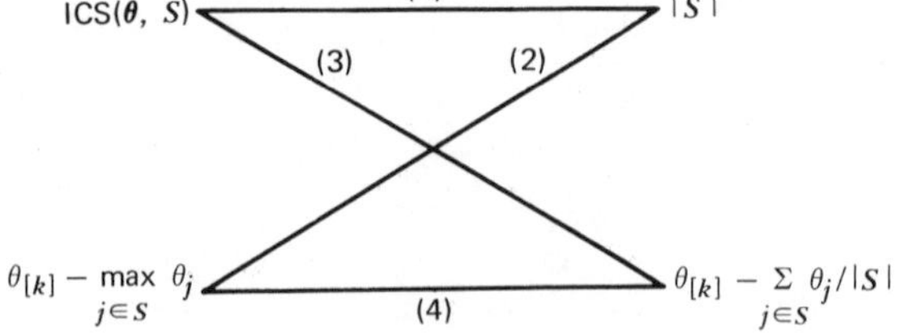

Figure 18.1.

Bjørnstad (1978) studied a class of procedures which are just and translation-invariant and whose selection probabilities are certain Schur-concave functions. For normal populations, these procedures and the maximum-type procedures are found to be quite comparable. Further, the average-type procedures are found to be clearly inferior to the other two. Under certain additive loss functions, Miescke (1978) has shown that in the normal case, with symmetrical priors, the Gupta-type procedure is the limit of Bayes rules as the sample size tends to infinity.

18.2.3 Selection Problem as a Restricted Product of Two-Decision Problems

Let us consider the loss function

$$L(S,\boldsymbol{\theta})=c_1\sum_{j\in S}\epsilon(\theta_{[k]}-\theta_j)+c_2\epsilon(\theta_{[k]}-\theta_{[S]}), \tag{18.21}$$

where $\epsilon(x)=1$ if $x>0$, and 0 if $x\leqslant 0$. Each action S of the selection problem (i.e., the subset of the k parameters) may be generated by taking the proper action from each of the k two-decision problems:

$$\begin{aligned} &d_+^i:\theta_i\in S,\\ &d_-^i:\theta_i\notin S, \end{aligned}\qquad i=1,2,\ldots,k. \tag{18.22}$$

The only sequence of decisions not acceptable is $d_-^1,d_-^2,\ldots,d_-^k$. Lehmann (1957a, b) has proved that an optimal solution for a restricted product problem is given by the compatible optimal solutions for the generating problems if the losses are additive. If the loss for each action in the restricted product problem is the sum of the losses of the corresponding generating actions, then we have additive losses. Bratcher and Bhalla (1974) have adopted this approach.

With the assumption of additive losses, the loss functions for the two-decision problems are appropriately defined by

$$\begin{aligned} L_+^i(\boldsymbol{\theta})&=\begin{cases}0 & \text{if } \theta_i=\theta_{[k]},\\ c_1 & \text{if } \theta_i\neq\theta_{[k]},\end{cases}\\ L_-^i(\boldsymbol{\theta})&=\begin{cases}c_2 & \text{if } \theta_i=\theta_{[k]},\\ 0 & \text{if } \theta_i\neq\theta_{[k]}.\end{cases} \end{aligned} \tag{18.23}$$

From the Bayes decision criterion we include π_i in the selected subset (i.e., take the action d^i_+) if

$$E\big(L^i_+(\boldsymbol{\theta})|\mathbf{y}\big) \leqslant E\big(L^i_-(\boldsymbol{\theta})|\mathbf{y}\big) \tag{18.24}$$

which is equivalent to

$$\Pr(\theta_i = \theta_{[k]}|\mathbf{y}) \geqslant (c+1)^{-1}, \tag{18.25}$$

where $c = c_2/c_1$. If there is exactly the same amount of information on each θ_i, then $\Pr(\theta_i = \theta_{[k]}|\mathbf{y}) = k^{-1}$, $i = 1,\ldots,k$. Hence, if $c < k-1$ then each π_i could be excluded from the selected subset. If $c \geqslant k-1$, the null set will never result. Bratcher and Bhalla (1974) have derived the expression for the PCS and considered the example of binomial populations with uniform prior.

18.3 EMPIRICAL BAYES PROCEDURES

In this section we continue with the general setup of §18.2. However, we assume only the existence of a prior distribution G of $\boldsymbol{\theta}$ and not any specific form of G. Deely (1965) has investigated selection procedures in this case using the empirical Bayes approach of Robbins [1955, 1963, 1964].

Let $\mathbf{Y} = (Y_1,\ldots,Y_k)$ where the joint distribution of the Y_j is $F(\cdot,\boldsymbol{\theta})$. Let $\mathbf{y}_i$, $i = 1,\ldots,r$, be r observations on $\mathbf{Y}$. Let $\mathbf{y}^* = (\mathbf{y}_1,\ldots,\mathbf{y}_r)'$. Thus $\mathbf{y}^*$ is an $r \times k$ matrix. Suppose that independent observations $(\mathbf{y}^*_1,\boldsymbol{\theta}_1),\ldots,(\mathbf{y}^*_n,\boldsymbol{\theta}_n)$ on the random variable $\mathbf{Y}$ are available with $\boldsymbol{\theta}_i$'s all being drawn from the same prior distribution G. The "prior observations" contain information about G and thus if a decision procedure t_n based upon $\mathbf{y}^*_1,\ldots,\mathbf{y}^*_n$ could be found such that the Bayes risk $R(t_n, G)$ converges to the Bayes risk $R(t_G, G)$ for any G in some family $\tilde{G}$, then the procedure t_n is *asymptotically optimal* with respect to t_G and is called an *empirical Bayes* procedure with respect to the unknown G. The following theorem of Deely (1965, pp. 33–34) gives a method for finding empirical Bayes procedures for the selection problem.

Theorem 18.4. Let $\mathbf{X}_1, \mathbf{X}_2, \ldots$ be independent random vectors, $\mathbf{X}_i$ having the density $f(\mathbf{x}|\boldsymbol{\theta}_i)$ and consisting of k components which are independent random variables. Let G_n be a distribution function on the parameter space Ω. Suppose G_n is a function of $\mathbf{X}_1, \mathbf{X}_2, \ldots, \mathbf{X}_n$ such that $\Pr(\lim_{n\to\infty} G_n(\boldsymbol{\theta}) = G(\boldsymbol{\theta})$ for every continuity point $\boldsymbol{\theta}$ of $G) = 1$, where G is a k-dimensional distribution on Ω and does not depend upon the $\mathbf{X}$'s and the probability is with respect to $\mathbf{X}_1, \mathbf{X}_2 \ldots$. Let the loss function $L(d_i, \boldsymbol{\theta})$ and the densities

$f(\mathbf{x}|\boldsymbol{\theta}_i)$ be such that $L(d_i,\boldsymbol{\theta})f(\mathbf{x}|\boldsymbol{\theta})$ is bounded and continuous in $\boldsymbol{\theta}$ for every $d_i \in \mathcal{Q}$ and $\mathbf{x} \in \mathcal{X}$. Let $L(\boldsymbol{\theta}) = \max\{L(d_i,\boldsymbol{\theta}): d_i \in \mathcal{Q}\}$. Then $\{t_{G_n}\}$ is asymptotically optimal with respect to G if $\int_\Omega L(\boldsymbol{\theta})\,dG(\boldsymbol{\theta}) < \infty$.

To find G_n, we assume G to be a member of some parametric family $\tilde{G}$ with parameter $\boldsymbol{\lambda} = (\lambda_1,\ldots,\lambda_k)$. Suppose now an estimate λ_{nj} of λ_j depending upon the prior observations from π_j can be found such that G_{nj} converges to G_j with probability one. Then $G_n = \prod_{j=1}^k G_{nj}$ converges to $G = \prod_{j=1}^k G_j$ with probability one. Further, G_n is also a member of $\tilde{G}$ and if the Bayes procedure with respect to any G in $\tilde{G}$ is available, then in particular the Bayes procedure with respect to G_n is available and thus an empirical Bayes procedure with respect to G is obtained by Theorem 18.4.

Deely (1965) has derived the empirical Bayes procedures in several special cases among which are (a) normal–normal, (b) normal–uniform, (c) binomial–beta, and (d) Poisson–gamma.

Using a result of Robbins [1964], Deely (1965) has also investigated the case where G is not assumed to be a member of some parametric family. The only assumption is that each G_i has a finite absolute first moment.

18.4 Γ-MINIMAX PROCEDURES

We discussed in §5.2.3 optimal sampling using Γ-minimax criterion. As we said earlier, the Γ-minimax criterion requires one to select a decision rule that minimizes the maximum expected risk over Γ, a class of prior distributions. If Γ consists of a single prior, then the Γ-minimax criterion is the Bayes criterion for that prior. On the other hand, if Γ consists of all possible priors then it becomes the usual minimax criterion. Gupta and D. Y. Huang (1977a) have discussed Γ-minimax selection procedures.

Let π_i have the associated probability density $f_{\theta_i}(\cdot)$, $i=1,\ldots,k$. We have one observation from each population. The vector of observations $\mathbf{x} = (x_1,\ldots,x_k)$ is assumed to have density $f_{\boldsymbol{\theta}}(\mathbf{x})$ (with respect to some σ-finite measure μ). Let Ω be the whole parameter space. We use τ_{ij} as a measure of distance between π_i and π_j. Define $\tau_i = \max_{\substack{1<j<k\\ j\neq i}} \tau_{ij}$. Let $\Omega_i = \{\boldsymbol{\theta}|\tau_i \leqslant \tau_0\}$, $i=1,\ldots,k$, where τ_0 is a given constant. The parameter space Ω is partitioned into $k+1$ mutually exclusive subsets, $\Omega = \Omega_0 \cup \Omega_1 \cup \ldots \cup \Omega_k$, where Ω_0 is an indifference zone. It is possible for Ω_0 to be an empty set.

Let $\delta_1,\ldots,\delta_k$ be individual selection probabilities. The population π_i associated with $\tau_i = \min_{1\leqslant j\leqslant k} \tau_j$ is called the best. A selection of the best population is called a correct selection. If $\boldsymbol{\theta} \in \Omega_0$, the selection of any one of the populations is a correct selection.

We define the loss function associated with $\delta=(\delta_1,\ldots,\delta_k)$ by

$$L(\boldsymbol{\theta},\delta)=\sum_{i=1}^{k}\sum_{j=1}^{k}L^{(i)}(\boldsymbol{\theta},\delta_j), \tag{18.26}$$

where, for any i $(1\leqslant i\leqslant k)$ and $\boldsymbol{\theta}\in\Omega_i$, $L^{(r)}(\boldsymbol{\theta},\delta_j)=0$ for all $r\neq i$,

$$L^{(i)}(\boldsymbol{\theta},\delta_j)=\begin{cases} w_{ij}\theta_j & \text{for } j\neq i,\\ 0 & \text{for } j=i,\end{cases} \tag{18.27}$$

and w_{ij} $(i\neq j)$ are given positive numbers and $w_{ii}=0$. For $\boldsymbol{\theta}\in\Omega_0$, the loss is zero. We assume that partial information is available so that we are able to specify $q_i=\Pr\{\boldsymbol{\theta}\in\Omega_i\}$, $\sum_{i=1}^{k}q_i\leqslant 1$. Define

$$\Gamma=\left\{\rho(\boldsymbol{\theta})\,|\,\textstyle\int_{\Omega_i}d\rho(\boldsymbol{\theta})=q_i,\ i=1,\ldots,k\right\}, \tag{18.28}$$

where $\rho(\boldsymbol{\theta})$ is a prior distribution over Ω. A decision rule δ^0 is called Γ-*minimax* if

$$\sup_{\rho\in\Gamma} r(\rho,\delta^0)=\inf_{\delta}\sup_{\rho\in\Gamma} r(\rho,\delta), \tag{18.29}$$

where $r(\rho,\delta)$ is the risk of a decision rule δ. Gupta and Huang (1977a) have proved the following result.

Theorem 18.5. For any i, $1\leqslant i\leqslant k$, let $\theta_i^*\in\Omega_i$ be such that

$$\sup_{\boldsymbol{\theta}\in\Omega_i}\int_{E^k}\sum_{j=1}^{k}w_{ij}\delta_j^0 f_{\boldsymbol{\theta}}(\mathbf{x})\,d\mu(\mathbf{x})=\int_{E^k}\sum_{j=1}^{k}w_{ij}\delta_j^0 f_{\boldsymbol{\theta}_i^*}(\mathbf{x})\,d\mu(\mathbf{x}),$$

where

$$\delta_i^0(\mathbf{x})=\begin{cases} 1 & \text{if } L_i(\mathbf{x})<\min\limits_{\substack{1\leqslant j\leqslant k\\ j\neq i}} L_j(\mathbf{x}),\\ a_i & \text{if } L_i(\mathbf{x})=\min\limits_{\substack{1\leqslant j\leqslant k\\ j\neq i}} L_j(\mathbf{x}),\\ 0 & \text{if } L_i(\mathbf{x})>\min\limits_{\substack{1\leqslant j\leqslant k\\ j\neq i}} L_j(\mathbf{x}),\end{cases}$$

$L_i(\mathbf{x})=\sum_{j=1}^{k}w_{ji}q_j f_{\boldsymbol{\theta}_j^*}(\mathbf{x})$, $1\leqslant i\leqslant k$, and $\sum_{i=1}^{k}a_i=1$. Then $\delta^0=(\delta_1^0,\ldots,\delta_k^0)$ is a Γ-minimax procedure.

Gupta and D. Y. Huang (1977a) have discussed an application of Theorem 18.5 to a problem of selecting "superior" populations. They have obtained similar results in the case of subset selection procedures under a slippage model. In another paper (1978a), Gupta and D. Y. Huang considered the class $\mathcal{C}$ of all rules δ such that

$$P_{\boldsymbol{\theta}}(CS|\delta) \geqslant \gamma \text{ for } \boldsymbol{\theta} \in \overline{\Omega} \tag{18.30}$$

where $\overline{\Omega} = \Omega - \Omega_0$. They have discussed how to construct an optimal procedure in $\mathcal{C}$ which minimizes the supremum of $E(S|\delta)$. Gupta and D. Y. Huang (1978b) have considered rules for which the expected subset size is controlled and the infimum of the PCS over Ω is maximized. Their main theorem deals with construction of an essentially complete class of selection rules of the preceding type. This class contains some classical rules.

In an ϵ-contaminated model, the form of the distribution is specified only with probability $1-\epsilon$, the probability being ϵ that the distribution is something totally different and unspecified. Berger (1977a) has shown that, in a finite parameter space multiple decision problem, the usual Bayes rule, ignoring any contamination, is robust in the sense that, for small ϵ, it is Γ-minimax when the subclass of priors is a class of priors on the family of ϵ-contaminations.

Berger and Gupta (1977) have discussed minimax rules in the class of nonrandomized, just and invariant rules when the risk is measured by the maximum probability of including a nonbest population. These minimax rules are of the form studied by Gupta (1956) in location and scale parameter problems. Berger and Gupta (1977) have applied their results to the normal means problem with known unequal variances obtaining comparison of several rules proposed earlier. The rule of Gupta and Wong (1976b) is shown to be minimax.

18.5 NOTES AND REMARKS

Some results regarding the monotonicity of Bayes subset selection procedures are given by Gupta and Hsu (1977a) and Miescke (1978). Broström (1976) has discussed admissibility of subset selection procedures. Bayes rules for subset selection under models involving series and parallel systems are studied by Broström (1977a) and Kim (1978).

CHAPTER 19

Some Modified Formulations and Other Related Problems

19.1 INTRODUCTION

In Chapter 11, while discussing the general subset selection theory, we discussed some modifications in the basic formulation and the goal. In the present chapter we continue this discussion with several other modifications and related problems among which is a subset selection formulation of the complete ranking problem.

19.2 SELECTING A SUBSET CONTAINING AT LEAST ONE OF THE t BEST POPULATIONS

Let π_i $(i=1,2,\ldots,k)$ have the cdf $F_i=F(x,\theta_i)$ where $\theta_i\in\Theta$, an open interval on the real line. Let $\delta(\theta_1,\theta_2)$ be a suitable nonnegative distance function measuring the distance between the populations with parameters $\theta_1 \leqslant \theta_2$ such that (i) $\delta(\theta_1, \theta_2) = 0$ if and only if $\theta_1 = \theta_2$ and (ii) $\delta(\theta_1, \theta_2)$ increases in θ_2 for fixed θ_1, and decreases in θ_1 for fixed θ_2. Let Ω denote the set of all $\boldsymbol{\theta}$ and $\Omega(\delta^*)$ denote the subset of Ω on which $\delta_t = \delta(\theta_{[k-t]}, \theta_{[k-t+1]}) \geqslant \delta^*$ where $\delta^* \geqslant 0$ is given.

Sobel (1969) investigated three procedures R, R', and R_M along with a truncated form of each for selecting a subset containing at least one of the t best populations (i.e., populations with the t largest θ-values) such that

$$\text{PCS} \geqslant P^* \quad \text{for all } \boldsymbol{\theta} \in \Omega(\delta^*). \tag{19.1}$$

These procedures select a subset which is either of fixed size or of random size depending on the values of P^*, δ^*, and so on. Hence, they can be regarded as an attempt to relate the fixed subset-size approach and the random subset-size approach.

Let X_{ij}, $j = 1, 2, \ldots, n$, be independent observations from π_i and $T_i = T(X_{i1}, \ldots, X_{in})$ a suitable statistic whose cdf is denoted by $G_{\theta_i}(y)$. We make the following two assumptions about the family $\{G_\theta(x) \colon \theta \in \Theta\}$.

Assumption 1. The family $\{G_\theta(x) \colon \theta \in \Theta\}$ is a stochastically increasing family of absolutely continuous distributions, and the derivative $G'_\theta(x)$ with respect to θ exists in Θ.

Assumption 2. For $x_1 \leqslant x_2$ and $\theta_1 \leqslant \theta_2$ (both in Θ)

$$G'_{\theta_1}(x_2)\, d_x G_{\theta_2}(x_1) - G'_{\theta_1}(x_1)\, d_x G_{\theta_2}(x_2) \geqslant 0, \tag{19.2}$$

where d_x denotes the differential element with respect to x.

It should be noted that for a location-parameter problem, the assumption (19.2) is equivalent to a monotone likelihood ratio property for $G_\theta(x)$. In this case Assumption 2 implies Assumption 1. Actually Assumption 2 is required only to show that the expected subset size is maximized when the parameters are equal.

Under the procedures R and R', we select a subset of size at least s, and to avoid the trivial case, it is assumed that $s \leqslant k - t$. For the truncated procedures the subset size is at most $k - t + 1$.

If only one value of s is considered, then we would assume that $P^* > P(s) = 1 - \binom{k-t}{s} \Big/ \binom{k}{s}$. But we consider different s-values ($1 \leqslant s \leqslant k - t$). It is sufficient to assume that $P^* > P(1) = t/k$; however, it is more convenient and practical to assume that $P^* > P(k-t) = 1 - \binom{k}{t}^{-1}$. This enables us to use the full range $1 \leqslant s \leqslant k - t$ for s. For any t ($1 \leqslant t < k$), it follows that $P^* > (k-1)k^{-1} \geqslant \frac{1}{2}$.

For convenience, we write $G_i(y)$ instead of $G_{\theta_{[i]}}(y)$ and omit the limits $-\infty$ to ∞ on integrals.

19.2.1 Procedure *R*

To define the procedure R of Sobel (1969), an ordered set of pairs (r,b_r), $r=1,2,\ldots,k-t$, with $0 \leqslant b_r < \infty$ for $r<k-t$ and $b_{k-t}=\infty$, is constructed as a function of k, t, P^*, and δ^*. We investigate the infimum $Q(r,b_r)$ of the PCS for $\boldsymbol{\theta} \in \Omega(\delta^*)$ for the rule based on selecting π_i if and only if $T_i \geqslant T_{[k-r+1]} - b_r$. Thus, each pair corresponds to a rule, and part of the procedure R is to decide on the value s of the integer r to be used.

R: Construct the ordered set (r,b_r), and use it to select an integer s, described as follows. Select π_i if and only if

$$T_i \geqslant T_{[k-s+1]} - a_s, \tag{19.3}$$

where a_s (with $0 \leqslant a_s \leqslant b_s$) is determined by using the equality sign in (19.1). If the pair $(s=1,\ a_1=0)$ already satisfies (19.1), then we use a fixed subset-size procedure and select only the population with the largest T-value.

For the normal means problem with common $\sigma^2=1$, the latter part of the procedure R applies when n is large enough so that

$$1-\int \Phi(x-\delta^*\sqrt{n}\,)\,d\Phi^{k-t}(x) \geqslant P^*. \tag{19.4}$$

For $t=1$, the left-hand side of (19.4) reduces to the well-known form

$$\int \Phi^{k-1}(x+\delta^*\sqrt{n}\,)\,d\Phi(x). \tag{19.5}$$

Sobel (1969) has shown that the infimum of the PCS over $\Omega(\delta^*)$ is attained for a configuration of the type

$$\theta_{[1]}=\theta_{[k-t]}=\theta', \qquad \theta_{[k-t+1]}=\theta_{[k]}=\theta \quad \text{with } \delta(\theta',\theta)=\delta^*. \tag{19.6}$$

Let $Q(s,a_s;\theta)$ denote the PCS for $\boldsymbol{\theta}$ satisfying (19.6) and let $Q(s,a_s)=\inf_{\theta\in\Theta} Q(s,a_s;\theta)$.

Similar to $Q(s,a_s;\theta)$, let $Q(r,a_r;\theta)$ denote the corresponding quantity for any integer $r (1 \leqslant r \leqslant k-t)$. The quantity $Q(r,a_r)=\inf_{\theta\in\Theta} Q(r,a_r;\theta)$ is used to define the constants b_r needed to make the procedure R explicit. For each r, define b_r as the smallest value of b for which

$$Q(r,b) \geqslant Q(r+1,0), \tag{19.7}$$

where $Q(k-t+1,0)\equiv 1$. Finally, we define the integer s as the smallest value of r for which

$$Q(r,b_r)\geqslant P^*. \tag{19.8}$$

If θ is a location parameter with $\delta(\theta',\theta)=\theta-\theta'=\delta$, and $G_0(x)=G(x)$ is symmetrical about zero, then

$$Q(s,a_s;\delta)=s\binom{k-t}{s}\int\{1-[1-G(x+a_s+\delta)]^t\}$$
$$\times G^{s-t}(x)[1-G(x)]^{k-s-t}dG(x). \tag{19.9}$$

It has been shown by Sobel that $E(S|R)$ is maximized over Ω at a vector $\boldsymbol{\theta}$ with equal components. Let $E\{S;\boldsymbol{\theta}\}$ denote the expected subset size when all the components of $\boldsymbol{\theta}$ are equal.

19.2.2 Procedure R'

Before we define the procedure R', let us define the fixed subset-size procedure R_0 with the same assumptions on δ^* and P^*. This is a special case of R and we need only set $a_s=0$. Under the procedure R_0, we find the smallest s for which $Q_0(s,t;\theta)\geqslant P^*$. We do not restrict ourselves to integer values of s but consider the average of two adjacent integers, denoted by $\bar{s}$ or by $(s,s+1;0)$. This corresponds to a randomized procedure R_0 which attains a PCS exactly equal to P^*.

To define the procedure R', we replace s by r and a_s by b' in the derivation of $E(S;\theta)$ and call the expression $E(S;r,\theta)$. For each $r(1\leqslant r\leqslant k-t)$, we define b'_r as the smallest b'-value for which

$$\max_{\theta\in\Theta} E(S;r,\theta)\geqslant r+1. \tag{19.10}$$

In most applications, the left-hand side of (19.10) is continuous and strictly increasing in b' for each r so that we can write (19.10) with equality sign and b'_r is the unique root of this equation.

Applying these new constants b'_r as in the procedure R, we again define s as the r-value in the first pair (r,b'_r) that satisfies (19.1). The procedure R' is now defined *either* as given by (19.3) with the preceding s-value and a'_s-value satisfying (19.1) with equality, *or* as the fixed subset-size procedure R_0 based on $(s,s+1;0)$, whichever gives the smallest maximum $E(S;s,\theta)$.

REMARK 19.1. Although the procedures R and R' are identical for many values of k, t, P^*, and δ^*, the tables of Sobel (1969) show that this is not always the case. It should also be noted that the constants b'_r do not depend on δ^* as in the case of the b_r. □

19.2.3 Procedure R_M

The procedure R_M arbitrarily sets $s=t$. Since $s \leqslant k-t$ for our goal, this procedure is defined only for $t \leqslant k-t$ or $t \leqslant k/2$. For $\delta^*=0$ and $t=1$, this procedure is identical with the procedure of Gupta (1965) which will be called R_G in this section. For $t=1$ and $\delta^*>0$, the improvement of R_M over R_G is due only to the introduction of the indifference zone. All the necessary formulas for the procedure R_M are the special cases of the previous results. As in the cases of R and R', we use a fixed subset size if the smallest constant a_t satisfying (19.1) is negative.

19.2.4 Truncated Versions of R, R', and R_M

Although we assumed that $s \leqslant k-t$, there was no assumption about an upper bound for the subset size S under the random subset-size (RSS) procedures R, R', and R_M. Corresponding to each of these RSS procedures, Sobel (1969) defined a truncated procedure that selects the populations associated with the $k-t+1$ largest T-values whenever $S \geqslant k-t+1$. Truncation is only concerned with the case $t>1$ since the procedure is unchanged for $t=1$. The truncated procedures are denoted by R_T, R'_T, and R_{MT}, respectively. Truncation does not affect the values of PCS or $Q(s,a_s;\theta)$. It can only reduce the value of $E(S;\boldsymbol{\theta})$, and the reduction is small unless $\lambda=\delta^*\sqrt{n}$ is close to zero or P^* is close to one. The maximum of $E(S;\boldsymbol{\theta})$ is at most $k-t+1$ under truncation. Let S_T denote the size of the subset selected under truncation. Sobel has shown that, for any integers $k,t,s(1 \leqslant s \leqslant k-t), E(S_T;\boldsymbol{\theta})$ under any one of the preceding truncated procedures is maximized at a vector $\boldsymbol{\theta}$ whose components are all equal.

19.2.5 Comparisons of Procedures

Sobel (1969) has given tables for a numerical comparison of the $E(S;\boldsymbol{\theta})$-values for (a) the untruncated procedures for $t=1$, and for (b) both the truncated and untruncated procedures for $t=2$ in the case of $k=5$ normal means problem with common $\sigma^2=1$. In the case of (a), the values are tabulated for the LFC and the equal parameter (EP) configurations when $P^*=0.95$, and $\lambda=\delta^*\sqrt{n}=0, 0.50, 1.00, 2.25, 3.00, \infty$. In the case of (b), the values are for the EP configuration with $P^*=0.95$, and $\lambda=0, 0.25, 0.50, 1.00, 3.00, \infty$.

For the normal means problem, Sobel (1969) has considered the asymptotic ($P^*\to 1$) relative efficiency of R with respect to R_M defined by

$$\mathrm{ARE}(R;R_M)=\lim_{P^*\to 1}\left[\frac{k-E(S|R,\mathrm{EP})}{k-E(S|R_M,\mathrm{EP})}\right]. \tag{19.11}$$

He has shown that for $t<k-t$, this limit is ∞, thus showing that, for P^*

close to 1, the max $E(S)$ is smaller for R than for R_M. In particular for $t=1$, this limit is ∞ for all $k \geqslant 3$.

For $t=1$ and $\delta^*=0$, the goal of the selection problem is the same as that considered by Gupta (1965). To compare R and R_G for a location parameter problem, we note that the $E(S)$ is maximized when the parameters are equal and this maximum is kP^* in both cases. Hence this criterion is not of much value here. Sobel and Yen (1972) have made asymptotic ($P^* \to 1$) comparisons of $E(S|\boldsymbol{\theta})$ for any vector $\boldsymbol{\theta} \in \Omega$ in the case of the normal means problem with common $\sigma^2=1$. They have shown that $E(S|\boldsymbol{\theta}, R)$ is smaller than $E(S|\boldsymbol{\theta}, R_G)$ for P^* close to one when

$$\sqrt{\frac{2}{k(k-1)}} \sum_{\alpha=1}^{k} \lambda_{\alpha 1} > \lambda_{k1}, \tag{19.12}$$

where $\lambda_{ij} = \sqrt{n}\,(\theta_{[i]} - \theta_{[j]})$, and it is larger when the inequality in (19.12) is reversed. For $k=2$, the procedures are identical and (19.12) is vacuous. For $k=3$, the inquality in (19.12) holds when

$$(\theta_{[2]} - \theta_{[1]}) > (1 + \sqrt{3}\,)(\theta_{[3]} - \theta_{[2]}). \tag{19.13}$$

If we define the configurations $C_j (j=1,2,\ldots,k)$ by setting

$$\theta_{[1]} = \cdots = \theta_{[k-j]};\ \theta_{[k-j+1]} = \cdots = \theta_{[k]} \tag{19.14}$$

then (19.12) takes the form

$$j > \sqrt{k(k-1)/2} \tag{19.15}$$

and we note that (19.12) always holds for C_{k-1} for $k>2$. On the other hand, for all $k>2$ the inequality in (19.12) is reversed for C_1 and also for the configuration in which the adjacent parameters are equally spaced. A table of values of $E(S|C_j), j=1,2,3,4$, for $k=5$ is given by Yen (1969) and it illustrates numerically the results proved by Sobel and Yen (1972).

Sobel has also studied the asymptotic ($P^* \to 1$) efficiency of a likelihood ratio procedure R_L with respect to R in the case of normal means problem with $t=1$. Under the procedure R_L [see Bechhofer, Kiefer, and Sobel (1968), p. 88], the size S of the selected subset is the smallest integer j for which

$$\frac{\sum_{i=1}^{k-j} \exp\{-\lambda(y_k - y_i)\}}{\sum_{i=k-j+1}^{k} \exp\{-\lambda(y_k - y_i)\}} \leqslant \frac{1-P^*}{P^*}, \tag{19.16}$$

where $y_1 < y_2 < \cdots < y_k$ are the ordered sample means. For $S=j$, the procedure selects the populations that give rise to the j largest sample means. The asymptotic efficiency of R_L with respect to R is zero.

19.3 SUBSET SELECTION WITH MODIFIED REQUIREMENTS

The usual goal in the subset selection formulation of Gupta (1956) is to select a nonempty subset of the k given populations so as to include the best (i.e., the population associated with $\theta_{[k]}$) with a minimum guaranteed probability P^*. In other words,

$$\inf_{\Omega} P(CS|R) \geqslant P^* \tag{19.17}$$

where $\Omega = \{\boldsymbol{\theta} : \boldsymbol{\theta} = (\theta_1, \ldots, \theta_k)\}$ is the set of all parametric configurations. A modified formulation was considered by Santner (1975) whose aim was to restrict the subset size by an upper bound $s \leqslant k-1$. In order to do this, he introduced an indifference zone (see §11.3). In this section, we will discuss a modified formulation studied by Huang and Panchapakesan (1976).

Let $*$ denote a continuous binary operation on the elements of the real line and c_0, the identity of the operation. Let $\delta(\theta_i, \theta_j)$ denote the distance between the populations π_i and π_j such that (i) $\delta(\theta_i, \theta_j) \geqslant c_0$ with equality if and only if $\theta_i = \theta_j$, (ii) $\delta(\theta_i, \theta_j) = \delta(\theta_j, \theta_i)$, and (iii) $(\theta_i \wedge \theta_j) * \delta(\theta_i, \theta_j) = \theta_i \vee \theta_j$, where $a \wedge b = \min(a,b)$, and $a \vee b = \max(a,b)$. For $\delta^* > c_0$, let

$$\Omega(\delta^*) = \{\boldsymbol{\theta} : \delta(\theta_{[k]}, \theta_{[k-1]}) \geqslant \delta^*\}. \tag{19.18}$$

The modified formulation requires that besides (19.17) the procedure R also satisfies

$$E(S'|R) \leqslant \beta \quad \text{whenever } \boldsymbol{\theta} \in \Omega(\delta^*), \tag{19.19}$$

where $\beta \in [0, k-1]$ is specified by the experimenter in advance, and S' denotes the number of nonbest populations included in the selected subset.

Let $\{X_{ij}\}, j = 1, 2, \ldots,$ be a sequence of independent observations from π_i and $T_{ni} = T_n(X_{i1}, \ldots, X_{in})$ be defined such that $\{T_{ni}\}$ is consistent for θ_i. Usually T_{ni} is a transitive sufficient sequence. Let $G_n(y|\theta_i)$ denote the cdf of T_{ni}. It is assumed that $\{G_n(y|\theta), \theta \in \Theta\}$ is a stochastically increasing family of distributions for each n having a common support E independent of n. For $n \geqslant 1$ and $\epsilon > c_0$, let $h \equiv h_{n,\epsilon}: E \to R$ (real line) be a function such that

(i) $h(x) = c_n * x * \epsilon$,

(ii) $c_n \downarrow 0$ as $n \to \infty$,

(iii) $h(x)$ is continuous in x for every $n \geqslant 1$ and $\epsilon > c_0$, and

(iv) $h(x) \geqslant x * \epsilon$. (19.20)

Finally we assume that

$$a < b \Rightarrow c * a < c * b \text{ for any } c \in E. \tag{19.21}$$

EXAMPLES

Location Parameter Case. $h(x) = c_n + x + \epsilon$, and $c_n \downarrow 0$ as $n \to \infty$. For $\theta_1 < \theta_2$, $\delta(\theta_1, \theta_2) = \theta_2 - \theta_1$.

Scale Parameter Case. $h(x) = c_n x \epsilon$, and $c_n \downarrow 1$ as $n \to \infty$. For $\theta_1 < \theta_2$, $\delta(\theta_1, \theta_2) = \theta_2 / \theta_1$.

For selecting a subset containing the population associated with $\theta_{[k]}$, Huang and Panchapakesan (1976) have proposed a class of procedures $R_{n,\epsilon} \equiv R(h_{n,\epsilon})$ defined as follows.

$\boldsymbol{R_{n,\epsilon}}$: Select π_i if and only if

$$h(T_{ni}) \geqslant T_{n[k]}, \tag{19.22}$$

where $T_{n[1]} \leqslant \cdots \leqslant T_{n[k]}$ denote the ordered T_{ni}.

Let $P_\theta(CS|R_{n,\epsilon})$ denote the PCS when $\boldsymbol{\theta} \in \Omega_0 = \{\boldsymbol{\theta}: \theta_1 = \cdots = \theta_k\}$. Let $g_n(y|\theta) = (\partial/\partial y) G_n(y|\theta)$ and $h'(y) = (\partial/\partial y) h_{n,\epsilon}(y)$. Further, let $C1$ and $C2$ denote the following two conditions.

$\boldsymbol{C1}$: For fixed n and ϵ

$$\frac{\partial}{\partial \theta} G_n(h(y)|\theta) g_n(y|\theta) - h'(y) \frac{\partial}{\partial \theta} G_n(y|\theta) g_n(h(y)|\theta) \geqslant 0 \tag{19.23}$$

almost everywhere on the support E.

$\boldsymbol{C2}$: For $\theta_1 \leqslant \theta_2$ both in Θ

$$\frac{\partial}{\partial \theta_1} G_n(h(y)|\theta_1) g_n(y|\theta_2) - h'(y) \frac{\partial}{\partial \theta_1} G_n(y|\theta_1) g_n(h(y)|\theta_2) \geqslant 0 \tag{19.24}$$

almost everywhere on E.

The general results of Huang and Panchapakesan (1976) are contained in the following theorem.

Theorem 19.1. The class of procedures $R_{n,\epsilon}$ has the following properties.

a. $\inf_{\Omega} P(CS|R_{n,\epsilon}) = \inf_{\Omega_0} P(CS|R_{n,\epsilon})$.

b. $P_{\theta}(CS|R_{n,\epsilon}) \to 1$ as $n \to \infty$ for any $\boldsymbol{\theta} \in \Omega_0$ and $\epsilon > c_0$.

c. For fixed n and ϵ, $P_{\theta}(CS|R_{n,\epsilon})$ is nondecreasing in $\theta \in \Theta$ provided that the condition $C1$ given by (19.23) is satisfied.

d. The supremum of $E(S'|R_{n,\epsilon})$ for $\boldsymbol{\theta} \in \Omega(\delta^*)$ is attained at a configuration given by $\theta_{[1]} = \cdots = \theta_{[k-1]} < \theta_{[k]}$ with $\delta(\theta_{[k]}, \theta_{[k-1]}) = \delta^*$.

Applications of these results to the location and scale parameter cases are straightforward. Huang and Panchapakesan (1976) have also considered the specific problems of selection from normal populations in terms of means (with common $\sigma^2 = 1$) and variances. The problem of finding the smallest n satisfying the requirements (19.17) and (19.19) has been discussed by obtaining suitable bounds on $\inf_{\Omega} P(CS|R)$ and $\sup_{\Omega(\delta^*)} E(S'|R)$. It should be noted that the procedure has to be modified in an obvious way for selecting the population with the smallest variance.

Huang and Panchapakesan have also considered selection in terms of the treatment effects in a two-way layout. Consider the complete block design with one observation per cell. Our observations are

$$X_{i\alpha} = \mu + \beta_{\alpha} + \tau_i + \epsilon_{i\alpha} \tag{19.25}$$

where μ is the mean effect, the $\beta_{\alpha}(\alpha = 1, 2, \ldots, n)$ are the block effects, the $\tau_i(i = 1, 2, \ldots, k)$ are the treatment effects, and the $\epsilon_{i\alpha}$ are the error components. It is assumed that $\boldsymbol{\epsilon}_{\alpha} = (\epsilon_{1\alpha}, \ldots, \epsilon_{k\alpha}), \alpha = 1, \ldots, n$, are independent and identically distributed having the k-variate normal distribution with zero mean and covariance matrix $\Sigma = \sigma^2 I + \sigma^2\rho(\mathbf{1}'\mathbf{1} - I)$, where I is a $k \times k$ identity matrix, $\mathbf{1}$ is a row vector with k unit components, and Σ is assumed to be positive definite. The best treatment is defined to be the one associated with $\tau_{[k]}$. The block effects are nuisance parameters in the fixed-effects model and are random variables in the mixed-effects model. The requirement (19.19) must now be met when $\boldsymbol{\tau} \in \Omega(\delta^*) = \{\boldsymbol{\tau}: \tau_{[k]} - \tau_{[k-1]} \geq \delta^*\}$. Let

$$\bar{X}_{ni} = n^{-1} \sum_{\alpha=1}^{n} X_{i\alpha}, i = 1, \ldots, n,$$

$$\bar{\bar{X}}_n = k^{-1} \sum_{i=1}^{k} \bar{X}_{ni}, \quad \text{and} \quad T_{ni} = \bar{X}_{ni} - \bar{\bar{X}}_n. \tag{19.26}$$

Then the procedure $R_{n,\epsilon}$ in this case is: Select π_i if and only if

$$T_{ni} \geqslant T_{n[k]} - d_n - \epsilon \tag{19.27}$$

where $d_n \downarrow 0$ as $n \to \infty$.

The following theorem of Huang and Panchapakesan (1976) shows that the problem can be reduced to the one-way model case.

Theorem 19.2. For the procedure $R_{n,\epsilon}$ defined by (19.27)

$$\inf_{\Omega} P(CS|R_{n,\epsilon}) = \Pr\left(U_{nk} \geqslant \max_{1 \leqslant j \leqslant k} U_{nj} - d_n - \epsilon \right) \tag{19.28}$$

where the U_{ni} are i.i.d. normal random variables with zero mean and variance equal to $(2k-1)(kn)^{-1}(1-\rho)\sigma^2$. Furthermore, the supremum of $E(S'|R_{n,\epsilon})$ over $\Omega(\delta^*)$ is attained when $\tau_{[1]} = \tau_{[k-1]} = \tau_{[k]} - \delta^*$.

19.4 SUBSET SELECTION FORMULATION OF THE COMPLETE RANKING PROBLEM

Let $\pi_i (i=1,\ldots,k)$ have the associated distribution $F_{\theta_i}(x)$ which is assumed to be absolutely continuous. The parameter θ_i is unknown and it belongs to an open interval Θ on the real line. Our interest is in ranking all the k populations according to their θ-values. Let us denote a typical permutation of the set of integers $\{1,2,\ldots,k\}$ by $\{i_1,i_2,\ldots,i_k\}$ which corresponds to a ranking of $\pi_1,\pi_2,\ldots,\pi_k$. Huang and Panchapakesan (1978) have considered the problem of selecting a random nonempty subset of all the $k!$ permutations to include the correct ranking of the populations. Selection of any subset containing the correct ranking is called a correct selection. We require any procedure R to satisfy the P^*-condition, namely, $P(CS|R) \geqslant P^*$ whatever be the unknown configuration of the θ_i.

Let $T_i = T_i(X_{i1},X_{i2},\ldots,X_{in})$ be a statistic based on n independent observations $X_{i1},X_{i2},\ldots,X_{in}$ from π_i and let $H_{\theta_i}(t)$ be the cdf of $T_i, i=1,2,\ldots,k$. Huang and Panchapakesan (1978) considered the following types of procedures, R_1 and R_2.

$\boldsymbol{R_1}$: Include the ranking $(i_1,i_2,\ldots,i_k)$ in the selected subset if and only if

$$\max_{2 \leqslant s \leqslant k} \max_{1 \leqslant r \leqslant s} \left(T_{i_r} - T_{i_s} \right) \leqslant d, \tag{19.29}$$

where $d = d(k,n,P^*) \geqslant 0$ is to be determined so that the P^*-condition is satisfied.

$\boldsymbol{R}_2$: Include the ranking $(i_1, i_2, \ldots, i_k)$ in the selected subset if and only if

$$\max_{2 \leqslant s \leqslant k} \max_{1 \leqslant r \leqslant s} \left(\frac{T_{i_r}}{T_{i_s}} \right) \leqslant c, \tag{19.30}$$

where $c = c(k, n, P^*) \geqslant 1$ is to be determined so that the P^*-condition is met.

It should be noted that R_1 as well as R_2 will always include the ranking corresponding to $T_{[1]} \leqslant \cdots \leqslant T_{[k]}$.

19.4.1 Location Parameter Case

If $H_{\theta_i}(t) = H(t - \theta_i), \Theta = (-\infty, \infty), -\infty < t < \infty$, we use the procedure R_1. In this case, the PCS is shown to be minimized for $\theta_1 = \cdots = \theta_k$. The value of the smallest d satisfying the P^*-condition is given by

$$\Pr\{ U_r - U_s \leqslant d,\ r, s = 1, 2, \ldots, k; r < s \} = P^*, \tag{19.31}$$

where the U_i are i.i.d. with cdf $H(t)$.

Let G be the group of permutations on the ranking $I = (1, 2, \ldots, k)$. Given any ranking $I_i = (i_1, i_2, \ldots, i_k)$, let $g_j I_i$ denote the permutation obtained from I_i by interchanging i_j and i_{j+1}. Huang and Panchapakesan (1978) have shown that, for $\boldsymbol{\theta} = (\theta_{i_1}, \ldots, \theta_{i_k})$ such that $\theta_{i_j} > \theta_{i_{j+1}}$, for some $j = 1, 2, \ldots, k$,

$$p(I_i) \leqslant p(g_j I_i), \tag{19.32}$$

where $p(I_i)$ denotes the probability that the ranking I_i is included in the selected subset. As a corollary to this, we get $P(CS|R_1) \geqslant p(I_i)$ where I_i is any ranking of the k populations. In other words, the procedure R_1 is unbiased.

For ranking the means of normal populations with a common known σ^2, T_i is the sample mean based on n observations. In this case, $D = d\sqrt{n}/\sigma$ is given by

$$\Pr\{ Z_r - Z_s \leqslant D,\ r, s = 1, 2, \ldots, k; r < s \} = P^*, \tag{19.33}$$

where the Z_i are independent $N(0, 1)$ variables. For $k > 2$, we can obtain a conservative value of D by equating to P^* a lower bound for the probability on the left-hand side of (19.33) given by

$$\Pr\left\{ W_i \leqslant \frac{D}{\sqrt{2}\,(k-1)}, i = 1, 2, \ldots, k-1 \right\}, \tag{19.34}$$

where the W_i are $N(0,1)$ random variables with correlations

$$\rho(W_i, W_j) = \begin{cases} 1, & i=j, \\ -\frac{1}{2}, & |i-j|=1, \\ 0, & \text{otherwise}. \end{cases} \tag{19.35}$$

19.4.2 Scale Parameter Case

When $H_{\theta_i}(t) = H(t/\theta_i), \Theta = (0, \infty), t > 0$, we use the procedure R_2. The PCS is minimized for $\theta_1 = \cdots = \theta_k$. The procedure R_2 is also unbiased. The results can be applied to the problem of ranking normal variances. In this case, $\theta_i = \sigma_i^2$ and T_i is the usual unbiased estimator s_i^2 based on $n(\geqslant 6)$ observations so that $\nu s_i^2/\sigma^2$ has a chi-square distribution with $\nu = n-1$ degrees of freedom. A conservative value of the constant $c = c(k, n, P^*)$ is given by

$$\Pr\{Y_i \leqslant c^{1/(k-1)}, \qquad i = 1, 2, \ldots, k-1\} = P^*, \tag{19.36}$$

where the Y_i are correlated F-variables with (ν, ν) degrees of freedom and correlations

$$\rho(Y_i, Y_j) = \begin{cases} 1, & i=j, \\ -\dfrac{(\nu-4)}{2(\nu-1)}, & |i-j|=1, \\ 0, & \text{otherwise}. \end{cases} \tag{19.37}$$

Huang and Panchapakesan (1978) have also discussed ranking according to stochastic ordering of populations that belong to a stochastically increasing family and that are star-shaped with respect to a known distribution G.

PART III

COMPARISON WITH A CONTROL, ESTIMATION, AND RELATED TOPICS

CHAPTER 20

Comparison of Several Populations with a Standard or a Control

---

20.1 INTRODUCTION

In the preceding chapters we discussed decision procedures devised for choosing the $t(1 \leqslant t < k)$ best from a given set of k populations. The problem of selection was considered under the indifference zone formulation as well as the subset selection approach. We have also discussed several modifications of the goals and also formulations that combine the features of both the approaches (see Chapters 11 and 19).

Although the experimenter is generally interested in selecting the best one of the competing categories, in certain classes of experiments even the best one of the competing categories may not be good enough to warrant the experimenter's choosing it. For example, if the competing categories are varieties of wheat, the best one will not be worthy of consideration unless its mean yield is at least a specified amount or is larger than the mean yield of a variety that may be in use already. Thus the problem of simultaneous comparison of k given experimental categories among themselves and with reference to a standard is of practical interest. This problem has been investigated by several authors under different types of formulations with different criteria to be satisfied by an acceptable procedure. When the true value of the parameter of the standard population is not known, it becomes necessary to take sample observations from it and in this case, it is called the control. Of course one can keep this distinction by referring to them as the "specified standard" and "variable control" cases. However, we do, at the expense of some precision, refer to the population as the control in either case, when it is convenient to do so.

20.2 $(k+1)$-DECISION PROCEDURES FOR COMPARING k EXPERIMENTAL CATEGORIES WITH A STANDARD OR CONTROL

Let π_0 be the control and $\pi_1,\ldots,\pi_k$ denote the experimental categories. Let $F(x,\theta_i)$ be the distribution function associated with $\pi_i, i=0,1,\ldots,k$. We say that the population π_i is better than π_0 if $\theta_i>\theta_0$. Of course, the number of populations that are better than the control is not known. The goal is to select the population associated with $\theta_{[k]}$, the largest of the θ_i, provided that $\theta_{[k]}>\theta_0$ and not to select any population if $\theta_i\leqslant\theta_0$ for all $i=1,\ldots,k$.

20.2.1 Comparison of Normal Means

The initial investigation was made by Paulson (1952) who considered the cases of normal means and binomial proportions of successes. Let us first discuss the normal means problem. Suppose π_i is a normal population with mean μ_i and variance σ^2. The $\mu_i (i=0,1,\ldots,k)$ are assumed to be unknown. Let $\mu_{[1]}\leqslant\cdots\leqslant\mu_{[k]}$ denote the ordered $\mu_i, i=1,\ldots,k$, and let

$$\Omega=\{\boldsymbol{\mu}:\boldsymbol{\mu}=(\mu_0,\mu_1,\ldots,\mu_k),\quad -\infty<\mu_i<\infty,\quad i=0,1,\ldots,k\},$$

$$\Omega_i=\{\boldsymbol{\mu}:\mu_i\geqslant\mu_j+\Delta,\quad j=0,1,\ldots,k;\quad j\neq i;\quad \Delta>0\},\quad i=1,\ldots,k \tag{20.1}$$

$$\Omega_0=\{\boldsymbol{\mu}:\mu_i<\mu_0,\quad i=1,\ldots,k\}.$$

REMARK 20.1. In the earlier chapters we have generally used Ω_0 to denote the part of the parameter space where all the populations have equal values for the parameter. The departure here from this convention must be kept in mind. □

Three constants (Δ,P_0^*,P_1^*) are specified in advance by the experimenter such that $\Delta>0, 2^{-k}<P_0^*<1$ and $(1-2^{-k})k^{-1}<P_1^*<1$. Paulson (1952) required that a rule R satisfied the following two conditions.

$$\Pr\{\pi_0 \text{ is selected}\}\geqslant P_0^* \text{ whenever } \boldsymbol{\mu}\in\Omega_0, \tag{20.2}$$

$$\Pr\{\pi_i \text{ is selected}\}\geqslant P_1^* \text{ whenever } \boldsymbol{\mu}\in\Omega_i,\qquad i=1,\ldots,k. \tag{20.3}$$

Case (i): σ^2 known. In this case, Paulson (1952) proposed the following rule R_1 defined in terms of the sample means $\bar{X}_i (i=0,1,\ldots,k)$ based on n independent observations from each of the $(k+1)$ populations. Let

$$\bar{X}^* = \max_{1 \leqslant j \leqslant k} \bar{X}_j \text{ and } \alpha = (1-P_0^*)/k.$$

R_1: Select π_i associated with $\bar{X}^*$ if

$$\bar{X}^* - \bar{X}_0 \geqslant \lambda\sigma\sqrt{2/n}\ ; \tag{20.4}$$

select π_0 otherwise.

The constant $\lambda > 0$ in rule R_1 is determined so that the requirement (20.2) is met.

For $\boldsymbol{\mu} \in \Omega_0$, the probability of choosing π_0 is minimized when $\mu_0 = \mu_1 = \cdots = \mu_k$ and thus λ is given by

$$\int_{-\infty}^{\infty} \Phi^k(y + \sqrt{2}\,\lambda)\, d\Phi(y) = P_0^*. \tag{20.5}$$

Paulson (1952) approximated the solution for λ by λ_0 given by $1 - \Phi(\lambda_0) = \alpha$. It should be noted that this approximation corresponds to the approximation n_2 for n discussed in §2.1.1 in connection with the procedure of Bechhofer (1954). The values of λ for selected values of k and P_0^* have been subsequently tabulated by Gupta [1963a] and Gupta, Nagel, and Panchapakesan [1973].

Now, for fixed λ, Δ, k, and n, the probability of selecting a particular experimental category, say π_k, when $\boldsymbol{\mu} \in \Omega_k$ is minimized when $\mu_0 = \mu_1 = \cdots = \mu_{k-1} = \mu_k - \Delta$. The minimum sample size required to meet the condition (20.3) is approximated by $\tilde{n}$ given by

$$\Phi\left(\Delta\sigma^{-1}\sqrt{\frac{\tilde{n}}{2}} - \lambda\right) = P_1^*. \tag{20.6}$$

Using the approximate solution for λ, we get

$$\tilde{n} = \left(\frac{2\sigma^2}{\Delta^2}\right)\{\Phi^{-1}(1-\alpha) + \Phi^{-1}(P_1^*)\}^2. \tag{20.7}$$

This approximation appears to be good when $P_0^* \geqslant .95$ and $P_1^* \geqslant .80$.

Case (ii): σ^2 unknown. In this case the procedure R_2 is the same as R_1 with σ^2 replaced by the usual pooled estimator with $\nu = (k+1)(n-1)$

degrees of freedom. The least favorable configurations for the probabilities in (20.2) and (20.3) are the same as before. An approximate solution of $\lambda=\lambda(k,P_0^*,n)$ is given by $\lambda^*=\lambda^*(k,P_0^*,n)$ which is defined by

$$\Pr\{t_\nu>\lambda^*\}=\frac{1-P_0^*}{k-1} \tag{20.8}$$

where t_ν is a Student t random variable with ν degrees of freedom. A good first approximation of n is usually given by (20.7) which can be modified if necessary.

20.2.2 Comparison of Normal Means: Modified Probability Requirements

Bechhofer and Turnbull (1978) also have considered the problem of comparing normal means with a specified standard; however, they have slightly modified the probability requirements. They have discussed both known and unknown σ^2 cases.

Let

$$\begin{aligned}
&\Omega=\{\boldsymbol{\mu}:\boldsymbol{\mu}=(\mu_1,\ldots,\mu_k):-\infty<\mu_i<\infty,\quad i=1,\ldots,k\},\\
&\Omega_0=\{\boldsymbol{\mu}:\mu_0\geqslant\mu_{[k]}+\delta_0^*\},\\
&\Omega'=\{\boldsymbol{\mu}:\mu_{[k]}\geqslant\mu_0+\delta_1^*\quad\text{and}\quad\mu_{[k]}\geqslant\mu_{[k-1]}+\delta_2^*\},
\end{aligned} \tag{20.9}$$

where δ_0^*, δ_1^*, and δ_2^* are specified constants such that $0<\delta_1^*,\delta_2^*<\infty$ and $-\delta_1^*<\delta_0^*<\infty$, and $\mu_{[1]}\leqslant\cdots\leqslant\mu_{[k]}$ are the ordered means of the experimental categories. For $\delta_0^*=0$ and $\delta_1^*=\delta_2^*$, we get the formulation of Paulson (1952). For specified constants P_0^* and $P_1^*(2^{-k}<P_0^*<1$ and $(1-2^{-k})/k<P_1^*<1)$, the probability requirements imposed on any rule are:

$$\begin{aligned}
&\Pr\{\pi_0\text{ is selected}\}\geqslant P_0^*\text{ whenever }\boldsymbol{\mu}\in\Omega_0,\text{ and}\\
&\Pr\{\text{the best experimental category is selected}\}\geqslant P_1^*\text{ whenever }\boldsymbol{\mu}\in\Omega'.
\end{aligned} \tag{20.10}$$

Case (i): Known σ^2. Let $\overline{X}_1,\ldots,\overline{X}_k$ be the sample means based on N independent observations from each of the experimental categories. The procedure R_3 proposed by Bechhofer and Turnbull (1978) is as follows.

$$\begin{aligned}
&\text{Choose }\pi_0\quad\text{if}\quad\overline{X}_{[k]}<\mu_0+c;\quad\text{otherwise choose the population}\\
&\text{that produced }\overline{X}_{[k]}\text{ as the one associated with }\mu_{[k]}.
\end{aligned} \tag{20.11}$$

The sample size N is the smallest integer which, when used with an appropriate c, guarantees (20.10). Let $N=[N_e]+1$ be the common sample size per population, where $[x]$ denotes the largest integer less than or equal to x. Then Bechhofer and Turnbull (1978) have shown that N_e and c which exactly guarantee (20.10) are those that satisfy the simultaneous equations:

$$\Phi^k\left\{\frac{\sqrt{N_e}\,(c+\delta_0^*)}{\sigma}\right\}=P_0^*,$$
$$\int_{\sqrt{N_e}\,(c-\delta_1^*)/\sigma}^{\infty}\Phi^{k-1}\left\{y+\frac{\sqrt{N_e}\,\delta_2^*}{\sigma}\right\}d\Phi(y)=P_1^*. \tag{20.12}$$

In many applications, we take $\delta_0^*=0$, and $\delta_1^*=\delta_2^*=\delta^*$ (say). Letting $h=\sqrt{N_e}\,c/\sigma$ and $g=\sqrt{N_e}\,\delta^*/\sigma$, we have $h^*=h^*(k,P_0^*)$ and $g^*=g^*(k,P_0^*,P_1^*)$ which satisfy the simultaneous equations:

$$\Phi^k(h)=P_0^*,$$
$$\int_{h-g}^{\infty}\Phi^{k-1}(y+g)\,d\Phi(y)=P_1^*. \tag{20.13}$$

Then $N=[(h^*\sigma/\delta^*)^2]+1$ and $c=h^*\delta^*/g^*$. Values of h^* and g^* are tabulated by Bechhofer and Turnbull (1978) for $k=2(1)15$, and selected values of P_0^* and P_1^*.

It has also been shown by Bechhofer and Turnbull that the same (N,c), which guarantee (20.10) when $\delta_1^*=\delta_2^*=\delta^*$, also guarantee that

$$\Pr\{\text{selecting any one of the } t \text{ best experimental categories}\}\geqslant P_1^*$$

whenever $\mu_{[k]}\geqslant\mu_0+\delta^*$, *and* $\mu_{[k-t+1]}\geqslant\mu_0\geqslant\mu_{[k-t]}$ for any $t(1\leqslant t\leqslant k-1)$, the probability inequality being strict if $t>1$. If $t=k$, we define $\mu_{[0]}=-\infty$.

Case (ii): σ^2 ***Unknown.*** In this case, Bechhofer and Turnbull (1978) have proposed the following two-stage procedure R_4. To define this procedure, let $\bar{X}_i(n_0)$ denote the mean of a sample of n_0 independent observations $X_{ij}, j=1,\ldots,n_0$, from π_i and define

$$s_\nu^2=\sum_{i=1}^{k}\sum_{j=1}^{n_0}\frac{\left(X_{ij}-\bar{X}_i(n_0)\right)^2}{k(n_0-1)}$$

which is an unbiased estimate of σ^2 based on $\nu = k(n_0 - 1)$ degrees of freedom. Let h and c be constants that satisfy the simultaneous equations:

$$\int_0^\infty \Phi^{k-1}(hz) f_\nu(z)\, dz = P_0^*,$$

$$\int_0^\infty \left[\int_{(c-\delta_1^*)hz/(c+\delta_0^*)}^\infty \Phi^{k-1}\left[(y+\delta_2^* hz)(c+\delta_0^*)^{-1} \right] \varphi(y)\, dy \right] f_\nu(z)\, dz = P_1^* \tag{20.14}$$

where f_ν is the density function of $\sqrt{\chi_\nu^2/\nu}$. Obviously, the pair (h,c) depends on k, ν, and the specified quantities $\{\delta_0^*, \delta_1^*, \delta_2^*, P_0^*, P_1^*\}$.

The procedure R_4 of Bechhofer and Turnbull (1978) is defined as follows:

a. In the first stage, take $n_0 > 1$ independent observations from π_i, $i = 1, \ldots, k$.

b. In the second stage, take $N - n_0$ additional independent observations from each of the k populations, where $N = \max\{[(hs_\nu/\delta^* + c)^2] + 1, n_0\}$ and $[x]$ denotes the largest integer less than x.

c. If $\bar{X}_{[k]}(N) < \mu_0 + c$, select no population; otherwise, select the population that produced $\bar{X}_{[k]}$ as the one associated with $\mu_{[k]}$.

As before, we are interested in the special case $\delta_0^* = 0, \delta_1^* = \delta_2^* = \delta^*$ (say). In this case, let (g^*, h^*) be the pair (g,h) which satisfies (20.14) where $g = h\delta^*/c$. Bechhofer and Turnbull (1978) have tabulated the values of (g^*, h^*) for $k = 2(1)15$ and selected ν and $\{P_0^*, P_1^*\}$ with an accuracy of $\pm 10^{-4}$ in the associated (P_0^*, P_1^*) values. For $k = 1$, g^* and h^* can be obtained from the usual tables of Student's t. If $G_\nu(t)$ denotes the cdf of the t-distribution with ν degrees of freedom, we get $G_\nu(h^*) = P_0^*$, and $G_\nu(h^* - g^*) = 1 - P_1^*$.

Bechhofer and Turnbull have also shown that the same (h,c) which guarantee (20.10) when $\delta_1^* = \delta_2^* = \delta^*$, also guarantee

$$\Pr\{\text{selecting one of the } t \text{ best experimental categories}\} \geqslant P_1^*$$

whenever $\mu_{[k]} \geqslant \mu_0 + \delta^*, \mu_{[k-t+1]} \geqslant \mu_0 \geqslant \mu_{[k-t]}$ for any $t (1 \leqslant t \leqslant k)$, the probability inequality being strict if $t > 1$; again $\mu_{[0]} \equiv -\infty$.

In the general case when we do not have $\delta_0^* = 0$ and $\delta_1^* = \delta_2^*$, one can find the value of c from the value of h^*. The h^*-values are also tabulated (to two decimal places) by Krishnaiah and Armitage [1966] for $\nu = 5(1)35, k = 1(1)10$, and $P^* = 0.95$, 0.99.

REMARK 20.2. In implementing the procedure R_4, the choice of n_0, the first-stage sample, is optional. Any prior information such as an interval for the possible values of σ can be used to choose n_0. With such information one can choose n_0 which minimizes the maximum expected loss in number of extra observations needed because of ignorance of σ. Also, increasing n_0 has the effect of reducing the chance of being required to take an extremely large N. □

20.2.3 Comparison of Processes with Regard to Conforming to Product Specifications

Let $\pi_1,\ldots,\pi_k$ be k industrial processes producing similar items. Each item can be classified as being conforming or nonconforming. Let $q_i = 1 - p_i$ $(1 \leqslant i \leqslant k)$ denote the fraction of conforming items produced by π_i. Then p_i is the familiar fraction defective in the process π_i. These processes are compared in terms of the fraction of conforming items among themselves and against a specified standard q_0. Bechhofer and Turnbull (1977) have considered this problem along the lines of their 1978 paper for comparing normal means. To be specific, the goal is to select the process associated with $q_{[k]}$ provided that $q_{[k]} > q_0$ and in case no process has a q-value larger than q_0 then to reject *all* the processes. As in §20.2.2, the probability requirements are:

$$\begin{aligned} &\Pr\{\text{rejecting } all \text{ processes}\} \geqslant P_0^* \text{ whenever } q_{[k]} \leqslant q_0^*, \text{ and} \\ &\Pr\{\text{selecting the best process}\} \geqslant P_1^* \text{ whenever } q_{[k-1]} \leqslant q_0^* \text{ and } q_1^* \leqslant q_{[k]}, \end{aligned} \tag{20.15}$$

where $\{q_0^*, q_1^*, P_0^*, P_1^*\}$ are specified constants all between 0 and 1.

Bechhofer and Turnbull (1977) have discussed two types of sampling plans: (1) variables sampling plans and (2) attributes sampling plans. For the *variables sampling plans*, we assume that an observation X_i from π_i is normally distributed with unknown mean μ_i and known variance σ^2 (common to all processes). Thus X_i is an observation on the quality characteristic of interest of a random item from the process π_i. Let A and $B (A < B)$ be lower and upper specification limits, respectively. Then

$$q_i = \Phi\left[\frac{B - \mu_i}{\sigma}\right] - \Phi\left[\frac{A - \mu_i}{\sigma}\right]. \tag{20.16}$$

First, let us take $-\infty < A$ and $B = \infty$. So we have a one-sided lower specification limit and $q_i = 1 - \Phi[(A - \mu_i)/\sigma]$. For given A and specified $\{q_0^*, q_1^*\}$, we define μ_0^* and μ_1^* by $q_r^* = 1 - \Phi[(A - \mu_r^*)/\sigma], r = 0, 1$. In this case, we use the procedure R_3 of Bechhofer and Turnbull (1978) defined by (20.11) with μ_0^* in the place of μ_0. Note that choosing π_0 means rejecting all the k processes.

For two-sided specification limits, define $M=(B+A)/2, D=(B-A)/2$, and let $\theta_i=|\mu_i-M|$. Then $q_i=\Phi[(D-\theta_i)/\sigma]-\Phi[(-D-\theta_i)/\sigma]$ which is maximized when $\theta_i=0$. The maximum value of q_i is $\Phi(D/\sigma)-\Phi(-D/\sigma)$. The process with the largest q-value is the one with the smallest θ-value. We assume that q_0^* and q_1^* are specified so that $q_0^*<q_1^*<\Phi(D/\sigma)-\Phi(-D/\sigma)$; otherwise, the probability requirements cannot be satisfied. For a given D, we define θ_0^* and θ_1^* as the positive solutions to the equations

$$q_r^*=\Phi\left[\frac{D-\theta_r^*}{\sigma}\right]-\Phi\left[\frac{-D-\theta_r^*}{\sigma}\right], \qquad r=0,1.$$

The procedure R_5 of Bechhofer and Turnbull, based on N independent observations from each π_i is as follows.

R_5: Compute $W_i=|\bar{X}_i-M|, i=1,\ldots,k$. If $W_{[1]}>\theta_0^*-d$, reject all processes; otherwise select the process that yielded $W_{[1]}$ as the one having the smallest θ-value (and hence the largest q-value).

(20.17)

The values of N and d/σ are tabulated by Bechhofer and Turnbull (1977) for $k=1(1)5, (\theta_0^*/\sigma, \theta_1^*/\sigma)=(0.6983,\ 0.2038),\ (1.3551,\ 0.6691),\ (2.3551,\ 1.6737)$, and $(P_0^*, P_1^*)=(.90, .90), (.95, .90), (.95, .95), (.99, .90), (.99, .95), (.99, .99)$.

Now let us consider *attributes sampling plan*. In this case, $X_{ij}=1$ or 0 according as the jth item from ith process is classified as conforming (nondefective) or as nonconforming (defective). We assume that all the X_{ij} are independent and that $\Pr\{X_{ij}=1\}=q_i, 1\leqslant i\leqslant k$. Then q_i is the fraction of conforming items produced by π_i. The procedure R_6 of Bechhofer and Turnbull (1977) is based on N independent items drawn from each π_i and $\hat{q}_i=\sum_{j=1}^{N}X_{ij}/N$. Their procedure is

R_6: If $\hat{q}_{[k]}\leqslant q^*$, reject all processes; otherwise select the process that yielded $\hat{q}_{[k]}$ as the one associated with $q_{[k]}$, using randomization to break ties.

(20.18)

Here N and q^* are to be chosen to guarantee (20.15). There are no readily available tables of these constants. When N is large, we can use the arc sine transformation and (asymptotic) normal approximations.

20.2.4 Some General Results for Comparison of Several Populations with a Specified Standard

Let $\pi_i(1 \leqslant i \leqslant k)$ be an experimental category or population whose associated distribution depends on an unknown real parameter θ_i. The possible values of θ_i are assumed to be some fixed subset Ω of the real line and we let $\underline{\theta}$ and $\bar{\theta}$ (which may be infinite) to be the infimum and supremum of Ω, respectively. Let X_i be a sufficient statistic for θ_i based on n independent observations from π_i, having an absolutely continuous distribution $F_n(x,\theta_i)$. We assume that the form of F is known, but the θ_i are unknown. For each x and $n, F_n(x,\theta)$ is assumed to be stochastically increasing in θ. The θ_i are to be compared with the specified standard $\theta_0(\underline{\theta}<\theta_0<\bar{\theta})$.

The goal is to select the population associated with $\theta_{[k]}$ unless $\theta_{[k]} \leqslant \theta_0$ in which case we select no population (i.e., we choose π_0, say). The procedure R_7 of Turnbull (1976) chooses π_0 if $X_{[k]}<c$; otherwise the population that produced $X_{[k]}$ is selected as associated with $\theta_{[k]}$. The sample size n and the constant c are chosen such that

$$\Pr\{\text{selecting } \pi_0\} \geqslant P_0^* \quad \text{whenever } \theta_{[k]} \leqslant \theta^* \tag{20.19}$$

and for *all* $t=1,2,\ldots,k$,

$$\Pr\{\text{selecting any one of the } t \text{ best populations}\} \geqslant P_1^* \tag{20.20}$$

whenever $\theta_{[k-t]} \leqslant \theta_L^* \leqslant \theta_{[k-t+1]}$ and $\theta_{[k]} \geqslant \theta_U^*$. Here $\theta^*, \theta_L^*, \theta_U^*, P_0^*, P_1^*$ are specified constants, with $\theta^* < \theta_U^*$, and $\theta_L^* < \theta_U^*$. When $t=k$, we define $\theta_{[0]} = -\infty$.

To guarantee (20.19) and (20.20), the values of n and c should be chosen to satisfy:

$$\begin{aligned} &[F_n(c,\theta^*)]^k \geqslant P_0^*, \\ &\int_c^{\infty} [F_n(x,\theta_L^*)]^{k-1} dF_n(x,\theta_U^*) \geqslant P_1^*. \end{aligned} \tag{20.21}$$

In practice, we solve for n and c with equality signs in (20.21), allowing first n to be real, and then round it up to the next larger integer.

When θ is a location or scale parameter, the pair (n,c), which guarantees the requirement (20.19) and (20.20), does in fact guarantee a requirement much stronger than (20.20), namely, for all $t(1 \leqslant t \leqslant k)$,

$$\Pr\{\text{one of the } t \text{ best populations is selected}\} \geqslant P_1^*$$

whenever $\theta_{[k]} \geqslant \theta_U^*$ and (a) $\theta_{[k-t+1]} - \theta_{[k-t]} \geqslant \delta^* = \theta_U^* - \theta_L^* > 0$ in the location parameter case, and (b) $\theta_{[k-t+1]}/\theta_{[k-t]} \geqslant \delta^* = \theta_U^*/\theta_L^* > 1$ in the scale parameter case.

20.3 ASYMPTOTICALLY OPTIMAL PROCEDURES

Let $\pi_1,\ldots,\pi_k$ be k experimental categories with the associated density functions $g(x,\tau_i), i=1,\ldots,k$, and π_0 be the control with the density $g(x,\tau_0)$. We say $\theta=0$ when $\tau_0=\tau_1=\cdots=\tau_k$ and say $\theta=i$ when $\tau_i=\tau_0+\Delta(\Delta>0)$ and $\tau_0=\tau_j, j\neq i$. The decision D_0 is preferred if $\tau_i \leqslant \tau_0$ for $i=1,\ldots,k$. The decision D_j is preferred if $\tau_j = \max_{1\leqslant i\leqslant k}\tau_i$ and $\tau_j > \tau_0$.

20.3.1 Fixed Sample Size Procedure

Let X_{ij} be the jth observation from $\pi_i, i=1,\ldots,k; j=1,\ldots,n_i$. Define

$$Z_{ij} = \log\left[\frac{g_1(X_{ij})}{g_0(X_{ij})}\right], \qquad i=1, \quad 2,\ldots,k, \tag{20.22}$$

where $g_0(x)=g(x,\tau_0)$ and $g_1(x)=g(x,\tau_0+\Delta)$. Define s to be the integer for which $\sum_{j=1}^{n_s} Z_{sj} = \max_{1\leqslant i\leqslant k}\sum_{j=1}^{n_i} Z_{ij}$. If s is not unique, select randomly one of those i for which the maximum is attained.

Roberts (1964) proposed the procedure δ_m^* which takes $n_1=\cdots=n_k=n$ and makes the decision $\theta=s$ if $\sum_{j=1}^{n} Z_{sj} > m$ and the decision $\theta=0$ otherwise. Let $c>0$ be the cost per observation. We use a simple zero–one loss function and assume a prior distribution ξ which assigns probability $\xi_j>0$ to $\theta=j$ with $\xi_0+\cdots+\xi_k=1$. The risk of any procedure δ when θ is the true state of nature is given by $r(\theta,\delta)=R(\theta,\delta)+cE_\theta N$, where $R(\theta,\delta)$ is the expected loss with procedure δ and $E_\theta N$ is the expected total sample size required. The price of the procedure δ is defined to be

$$\rho(\delta) = \limsup_{c\to 0}\left[\frac{-r(\delta)}{c\log c}\right],$$

where $r(\delta)$ is the expected risk given by $r(\delta)=\sum_{i=0}^{k}\xi_i r(i,\delta)$.

Define $A_0=\inf_t E_0(e^{tZ_{11}})$ and $B_0=\inf_t E_1(e^{-tZ_{11}})$, t real. Then A_0 and B_0 are both finite and positive. The following results have been obtained by Roberts (1964).

Theorem 20.1.

1. Any fixed sample size procedure δ (whose sample sizes may depend on the cost c) has $\rho(\delta) \geqslant -k/\log\{\max(A_0, B_0)\}$.
2. If $n_1 = \cdots = n_k = \log c/\log\{\max(A_0, B_0)\}$, then for each fixed m, $\rho(\delta_m^*) = -k/\log\{\max(A_0, B_0)\}$.

The first part of the preceding theorem states that there is a certain minimal price possible for fixed sample size procedures and the second part shows that for a particular choice of the sample sizes the procedure δ_m^* is asymptotically ($c \to 0$) the best fixed sample size procedure.

The constant m is chosen so that the probability of making the decision D_0 when $\theta = 0$ is at least P^*.

20.3.2 Sequential Procedures

For the problem of comparing several experimental categories with a control discussed in §20.3.1 Roberts (1963) has compared three sequential procedures defined in this subsection. The three procedures differ in the sampling rule; they have the same stopping rule and terminal decision rule. The third procedure δ_3 was proposed by Paulson (1962).

PROCEDURE δ_1. Take one observation from each of $\pi_1, \ldots, \pi_k$. Then select after each single observation a category on which to sample next. The rule is to take one observation next from π_s (the integer s is defined in §20.3.1).

PROCEDURE δ_2. Select at random an order to examine the categories, and then one-by-one decide if a category is better than the control. If the order $(i_1, i_2, \ldots, i_k)$ is chosen, sample first from π_{i_1}, then from π_{i_2}, and so on. Once sampling from $\pi_{i_{j+1}}$ starts no more observations are taken from π_{i_j}.

PROCEDURE δ_3. Sample in k (or less)-tuples of one observation from each experimental population beginning with a k-tuple. After each observation, either decide a population is better than the control and stop further sampling, or continue sampling after (possibly) eliminating populations that appear no better than the control.

Let $b < 0 < a$. For δ_1, δ_2, and δ_3 the following stopping and decision rules are used.

1. Stop sampling from π_i as soon as $b < \Sigma_{j=1}^{n_i} Z_{ij} < a$ is violated for some n_i.
2. If for some $n_i, \Sigma_{j=1}^{n_i} Z_{ij} \geqslant a$ stop further sampling and make decision D_s.

3. As soon as $\sum_{j=1}^{n_s} Z_{sj} \leqslant b$ and observations have been taken on all k categories, stop further sampling and make decision D_0.

We assume the cost per observation and zero–one loss function as before. We also assume that $I_0 = -E_0 Z_{11}$ and $I_1 = E_1 Z_{11}$ exist and are positive.

Roberts (1963) has shown that any procedure δ has $\rho(\delta) \geqslant k\xi_0/I_0 + (1-\xi_0)/I_1$, where ξ_0 is the probability assigned to $\theta = 0$ by the prior distribution ξ. By obtaining lower bounds for $\rho(\delta_2)$ and $\rho(\delta_3)$, he has also shown that the minimum value for $\rho(\delta)$ cannot be achieved by δ_2 and δ_3. With the choice of $a = -b = -\log c$, the procedure δ_1 is shown to be optimal.

Comparing δ_2 and δ_3, we find that δ_2 is better than δ_3 if $I_1 < I_0$ and δ_3 is better than δ_2 if $2I_0 < I_1$. Also δ_2 is approximately optimal if I_1/I_0 is small and δ_3 is approximately optimal if I_1/I_0 or I_0/I_1 is small.

To meet the requirements that the probability of selecting D_0 when $\theta = 0$ is at least P_0^* and that the probability of selecting D_j when $\theta = j$ is at least P_1^* for $j = 1, \ldots, k$, Roberts (1963) suggested the choices of the constants a and b given by

$$\lambda = \min\left\{1, \frac{k(1-P_1^*)I_0}{(k-1)(1-P_0^*)(I_0+(k-1)I_1)}\right\},$$

$$a = \log\frac{k}{\lambda(1-P_0^*)},$$

$$b = \log\left[(1-P_1^*) - \frac{(k-1)(1-P_0^*)\lambda}{k}\right].$$

The preceding choices of a and b were earlier suggested by Paulson (1962) who has made some numerical comparisons between the sequential procedure δ_3 and the nonsequential procedure of his defined in (20.4) in the normal means case when $\mu_0 = 0$, and $\sigma = 1$.

Paulson (1962) has also briefly discussed a procedure similar to δ_3 when the μ_i are normal means and μ_0 is unknown. Roberts (1962) has shown that each of the three sequential procedures δ_1, δ_2, and δ_3 is strictly better than any fixed sample size procedure δ (whose sample size may depend upon the cost c).

20.4 CHOOSING ALL POPULATIONS THAT ARE BETTER THAN A CONTROL

Dunnett (1955) considered a procedure to choose a subset (possibly empty) of the experimental categories that consists only of all those that are better than the standard. In other words, the procedure selects one of the 2^k possible decisions: D_0 corresponding to selecting the control as the best, and the remaining 2^k-1 decisions corresponding to all possible nonempty subsets of the set of experimental categories. His requirement on the procedure is

$$\Pr\{\text{making the decision } D_0\} \geqslant P^* \tag{20.23}$$

whenever all the experimental categories are equivalent to the control.

Let $\pi_i(i=0,1,\ldots,k)$ be a normal population with mean μ_i and variance σ^2. The common variance σ^2, and the mean μ_0 of the control population are assumed to be unknown. Let $\overline{X}_i$ be the mean of n independent observations from π_i and s^2 be the pooled estimate of σ^2 based on $\nu=(k+1)(n-1)$ degrees of freedom. Dunnet's rule R includes $\pi_i(i=1,\ldots,k)$ in the selected subset of the experimental categories if and only if

$$\overline{X}_i-\overline{X}_0 \geqslant ds\sqrt{\frac{2}{n}}\ . \tag{20.24}$$

If the subset is empty, then the control is selected as the best. The smallest constant d satisfying (20.23) is given by

$$\Pr\{t_1<d,\ldots,t_k<d\}=P^*, \tag{20.25}$$

where the joint distribution of the t_i is the multivariate t (see Dunnett and Sobel [1954]). The d-values are tabulated by Dunnett (1955) for $k=1(1)9, \nu=5(1)20$, 24, 30, 40, 60, 120, ∞, and $P^*=0.95$, 0.99.

The sample size n may be determined as the minimum sample size required to achieve a specified probability of accepting a particular one of the experimental categories, say π_1, as the best when $\mu_1-\delta=\mu_0=\mu_2=\cdots=\mu_k, \delta>0$.

20.5 SELECTING A SUBSET CONTAINING ALL POPULATIONS BETTER THAN A CONTROL

Analogous to the subset selection formulation of Gupta (1956) for selecting the best population, Gupta and Sobel (1958a) investigated procedures for selecting a subset of random size of the experimental categories to include all those that are better than a control with a minimum guaranteed probability P^* whatever be the true configurations of the parameters. Thus a correct selection is selecting any subset that contains all the populations that are better than the standard.

20.5.1 Parametric Procedures

Gupta and Sobel (1958a) considered the problems of normal means, gamma scale parameters, and binomial success probabilities.

Normal Populations. Let π_i be $N(\mu_i, \sigma^2)$ where $\mu_i, i=1,\ldots,k$, are unknown. Several cases arise depending on whether the common variance σ^2 and the control mean μ_0 are known or not. Let $\bar{X}_i$ be the sample mean based on n_i independent observations from π_i. When μ_0 and σ^2 are known, the rule R_1 of Gupta and Sobel (1958a) selects $\pi_i (i=1,\ldots,k)$ if and only if

$$\bar{X}_i \geqslant \mu_0 - \frac{d\sigma}{\sqrt{n_i}}, \tag{20.26}$$

where $d>0$ is determined so that the probability requirement on a correct selection is met. It is easy to see that

$$P(CS|R_1) = \prod_{i \in I} \left\{ 1 - \Phi\left(\frac{-d + \sqrt{n_i}\,(\mu_0 - \mu_i)}{\sigma} \right) \right\}, \tag{20.27}$$

where I is a subset of $\{1,2,\ldots,k\}$ whose elements correspond to those π_i for which $\mu_i \geqslant \mu_0$. However the number of such populations is not known. The infimum of $P(CS|R_1)$ is attained when $I = \{1,2,\ldots,k\}$ and $\mu_0 = \mu_1 = \cdots = \mu_k$. Thus the constant d is given by

$$\Phi(d) = (P^*)^{1/k}. \tag{20.28}$$

For other cases, Gupta and Sobel (1958a) proposed rules of the form R_1 with μ_0 (unknown) replaced by $\bar{X}_0$ and σ^2 (unknown) replaced by the pooled estimate s_ν^2 based on ν degrees of freedom, where $\nu = \sum_{i=1}^k (n_i - 1)$ when μ_0 is known and $\nu = \sum_{i=0}^k (n_i - 1)$ when μ_0 is unknown. The value of d in these cases is given by

$$P^* = \begin{cases} \Phi^k(d), & \mu_0 \text{ and } \sigma \text{ both known} \\ \int_{-\infty}^{\infty} \Phi^k(u+d)\,d\Phi(u), & \mu_0 \text{ unknown, } \sigma \text{ known} \\ \int_0^{\infty} \Phi^k(yd)\,q_\nu(y)\,dy, & \mu_0 \text{ known, } \sigma \text{ unknown} \\ \int_0^{\infty}\int_{-\infty}^{\infty} \left[\prod_{i=1}^{k} \Phi\left(u\sqrt{\frac{n_i}{n_0}} + yd \right) \right] \varphi(u)\,q_\nu(y)\,du\,dy, & \mu_0 \text{ and } \sigma \text{ both unknown,} \end{cases} \tag{20.29}$$

where $q_\nu(y)$ is the density of $y = s_\nu/\sigma$.

The d-values satisfying the second equation in (20.29) are tabulated for several selected values of k and P^* by Bechhofer (1954), Gupta [1963a], and Gupta, Nagel, and Panchapakesan [1973]. When $n_0 = n_1 = \cdots = n_k$, methods of evaluating the double integral in (20.29) are given by Gupta and Sobel (1957) who have tabulated the d-values for $k = 1$, 4, 9(1)15(2)19 (5)39, 49, $\nu = 15(1)20$, 24(6)36, 40, 48, 60(20)120, 360, ∞, and $P^* = 0.75$, 0.90, 0.95, 0.975, 0.99. Values of $d/\sqrt{2}$ for other values of k and ν are given by Dunnett (1955), whose tables were referred to in §20.4.

Gamma Populations. The population π_i has the density

$$\frac{1}{\Gamma\left(\frac{\alpha_i}{2}\right)} \theta_i^{-\alpha_i/2} x^{\alpha_i/2-1} e^{-x/\theta_i} \qquad i = 0, 1, \ldots, k. \tag{20.30}$$

It will be more natural to define π_i as better than π_0 if $\theta_i < \theta_0$. Let t_i be the sum of n_i independent observations from π_i. Then the density of t_i is given by (20.30) with α_i replaced by $\nu_i = n_i \alpha_i$. The procedure R of Gupta and Sobel (1958a) includes $\pi_i (i = 1, \ldots, k)$ in the selected subset if

$$\frac{t_i}{\nu_i} \leqslant \begin{cases} (1+d)\theta_0, & \theta_0 \text{ known,} \\ (1+d)t_0/\nu_0, & \theta_0 \text{ unknown,} \end{cases} \tag{20.31}$$

where $d > 0$ is determined subject to the probability requirement. Application of this procedure to normal populations when comparison is in terms of their variances is straightforward. We can get an approximate value of d by applying the transformation $y_i = \log(t_i/\nu_i)$ and using the normal approximation.

Binomial Populations. Here π_i is a binomial population with parameter θ_i which denotes the probability of a unit being defective in the population π_i. We define π_i to be better than π_0 when $\theta_i < \theta_0$. Let x_i denote the number of defectives observed in the sample of n_i observations from π_i. When θ_0 is known, the procedure R of Gupta and Sobel (1958a) includes $\pi_i (i = 1, \ldots, k)$ in the selected subset if

$$\frac{x_i}{n_i} \leqslant \theta_0 + d\sqrt{\theta_0(1-\theta_0)/n_i}\,, \tag{20.32}$$

where $d > 0$ satisfying the probability requirement is given by

$$\prod_{i=1}^{k} \left[\sum_{j=0}^{[m_i(d)]} \binom{n}{j} \theta_0^j (1-\theta_0)^{n_i - j} \right] \geqslant P^*, \tag{20.33}$$

where $[m_i(d)]$ denotes the largest integer contained in

$$m_i(d) = n_i\theta_0 + d\sqrt{n_i\theta_0(1-\theta_0)}\ , \qquad i=1,\dots,k. \tag{20.34}$$

When θ_0 is unknown, the procedure R' proposed by Gupta and Sobel (1958a) selects π_i if

$$\frac{x_i}{n_i} \leqslant \frac{x_0}{n_0} + \left(\frac{d}{2}\right)\sqrt{n_i^{-1} + n_0^{-1}}\ . \tag{20.35}$$

The probability of a correct selection is minimized when the number of populations better than π_0 is k and $\theta_0 = \theta_1 = \cdots = \theta_k = \theta$ (say). If this probability is denoted by $P(\theta, d)$ then the desired value of d is the smallest number for which

$$\min_{0<\theta\leqslant 1} P(\theta,d) \geqslant P^*. \tag{20.36}$$

The exact value of θ for which $P(\theta,d)$ is minimized is not known. Since, except for very small n_i or very large k, the minimum occurs near $\theta = 0.5$, we can obtain an approximate solution by finding the smallest number for which $P(0.5,d) \geqslant P^*$.

Naik (1975) considered a slight modification in the usual procedure. Suppose π_i has an associated distribution $F(x;\theta_i), i=0,1,\dots,1$. We assume that θ_0 is known. Let $T_i(i=1,\dots,k)$ be a suitable statistic based on n independent observations from π_i whose density is $g(t_i|\theta_i)$. We assume that the densities of T_i differ only in θ-values and that the distribution of T_i is stochastically increasing in θ_i. The rule R considered by Naik (1975) uses k constants $a_{[1]} < \cdots < a_{[k]}$. The T_i are ordered and $T_{[i]}$ is compared with $a_{[i]}$ starting with $i=1$ and proceeding until the first time for some $t(1 \leqslant t \leqslant k) T_{[t]} \geqslant a_{[t]}$. At this stage select all the populations associated with $T_{[t]}, \dots, T_{[k]}$. The probability requirement for a correct selection is met by choosing the constants such that

$$\Pr\{T_1 \geqslant a_{[l]} | \theta_1 = \theta_0\} = (P^*)^{1/(k+1-l)}, \qquad l=1,\dots,k.$$

20.5.2 Nonparametric Procedures

In this section we discuss several nonparametric procedures. Rizvi, Sobel, and Woodworth (1968) discussed comparison of populations in terms of α-quantiles. Puri and Puri (1968, 1969) considered procedures based on ranks in the cases of location and scale parameters. Sen and Puri (1972) discussed procedures based on rank order estimators in a two-way layout model.

A. COMPARISON IN TERMS OF α-QUANTILES. Let π_0 be the control population and $\pi_1,\ldots,\pi_k$ be k experimental populations. We can think of these as $(k+1)$ lots of items available for purchase. On the basis of random samples of common size n, we want to select those that are at least as good as π_0. We suppose that items are judged on the basis of a continuously distributed attribute X and a known fraction $\alpha(0<\alpha<1)$ of the items in the control population are deficient (i.e., their X-values are too small). A population is considered to be better than the control if it has a smaller proportion of deficient items. Let $F_j(j=0,\ldots,k)$ denote the distribution function of X for the population π_j and $x_\alpha(F_j)$ its αth quantile. Then π_j is better than π_0 if $x_\alpha(F_j)\geqslant x_\alpha(F_0)$.

Rizvi, Sobel, and Woodworth (1968) considered the problem of selecting a subset containing all populations better than π_0. To make this more precise, we say that F_i is *as good as F_0 uniformly* if and only if $F_i(x)\leqslant F_0(x)$ for all x and that F_i is *worse than F_0 uniformly* if and only if $x_\alpha(F_i)<x_\alpha(F_0)$ and $F_i(x)\geqslant F_0(x)$ for all x. Let Ω denote the space of all $(k+1)$-tuples $\omega=(F_0,F_1,\ldots,F_k)$ and let Ω_1 denote the subspace of Ω consisting of those ω such that for each $i(i=1,\ldots,k)$ either F_i is as good as F_0 uniformly or F_i is worse than F_0 uniformly.

For any preassigned P^* with $2^{-k}<P^*<1$, we want a procedure R such that

$$P(CS|R)\geqslant P^* \quad \text{whenever } \omega\in\Omega_1. \tag{20.37}$$

For any fixed α with $0<\alpha<1$, we assume that $1\leqslant(n+1)\alpha\leqslant n$ and define the integer r by the inequalities $r\leqslant(n+1)\alpha<r+1$ (see §15.2).

The procedure $R=R(c)$ is defined in terms of an integer c and the order statistics Y_{ji} where Y_{ji} is the jth order statistic in a sample of size n from π_i. Since the F_i are unknown we take Y_{0i} to mean $-\infty$ for each i.

Case (1): Unknown F_0. The procedure $R(c)$ includes $\pi_i(i=1,\ldots,k)$ in the selected subset if and only if

$$Y_{ri}\geqslant Y_{r-c,0}. \tag{20.38}$$

The procedure R is defined as that $R(c)$ for which c is the smallest integer $(0\leqslant c\leqslant r)$ such that $R(c)$ satisfies (20.37).

For any α and k, a value of $c\leqslant r-1$ (for which the procedure is nondegenerate) may not exist for all pairs (n,P^*). If P^* is chosen not greater than some $\overline{P}_0=\overline{P}_0(n,\alpha,k)$, then a value of $c\leqslant r-1$ satisfying (20.37) does not exist. It is shown that

$$\bar{P}_0 = n\int_0^1 [G_{n-r+1}(v)]^k v^{n-1}\,dv, \tag{20.39}$$

where $G_r(p) = I_p(r, n-r+1)$, the standard incomplete beta function.

As a secondary problem we suppose that for some preassigned fraction δ^* the decision-maker considers a population π_i to be δ^*-inferior to π_0 if more than $100(\alpha+\delta^*)$ percent of the items in π_i are as bad as at least one of the worst $100(\alpha-\delta^*)$ percent of items in π_0; that is, π_i is δ^*-inferior if $x_{\alpha-\delta^*}(F_0) \geqslant x_{\alpha+\delta^*}(F_i)$. Rizvi, Sobel, and Woodworth (1968) have obtained asymptotic expressions for the smallest sample size needed to guarantee that the expected proportion of δ^*-inferior populations selected by R will be less than a preassigned number β^*.

The procedure R is compared with a competing nonparametric procedure S based on rank sums and a competing asymptotically nonparametric procedure M based on sample means. For small values of δ^*, these two procedures require sample sizes proportional to the square of that required by R to achieve the same degree of rejection of δ^*-inferiors. For moderate δ^*-values, the procedure S is shown to require a sample size that has the same order of magnitude as that required by R.

Case (2): Known F_0. Let $x_{\alpha-\beta}(F_0)$ denote the $(\alpha-\beta)$-th quantile of F_0. The procedure R_1 defined by Rizvi, Sobel, and Woodworth (1968) includes $\pi_i (i=1,\ldots,k)$ in the selected subset if and only if

$$Y_{ri} \geqslant x_{\alpha-\beta} \tag{20.40}$$

where β is the smallest number between 0 and α for which (20.37) holds. The results obtained in this case for investigating the expected number of misclassifications are exact.

B. PROCEDURES BASED ON RANKS. Puri and Puri (1968, 1969) have considered the problem of selecting a subset containing all populations better than a control when location and scale parameters, are involved in comparison. We first consider the case of location parameters.

Case (1): Location Parameters. The population π_i has the continuous distribution $F(x-\theta_i), i=0,1,\ldots,k$. A population π_i is defined to be as good as or better than π_0 if $\theta_i \geqslant \theta_0 + \Delta^{(n)}$ where $\Delta^{(n)} = \Delta\sigma/\sqrt{n} + 0(1/\sqrt{n})$. Here σ^2 is the variance of F, and Δ is determined by the condition

$$\Pr\{\max(U_1,\ldots,U_k) \leqslant \Delta/\sqrt{2}\} = P^*, \tag{20.41}$$

where the U_i are equally correlated $N(0,1)$ variables with correlation equal to $\frac{1}{2}$. Puri and Puri (1969) have studied two procedures based on ranks.

Let $X_{ij}, j=1,2,\ldots,n$, be independent observations from $\pi_i, i=0,1,\ldots,k$. Let $N=(k+1)n$ and define

$$Z_{r,N}^{(i)}=\begin{cases}1, & \text{if the } r\text{th smallest observation in the pooled sample is from } \pi_i,\\ 0, & \text{otherwise.}\end{cases} \tag{20.42}$$

Let

$$T_i=n^{-1}\sum_{r=1}^{N} E(V_{[r]})Z_{r,N}^{(i)}, i=0,1,\ldots,k, \tag{20.43}$$

where $V_{[1]}<\cdots<V_{[N]}$ is an ordered sample from a given distribution F_0. Then the first procedure R_1 selects π_i if and only if

$$T_i \geqslant T_0. \tag{20.44}$$

To find the desired sample size n, we restrict ourselves to a certain part of Ω and obtain the LFC over this part. The asymptotic efficiency of the means procedure relative to R_1 is same as in the case of their procedure for selecting the t best populations discussed in §8.3.2. The two procedures have the same asymptotic performance.

For the alternative procedure R_2 we combine separately each of the k samples from $\pi_1,\ldots,\pi_k$ with the sample from π_0. We define

$$Z_{r,n}^{(i,0)}=\begin{cases}1, & \text{if } r\text{th smallest of } 2n \text{ observations from } \pi_i \text{ and } \pi_0 \text{ belongs to } \pi_i,\\ 0, & \text{otherwise.}\end{cases}$$

Let

$$\begin{aligned}\overline{H}_i&=n^{-1}\sum_{r=1}^{2n} E(V_{[r]})Z_{r,n}^{(i,0)}\\ \overline{H}_{i,0}&=n^{-1}\sum_{r=1}^{2n} E(V_{[r]})(1-Z_{n,r}^{(i,0)})\end{aligned} \tag{20.45}$$

where $V_{[1]}<\cdots<V_{[2n]}$ is an ordered sample from a given distribution F_0. Then the procedure R_2 selects π_i if and only if

$$\overline{H}_i \geqslant \overline{H}_{i,0} \tag{20.46}$$

which is equivalent to $\bar{H}_i \geqslant$ constant. The two procedures R_1 and R_2 are asymptotically equi-efficient.

Case (2): Scale Parameters. Let π_i have a continuous distribution function $F(x/\sigma_i), i=0,1,\ldots,k$. The population π_i is defined to be better than π_0 if $\theta_{i0} \leqslant \theta^{(n)}$, where $\theta_{i0} = \sigma_i^2/\sigma_0^2$. Here $\theta^{(n)}$ is known to be less than or equal to unity and in actual practice is identified with a constant $0 < \theta^* \leqslant 1$ prescribed by the experimenter. Let $X_{ij}, j=1,\ldots,n$, be independent observations from $\pi_i, i=0,1,\ldots,k$. Define $Z_{r,N}^{(i)} = 1$ if the rth smallest of $N = n(k+1)$ absolute values $|X_{ij}|$ is from π_i and $Z_{r,N}^{(i)} = 0$ otherwise. Alternatively, consider for each $i = 1,2,\ldots,k$, the $2n$ absolute values $|X_{ij}|$, from π_i and π_0. Let $Z_{r,n}^{(i,0)} = 1$ if the rth smallest in the above $2n$ absolute values is from π_i; otherwise, let $Z_{r,n}^{(i,0)} = 0$. Define

$$T_i = n^{-1} \sum_{r=1}^{N} E\left(V_{[r]}^2\right) Z_{r,N}^{(i)}, i=0,1,\ldots,k,$$

$$\bar{T}_i = n^{-1} \sum_{r=1}^{2n} E\left(V_{[r]}^2\right) Z_{r,n}^{(i,0)}, i=1,\ldots,k, \tag{20.47}$$

$$\bar{T}_{i,0} = n^{-1} \sum_{r=1}^{2n} E\left(V_{[r]}^2\right)\left(1 - Z_{r,n}^{(i,0)}\right), i=1,\ldots,k,$$

where $V_{[1]} < \cdots < V_{[N]}$ (in the definition of T_i) and $V_{[1]} < \cdots < V_{[2n]}$ (in the definition of $\bar{T}_i$) are ordered samples of sizes N and $2n$, respectively, from a given distribution F_0.

For selecting a subset containing all the populations that are better than the control Puri and Puri (1968) defined the following two rules R_3 and R_4. The rule R_3 selects π_i if and only if $T_i \leqslant T_0$ and the rule R_4 selects π_i if and only if $\bar{T}_i \leqslant \bar{T}_{i,0}$ (which is equivalent to $\bar{T}_i \leqslant$ constant).

The two procedures R_3 and R_4 are asymptotically equi-efficient. The asymptotic efficiency of the procedure R_3 relative to the normal theory procedure is studied in a manner similar to the procedures of Puri and Puri (1968) discussed in §8.3.2.

C. PROCEDURES BASED ON RANK ORDER ESTIMATORS. Sen and Puri (1972) have discussed the problem of selecting a subset of treatments better than a control in a two-factor complete block design with one observation per cell. The observable random variable $X_{i\alpha}(i=0,1,\ldots,k; \alpha=1,\ldots,n)$ is expressed as

$$X_{i\alpha} = \mu + \beta_\alpha + \tau_i + \epsilon_{i\alpha}, \tag{20.48}$$

where μ is the mean effect, $\beta_1,\ldots,\beta_n$ are the block effects (nuisance parameters for the fixed effects model and random variables for the mixed effects model), $\tau_1,\ldots,\tau_k$ are the experimental treatments, τ_0 is the control treatment, and the $\epsilon_{i\alpha}$ are the error components. It is assumed that $\boldsymbol{\epsilon}_\alpha=(\epsilon_{0\alpha},\epsilon_{1\alpha},\ldots,\epsilon_{k\alpha}),\alpha=1,\ldots,n$, are independent and identically distributed random vectors with a continuous cumulative distribution function $F(\boldsymbol{\varepsilon})$, where $\boldsymbol{\varepsilon}\in R^{k+1}$ (the $(k+1)$-dimensional Euclidean space) and $F(\boldsymbol{\varepsilon})$ is symmetric in its $(k+1)$ arguments.

Let $X^*_{i0,\alpha}=X_{i\alpha}-X_{0\alpha}$ and $\mathbf{X}^*_{i0}=(X^*_{i0,1},\ldots,X^*_{i0,n})$. Define

$$h_n(\mathbf{X}^*_{i0})=n^{-1}\sum_{\alpha=1}^{n}J_n\big((n+1)^{-1}R_{i0,\alpha}\big)\operatorname{sgn}X^*_{i0,\alpha}$$

where $R_{i0,\alpha}=$rank of $|X^*_{i0,\alpha}|$ among $|X^*_{i0,1}|,\ldots,|X^*_{i0,n}|$, sgn u is equal to 1, 0, or -1 according as u is $>$, $=$, or <0, and $J_n((n+1)^{-1}i)$ is the expected value of the ith order statistic of a sample of size n from the distribution $\psi^*(x)=\psi(x)-\psi(-x)$, $x\geqslant 0$, $i=1,\ldots,n$, where $\psi(x)$ is a symmetric (about the origin) nondegenerate distribution function satisfying certain assumptions (see assumptions I, II, and III in Puri and Sen [1967]).

Now, for $i=1,\ldots,k$, let

$$\hat{\Delta}^{(n)}_{i0,1}=\sup\{t:h_n(\mathbf{X}^*_{i0}-t\mathbf{1}_n)>0\},$$

$$\hat{\Delta}^{(n)}_{i0,2}=\inf\{t:h_n(\mathbf{X}^*_{i0}-t\mathbf{1}_n)<0\},$$

$$\hat{\Delta}^{(n)}_{i0}=\frac{\big[\hat{\Delta}^{(n)}_{i0,1}+\hat{\Delta}^{(n)}_{i0,2}\big]}{2}.$$

The rule R_5 proposed by Sen and Puri (1972) selects all the treatments for which $\hat{\Delta}^{(n)}_{i0}>0$.

The procedure of Gupta and Sobel (1958a) can be modified to use in the two-way layout. The only change needed is to replace σ^2 by $\sigma^2(1-\rho)$. Here σ^2 is the variance of $\epsilon_{i\alpha}$ and ρ is the common correlation of $\epsilon_{i\alpha}$ and $\epsilon_{i\beta}(\alpha\neq\beta)$. By virtue of the central limit theorem, the same solution holds asymptotically for the entire class of cdf's with finite second moments. The Gupta–Sobel solution also asymptotically holds for the rank scores procedure provided σ^2 is replaced by certain appropriate quantity σ_0^2 [see (3.7) in Sen and Puri (1972)]. Hence the asymptotic relative efficiency of the rank scores procedure with respect to the Gupta–Sobel procedure can be measured by $\sigma^2(1-\rho)/\sigma_0^2$.

20.5.3 Procedures with Modified Requirements

Turnbull (1976) also considered the problem of choosing a subset consisting of those populations that are better than the specified standard; however, his probability requirements incorporated an indifference zone feature as explained in this subsection. We have the same assumptions regarding the populations and the distribution of the statistics X_i as in §20.2.4. The procedure R_6 of Turnbull is as follows.

R_6: If $X_{[k]} \leqslant B$, select no population (i.e., choose π_0); otherwise select all populations π_i for which $X_i > A$. (20.49)

The real numbers $A, B, (A < B)$ are chosen to satisfy:

Pr{selecting π_0} $\geqslant P_0^*$ whenever $\theta_{[k]} \leqslant \theta_L^*$, and Pr{selecting all populations π_i for which $\theta_i \geqslant \theta_U^*$} $\geqslant P_1^*$ whenever $\theta_{[k]} \geqslant \theta_U^*$. (20.50)

Here $\{\theta_L^*, \theta_U^*, P_0^*, P_1^*\}$ are specified constants with $\theta_L^* < \theta_U^*, 0 < P_0^* < 1$, and $[1 - F_n(B, \theta_U^*)]^k \leqslant P_1^* \leqslant 1 - F_n(B, \theta_U^*)$, where B satisfies the relation $P_0^* = [F_n(B, \theta_L^*)]^k$. The procedure R_6 will guarantee the probability requirements for *all* sample sizes n. In general, $F_n(B, \theta_U^*) \to 0$ as $n \to \infty$ for fixed $\theta_L^* < \theta_U^*$. Therefore a P_1^*-value may be specified as close to unity as desired, provided that the sample size is large enough.

To guarantee the probability requirements, Turnbull (1976) has shown that A and B should be chosen to satisfy:

$$P_0^* = \left[F_n(B, \theta_L^*) \right]^k,$$

$$P_1^* = \min\left\{ 1 - F_n(A, \theta_U^*) - \alpha^{k-1}\left[F_n(B, \theta_U^*) - F_n(A, \theta_U^*) \right], \right.$$

$$\left. \left[1 - F_n(A, \theta_U^*) \right]^k - \left[F_n(B, \theta_U^*) - F_n(A, \theta_U^*) \right]^k \right\} \quad (20.51)$$

where $\alpha = F_n(B, \underline{\theta})$. Usually $\underline{\theta} = -\infty$ and $\alpha = 1$.

20.6 PARTITIONING A SET OF POPULATIONS WITH RESPECT TO A CONTROL

In the preceding two sections we discussed the formulations of Dunnett (1955) and Gupta and Sobel (1958a). The goal in these formulations is to divide the set of k populations $\pi_1, \ldots, \pi_k$ into two sets, say, S_G and S_B. The goal is either to partition such that S_G contains all the populations that are better than a control population π_0 and S_B contains the rest, or only such that S_G will include all the better populations satisfying in either case certain probability requirements imposed on a correct selection. It is possible that the experimenter is really concerned about only misclassifying populations that are "sufficiently better" or "sufficiently worse" than the control population.

20.6.1 Normal Populations

Let us consider $(k+1)$ normal populations where $\pi_i (i=0,1,\ldots,k)$ is normal with mean μ_i and variance σ^2. For arbitrary but fixed constants δ_1^* and δ_2^* such that $\delta_1^* < \delta_2^*$, we define three disjoint and exhaustive subsets $\mathcal{P}_B, \mathcal{P}_I$, and $\mathcal{P}_G$ of the set $\mathcal{P} = \{\pi_1, \ldots, \pi_k\}$:

$$\mathcal{P}_B = \{\pi_i : \mu_i \leqslant \mu_0 + \delta_1^*\},$$

$$\mathcal{P}_I = \{\pi_i : \mu_0 + \delta_1^* < \mu_i < \mu_0 + \delta_2^*\},$$

$$\mathcal{P}_G = \{\pi_i : \mu_i \geqslant \mu_0 + \delta_2^*\}. \tag{20.52}$$

Any population $\pi_i \in \mathcal{P}_G$ is sufficiently better than the control and can be called a good population. Similarly any population $\pi_i \in \mathcal{P}_B$ is sufficiently worse than the control and is called a bad population. After observations are taken, the set $\mathcal{P}$ is partitioned into two disjoint sets S_B and S_G which are declared as the sets of bad and good populations, respectively. A correct decision (CD) is made if $\mathcal{P}_B \subset S_B$ and $\mathcal{P}_G \subset S_G$. In other words, a correct decision requires no misclassification of the populations in $\mathcal{P}_B$ and $\mathcal{P}_G$. The experimenter is not concerned with how the populations in P_I are classified. Thus, a population $\pi_i \in \mathcal{P}$ is said to be misclassified if $\pi_i \in (\mathcal{P}_B \cap S_G) \cup (\mathcal{P}_G \cap S_B)$. The interest is in finding a procedure R for which

$$P(\mathrm{CD}|R) \geqslant P^* \quad \text{whenever } \boldsymbol{\mu} \in \Omega, \tag{20.53}$$

where $\Omega = \{\boldsymbol{\mu} : \boldsymbol{\mu} = (\mu_0, \mu_1, \ldots, \mu_k)\}$ when σ^2 is known and $\Omega = \{\boldsymbol{\mu} : \boldsymbol{\mu} = (\mu_0, \ldots, \mu_k, \sigma^2)\}$ when σ^2 is unknown. Tong (1969) has discussed a single-stage (σ^2 known), a two-stage, and a sequential procedure for this problem. Sobel and Tong (1971) have discussed the optimal allocation of observations using two optimal rules. These problems and a slightly different formulation of Bhattacharya (1956, 1958) are discussed in this section. However, the formulation of Lehmann (1961) considering several possible optimality requirements and related results are discussed in the next section.

A. A SINGLE-STAGE PROCEDURE. Let $\bar{X}_i\,(i=0,1,\ldots,k)$ be the mean of a sample of size N from π_i. The decision rule R_1 of Tong (1969) is:

$$
\begin{aligned}
S_G &: \left\{\pi_i : \bar{X}_i - \bar{X}_0 < d\right\} \\
S_B &: \left\{\pi_i : \bar{X}_i - \bar{X}_0 \geqslant d\right\}
\end{aligned}
\tag{20.54}
$$

where $d=(\delta_1^*+\delta_2^*)/2$. The sample size N is to be determined such that (20.53) is satisfied.

Let $\lambda=2\sigma/(\delta_2^*-\delta_1^*)$. Then, for every λ and N, the LFC under the procedure R_1 is given by

$$
\begin{aligned}
\mu_1=\mu_2=\cdots=\mu_m&=\mu_0+\delta_1^*, \\
\mu_{m+1}=\mu_{m+2}=\cdots=\mu_k&=\mu_0+\delta_2^*,
\end{aligned}
\tag{20.55}
$$

where $m=k/2$ if k is even, and $m=(k+1)/2$ if k is odd. The minimum sample size N required to satisfy (20.53) is the smallest integer satisfying

$$N \geqslant 2\lambda^2 b^2, \tag{20.56}$$

where $b=b(P^*,k)$ is given by

$$\Pr\{Y_1 \leqslant b,\ldots,Y_k \leqslant b\}=P^*, \tag{20.57}$$

where $Y_1,\ldots,Y_k$ are correlated $N(0,1)$ variables with the covariance matrix

$$
\Sigma=
\begin{array}{c}
\begin{array}{ccc|ccc}
1 & \cdots & m & m+1 & \cdots & k
\end{array} \\
\left[\begin{array}{ccc|ccc}
1 & & \frac{1}{2} & & & \\
 & \ddots & & & -\frac{1}{2} & \\
\frac{1}{2} & & 1 & & & \\
\hline
 & & & 1 & & \frac{1}{2} \\
 & -\frac{1}{2} & & & \ddots & \\
 & & & \frac{1}{2} & & 1
\end{array}\right]
\end{array}
\begin{array}{c}
1 \\ \vdots \\ m \\ \hline m+1 \\ \vdots \\ k
\end{array}.
\tag{20.58}
$$

Tong (1969) has tabulated the equicoordinate percentage point b for $k=1(1)10(2)20$ and $P^*=0.50,\ 0.75,\ 0.90,\ 0.95,\ 0.975,\ 0.99$.

An upper bound N' for the minimum sample size required is given by the smallest integer N' satisfying

$$N' \geqslant 2\lambda^2 b'^2, \tag{20.59}$$

where

$$\begin{aligned} H_m(b') + H_{k-m}(b') &= 1 + P^*, \\ H_q(c) &= \Pr\{U_1 \leqslant c, \ldots, U_q \leqslant c\}, \end{aligned} \tag{20.60}$$

and the U_i are equicorrelated $N(0,1)$ variables with correlation $\frac{1}{2}$. Numerical computations of $\gamma = (b'/b)^2$ carried out for selected values of k and P^* lead to the conjecture that $\gamma = \gamma(P^*, k)$ is monotonically decreasing both in P^* and k.

It has also been shown by Tong (1969) that the procedure R_1 is Bayes, minimax, and admissible among the class of translation invariant decision rules. He has tacitly assumed, apparently, that the decision in the ith component problem depends only on $\overline{X}_i$ and $\overline{X}_0$. Randles and Hollander (1971) have shown through an example that this restriction is necessary for the minimax property to be true.

When σ^2 is unknown, there is no single-stage procedure that can solve the problem. In this case, one can use the two-stage or the sequential procedure discussed as follows.

B. A TWO-STAGE PROCEDURE. We assume that σ^2 is unknown. The following procedure R_2 is proposed by Tong (1969)

$\boldsymbol{R_2}$: 1. Let $n_0 \geqslant 2$ be a preassigned positive integer. Let s_ν^2 be the pooled estimator of σ^2 with $\nu = (k+1)(n_0 - 1)$ degrees of freedom based on a sample of size n_0 from each population.

2. Take an additional sample of size $(N - n_0)$ from each population, where N is to be determined as follows.

3. Let $\overline{X}_i, i = 0, 1, \ldots, k$, be sample means based on the complete samples of size N.

4. Apply the decision rule in (20.54).

Let $h_\nu = h_\nu(P^*, k)$ be the solution of the equation

$$\Pr\{V_1 \leqslant h_\nu, \ldots, V_k \leqslant h_\nu\} = P^*, \tag{20.61}$$

where the V_i are Student's t variables with ν degrees of freedom, and they are correlated with correlation matrix Σ given in (20.58). Then the smallest sample size N required by R_2 to satisfy (20.53) is the smallest integer N satisfying

$$N \geqslant \max\left\{ n_0, \frac{2h_\nu^2 s_\nu^2}{a^2} \right\} \tag{20.62}$$

where $a=(\delta_2^* - \delta_1^*)/2$. The values of h_ν are tablulated by Tong (1969) for $k=2(1)6(2)12(4)20, P^*=0.50$, 0.75, 0.90, 0.95, 0.975, 0.99, and $\nu=5(1)10(2)20(4)60(30)120, \infty$.

Let N be the random sample size defined by (20.62) and N_0 be the sample size required under the single-stage procedure for the LFC. Then it is shown by Tong (1969) that, for every P^*, k and the first-stage sample size n_0,

$$\begin{aligned} &\text{(i)} \quad \frac{E(N)}{N_0} \geqslant \left(\frac{h_\nu}{b}\right)^2 \text{ for every } \lambda, \\ &\text{(ii)} \quad \lim_{\lambda\to\infty} \frac{E(N)}{N_0} = \left(\frac{h_\nu}{b}\right)^2. \end{aligned} \tag{20.63}$$

Tong has also obtained lower and upper bounds for the expected misclassification size and studied its asymptotic ($\lambda\to\infty$) behavior.

C. A SEQUENTIAL PROCEDURE. The observations are taken a vector at a time. Let $\mathbf{X}_j=(X_{0j}, X_{1j}, \ldots, X_{kj})$. Thus the observation sequence is $\{\mathbf{X}_j\}_{j=1}^{\infty}$. The procedure R_3 defined by Tong (1969) has the following stopping rule: Stop with $\mathbf{X}_N$ where N is the first integer $n \geqslant 2$ such that $s_\nu^2 = na^2/(2h_\nu^2)$, where $a=(\delta_2^* - \delta_1^*)/2, \nu=(k+1)(n-1)$, h_ν satisfies (20.61), and s_ν^2 is the pooled estimator of σ^2 based on the observations $\mathbf{X}_j, j=1,2,\ldots,n$.

Let the observed value of the stopping time N be n. Then the decision rule of R_3 is same as (20.54), where $\overline{X}_i$ is the mean based on n observations from π_i.

As before, let N_0 be the sample size needed for the single-stage procedure. Then it is shown that

$$\begin{aligned} &\text{(i)} \quad \lim_{\lambda\to\infty} \frac{N}{N_0} = 1 \text{ a.s.}; \\ &\text{(ii)} \quad \lim_{\lambda\to\infty} \frac{E(N)}{N_0} = 1. \end{aligned} \tag{20.64}$$

In other words, the procedure R_3 is asymptotically relatively efficient.

The sequential procedure provides only an "asymptotic" solution to the problem in the sense that the probability of a correct decision under this procedure may be slightly less than P^* for some values of the unknown parameter σ^2, or equivalently λ. Tong (1969) has shown that, for every mean vector $\boldsymbol{\mu}=(\mu_0,\ldots,\mu_k)$,

$$\lim_{\lambda\to\infty} P(CD|\boldsymbol{\mu},\lambda;R_3)=1,$$

$$\lim_{\lambda\to\infty} P(CD|\boldsymbol{\mu},\lambda;R_3)\geqslant P^*.$$

D. Optimal Allocation of Observations. Here we assume that π_0 is $N(\mu_0,\sigma_0^2)$ and $\pi_i(i=1,\ldots,k)$ is $N(\mu_i,\sigma_1^2)$. We consider a single-stage procedure based on n_0 observations from π_0 and n_1 observations from each of $\pi_1,\ldots,\pi_k$. Let $\overline{X}_i(i=0,1,\ldots,k)$ be the sample mean associated with π_i. We use the decision rule R defined by (20.54). Let $\alpha=n_0/n_1$. For a fixed $n=n_0+kn_1$, let $C_{\boldsymbol{\mu}}(\alpha)$ and $M_{\boldsymbol{\mu}}(\alpha)$ denote the probability of a correct decision and the expected number of populations misclassified, respectively. These two quantities depend on n_0 and n_1 only through α. For every fixed α the LFC $\boldsymbol{\mu}^*$ for $C_{\boldsymbol{\mu}}(\alpha)$ is given by (20.55). Let $C(\alpha)=C_{\boldsymbol{\mu}^*}(\alpha)$. Sobel and Tong (1971) have shown that $\inf_\alpha M_{\boldsymbol{\mu}}(\alpha)=M_{\boldsymbol{\mu}}(\alpha_1)$, where $\alpha_1=\sqrt{k}\,\sigma_0/\sigma_1$. Let α_2 satisfy $\sup_\alpha C(\alpha)=C(\alpha_2)$. Then the behavior of α_2 is summarized in the following theorem of Sobel and Tong (1971).

Theorem 20.2.

a. When $k=2$, $\alpha_2>\alpha_1$ for every λ and every $b=(\delta_2^*-\delta_1^*)\sqrt{n}\,/2\sigma_1$;
b. when $k>2$, $\alpha_2<\alpha_1$ for every b and large λ;
c. for every $k\geqslant 2$ and every λ, $\alpha_2\to\alpha_1$ as $b\to\infty$.

So we have two optimal α values, α_1 minimizing the expected number of populations misclassified and α_2 maximizing the probability of a correct decision. These two optimal rules are not the same for finite n, but they are asymptotically equivalent.

When σ_0^2,σ_1^2, and λ are completely unknown, Sobel and Tong (1971) have also defined an asymptotically optimal sequential sampling rule for the purpose of optimal allocation when the total sample size n is large. This sequential rule has the nature that, at each stage, a decision is made on whether we should take one observation from the control or take a vector of observations from the k experimental populations. This sequential rule R' is as follows.

R′: a. Let $n^* \geqslant 2$ be a predetermined positive integer. We first take n^* observations from each one of the $(k+1)$ populations.

b. At each stage after observing r observations from π_0 and s observations from each one of $\pi_1,\ldots,\pi_k$, compute the sample standard deviations u_r and v_s, where

$$(r-1)u_r^2 = \sum_{j=1}^{r}\left(X_{0j}-\bar{X}_0\right)^2 \quad \text{and} \quad k(s-1)v_s^2 = \sum_{i=1}^{k}\sum_{j=1}^{s}\left(X_{ij}-\bar{X}_i\right)^2.$$

c. For the next stage take one observation from π_0 if $r/s \leqslant \sqrt{k}\; u_r/v_s$, or a vector of observations from $\pi_1,\ldots,\pi_k$ if $r/s > \sqrt{k}\; u_r/v_s$.

d. Stop sampling when $r+ks=n$ and apply the decision rule in (20.54).

E. A Modified Formulation In the problem of partitioning a set of k normal populations with respect to a control discussed so far, the set $\mathcal{P}$ consisting of the experimental populations is partitioned into three sets $\mathcal{P}_B$, $\mathcal{P}_I$, and $\mathcal{P}_G$. The procedure sought partitions the set $\mathcal{P}$ on the basis of observations into two sets S_B and S_G. We now discuss a formulation of Bhattacharya (1956, 1958) whose aim is to find a rule for partitioning the set $\mathcal{P}$ into three sets S_B, S_I, and S_G assuming the control mean μ_0 to be known with $\delta_2^* = \delta$ and $\delta_1^* = -\delta$. Restating the problem, we have $\pi_0,\ldots,\pi_k$, where π_i is $N(\mu_i,\sigma^2)$. Based on n independent observations the decision rule D_λ is defined by

$$\begin{aligned} S_G &= \left\{\pi_i\colon \bar{X}_i-\mu_0 \geqslant \lambda\right\},\\ S_I &= \left\{\pi_i\colon |\bar{X}_i-\mu_0| < \lambda\right\},\\ S_B &= \left\{\pi_i\colon \bar{X}_i-\mu_0 < -\lambda\right\}, \end{aligned} \tag{20.65}$$

where λ is chosen to be λ^*, the value for which $\inf_{\boldsymbol{\mu}} P(\text{CD};\lambda)$ is maximized. It should be noted that a correct decision occurs if and only if $S_B=\mathcal{P}_B$, $S_I=\mathcal{P}_I$ and $S_G=\mathcal{P}_G$. There is no significance of an indifference zone. Assuming a simple loss function for the classification of a population, it is shown that D_{λ^*} is an admissible minimax rule. However, it is inconsistent in the sense that the supremum of the risk function does not tend to zero even if the sample sizes increase indefinitely. This inconsistency arises out of the simple loss function. The problem was therefore studied afresh by Bhattacharya (1958) with a loss function which is identical to simple loss function except in some neighborhoods of $\mu_0 \pm \delta$. In other words, for some $\epsilon>0$, the decisions $\pi_i \in S_G$ and $\pi_i \in S_I$ are equally good when $\delta-\epsilon<\mu_i-\mu_0$

$<\delta+\epsilon$ and decisions $\pi_i \in S_I$ and $\pi_i \in S_B$ are equally good when $-\delta-\epsilon < \mu_i - \mu_0 < -\delta+\epsilon$. This was essentially done also by Seeger (1972) who formulated the problem as one of comparing the normal populations π_i: $N(\mu_i, \sigma^2)$, $i=1,\dots,k$, with two controls π_{0j}: $N(\mu_{0j}, \sigma^2)$, $j=1,2$, where $\mu_{01} < \mu_{02}$. All the means are unknown and σ^2 is known. For arbitrary but fixed constants δ_{11}^*, δ_{21}^*, δ_{12}^*, and δ_{22}^*, such that $\delta_{11}^* < \delta_{21}^*$ and $\delta_{12}^* < \delta_{22}^*$, the set $\mathcal{P} = \{\pi_1, \dots, \pi_k\}$ is partitioned into five disjoint subsets: $\mathcal{P}_B = \{\pi_i\colon \mu_i \leq \mu_{01} + \delta_{11}^*\}$, $\mathcal{P}_{I1} = \{\pi_i\colon \mu_{01} + \delta_{11}^* < \mu_i < \mu_{01} + \delta_{21}^*\}$, $\mathcal{P}_M = \{\pi_i\colon \mu_{01} + \delta_{21}^* \leq \mu_i \leq \mu_{02} + \delta_{12}^*\}$, $\mathcal{P}_{I2} = \{\pi_i\colon \mu_{02} + \delta_{12}^* < \mu_i < \mu_{02} + \delta_{22}^*\}$, and $\mathcal{P}_G = \{\pi_i\colon \mu_i \geq \mu_{02} + \delta_{22}^*\}$. On the basis of the observed sample means $\bar{X}_{01}, \bar{X}_{02}, \bar{X}_1, \dots, \bar{X}_k$, the set $\mathcal{P}$ is partitioned into three subsets S_B, S_M, and S_G. A correct decision occurs if

$$\mathcal{P}_B \subset S_B;\ \mathcal{P}_M \subset S_M;\ \mathcal{P}_G \subset S_G;$$

$$\mathcal{P}_{I1} \subset S_M \cup S_B;\ \mathcal{P}_{I2} \subset S_M \cup S_G.$$

When $\mu_{01} + \delta_{21}^* > \mu_{02} + \delta_{12}^*$, a correct decision occurs if $\mathcal{P}_B \subset S_B$ and $\mathcal{P}_G \subset S_G$.

Following Tong (1969), Seeger (1972) proposed the procedure:

$$\begin{aligned} S_B &= \{\pi_i\colon \bar{X}_i \leq \bar{X}_{01} + d_1\}, \\ S_M &= \{\pi_i\colon \bar{X}_{01} + d_1 < \bar{X}_i \leq \bar{X}_{02} + d_2\}, \\ S_G &= \{\pi_i\colon \bar{X}_i > \bar{X}_{02} + d_2\}, \end{aligned} \tag{20.66}$$

where $d_1 = (\delta_{12}^* + \delta_{21}^*)/2$ and $d_2 = (\delta_{12}^* + \delta_{22}^*)/2$. It is assumed that the probability for $\bar{X}_{01} + d_1 > \bar{X}_{02} + d_2$ is very low. However, Seeger's minimization of the probability of a correct decision is in error. See Cane (1979) for a discussion of this.

20.6.2 Exponential Populations

Let $\pi_0, \pi_1, \dots, \pi_k$ be $(k+1)$ one-parameter exponential populations. Let π_i have the associated density function

$$f(x,\theta) = \exp\{P(x)Q(\theta) + R(x) + S(\theta)\}, \qquad i=0,1,\dots,k.$$

As usual, π_0 is the control and the θ_i $(i=0,1,\dots,k)$ are all unknown. We consider the partition in terms of $\tau_i = Q(\theta_i)$.

Perng and Groves (1977) have discussed two sequential procedures. Their formulation of the problem is similar to that of Tong (1969) in the case of normal populations which was discussed earlier in this section.

For arbitrary but fixed constants δ_1 and δ_2 specified by the experimenter, the set $\mathcal{P}=\{\pi_1,\ldots,\pi_k\}$ is partitioned into three subsets,

$$\mathcal{P}_B=\{\pi_i\colon \tau_i \leqslant \tau_0-\delta_1\},$$

$$\mathcal{P}_I=\{\pi_i\colon \tau_0-\delta_1<\tau_i<\tau_0+\delta_2\},$$

$$\mathcal{P}_G=\{\pi_i\colon \tau_i \geqslant \tau_0+\delta_2\}.$$

On the basis of the observations we want to partition $\mathcal{P}$ into two disjoint sets S_B and S_G. As before, a correct decision (CD) is made if $\mathcal{P}_B \subset S_B$ and $\mathcal{P}_G \subset S_G$. We are interested in procedures which will guarantee that $P(\text{CD}) \geqslant P^*$, where $2^{-k}<P^*<1$.

Let X_{ij} denote the jth observation from π_i, $i=0,1,\ldots,k$; $j=1,2,\ldots$. Define

$$Y_{im}=\sum_{j=1}^{m} P(X_{ij}), \qquad i=0,1,\ldots,k; \qquad m=1,2,\ldots.$$

The sampling for the procedure R_1 of Perng and Groves (1977) starts with taking one observation from each of the $k+1$ populations. At any stage m $(m=1,2,\ldots)$,

$$\begin{aligned}&\text{put } \pi_i \text{ in } S_G \text{ if } Y_{im}-Y_{om} \geqslant C_1,\\ &\text{put } \pi_i \text{ in } S_B \text{ if } Y_{im}-Y_{om} \leqslant -C_2,\\ &\text{make no decision about } \pi_i \text{ if } -C_2<Y_{im}-Y_{om}<C_1,\end{aligned} \tag{20.67}$$

where $C_i=\delta_i^{-1}\log[k/(1-P^*)]$, $i=1,2$. At stage $m+1$, we take one observation from each of π_0 and those other populations for which no decision has been made. The procedure is terminated when all the populations are classified.

The previous procedure R_1 terminates with probability 1 and $P(\text{CD}|R_1) \geqslant P^*$ for all $\tau=(\tau_1,\tau_2,\ldots,\tau_k)$. Perng and Groves (1977) have proposed another sequential procedure R_2 for the same problem based on the approach of Girshick [1946]. The procedure R_2 is obtained from R_1 by replacing C_1 and C_2 in (20.67) with $C=\delta^{-1}\log[(k-1+P^*)/(1-P^*)]$ where $\delta=\min(\delta_1,\delta_2)$. This procedure also guarantees the probability requirement. If $\delta_1=\delta_2$ $(=\delta)$, the procedure R_2 is superior to R_1 in the sense that it requires a smaller number of observations for its termination.

20.7 LEHMANN'S FORMULATION

Let $\pi_1,\ldots,\pi_k$ be k given populations where the quality of π_i is characterized by a real-valued parameter θ_i, $i=1,\ldots,k$. A population is said to be *positive* (or *good*) if $\theta_i \geqslant \theta_0+\Delta$, and *negative* (or *bad*) if $\theta_i \leqslant \theta_0$, where Δ is a given positive constant and θ_0 is either a given number or a parameter that may be estimated. We wish to select a subset of the k populations in some desirable way. A negative population included in the selected subset is called a *false positive*, whereas a positive population included in the set is called a *true positive*. Roughly speaking, the aim of a selection procedure is to seek out the true positive while holding false positives to a minimum.

For measuring the effectiveness of a procedure in identifying the positive populations, a number of criteria can be used, namely,

a. the expected number of true positives,
b. the expected proportion of true positives,
c. the probability of at least one true positive,
d. the probability of including in the selected group the best population (i.e., the population associated with the largest θ_i), provided it is positive,
e. the probability of including all good populations.

The criteria (a) and (b) are appropriate if the aim is to include in the selected subset as many of the positive populations as possible. If the selection is only a step in a scheme, of which the eventual aim is the selection of a single population, then (c) and (d) may be appropriate.

As a measure of the performance of a procedure with respect to false positives we shall take either (i) the expected number of false positives or (ii) the expected proportion of false positives.

For a given $\boldsymbol{\theta}=(\theta_1,\ldots,\theta_k)$ and a selection procedure δ, let $S(\boldsymbol{\theta},\delta)$ denote any one of the quantities (a) through (e) and $R(\boldsymbol{\theta},\delta)$ denote either (i) or (ii). It should be noted that S is defined only for the subset Ω' of the parameter space Ω for which at least one of the populations is positive. Then the problem is to find a procedure δ for which $\sup_{\Omega} R(\boldsymbol{\theta},\delta)$ is minimized subject to

$$\inf_{\Omega'} S(\boldsymbol{\theta},\delta) \geqslant \gamma. \tag{20.68}$$

In the dual problem, $\inf S(\boldsymbol{\theta},\delta)$ is maximized subject to an upper bound on $\sup R(\boldsymbol{\theta},\delta)$. It is interesting to note that the condition (20.68) with S given by (b), that is,

$$\inf_{\Omega'} \{\text{expected proportion of true positives}\} \geqslant \gamma,$$

implies

$$\inf_{\Omega'} \Pr\{\text{at least one true positive}\} \geqslant \gamma.$$

20.7.1 A Minimax Solution

The following theorem of Lehmann (1961) gives a minimax solution in the case of certain families of distributions.

Theorem 20.3. Let the probability density of $\mathbf{X}$ be denoted by p_0 when $\theta_1 = \cdots = \theta_k = \theta_0$, and by p_i when $\theta_i = \theta_0 + \Delta$ and the parameters θ_j for $j \neq i$ have a common value $\theta' \leqslant \theta_0 + \Delta$ determined so that the following conditions are satisfied. Suppose that $p_i(\mathbf{x})/p_0(\mathbf{x})$ is a nondecreasing function of a real-valued statistic T_i and that the distribution of T_i depends only on θ_i, is stochastically increasing in θ_i, and is independent of i. Then the procedure δ_0 that minimizes $\sup_\Omega R(\boldsymbol{\theta}, \delta)$ subject to (20.68) with S equal to any one of the quantities (a)–(d) and R defined by (i) or (ii), is given by

$$\psi_i = 1, \lambda_0, 0 \quad \text{as} \quad T_i >, =, < C, \tag{20.69}$$

where λ_0 and C are determined by

$$E_{\theta_0 + \Delta} \psi_i = \gamma. \tag{20.70}$$

The solution to the dual problem in which (20.68) is replaced by

$$\sup_{\Omega} R(\boldsymbol{\theta}, \delta) \leqslant \gamma' \tag{20.71}$$

is also given by (20.69), with λ_0 and C now determined by

$$R(\boldsymbol{\theta}_0, \delta_0) = \gamma' \tag{20.72}$$

where $\boldsymbol{\theta}_0 = (\theta_0, \ldots, \theta_0)$.

The preceding theorem readily applies to the case of independent samples $X_{i1}, \ldots, X_{in}$ from populations with probability density f_{θ_i} depending only on the real-valued parameter θ_i with respect to which we wish to select, if there exists a sufficient statistic T_i for $(X_{i1}, \ldots, X_{in})$ with monotone likelihood ratio. Lehmann (1961) has also shown that, if the criterion (e) is used instead of one of the criteria (a)–(d), then the minimax solution is no longer given by (20.69). The case of unequal sample sizes requires only a slight generalization of Theorem 20.3.

20.7.2 Normal Populations with Common Unknown Variance

Let $X_{ij}, j=1,\ldots,n_i$, be independent observations from π_i which is $N(\mu_i,\sigma^2)$, $i=1,\ldots,k$. We wish to select the populations with large values of $\theta_i=\mu_i/\sigma$. More specifically, we consider π_i to be negative if $\theta_i \leqslant 0$ and positive if $\theta_i \geqslant \Delta$. In this case we do not get a minimax solution. However, Lehmann (1961) has shown that the intuitive procedure which selects π_i if $\bar{X}_i/[\Sigma\Sigma(X_{ij}-\bar{X}_i)^2]^{1/2} \geqslant c_i$ is approximately minimax under the assumption that $\theta_i \geqslant 0$ for all i in the sense that subject to (20.68), the maximum expected number of false positives can at least be improved only slightly over its value for the intuitive procedure.

D. Y. Huang (1974) has considered the problem of finding an optimal procedure to seek out the true positives while controlling false positives. Let $\delta_i(\mathbf{x})$ denote the individual selection probability for π_i, $i=1,\ldots,k$. As in Theorem 20.3, we assume that $p_i(\mathbf{x})/p_0(\mathbf{x})$ is a nondecreasing function of a real-valued statistic T_i whose distribution depends only on θ_i and is stochastically increasing in θ_i. Then Huang (1974) has shown that the procedure given by $\delta_i = 1, \lambda_i, 0$ as $T_i >, =, < c_i$ maximizes $\inf_{\boldsymbol{\theta}\in\Omega_G}\Sigma_{i=1}^k E_{\boldsymbol{\theta}^{(i)}}\delta_i(\mathbf{X})$ among all rules for which $\Sigma_{i=1}^k E_{\boldsymbol{\theta}_0}\delta_i(\mathbf{X})=\gamma$, where $\Omega_G=\{\boldsymbol{\theta}:\ \theta_i \geqslant \theta_0+\Delta,\ i=1,\ldots,k\}$, $\boldsymbol{\theta}_0=(\theta_0,\ldots,\theta_0)$, and $\boldsymbol{\theta}^{(i)}$ is a parameter vector $\boldsymbol{\theta}$ which has $\theta_i \geqslant \theta_0+\Delta$ and $\theta_j<\theta_0+\Delta$ for all $j\neq i$.

20.8 COMPARISON OF MULTIVARIATE NORMAL POPULATIONS

Let π_i $(i=1,\ldots,k)$ be a p-variate normal population with mean vector $\boldsymbol{\mu}_i$ and covariance matrix Σ_i. The control population π_0 is $N_p(\boldsymbol{\mu}_0,\Sigma_0)$. In this section we discuss selection of a subset of the k populations which contains all positive populations (suitably defined) with a guaranteed minimum probability. We also consider partitioning the set of k populations with respect to a control in terms of generalized variance under the indifference zone setup used by Tong (1969).

20.8.1 Comparisons in Terms of Linear Combinations of the Elements of the Mean Vectors

Krishnaiah and Rizvi (1966) have considered comparison of k multivariate normal populations with a control defining positive and negative populations using different criteria based on linear combinations of the elements of the mean vectors. In all these cases, let $P(\omega,\delta)$ denote the probability of including all positive populations, $S(\omega,\delta)$ the expected proportion of true positives, and $R(\omega,\delta)$ the expected proportion of false positives in the selected subset; ω and δ are used as generic notations for a point in the appropriate parameter space and the selection procedure applied, respectively. It should be noted that $P(\omega,\delta)$ and $S(\omega,\delta)$ are defined only for the set of parameters for which there is at least one positive population. We seek a procedure δ subject to

$$\inf_{\omega} P(\omega,\delta) \geqslant P^*, \tag{20.73}$$

or

$$\inf_{\omega} S(\omega,\delta) \geqslant p^*, \tag{20.74}$$

where P^* and p^* are given constants. The performance of δ is measured by $\sup_{\omega} R(\omega,\delta)$. The selection procedures considered are of the form: Select π_i if $T_{ic} \geqslant d$ for $c=1,\ldots,r$, where T_{ic} is a suitable function of the sufficient statistics from π_i and π_0, and d is a constant determined subject to (20.73) or (20.74).

Let $\theta_{ic} = \mathbf{a}_c'\boldsymbol{\mu}_i$ $(c=1,\ldots,r;\ i=0,1,\ldots,k)$, where $\mathbf{a}_1,\ldots,\mathbf{a}_r$ are specified vectors whose elements can be considered as economic weights assigned to the elements of the mean vectors. The different definitions of positive and negative populations adopted by Krishnaiah and Rizvi (1966) are as follows.

	Positive	*Negative*
A.	$\theta_{ic} \geqslant \theta_{0c} + \Delta_c,\ c=1,\ldots,r$	$\theta_{ic} \leqslant \theta_{0c},\ c=1,\ldots,r$
	where Δ_c are given positive constants.	
B.	$(\theta_{ic} - \theta_{0c})^2 \geqslant \Delta_{1c},\ c=1,\ldots,r$	$(\theta_{ic} - \theta_{0c})^2 \leqslant \Delta_{2c},\ c=1,\ldots,r,$
	where $\Delta_{1c} \geqslant \Delta_{2c} > 0$ are known constants.	
C.	$\lvert\theta_{ic}\rvert \geqslant \lvert\theta_{0c}\rvert,\ c=1,\ldots,r$	$\lvert\theta_{ic}\rvert < \lvert\theta_{0c}\rvert,\ c=1,\ldots,r.$

As we have said earlier, the rule δ proposed by Krishnaiah and Rizvi (1966) in each of these cases includes π_i $(i=1,\ldots,k)$ in the selected subset if $T_{ic} \geqslant d$ for $c=1,\ldots,r$, where the appropriate choice for T_{ic} in each case is described as follows.

Let a sample of size n_i be taken from π_i $(i=0,1,\ldots,k)$ and let $\overline{\mathbf{X}}_i$ denote the sample mean. Define

$$\begin{aligned} U_{ic} &= \frac{\mathbf{a}_c'(\overline{\mathbf{X}}_i - \overline{\mathbf{X}}_0)}{\left[\mathbf{a}_c'(n_i^{-1}\Sigma_i + n_0^{-1}\Sigma_0)\mathbf{a}_c\right]^{1/2}} \\ W_{ic} &= \frac{n^{1/2}\left[\,\lvert\mathbf{a}_c'\overline{\mathbf{X}}_i\rvert - \lvert\mathbf{a}_c'\overline{\mathbf{X}}_0\rvert\,\right]}{\left[\mathbf{a}_c'\Sigma\mathbf{a}_c\right]^{1/2}} \end{aligned} \tag{20.75}$$

where W_{ic} is defined when $n_i=n$ and $\Sigma_i=\Sigma$ for all $i=0,1,\ldots,k$. Then the statistic T_{ic} is given by

$$T_{ic}=\begin{cases} U_{ic} \text{ in Case } A, \\ U_{ic}^2 \text{ in Case } B, \\ W_{ic} \text{ in Case } C. \end{cases} \tag{20.76}$$

In Case A, a lower bound for $\inf P(\omega,\delta)$ is given; solving for d by equating this lower bound to P^* is difficult unless the covariance matrices are assumed to have special structures. In Cases B and C, lower bounds for $\inf P(\omega,\delta)$ have been obtained by using Bonferroni's inequality. Krishnaiah and Rizvi (1966) have no results regarding $\sup R(\omega,\delta)$.

In another paper, Krishnaiah (1967) has considered the problem of selecting multivariate normal populations better than a control on the basis of linear combinations of the elements of covariance matrices.

20.8.2 Comparisons Based on Distance Functions

Krishnaiah and Rizvi (1966) have also considered positive and negative populations defined through distance functions as follows.

	Positive	*Negative*
D.	$(\mathbf{a}'\boldsymbol{\mu}_i-\mathbf{a}'\boldsymbol{\mu}_0)^2\geqslant\Delta_1$	$(\mathbf{a}'\boldsymbol{\mu}_i-\mathbf{a}'\boldsymbol{\mu}_0)^2\leqslant\Delta_2,$
	for $\mathbf{a}\neq\mathbf{0}$, where $\Delta_1\geqslant\Delta_2>0$ are known constants.	
E.	$\boldsymbol{\mu}_i'\Sigma^{-1}\boldsymbol{\mu}_i\geqslant\boldsymbol{\mu}_0'\Sigma^{-1}\boldsymbol{\mu}_0$	$\boldsymbol{\mu}_i'\Sigma^{-1}\boldsymbol{\mu}_i<\boldsymbol{\mu}_0'\Sigma^{-1}\boldsymbol{\mu}_0,$
	assuming that $\Sigma_i=\Sigma$ for all i.	

Their procedure includes π_i in the selected subset if

$$(\overline{\mathbf{X}}_i-\overline{\mathbf{X}}_0)'\left[n_i^{-1}\Sigma_i+n_0^{-1}\Sigma_0\right]^{-1}(\overline{\mathbf{X}}_i-\overline{\mathbf{X}}_0)\geqslant d \text{ in Case D, and if}$$

$$\overline{\mathbf{X}}_i'\Sigma^{-1}\overline{\mathbf{X}}_i\geqslant d\overline{\mathbf{X}}_0'\Sigma^{-1}\overline{\mathbf{X}}_0,\ d>1, \text{ in Case E.}$$

These procedures have not been investigated thoroughly. In Case E, it is also assumed that $\boldsymbol{\mu}_0'\Sigma^{-1}\boldsymbol{\mu}_0<\tau$, where τ is known. In these, as well as the earlier cases, the procedures are modified in the usual way if the Σ_i are unknown.

20.8.3 Partitioning a Set of Multivariate Normal Populations

W. T. Huang (1973a) considered the problem of partitioning a set of p-variate normal populations with respect to a control when the comparison is made in terms of their generalized variances. The formulation is similar to that of Tong (1969). Given $0<\rho_1<1$, and $\rho_2>\rho_1$, the set $\mathcal{P}=\{\pi_1,\ldots,\pi_k\}$ is partitioned into the subsets $\mathcal{P}_G=\{\pi_i\colon |\Sigma_i|\leqslant\rho_1|\Sigma_0|\}$, $\mathcal{P}_I=\{\pi_i\colon \rho_1|\Sigma_0|<|\Sigma_i|<\rho_2|\Sigma_0|\}$, and $\mathcal{P}_B=\{\pi_i\colon |\Sigma_i|\geqslant\rho_2|\Sigma_0|\}$. A typical decision corresponds to a partition of $\mathcal{P}$ into two sets S_G and S_B based on samples of equal size n from all the $(k+1)$ populations. A correct decision occurs when $\mathcal{P}_B\subset S_B$ and $\mathcal{P}_G\subset S_G$. Huang (1973a) considered a single-stage as well as a sequential procedure.

Let S_i denote the sample covariance matrix based on the sample observations from π_i, $i=0,1,\ldots,k$. Then the procedure R_1 is given by

$$\begin{aligned} S_G&=\{\pi_i\colon |S_i|<c|S_0|\},\\ S_B&=\{\pi_i\colon |S_i|\geqslant c|S_0|\}, \end{aligned} \tag{20.77}$$

where $c=c(n,k,p,P^*)>0$ is to be determined so that $P(\mathrm{CD}|R_1)\geqslant P^*$. The LFC is of the type

$$|\Sigma_i|=\begin{cases}\rho_1|\Sigma_0|, & i=1,\ldots,q,\\ \rho_2|\Sigma_0|, & i=q+1,\ldots,k.\end{cases} \tag{20.78}$$

Let $\Sigma_0(q)$ denote the preceding configuration. Of course, $q=0$ means $|\Sigma_i|=\rho_2|\Sigma_0|$, $i=1,\ldots,k$. The value of $P(\mathrm{CD}|R_1)$ for the LFC can be evaluated using the fact that $A_i=(n-1)^p|S_i|/|\Sigma_i|$ is distributed as the product of p independent random variables which have a chi-square distribution with $(n-1),(n-2),\ldots,$ and $(n-p)$ degrees of freedom. The exact distribution of A_i is known when $p=1$ or 2 (see §14.3.2). In these two cases, Huang (1973a) has obtained a lower bound for $\min_{0\leqslant q\leqslant k}P(\mathrm{CD}|R_1;\Sigma_0(q))$. For $p\geqslant 3$, we can use, as noted earlier in §14.3.2, the approximation for the distribution of $A_i^{1/p}$ suggested by Hoel [1937]. This approximation becomes less accurate as p increases. When p is large, we can use the fact that $\sqrt{n}\,[(|S_i|/|\Sigma_i|)-1]$ is asymptotically normally distributed with mean zero and variance $2p$. Thus, for large n,

$$\inf_{\Omega} P(\mathrm{CD}|R_1)\approx$$

$$\min_{0\leqslant q\leqslant k}\int_{-\infty}^{\infty}\Phi^q\left[\lambda x+\sqrt{\frac{n}{2p}}\,(\lambda-1)\right]\left[1-\Phi\left(\frac{x}{\lambda}+\frac{\sqrt{n}\,(\lambda-1)}{\lambda^2\sqrt{2p}}\right)\right]^{k-p} d\Phi(x), \tag{20.79}$$

where $\lambda^2 = \rho_2/\rho_1$. It is only conjectured that the minimum on the right-hand side of (20.79) is attained for $q = k$. The procedure R_1 is minimax in the class of invariant rules.

For the preceding problem, Huang (1973a) has also considered truncated sequential procedures when $p = 1, 2$. For given ρ_1, ρ_2 and P^* such that $0 < \rho_1 < 1$, $\rho_2 > \rho_1$, and $2^{-k} < P^* < 1$, let τ be a value such that $1 < \tau < \min(\rho_2^2, \rho_1^{-2})$, $\alpha = (1 - P^*)/2k$, and $\beta_n = \alpha^{1/(n-1)}$, $n \geqslant 2$. Define

$$
\begin{aligned}
A_n &= \rho_2^2(\lambda\beta_n - 1)/\lambda(\lambda - \beta_n),\\
B_n &= \rho_1^2\lambda(\lambda - \beta_n)/(\lambda\beta_n - 1),\\
m_0 &= \left[\log\alpha^{-1}/\log\lambda\right] + 2,\\
m(A) &= 1 + \left[\log\alpha^{-1}/\log\{\lambda(\rho_2^2 + 1)/(\lambda^2 + \rho_2^2)\}\right],\\
m(B) &= 1 + \left[\log\alpha^{-1}/\log\{(\lambda + \rho_1^2\lambda)/(\rho_1^2\lambda^2 + 1)\}\right],\\
m_1 &= \min\{m(A), m(B)\},\\
m_2 &= \min\{n | A_n \geqslant B_n \quad \text{for} \quad m_1 \leqslant n \leqslant m_3\},\\
m_3 &= \max\{m(A), m(B)\},
\end{aligned}
\tag{20.80}
$$

where $[\cdot]$ denotes the integral part.

Suppose that $m_1 = m(A)$. Then, the procedure $R_2(\lambda)$ defined by Huang (1973a) is as follows.

1. At each stage of sampling, we always sample from π_0. At first stage, draw m_0 observations from each of the $(k+1)$ populations. Let $S_{im_0}^2$ denote the sample variance for π_i. Put π_i in S_G or S_B according as $S_{im_0}^2 < A_{m_0} S_{0m_0}^2$ or $S_{im_0}^2 > B_{m_0} S_{0m_0}^2$. If all populations are classified, the sampling is terminated and the disjoint sets S_G and S_B are obtained. Otherwise, draw one more observation from those populations that are not yet classified. Any π_i which is not classified in the first stage is now put in S_G or S_B according as $S_{i,m_0+1}^2 < A_{m_0+1} S_{0,m_0+1}^2$ or $S_{i,m_0+1}^2 > B_{m_0+1} S_{m_0+1}^2$. If there are populations yet to be classified, we continue sampling from those remaining ones. This procedure is continued until all the populations are classified or until $(m_1 - m_0)$ stages, whichever occurs first.

2. At stage $(m_1 - m_0 + 1)$, take one more observation from those populations that are not yet classified. If π_i is sampled at this stage, include it in S_G or S_B according as $S_{im_1}^2 \leqslant S_{0m_1}^2$ or $S_{im_1}^2 > B_{m_1} S_{0m_1}^2$.

3. If all the populations are not yet classified after the stage (m_1-m_0+1), then take (m_2-m_1) observations from the remaining populations and assign π_i to S_B if $S^2_{im_2}>B_{m_2}S^2_{0m_2}$. This is continued in an obvious manner taking one observation at a time from each remaining population until the stage (m_3-m_0+1).

4. At stage (m_3-m_0+1), one more observation is drawn from those populations that are yet to be classified. Any π_i which is sampled now is put in S_B if $S^2_{im_3}>S^2_{0m_3}$ and in S_G otherwise.

Some obvious modifications are to be made in the description of $R_2(\lambda)$ when $m_1=m(B)$. Huang (1973a) has shown that the probability requirement on the correct decision is met when $p=1$ and 2. It is also shown that the procedure has desirable monotonicity properties.

20.9 Γ-MINIMAX AND BAYES PROCEDURES

In this section we first discuss a Γ-minimax procedure investigated by Randles and Hollander (1971) for selecting treatment populations that have larger translation parameters than that of a control population assuming that the associated density functions possess the monotone likelihood ratio property. Both known and unknown control populations are considered. The Γ-minimax rule has been compared with a Bayes competitor based on independent normal priors for the case of normal populations with common known variance. We also discuss other Bayes procedures for comparison with a standard and for partitioning a set of normal populations.

20.9.1 Γ-Minimax Procedures for Translation Parameter Family

Let π_i $(i=0,1,\ldots,k)$ have the associated density function $f_i(x-\theta_i)$ which is assumed to have a monotone likelihood ratio in its parameter. We define π_i $(i=1,\ldots,k)$ to be positive if $\theta_i \geqslant \theta_0+\Delta$ and negative if $\theta_i \leqslant \theta_0$ where Δ is a specified positive constant. The objective is to select all positive populations while rejecting all negative ones. Let L_1 denote the loss incurred if we fail to select a positive population and L_2, the loss for each negative population selected. We consider decision rules of the form $\delta(\mathbf{x})=(\delta_1(\mathbf{x}),\ldots,\delta_k(\mathbf{x}))$ where $\delta_i(\mathbf{x})$ denotes the conditional probability of selecting π_i given the observation $\mathbf{x}=(x_0,x_1,\ldots,x_k)$. The loss function is then

$$L(\boldsymbol{\theta},\delta)=\sum_{i=1}^{k} L^{(i)}(\boldsymbol{\theta},\delta), \tag{20.81}$$

where

$$L^{(i)}(\boldsymbol{\theta},\delta)=\begin{cases} L_1(1-\delta_i) & \text{if } \theta_i \geqslant \theta_0+\Delta, \\ L_2\delta_i & \text{if } \theta_i \leqslant \theta_0, \\ 0 & \text{otherwise.} \end{cases} \tag{20.82}$$

The risk function $R(\boldsymbol{\theta},\delta)=L_1N_1+L_2N_2$, where $N_1(N_2)$ is the expected number of positive (negative) populations rejected (selected). Randles and Hollander (1971) assumed the following partial prior information regarding the family of prior distributions of $\boldsymbol{\theta}$. Let $\Omega_P(i)=\{\boldsymbol{\theta}|\theta_i \geqslant \theta_0+\Delta\}$ and $\Omega_N(i)=\{\boldsymbol{\theta}|\theta_i \leqslant \theta_0\}$. It is assumed that we are able to specify $q_i=\Pr\{\boldsymbol{\theta}\in\Omega_P(i)\}$ and $q_i'=\Pr\{\boldsymbol{\theta}\in\Omega_N(i)\}$ so that $q_i+q_i'\leqslant 1$ for $i=1,\ldots,k$. The subset Γ of the set of all prior distributions $\tau(\boldsymbol{\theta})$ over Ω is defined by

$$\Gamma=\left\{\tau(\boldsymbol{\theta})|\int_{\Omega_P(i)} d\tau(\boldsymbol{\theta})=q_i \quad\text{and}\quad \int_{\Omega_N(i)} d\tau(\boldsymbol{\theta})=q_i' \quad\text{for}\quad i=1,\ldots,k\right\}. \tag{20.83}$$

The Γ-minimax rule minimizes the maximum of the Bayes risk over Γ.

Case I: Known Control Population. In this case Randles and Hollander (1971) have shown that the Γ-minimax rule $\delta^0=(\delta_1^0,\ldots,\delta_k^0)$ is of the form $\delta_i^0=1$ or 0 as $x_i \geqslant$ or $<d_i$ for $i=1,\ldots,k$, where each d_i is determined so that

$$L_2q_i'f_i(x-\theta_0)-L_1q_if_i(x-\theta_0-\Delta)\leqslant \text{ or } >0 \tag{20.84}$$

according as $x \geqslant$ or $<d_i$.

It should be noted that the procedure depends on the prior probability $q_i(q_i')$ that π_i is positive (negative) only through the ratio q_i/q_i'.

Suppose π_i is a normal population with mean θ_i and variance σ^2. Assume that θ_0 and σ^2 are known. In this case, the Γ-minimax rule selects π_i if and only if

$$\overline{X}_i \geqslant \theta_0+\frac{\Delta}{2}+\sigma^2(\Delta n_i)^{-1}\log\left(\frac{L_2q_i'}{L_1q_i}\right), \tag{20.85}$$

where $\overline{X}_i$ is the mean of a sample of size n_i from π_i, $i=1,\ldots,k$.

Case II: Unknown Control Population. Let X_i be an observation from π_i, $i=0,1,\ldots,k$, and let $g_i(y-(\theta_i-\theta_0))$ denote the density of $Y_i=X_i-X_0$. The Γ-minimax rule in this case is of the form: $\delta_i^0=1$ or 0 according as

$Y_i \geqslant$ or $<d_i'$ where the d_i' are determined so that

$$L_2 q_i' g_i(y) - L_1 q_i g_i(y-\Delta) \leqslant \text{ or } >0 \tag{20.86}$$

according as $y \geqslant$ or $<d_i'$. See also Miescke (1979b) regarding the proof.

For the normal means problem discussed earlier, the Γ-minimax rule selects π_i if and only if

$$\bar{X}_i - \bar{X}_0 \geqslant \frac{\Delta}{2} + \frac{\sigma^2}{\Delta}(n_i^{-1} + n_0^{-1})\log\left(\frac{L_2 q_i'}{L_1 q_i}\right), \tag{20.87}$$

where $\bar{X}_i$ is the mean of a sample of size n_i from π_i, $i=0,1,\ldots,n$.

If a priori considerations yield a class of prior distributions over Ω, one can utilize such information by selecting a member of the class and using the corresponding Bayes decision rule. Alternatively, one can find a rule that is Γ-minimax with respect to the class of priors given. Thus Bayes rules corresponding to prior distributions in Γ are natural competitors for a Γ-minimax rule.

Suppose π_i $(i=0,1,\ldots,k)$ is a normal population with mean θ_i and variance σ^2. We assume σ^2 to be known. Let $\theta_0, \theta_1, \ldots, \theta_k$ have independent normal priors $\tau_i(\theta_i)$ with known mean α_i and variance β_i^2. Let us also assume that $L_1 = L_2 = 1$. Then the Bayes procedure $\delta^* = (\delta_1^*, \ldots, \delta_k^*)$ is given by

$$\delta_i^* = \begin{cases} 1 & \text{if } a_i - a_0 \geqslant \Delta/2, \\ 0 & \text{if } a_i - a_0 < \Delta/2, \end{cases} \tag{20.88}$$

where

$$a_i = \frac{(n_i \bar{x}_i \sigma^{-2} + \alpha_i \beta_i^{-2})}{(n_i \sigma^{-2} + \beta_i^{-2})}.$$

Under these independent normal priors, we know q_i and q_i'. Randles and Hollander (1971) have compared the Γ-minimax rule and the Bayes procedure given in (20.88) for three distributions in Γ in terms of the expected risks and have found the performance of the Γ-minimax rule satisfactory.

20.9.2 Bayes Procedures for Comparison with a Standard

Here we continue the general setup of §20.2.4. The comparison of the populations with the standard (θ_0) is in terms of the parameters θ_i. The X_i are suitably chosen statistics. Let $B(\boldsymbol{\theta})$ denote the joint prior distribution for $\boldsymbol{\theta}=(\theta_1,\ldots,\theta_k)$ defined on Ω^k. We also let $L(\boldsymbol{\theta})=(L_0(\boldsymbol{\theta}),L_1(\boldsymbol{\theta}),\ldots,L_k(\boldsymbol{\theta}))$, where $L_i(\boldsymbol{\theta})$ is the loss incurred when population π_i is selected and $\boldsymbol{\theta}$ is the true configuration. We assume that, given $\boldsymbol{\theta}$, the X_i are independent with marginal density functions, $f(\cdot,\theta_i)$, with respect to some measure. For $0 \leqslant j \leqslant k$, we define

$$R_j(\mathbf{x})=\int_{\Omega^k} L_j(\boldsymbol{\theta}) \prod_{i=1}^{k} f(x_i,\theta_i)\,dB(\boldsymbol{\theta})$$

where $\mathbf{x}=(x_1,\ldots,x_n)$. A Bayes rule chooses π_i $(0 \leqslant i \leqslant k)$ if only if

$$R_i(\mathbf{x})=\min_{0 \leqslant j \leqslant k} R_j(\mathbf{x}). \tag{20.89}$$

In case of ties, we use randomization.

Let us consider loss functions which satisfy the following conditions:

A. $L_j(\theta_{\pi 1},\ldots,\theta_{\pi k})=L_{\pi j}(\theta_1,\ldots,\theta_k)$ for $0 \leqslant j \leqslant k$, where π is any permutation of $\{1,2,\ldots,k\}$ and $\pi 0=0$.

B. $L_0(\boldsymbol{\theta})<L_i(\boldsymbol{\theta})$ for $1 \leqslant i \leqslant k$, if $\theta_j=\lambda$ for all j.

C. $L_j(\boldsymbol{\theta})<L_i(\boldsymbol{\theta})=L_0(\boldsymbol{\theta})$, $1 \leqslant i,j \leqslant k$ $(i \neq j)$, if $\theta_t=\lambda$ for $t \neq j$, and $\theta_j=\lambda+\Delta$, where $\Delta \geqslant 0$.

An example (Turnbull (1976)) of such a loss function is:

$$L_0(\boldsymbol{\theta})=\begin{cases} -1 & \text{if } \theta_{[k]} \leqslant \lambda \\ 0 & \text{otherwise,} \end{cases}$$

and

$$L_i(\boldsymbol{\theta})=\begin{cases} -1 & \text{if } \theta_{[k]} > \lambda \\ 0 & \text{otherwise.} \end{cases}$$

Let us now consider symmetric priors B which give weight only to slippage configurations. Specifically, let $B_{\lambda,p,G}$ denote the prior that assigns probability $(1-kp)$ to the configuration with all $\theta_i=\lambda$ and probability p for the configuration $\boldsymbol{\theta}$ with $\theta_j=\lambda+\Delta$ for some $\Delta>0$ and the remaining θ_i equal to λ. Let $G(\Delta)$ be the conditional distribution of Δ given that $\Delta>0$. Recall the natural rule R_7 of Turnbull (1976) discussed in §20.2.4. Turnbull has shown that if $f(x,\theta)$ has a monotone nondecreasing likelihood ratio, then the natural decision rule is admissible when the loss function satisfies the conditions A, B, and C. He has also established that the rule R_7 is Bayes with respect to some symmetric slippage prior $B_{\lambda,p,G}$.

20.9.3 Bayes Procedures for Partitioning a Set of Normal Populations

Let π_i $(i=0,1,\ldots,k)$ be a normal population with mean θ_i and variance σ_i^2. Given $0<\delta_2<\delta_1$, the set $\mathcal{P}=\{\pi_1,\ldots,\pi_k\}$ is partitioned into three subsets: $\mathcal{P}_G=\{\pi_i\colon \theta_i-\theta_0\geqslant\delta_1\}$, $\mathcal{P}_B=\{\pi_i\colon \theta_i-\theta_0\leqslant-\delta_2\}$, and $\mathcal{P}_I=\{\pi_i\colon -\delta_2<\theta_i-\theta_0<\delta_1\}$. Based on the sample observations, the decision is a partition of the set $\mathcal{P}$ into S_G and S_B. A misclassification occurs if a population in $\mathcal{P}_G$ gets assigned to S_B or if a population in $\mathcal{P}_B$ gets assigned to S_G. W. T. Huang (1975a) has obtained a Bayes procedure assuming independent normal priors $N(\lambda_i,\tau_i^2)$, $i=1,\ldots,k$, where the λ_i and τ_i^2 are known and using two different loss functions.

Let $L_{1i}(S_G,\boldsymbol{\theta})$ denote the loss associated with π_i when $\boldsymbol{\theta}$ is the true parameter vector and the decision results in the partition $\mathcal{P}=\{S_G,S_B\}$. Then the loss $L(S_G,\boldsymbol{\theta})$ is given by

$$L(S_G,\boldsymbol{\theta})=\sum_{i=1}^{k} L_{1i}(S_G,\boldsymbol{\theta}), \tag{20.90}$$

where

$$L_{1i}(S_G,\boldsymbol{\theta})=\begin{cases}\alpha & \text{if } \pi_i\in\mathcal{P}_G\cap S_B,\\ \beta & \text{if } \pi_i\in\mathcal{P}_B\cap S_G,\\ 0 & \text{otherwise.}\end{cases} \tag{20.91}$$

An alternative loss function considered by W. T. Huang (1975a) is with $L_{2i}(S_G,\boldsymbol{\theta})$ instead of $L_{1i}(S_G,\boldsymbol{\theta})$ in (20.90) where

$$L_{2i}(S_G,\boldsymbol{\theta})=\begin{cases}\alpha(\theta_i-\theta_0) & \text{if } \pi_i\in\mathcal{P}_G\cap S_B,\\ \beta(\theta_0-\theta_i) & \text{if } \pi_i\in\mathcal{P}_B\cap S_G,\\ 0 & \text{otherwise.}\end{cases} \tag{20.92}$$

The Bayes procedure is admissible. Huang has also considered an empirical Bayes procedure.

20.10 MULTIPLE COMPARISONS WITH A CONTROL: ONE-SIDED AND TWO-SIDED INTERVALS, AND OPTIMAL ALLOCATION OF OBSERVATIONS

In this section we consider $(k+1)$ normal populations, where π_i is $N(\mu_i,\sigma_i^2)$, $i=0,1,\ldots,k$. As before, π_0 is the control population. The problem here is to obtain confidence intervals for $\mu_i-\mu_0$, $i=1,\ldots,k$, such that the joint confidence coefficient is $P^*(0<P^*<1)$. The earliest paper in this direction is that of Dunnett (1955) who considered one-sided and two-sided intervals when $\sigma_0^2=\sigma_1^2=\cdots=\sigma_k^2=\sigma^2$. He posed the problem of optimal allocation of the observations between the control and the treatments but did not really solve it. Bechhofer (1969) and Bechhofer and Nocturne (1972) considered the optimal allocation problem for one-sided and two-sided comparisons, respectively, assuming that the σ_i^2 are known. Their allocation procedure is globally optimal when $\sigma_1^2=\cdots=\sigma_k^2$ and is suboptimal when σ_i^2, $i=1,\ldots,k$, are not all equal. In the latter case, Bechhofer and Turnbull (1971) have obtained the globally optimal procedure for one-sided comparisons. We discuss in this section these procedures as well as those of Sobel and Tong (1971) for two-sided comparisons and Bechhofer (1968b) who considered one-sided comparisons for multiply classified variances of normal populations.

20.10.1 One- and Two-Sided Comparisons: Equal Variances Case

Dunnett (1955) considered the problem of obtaining one- and two-sided confidence intervals for $\mu_i-\mu_0$, $i=1,\ldots,k$, so that the joint confidence coefficient is at least $P^*(0<P^*<1)$. Let $\bar{X}_i$ be the mean of a sample of n_i observations from π_i, $i=0,1,\ldots,k$. The lower joint confidence limits are given by

$$\bar{X}_i-\bar{X}_0-d_i s\sqrt{\frac{1}{n_i}+\frac{1}{n_0}}\,, \qquad i=1,\ldots,k, \tag{20.93}$$

where s^2 is the usual unbiased estimate of σ^2 based on $m=\Sigma_{i=0}^k n_i-(k+1)$ degrees of freedom and the constants d_i are chosen such that

$$\Pr\{t_1<d_1,\ldots,t_k<d_k\}=P^*. \tag{20.94}$$

The joint distribution of the t_i is the multivariate t distribution (see Dunnett and Sobel [1954]). Similar statements can be made with respect to upper confidence limits and two-sided limits. If $n_1=\cdots=n_k=n$, then the t_i are equicorrelated with correlation $\rho=n/(n+n_0)$. Further if $n=n_0$, the correlation is 0.5. Percentage points of the bivariate t are given by Dunnett and Sobel [1954] for $\rho=\pm 0.5$. Tables of Krishnaiah and Armitage [1966] give the equi-percentage point of the multivariate t for several values of the parameters and constants involved. These tables are discussed in §23.3.

Dunnett (1955) has tabulated the values of d, the common value of $d_1,\ldots,d_k$ satisfying (20.94) for $k=1(1)9$, $m=5(1)20$, 24, 30, 40, 60, 120, ∞ and $P^*=0.95$, 0.99. In another set of tables, Dunnett has tabulated the values of the constant for two-sided comparisons for the same values of k,

m, and P^* as previously. But the values are not exact, as a result of an approximation, for more than two comparisons. In a later paper, Dunnett (1964) has given the exact values for $k=1(1)12$, 15, 20 and the same values of m and P^* as before.

Dunnett (1955) also posed the problem of optimal allocation of observations between the control and treatments. The optimality is in the sense of maximizing the confidence coefficient for fixed $N=\Sigma_{i=0}^{k} n_i$. For the case of one-sided comparisons with $n_1=\cdots=n_k$, Dunnett's numerical calculations of the effect of different allocations on the confidence coefficient show that if the experimenter is working with P^* in the neighborhood of 0.95 or more, then he should design such that $n_0=n_1\sqrt{k}$, a result recommended earlier by Fieller [1944] and Finney [1952].

When σ^2 is known, we replace s by σ in (20.93). If $d_1=\cdots=d_k=d$, then d is an equi-percentage point of a k-variate normal distribution. For the equicorrelation case, the values of d can be obtained from the tables of Gupta, Nagel, and Panchapakesan [1973].

20.10.2 One- and Two-Sided Comparisons: Unequal Known Variances

Bechhofer (1969) considered the optimal allocation problem for one-sided comparisons. Let $\bar{X}_i$ denote the mean of a sample of size n_i from π_i $(i=0,1,\ldots,k)$. Define

$$\theta_i=\frac{\sigma_i^2}{\sigma_0^2}, \qquad 1\leqslant i\leqslant k,$$

$$\gamma_i=\frac{n_i}{N}, \qquad 0\leqslant i\leqslant k, \tag{20.95}$$

where $N=\Sigma_{i=0}^{k} n_i$. For $1\leqslant i\leqslant k$, we also define

$$Z_i=\left[(\bar{X}_0-\bar{X}_i)-(\mu_0-\mu_i)\right]\left(\frac{\sigma_0^2}{n_0}+\frac{\sigma_i^2}{n_i}\right)^{-1/2} \tag{20.96}$$

and consider

$$\Pr\{Z_i<d', i=1,\ldots,k\}=P^*. \tag{20.97}$$

The Z_i are correlated $N(0,1)$ random variables with

$$\operatorname{cov}(Z_i,Z_j)=\left[\left(1+\frac{\theta_i\gamma_0}{\gamma_i}\right)\left(1+\frac{\theta_j\gamma_0}{\gamma_j}\right)\right]^{-1/2}.$$

Assuming that $d'=d'(k,P^*)$ has been calculated, equations (20.96) and (20.97) permit one to make joint lower (or equivalently) upper confidence interval estimates, with confidence coefficient P^*, of the k differences $\mu_0-\mu_i$ $(i=1,\ldots,k)$.

Suppose we now set

$$d'\left(\frac{\sigma_0^2}{n_0}+\frac{\sigma_i^2}{n_i}\right)^{1/2}=d, \tag{20.98}$$

where d is regarded as a specified "yardstick" and define $\lambda=d\sqrt{N}/\sigma_0$. Then the left-hand side of (20.97) becomes

$$\Pr\left\{Z_i<\lambda\left(\gamma_0^{-1}+\theta_i\gamma_i^{-1}\right)^{-1/2}, \qquad i=1,\ldots,k\right\} \tag{20.99}$$

where λ is a pure number, the θ_i are the known variance ratios, and γ_i $(1\leqslant i\leqslant k)$ is the proportion of the total number of observations taken from π_i.

For fixed k, λ, and values of the θ_i $(1\leqslant i\leqslant k)$, the problem is to find the optimal allocation $\hat{\gamma}=(\hat{\gamma}_0,\hat{\gamma}_1,\ldots,\hat{\gamma}_k)$, that is, the choice of the γ_i $(\Sigma_{i=0}^k\gamma_i=1)$ for which (20.99) is maximum. Bechhofer (1969) has obtained an explicit formula for determining $\hat{\gamma}$ and studied the behavior of $\hat{\gamma}$ as a function of λ. Some special cases are of interest. In the limiting (as $\lambda\to\infty$) case, we have $\hat{\gamma}_0=(1+\sqrt{\beta}\,)^{-1}$, and $\hat{\gamma}_i=\theta_i/\sqrt{\beta}\,(1+\sqrt{\beta}\,)$, $i=1,\ldots,k$, where $\beta=\Sigma_{i=1}^k\theta_i$. As a consequence, $\hat{\gamma}_0/\hat{\gamma}_i=\sqrt{\beta}\,/\theta_i$. If $\theta_i=1$ $(1\leqslant i\leqslant k)$, that is, if $\sigma_0^2=\sigma_1^2=\cdots=\sigma_k^2$, then $\hat{\gamma}_0/\hat{\gamma}_i=\sqrt{k}$, which leads to the recommendation of Dunnett (1955). Finally, for $k=1$ we get the well-known optimal allocation $\hat{\gamma}_0=\hat{\gamma}_1/\sqrt{\theta_1}$ which holds uniformly in $\lambda(0<\lambda<\infty)$.

One could consider a more general formulation of the problem which would involve replacing d' in (20.98) by d_i' $(1\leqslant i\leqslant k)$ and optimizing simultaneously with respect to the d_i' and the γ_i. An exact solution for this general formulation would be quite difficult to obtain. It can be shown that the exact solution for the general formulation coincides with the exact solution for the restricted formulation only if $\theta_i=\theta$ $(1\leqslant i\leqslant k)$. Thus the results of Bechhofer (1969) are optimal for the restricted formulation (20.98), but are not globally optimal for the more general formulation unless the $\theta_i=\theta$ condition holds.

A similar investigation was carried out by Bechhofer and Nocturne (1972) for the case of two-sided comparisons. It is desired to make an exact confidence statement of the form

$$\overline{X}_0-\overline{X}_i-d<\mu_0-\mu_1<\overline{X}_0-\overline{X}_i+rd, \qquad i=1,\ldots,k, \tag{20.100}$$

where the constants $r>-1$ and $d>0$ are specified prior to experimentation. Bechhofer and Nocturne (1972) have obtained the optimal allocation

$\hat{\gamma}$. Their results will be globally optimal only if $\sigma_1^2 = \cdots = \sigma_k^2$, and will be optimal for the situation of possibly unequal σ_i^2 $(1 \leqslant i \leqslant k)$ only within the class of procedures for which the sample sizes n_i $(1 \leqslant i \leqslant k)$ are chosen so that $\sigma_i^2/n_i = \text{constant}$ for $i = 1,\ldots,k$. The proportion $\hat{\gamma}_i$ to be taken from π_i depends only on θ_i and λ. Bechhofer and Nocturne (1972) have studied analytically the behavior of the $\hat{\gamma}_i$ as functions of r and λ.

It should be noted that the problem of Bechhofer (1969) is obtained as a special case of the general problem by taking $r = \infty$. The results stated earlier concerning the optimal allocation when $\lambda \to \infty$ are true in the general case.

Bechhofer and Nocturne (1972) have tabulated the optimal allocation on the control ($\hat{\gamma}_0$), and the associated maximum probability ($\hat{P}$) for the following values of r and λ when $k = \beta = 2$:

a. $\lambda = 0.5$, 1.0; $r = 0(1)10(5)30$, ∞;
b. $r = 0.5$, 1.0, 1.5, 4, ∞; $\lambda = 0(0.2)7.0(0.5)9.5$, $\sqrt{2/\pi}$, ∞.

Sobel and Tong (1971) have also considered the optimal allocation of observations for the construction of two-sided confidence intervals, assuming that $\sigma_1^2 = \cdots = \sigma_k^2$ and $n_1 = \cdots = n_k$. It is desired to make a confidence statement of the form

$$\bar{X}_0 - \bar{X}_i - d_i \leqslant \mu_0 - \mu_i \leqslant \bar{X}_0 - \bar{X}_i + d_i, \qquad i = 1,\ldots,k. \tag{20.101}$$

The confidence coefficient associated with (20.101) is

$$G(\alpha) = \Pr\{|Z_i| \leqslant d_i, \qquad i = 1,\ldots,k\}, \tag{20.102}$$

where the Z_i are equicorrelated standard normal random variables with correlation $\rho = \rho(\alpha) = \lambda/(\alpha + \lambda)$, $\lambda = \sigma_0^2/\sigma_1^2$ and $\alpha = n_0/n_1$. Let α^* satisfy $G(\alpha^*) = \sup G(\alpha)$. Sobel and Tong (1971) have shown that, for every $k \geqslant 2$, (a) $\alpha^* < \sqrt{k\lambda}$ for every $\mathbf{d} = (d_1,\ldots,d_k)$ and every N, and (b) $\alpha^* \to \sqrt{k\lambda}$ as $N \to \infty$ if $d_1 = \cdots = d_k$. Part (b) of the preceding statement is consistent with the results of Bechhofer and Nocturne (1972). It should, however, be noted that Bechhofer and Nocturne (1972) have followed the multiple comparisons approach of Dunnett (1955, 1964) whereas Sobel and Tong have used the formulation of partitioning into random subsets adopted by Tong (1969) [see §20.6].

As we have remarked earlier, the solution of Bechhofer and Nocturne (1972) is globally optimal if $\sigma_1^2 = \cdots = \sigma_k^2$. When the variances of the "test" populations are not equal, their allocation procedure is optimal within a more restricted class of procedures. The results of Bechhofer and Nocturne have been generalized by Bechhofer and Turnbull (1971) who have obtained an allocation procedure for one-sided comparisons that is globally optimal.

It is desired to make a confidence statement of the form

$$\bar{X}_0 - \bar{X}_i - d < \mu_0 - \mu_i, \qquad i = 1, \ldots, k. \tag{20.103}$$

The formulation of the optimal allocation problem is a discrete one and its solution involves the solution of an integer programming problem. Bechhofer and Turnbull (1971) have given a continuous version of the same problem. For given k, σ_i^2 $(0 \leqslant i \leqslant k)$ and specified N and d, the globally optimal proportion $\hat{\gamma}_i$ depends only on the θ_i and λ. In the special case of $k=2$, it is interesting to note that $\hat{\gamma}_1/\hat{\gamma}_2 \approx \sqrt{\theta_1/\theta_2}$ for λ sufficiently small, whereas $\hat{\gamma}_1/\hat{\gamma}_2 \approx \theta_1/\theta_2$ for λ arbitrarily large.

In the case of $k=2$, Bechhofer and Turnbull (1971) have tabulated the values of the optimal allocation proportions $\hat{\gamma}_0, \hat{\gamma}_1, \hat{\gamma}_2$, and the associated maximum probability for the following combinations of θ_1 and θ_2 and the selected values of λ as shown in the table besides $\lambda = (\sqrt{\theta_1} + \sqrt{\theta_2})/\sqrt{2\pi}$, and ∞.

θ_1	θ_2	λ
1/10	1/10	0.3(0.1)1(0.2)4, 5
1/10	8/10	0.7(0.1)1(0.2)5
1/10	1	0.7(0.1)3(1)8
1/10	10/8	0.7(0.1)3(1)8
1/10	10	1.6(0.2)4(0.5)13
8/10	8/10	0.8(0.1)2(0.2)3(0.5)8
8/10	1	0.9(0.1)2(0.2)3(0.5)8
8/10	10/8	0.9(0.1)2(0.2)3(0.5)8
8/10	10	1.8(0.2)6(1)14
1	1	1(0.2)6(1)8
1	10/8	1(0.2)5(0.4)7(1)9
1	10	1.8(0.2)5(0.4)7(1)14
10/8	10/8	1(0.2)5(0.4)7(0.5)9
10/8	10	1.8(0.2)5(0.4)7(1)14
10	10	3(0.2)5(0.4)7(1)20

20.10.3 Multiple Comparisons with a Control for Multiply Classified Variances of Normal Populations

Bechhofer (1968a) introduced a multiplicative model for variances of normal populations and studied the problem of ranking multiply classified variances, which has been discussed in §2.5. In a companion paper, Bechhofer (1968b) used the multiplicative model as the basis for making multiple comparisons with a control for multifactor experiments involving variances of normal populations. We use here the notation of §2.5. Suppose that the first level of Factor A is the "control" level, and the remaining are "test" levels, that is, α_1 is the effect associated with the control level, and $\alpha_2, \alpha_3, \ldots, \alpha_a$ are the effects associated with the test levels. It is desired to make a joint confidence statement concerning the effect-ratios α_i/α_1 $(i=2,3,\ldots,a)$.

For fixed $a \geqslant 2$, $b \geqslant 1$, $n \geqslant 1$, and for $i=2,3,\ldots,a$, let us define the $a-1$ random variables

$$G_n^{(i),b} = \prod_{j=1}^{b} \left\{ \frac{\chi_n^2(i,j)}{\chi_n^2(1,j)} \right\}, \tag{20.104}$$

where the $\chi_n^2(i,j)$ $(1 \leqslant i \leqslant a;\ 1 \leqslant j \leqslant b)$ are independent chi-square variates, each based on n degrees of freedom. Thus, for fixed i, $G_n^{(i),b}$ is the product of b independent central F-statistics, each based on (n,n) degrees of freedom. Note that for $i \neq i'$, $G_n^{(i),b}$ and $G_n^{(i'),b}$ are correlated. Now, for $0<\lambda<1$, let $G_n^{a,b}(1-\lambda)$ be defined by the equation

$$\Pr\left\{ G_n^{a,b}(1-\lambda) < G_n^{(i),b}, \qquad i=2,3,\ldots,a \right\} = 1-\lambda. \tag{20.105}$$

Let s_{ij}^2 denote the sample variance based on n degrees of freedom associated with the ith level of Factor A and the jth level of Factor B. Then the following statement holds with confidence coefficient $1-\lambda$:

$$\frac{\alpha_i}{\alpha_1} < \left[\frac{1}{G_n^{a,b}(1-\lambda)} \prod_{j=1}^{b} \frac{s_{ij}^2}{s_{1j}^2} \right]^{1/b}, \qquad i=2,\ldots,a. \tag{20.106}$$

Bechhofer (1968b) has also given a two-sided interval estimate of α_2/α_1 when $a=2$. For $0<\lambda_1,\lambda_2;\ \lambda_1+\lambda_2<1$, the following statement holds with confidence coefficient $1-\lambda_1-\lambda_2$:

$$\left[\frac{1}{G_n^{2,b}(\lambda_2)} \prod_{j=1}^{b} \left\{ \frac{s_{2j}^2}{s_{1j}^2} \right\} \right]^{1/b} < \frac{\alpha_2}{\alpha_1} < \left[\frac{1}{G_n^{2,b}(1-\lambda_1)} \prod_{j=1}^{b} \left\{ \frac{s_{2j}^2}{s_{1j}^2} \right\} \right]^{1/b}. \tag{20.107}$$

Bechhofer (1968b) has tabulated the values $G_n^{2,b}(\lambda)$ for $\lambda=0.0005$, 0.001, 0.005, 0.01, 0.025, 0.05, 0.10, 0.15, 0.25 and (i) $b=2$, $n=1(1)6$, 8; (ii) $b=3$, $n=2$, 4.

CHAPTER 21

Estimation of Ordered Parameters

21.1 INTRODUCTION

Let $\pi_i(i=1,2,\ldots,k)$ be a population with the associated distribution function $F(x,\theta_i)$ where θ_i is an unknown real-valued parameter. The ordered θ_i are denoted by $\theta_{[1]}\leqslant\cdots\leqslant\theta_{[k]}$. We are interested in obtaining point and interval estimates of $\theta_{[i]}(1\leqslant i\leqslant k)$. We assume that there is no prior information regarding the correct pairing of the ordered and the unordered parameters. Under different assumptions regarding the distribution functions and the nature of the parameters involved we discuss the problems of maximum likelihood estimators, maximum probability estimators, minimax estimators, and optimal confidence intervals. We also discuss a combined goal of selecting the population associated with $\theta_{[k]}$ and estimating $\theta_{[k]}$.

In some contexts we may assume that the correct pairing of the ordered and the unordered parameters is known but not the values of the parameters themselves. In this case we briefly deal with the theory of isotonic regression and the maximum likelihood estimation.

21.2 POINT ESTIMATION

We first discuss the estimation of the largest parameter $\theta_{[k]}$ when $k=2$ and the θ_i are translation parameters.

21.2.1 Estimation of the Larger Translation Parameter

Let $f(x-\theta_i)$ be the density (with respect to Lebesgue measure) of $\pi_i, i=1,2$. We assume, with no loss of generality, that $\int_{-\infty}^{\infty}xf(x)=0$. Let $X_{i1},\ldots,X_{in}$ be independent observations from $\pi_i, i=1,2$. Let $\psi(a,b)=\max(a,b)$. If $n=1$, a natural estimator of $\psi(\theta_1,\theta_2)$ is $\psi(X_{11},X_{21})$. This estimator is symmetric and invariant under translations which take (θ_1,θ_2) to $(\theta_1+\alpha,\theta_2+\alpha)$ for any real constant α. Under suitable conditions (one of which is that f be symmetric), it can be shown that $\psi(X_{11},X_{21})$ is minimax with a squared error loss. However, it is not in general admissible. Further, when f is not symmetric, it need not be minimax.

Another estimator with the invariance property stated previously is the a posteriori expected value of $\psi(\theta_1,\theta_2)$, given (X_{11},X_{21}), when the generalized prior distribution of (θ_1,θ_2) is taken to be the uniform distribution over the two-dimensional plane. An adaptation such as this of the Pitman estimator is known to be minimax for many estimation problems. However, for the present problem, under suitable conditions (again including the symmetry of f), this estimate need not be minimax. Yet this estimator is admissible.

Blumenthal and Cohen (1968a) have obtained the following results for general n.

Let X_i be the a posteriori expected value of θ_i, given $X_{ij}, j=1,\ldots,n$, when θ_i has the generalized uniform distribution as a prior distribution, that is,

$$X_i = \frac{\int \theta_i \prod_{j=1}^{n} f(X_{ij}-\theta_i)\,d\theta_i}{\int \prod_{j=1}^{n} f(X_{ij}-\theta_i)\,d\theta_i}, \qquad i=1,2. \tag{21.1}$$

It should be noted that X_i is the usual Pitman estimataor of θ_i. Thus, if f is the normal density, then X_i is the sample mean.

Let $\psi(X_{11},\ldots,X_{1n},X_{21},\ldots,X_{2n}) = \psi(X_1,X_2) = \max(X_1,X_2)$. Define δ^* to be the a posteriori expected value of $\psi(\theta_1,\theta_2)$, given $X_{ij}, i=1,2;\ j=1,\ldots,n$, when the general prior distribution of (θ_1,θ_2) is the uniform distribution on the two-dimensional plane, that is,

$$\delta^*(X_{11},\ldots,X_{2n}) = \frac{\int\int \psi(\theta_1,\theta_2) \prod_{j=1}^{n} \left[f(X_{1j}-\theta_1) f(X_{2j}-\theta_2) \right] d\theta_1 d\theta_2}{\int\int \prod_{j=1}^{n} \left[f(X_{1j}-\theta_1) f(X_{2j}-\theta_2) \right] d\theta_1 d\theta_2}. \tag{21.2}$$

Let $Y_i = (Y_{i1}, Y_{i2},\ldots, Y_{i,n-1})$ where $Y_{ij} = X_{i,j+1} - X_{i,j}$. Let $p(x,y)$ denote the conditional density of X_i given Y_i when $\theta_i = 0$. The main results of Blumenthal and Cohen (1968a) are summarized in the following theorem.

Theorem 21.1

a. If $p(x,y) = p(-x,y)$ and $E\{E((X_1^2+X_2^2)|Y_1,Y_2)\} < \infty$, then $\psi(X_1,X_2)$ is minimax.

b. The estimator $\psi(X_1,X_2)$ need not in general be admissible.

c. If $E[E^2\{(X_1^2+X_2^2)|\log(X_1^2+X_2^2)|^{\beta}|Y_1,Y_2\}] < \infty$ for some $\beta > 0$, then δ^* is admissible.

d. δ^* need not be minimax.

REMARK 21.1 If $p(x,y) \neq p(-x,y)$, then the problem is complicated. Blumenthal and Cohen (1968a) have given an example in which $\psi(X_1,X_2)$ is not minimax. They have also shown that $\psi(X_1,X_2)$ is admissible when f is normal. □

Despite the inadmissibility of $\psi(X_1,X_2)$, this estimator has great intuitive appeal and it is easy to use. There is apparently no estimator whose risk improves on that of $\psi(X_1,X_2)$. The estimator δ^*, though admissible, is more difficult to compute and has a larger bias than $\psi(X_1,X_2)$.

Blumenthal and Cohen (1968b) have considered the special case of estimating the larger of two normal means and compared five different estimators (or sets of estimators) including ψ and δ^* discussed previously. Suppose that π_i is $N(\theta_i,\tau^2), i=1,2$, where τ is known. The problem of estimating the larger θ_i can arise in practical problems. Given two equal volume constant pitch signals, some physiological models assume that a subject will hear only the higher pitched one. Here θ_i will denote the pitch of signal. Using repeated readings from unbiased but imperfect measuring instruments regarding the two pitch values, we wish to estimate the pitch which the subject hears.

As pointed out earlier, X_1 and X_2 are now the sample means. We have

$$\psi(X_1,X_2)=\frac{X_1+X_2}{2}+\frac{|X_1-X_2|}{2} \tag{21.3}$$

and

$$\delta^*=\frac{X_1+X_2}{2}+|X_1-X_2|\left[\Phi\left(\sqrt{n}\,|X_1-X_2|/\tau\sqrt{2}\,\right)-\tfrac{1}{2}\right]$$
$$+\frac{\tau}{\sqrt{n\pi}}\exp\left\{\frac{-n(X_1-X_2)^2}{4\tau^2}\right\}. \tag{21.4}$$

As regards the other estimators, let δ_M denote the maximum likelihood estimator, and $\delta_\theta(d)$ the generalized Bayes estimator with respect to a generalized prior distribution which is the product of priors; one of which is uniform for $(\theta_1+\theta_2)/2$ and the other is normal with mean zero and variance ν^2 for $(\theta_1-\theta_2)/2$. Here $d=(2\nu^2)^{-1}[2\nu^2+(\tau^2/n)]$. The last is a set of hybrid estimators. For $c \geqslant 0$, let

$$\delta_H(c)=\begin{cases}(X_1+X_2)/2 & \text{if } |X_1-X_2|<2c\tau\sqrt{n}\,,\\ \psi(X_1,X_2) & \text{otherwise.}\end{cases} \tag{21.5}$$

For $c=0, \delta_H(c)=\psi(X_1,X_2)$.

All the estimators described previously are symmetric in (X_1, X_2) and translation invariant. Blumenthal and Cohen (1968b) have the bias and MSE (mean squared error) formulas for each of these estimators. They have made numerical comparisons in several special cases. The findings can be summarized as follows.

The intuitively appealing and easy-to-compute estimators $\psi(X_1, X_2)$ and $\delta_H(c)$ compare reasonably well with the others. Although these two estimators are inadmissible, it is not known which estimators have smaller mean squared errors than these. The estimator $\psi(X_1, X_2)$ is minimax. Both are consistent. In terms of MSE it appears that δ^* has some advantage.

21.2.2 A Natural Estimator of $\theta_{[i]}$

Let $\mathbf{X}_i = (X_{i1}, X_{i2}, \ldots, X_{in_i})$ be a random sample of size n_i from π_i. Let $T_i = T_{n_i}(\mathbf{X}_i), i = 1, \ldots, k$, be estimators for $\theta_1, \ldots, \theta_k$, respectively. Let $T_{[1]} \leqslant \cdots \leqslant T_{[k]}$ denote the ordered T_i.

Definition 21.1. If ω is a constant with probability one (w.p. 1), a sequence of estimators $\{\hat{\omega}_n : n \geqslant 1\}$ is said to be *strongly consistent* for ω if $\hat{\omega}_n$ converges to ω (w.p. 1).

Chen and Dudewicz (1973a) have shown that, for any $i (1 \leqslant i \leqslant k), T_{[i]}$ is strongly consistent as an estimator of $\theta_{[i]}$ if T_j is strongly consistent as an estimator of θ_j for $j = 1, 2, \ldots, k$. They have also shown that $T_{[i]}$ is an asymptotically unbiased estimator of $\theta_{[i]}$ if T_j converges to θ_j in L_1 (i.e., $E|T_j - \theta_j| \to 0$ as $n \to \infty$) for $j = 1, \ldots, k$.

21.2.3 Maximum Likelihood Estimators for Ranked Means

Let $\pi_i (i = 1, \ldots, k)$ be a normal population with mean μ_i and variance σ^2. The μ_i are unknown and the common variance σ^2 is assumed to be known. Let $X_{ij}, j = 1, \ldots, n$, denote n independent observations from π_i and let $\bar{X}_i$ be their mean. If all the μ_i are distinct, the maximum likelihood estimators (MLEs) of the μ_i based on the $\bar{X}_i$ exist and are uniquely given by $\hat{\mu}_i = \bar{X}_i, i = 1, \ldots, k$. If $g(\mu_1, \ldots, \mu_k)$ is a one-to-one function, then it is well known that $g(\hat{\mu}_1, \ldots, \hat{\mu}_k) = \hat{g}$ (say) furnishes a solution to the problem of finding an MLE of g. If g is not one-to-one, Zehna [1966] and Berk [1967] have given different justifications for picking only the "right" point $\mu_1 = \hat{\mu}_1, \ldots, \mu_k = \hat{\mu}_k$ for attention and calling it an MLE. Thus $\hat{\mu}_i = \bar{X}_{[i]}, i = 1, \ldots, k$, is the Berk–Zehna MLE of $(\mu_{[1]}, \ldots, \mu_{[k]})$. Dudewicz (1971c) has discussed the problem of MLE-type estimators of $(\mu_{[1]}, \ldots, \mu_{[k]})$ based on $\bar{X}_{[1]}, \ldots, \bar{X}_{[k]}$.

Let $f_{\bar{X}_1, \ldots, \bar{X}_k}(y_1, \ldots, y_k)$ denote the joint density of $\bar{X}_1, \ldots, \bar{X}_k$. Also, let β be any element of the permutation group S_k on k elements. Then the joint density of $\overline{X}_{[1]},\ldots,\overline{X}_{[k]}$ is given by

$$f_{\overline{X}_{[1]},\ldots,\overline{X}_{[k]}}(x_1,\ldots,x_k) = \begin{cases} \sum_{\beta\in S_k} f_{\overline{X}_1,\ldots,\overline{X}_k}(x_{\beta(1)},\ldots,x_{\beta(k)}), & x_1 \leqslant \cdots \leqslant x_k \\ 0 & \text{otherwise.} \end{cases} \tag{21.6}$$

Because of the symmetry of the preceding likelihood function in $\mu_{[1]},\ldots,\mu_{[k]}$, it is clear that if $(\hat{\mu}_{[1]},\ldots,\hat{\mu}_{[k]})$ is an MLE so is any permutation of it. Thus it is not necessary that $\hat{\mu}_{[1]} \leqslant \cdots \leqslant \hat{\mu}_{[k]}$. Such a possibility is eliminated by requiring the *consistency criterion* which allows only that permutation for which $\hat{\mu}_{[1]} \leqslant \hat{\mu}_{[2]} \leqslant \cdots \leqslant \hat{\mu}_{[k]}$. (See Dudewicz (1971c).)

Dudewicz (1971c) has shown that the MLE is given by

$$\hat{\mu}_{[i]} = \overline{X}_{[i]} + a_i(\overline{X}_{[1]},\ldots,\overline{X}_{[k]}), \qquad i=1,\ldots,k, \tag{21.7}$$

where $a_1,\ldots,a_k$ are some solution of a certain system of equations and must satisfy

$$-(x_i - x_1) \leqslant a_i \leqslant (x_k - x_i), \qquad i=1,\ldots,k$$

and $a_1 + \cdots + a_k = 0$. When $k=2, a_1 = -a_2$ and $0 \leqslant a_1 \leqslant x_2 - x_1$. Let $d = x_2 - x_1$ and $\bar{x} = (x_1 + x_2)/2$. Then it has been shown by Dudewicz (1971c) that the MLE is given by

a. $(\bar{x},\bar{x})$ when $0 \leqslant d \leqslant \sigma\sqrt{2/n}$ and

b. $(\bar{x} - \epsilon_0\sigma^2/2dn, \bar{x} + \epsilon_0\sigma^2/2dn)$ when $d > \sigma\sqrt{2/n}$, where ϵ_0 is the positive solution of $d^2 n\sigma^{-2} = \epsilon \coth(\epsilon/2)$.

21.2.4 Maximum Probability Estimators for Ranked Means

Weiss and Wolfowitz [1966] introduced the concept of generalized maximum likelihood estimators (GMLEs). The GMLEs provide (where available) asymptotically efficient estimators whereas this is not always true for MLEs even if thé latter can be found. The theory of Weiss and Wolfowitz [1966] is applicable in more general situations even though most of their applications are to i.i.d. "nonregular" cases. The concept of maximum probability estimators (MPEs) was introduced by Weiss and Wolfowitz [1967] who have shown that, for the case of $m=1$ parameter, every GMLE is an MPE. Dudewicz (1963) has extended this result to $m \geqslant 2$ parameters.

To define an MPE, let Θ be a closed region in R^m [m-dimensional Euclidean space] and $\Theta \subseteq \overline{\Theta}$, a closed region such that every finite boundary point of Θ is an inner point of $\overline{\Theta}$. For each n, let $X(n)$ denote the (finite) vector of random variables of which the estimator is to be a function. Let $K_n(x|\theta)$ be the density (with respect to a σ-finite measure μ_n) of $X(n)$ at the point x (of the appropriate space) when θ is the "true" value of the unknown parameter. Finally, let E be a fixed bounded set in R^m, and define

$$d-[k(n)]^{-1}E=\left\{(z_1,\ldots,z_m)\in\overline{\Theta}: d_i-\frac{y_i}{k_i(n)}=z_i, \qquad i=1,\ldots,m,\right.$$
$$\left.(y_1,\ldots,y_m)\in E\right\}, \tag{21.8}$$

where $d=(d_1,\ldots,d_m)$ and $k(n)=(k_1(n),\ldots,k_m(n))$ is such that $k(n)\to\infty$ as $n\to\infty$. The following definition is from Dudewicz (1973).

Definition 21.2. Z_n is a *maximum probability estimator* with respect to E and $k(n)$ if (for a.e. (μ_n) value x of $X(n)$) $Z_n(x)$ equals a $d\in\Theta$ such that

$$\int\cdots\int_{d-[k(n)]^{-1}E} K_n(x|\theta)\,d\theta_1\ldots d\theta_m = \sup_{t\in\Theta}\int\cdots\int_{t-[k(n)]^{-1}E} K_n(x|\theta)\,d\theta_1\ldots d\theta_m. \tag{21.9}$$

Dudewicz (1973) has discussed MPEs of ranked normal means where the populations concerned are assumed to have a common known variance. He has derived the MPE in the case of $m=2$ and shown it to have all the good properties of the GMLE.

Some corrections to Weiss and Wolfowitz [1966] are given in their later papers of 1967 and 1968. For more information on MPEs and related references the reader is referred to Weiss and Wolfowitz [1974].

21.2.5 A Two-Sample Estimator of the Largest Normal Mean

Let π_i be $N(\mu_i,\sigma^2), i=1,\ldots,k$. It is required to estimate $\mu_{[k]}$. The loss function is defined by

$$L=\frac{n(T-\mu_{[k]})^2}{\sigma^2} \tag{21.10}$$

where T denotes the estimate of $\mu_{[k]}$ and n the total number of observations from all the populations, which is fixed. Alam (1967) has considered the following procedure which requires experimentation in two stages.

Draw a sample of equal size m, say, from each of the populations in the first experiment, where $mk<n$. Compute the sample means and take the remaining $n-mk$ observations in the second experiment from the population that yielded the largest sample mean breaking any possible tie by randomization. Let $\bar{X}$ be the largest sample mean from the first stage and $\bar{Y}$ be the mean of the sample taken at the second stage. Then an estimator of $\mu_{[k]}$ is given by

$$T=\frac{m\bar{X}+(n-mk)\bar{Y}}{n-mk+m}. \tag{21.11}$$

The problem is to determine the optimal choice of m, given n. Alam (1967) has considered a restricted minimax solution, that is, the value m_k of m which minimizes $\max R_k$ where R_k is the risk and the maximum is taken over the subset H of parameter space where $k-1$ populations are equally shifted from the largest mean.

Let $u_k=m_k/(n-km_k+m_k)$. Alam (1967) has shown that u_k is uniquely determined for $k=2$. The same result is seen to hold for $k=3$ from the computed values of R_k. We get $u_2=0.645$ and $u_3=0.68$ (approx.).

If we let $u=m/(n-mk+m)$, then $u=1$ corresponds to experimentation restricted to one stage with $km=n$. Let $R_k(u)=\max_H R_k$. Comparison of the values of $R_k(u_k)$ and $R_k(1)$ would indicate the advantage of the two-stage sampling procedure. It is shown that $R_2(1)=2$ and $R_3(1)=3+3\sqrt{3}/2\pi$, whereas $R_2(u_2)=1.78$ and $R_3(u_3)=2.59$ (approx.). Approximate values of $R_k(u)$ are given by Alam (1967) for $k=2,3$ and $u=0\,(0.1)1$. He has further shown that, for $u=1$, T is admissible in a class of invariant estimators and that it is also minimax for $k=2$.

21.2.6 Estimation of Ordered Parameters with Prior Information: Isotonic Regression

In our discussion so far we assumed that there is no prior information available regarding the pairing of the ordered and unordered θ_i. Now we discuss the problem of estimation of ordered parameters where complete or partial information is available regarding this pairing. Such problems arise in practice. For example, consider the two-sample problem in which the mean responses for two treatments are compared. If Treatment 1 is the application of a fertilizer to the soil in which a crop is grown and if Treatment 2 is simply to grow the crop without fertilizer then it may be possible to assert a priori that $\mu_1 \geqslant \mu_2$. A similar situation arises when several means are being considered. Suppose that Y is a measure of performance or a test of ability carried out under varying degrees of stress, x. It may be expected that $\mu(x)=E(Y|x)$ is a decreasing function of x. It may be impossible to quantify stress; however, if the experiment is carried out at levels of stress $x_1, x_2, \ldots, x_k$, then it could be asserted that $\mu(x_1) \geqslant \mu(x_2) \geqslant \cdots \geqslant \mu(x_k)$.

It may not be possible always to have a total, or simple, ordering as in the preceding example. To be specific, the conditions $\mu(x_1) \geqslant \mu(x_2), \mu(x_1) \geqslant \mu(x_3)$ may hold with no possible ranking of $\mu(x_2)$ and $\mu(x_3)$ with respect to each other. This kind of partial orderings may also arise when x is vector valued. For example, it might be possible to assert only that $\mu(x_1', x_1'') \geqslant \mu(x_2', x_2'')$ *if* $x_1' \leqslant x_2'$ and $x_1'' \leqslant x_2''$. The ranking of the μ-values when only one of these inequalities holds may be unknown.

The subject of statistical inference under order restriction has been elaborately treated by Barlow, Bartholomew, Bremner, and Brunk [1972] (referred to hereafter as BBBB [1972]) who have given a unified account of the statistical theory which has grown up to deal with problems in which conditional expectations are subject to order restrictions. Underlying the diverse applications is the theory of isotonic regression over an ordered set. This provides maximum likelihood estimators of ordered parameters. We briefly outline the basic theory of isotonic regression to indicate how it is applied to obtain maximum likelihood estimators.

Definition 21.3. A binary relation "$\lesssim$" on a set X establishes a *simple order* on X if

1. it is reflexive: $x \lesssim x$ for $x \in X$;
2. it is transitive: $x, y, z \in X$, $x \lesssim y, y \lesssim z \Rightarrow x \lesssim z$;
3. it is antisymmetric: $x, y \in X$, $x \lesssim y, y \lesssim x \Rightarrow x = y$;
4. every two elements are comparable: $x, y \in X \Rightarrow$ either $x \lesssim y$ or $y \lesssim x$.

A *partial order* is reflexive, transitive and antisymmetric, but there may be noncomparable elements. A *quasi-order* is reflexive and transitive but not necessarily antisymmetric. Thus, every simple order is a partial order and every partial order is a quasi-order.

Definition 21.4. A real-valued function f on X is *isotonic* with respect to a quasi-ordering "$\lesssim$" on X if $x, y \in X, x \lesssim y$ imply $f(x) \leqslant f(y)$.

Definition 21.5. Let g be a given function on X and w a given positive function on X. An isotonic function g^* on X is an *isotonic regression of g with weights w* if and only if it minimizes in the class of isotonic functions f on X the sum $\Sigma_{x \in X}[g(x) - f(x)]^2 w(x)$.

The existence of an isotonic regression of g can be demonstrated (Theorem 1.6 in BBBB [1972]). Theorem 1.3 in BBBB states that the isotonic regression is unique, and gives a necessary and sufficient condition that an isotonic function be the isotonic regression.

Let Φ be a convex function that is finite on an interval I containing the range of the function g and infinite elsewhere, and let φ be an arbitrary determination of its derivative (i.e., having at a "corner" any value between the left and right derivatives), defined and finite on I. Then φ is nondecreasing. For numbers u and v set

$$\Delta(u,v)=\begin{cases}\Phi(u)-\Phi(v)-(u-v)\varphi(v), & \text{if } u,v\in I,\\ \infty, & \text{if } v\in I, u\notin I.\end{cases}$$

The number Δ is nonnegative, and may be interpreted as the excess of the rise of the graph of Φ between v and u over the rise of the line tangent to the graph at v. Then the generalized isotonic regression can be defined in terms of minimizing $\Sigma_x \Delta[g(x),f(x)]w(x)$. As we can see, $\Phi(u)=u^2$ in Definition 21.5.

The estimation of ordered parameters associated with binomial, geometric, Poisson, and gamma populations have been discussed in Section 1.4 of BBBB [1972]. In all these cases, the problem of maximizing the likelihood function subject to the order restrictions on the parameters is equivalent to that of generalized isotonic regression estimators for suitable choices of Φ. Further general theory and examples relating to maximum likelihood estimation of ordered multinomial parameters, and ordered means of populations belonging to a common exponential family are discussed in Chapter 2 of BBBB [1972].

21.3 CONFIDENCE INTERVALS FOR ORDERED PARAMETERS: SOME GENERAL RESULTS

Several authors have investigated the problem of constructing one-sided confidence intervals for ordered parameters. Most of these investigations relate to the largest (or the smallest) parameter in the cases of location and scale parameter families. We discuss these results in detail in the next section. Our present discussion is confined to certain general results.

Let $\pi_i (i=1,\ldots,k)$ have the associated distribution function $F(x,\theta_i)$, where θ_i is a real-valued parameter belonging to the open interval (θ_*,θ^*), $-\infty\leqslant\theta_*<\theta^*\leqslant\infty$.

21.3.1 Confidence Intervals for $\theta_{[i]}$

Let $\mathbf{X}_i=(X_{i1},\ldots,X_{in})$ be a random sample of n observations from π_i and let $T_i=T_n(\mathbf{X}_i)$ be an appropriate estimator of θ_i with cdf $G_n(t,\theta_i)$. It is assumed that the family $\{G_n(t,\theta)\}$ is stochastically increasing in θ. It is also assumed that $G_n(t,\theta)$ is *degenerate at* θ_* *and* θ^*, that is,

$$G_n(t,\theta)=\begin{cases}1 & \text{as } \theta=\theta_*\\ 0 & \text{as } \theta=\theta^*\end{cases} \tag{21.12}$$

for all $t\in(T_*,T^*)$, the support of T. Let $g_n(t,\theta_i)$ be the density of T_i. We want a procedure for constructing a random interval I so that

$$\inf_{\Omega(\theta_{[i]})} \Pr(\theta_{[i]}\in I)\geqslant\gamma \tag{21.13}$$

where $0<\gamma<1$ is a preassigned number and $\Omega(\theta_{[i]})=\{\boldsymbol{\theta}=(\theta_1,\ldots,\theta_k)$ for which $\theta_{[i]}$ has a fixed value$\}$.

Chen (1977a) has obtained a class of one-sided (upper and lower) intervals and used these to construct two-sided intervals. These results are described as follows.

A. Lower Confidence Intervals. Let h_1 be a continuous increasing function with inverse g_2. Define

$$I_L=(h_1(T_{[i]}|\gamma_1),\theta^*) \tag{21.14}$$

to be lower confidence interval for $\theta_{[i]}$ with coverage probability γ_1. Chen (1977a) has shown that

$$\inf_{\Omega(\theta_{[i]})} \Pr\{\theta_{[i]}\in I_L\}=\left[G_n(g_2(\theta_{[i]}),\theta_{[i]})\right]^i, \tag{21.15}$$

and given $\gamma_1\in(0,1),k,n,$ and $i(1\leqslant i\leqslant k)$, the end point $h_1(T_{[i]}|\gamma_1)$ can be obtained as the unique solution in $\theta_{[i]}$ of

$$G_n(g_2(\theta_{[i]}),\theta_{[i]})=\gamma_1^{1/i}. \tag{21.16}$$

B. Upper Confidence Intervals. Let h_2 be a continuous increasing function with inverse g_1. Define

$$I_U=(\theta_*,h_2(T_{[i]}|\gamma_2)) \tag{21.17}$$

to be an upper confidence interval for $\theta_{[i]}$ with coverage probability γ_2.

Then

$$\inf_{\Omega(\theta_{[i]})} \Pr\{\theta_{[i]} \in I_U\} = \left[1 - G_n(g_1(\theta_{[i]}), \theta_{[i]})\right]^{k-i+1} \tag{21.18}$$

and given $\gamma_2 \in (0,1)$, k, n, and i, the end point $h_2(T_{[i]}|\gamma_2)$ can be obtained as the unique solution in $\theta_{[i]}$ of

$$G_n(g_1(\theta_{[i]}), \theta_{[i]}) = 1 - \gamma_2^{1/(k-i+1)} \tag{21.19}$$

C. TWO-SIDED CONFIDENCE INTERVALS. Let

$$I_T = (h_1(T_{[i]}|\gamma_1), h_2(T_{[i]}|\gamma_2)) \tag{21.20}$$

be the intersection of the lower interval I_L and the upper interval I_U with coverage probabilities γ_1 and γ_2, respectively. Let γ_1 and γ_2 be such that $\gamma_1 + \gamma_2 - 1 = \alpha$ (α fixed and $0 < \alpha < 1$). Then the interval I_T is a two-sided confidence interval for $\theta_{[i]}$ with coverage probability at least α. This method was earlier used by Dudewicz (1972a) in the case of location parameters.

The lower confidence interval I_L and the upper confidence interval I_U have minimal coverage probability γ_1 and γ_2, respectively. Chen (1977a) has shown that I_L and I_U have maximal coverage probability $1-(1-\gamma_1^{1/i})^{k-i+1}$ and $1-(1-\gamma_2^{1/(k-i+1)})^i$, respectively.

21.3.2 An Alternative Two-Sided Interval

Alam, Lal Saxena, and Tong (1973) have discussed a different class of two-sided intervals for $\theta_{[k]}$. It is assumed as before that, for each n, the family $\{G_n(t,\theta)\}$ is stochastically increasing in θ and $G_n(t,\theta)$ satisfies (21.12). It is further assumed that $u_n(t,\theta) = -(\partial/\partial\theta)(G_n(t,\theta))$ exists for $\theta \in (\theta_*, \theta^*)$ and that

$$u_n(t,\theta)u_n(t',\theta') - u_n(t,\theta')u_n(t',\theta) \geqslant 0 \tag{21.21}$$

for $t \leqslant t'$ and $\theta \leqslant \theta'$. The condition (21.21) reduces to that of monotone likelihood ratio when θ is a location or a scale parameter.

Let h_1 and h_2 be strictly increasing functions such that $h_1(t) \geqslant h_2(t)$ for all t and let g_1 and g_2 be the corresponding inverse functions. Alam, Lal Saxena, and Tong (1973) considered

$$I = [h_2(T_{[k]}), h_1(T_{[k]})] \tag{21.22}$$

as a two-sided confidence interval for $\theta_{[k]}$. Let

$$H_r(\theta)=G_n^r(g_2(\theta),\theta)-G_n^r(g_1(\theta),\theta), \qquad r=1,\ldots,k. \tag{21.23}$$

Under the assumptions about $G_n(t,\theta)$ stated earlier, Alam, Lal Saxena, and Tong (1973) have shown that

$$\inf_{\Omega(\theta_{[k]})} \Pr\{\theta_{[k]}\in I\}=\min\left[H_1(\theta_{[k]}),H_k(\theta_{[k]})\right]. \tag{21.24}$$

Similary, if we take $J=[h_2(T_{[1]}),h_1(T_{[1]})]$, then

$$\inf_{\Omega(\theta_{[1]})} \Pr\{\theta_{[1]}\in J\}=\min\left[H_1'(\theta_{[1]}),H_k'(\theta_{[1]})\right] \tag{21.25}$$

where $H_r'(\theta)=[1-G_n(g_1(\theta),\theta)]^r-[1-G_n(g_2(\theta),\theta)]^r, r=1,\ldots,k$.

Now, if the functions g_1 and g_2 are such that $G_n(g_1(\theta),\theta)$ and $G_n(g_2(\theta),\theta)$ are independent of θ, it is possible to satisfy the condition

$$\inf_{\Omega} \Pr\{\theta_{[k]}\in I\}\geqslant\gamma \tag{21.26}$$

where $0<\gamma<1$ is specified in advance. Thus we will have a class of confidence intervals given by (21.22) satisfying (21.26) and we can select an optimal interval from this class. Similar statements hold in the case of J.

Results along these lines in the case of $\theta_{[i]}$ are given by Alam and Lal Saxena (1974).

21.4 CONFIDENCE INTERVALS FOR ORDERED PARAMETERS: LOCATION AND SCALE PARAMETERS

In this section we discuss the applications of the general results of the previous section to the cases of location and scale parameters. Some special results relating to the normal means and variances are also discussed.

21.4.1 Location Parameters

In the notations of §21.3, $G_n(t,\theta)=G_n(t-\theta)$ for $t\in(T_*,T^*)$ and $\theta\in(-\infty,\infty)$. In this case, we have

$$I_L=(T_{[i]}-G_n^{-1}(\gamma_1^{1/i}),\infty),$$

$$I_U=(-\infty,T_{[i]}-G_n^{-1}(1-\gamma_2^{1/k-i+1})),$$

$$I_T=(T_{[i]}-G_n^{-1}(\gamma_1^{1/i}),T_{[i]}-G_n^{-1}(1-\gamma_2^{1/k-i+1})). \tag{21.27}$$

The one-sided intervals in (21.27) have been discussed earlier by Dudewicz (1970a) and the two-sided intervals by Dudewicz (1972a). Both assume that $F(x,\theta)=F(x-\theta)$ and $\mu_0=\int_{-\infty}^{\infty}x\,dF(x)$ is finite and known. Thus the mean of π_i is $\mu_i=\mu_0+\theta_i$. Let $G_n(y)$ denote the cdf of $\overline{X}_i-\theta_i$ so that $T_i=\overline{X}_i$ has the cdf $G_n(y-\theta_i)$. The two-sided interval in this case is

$$I_T=\left(\overline{X}_{[i]}-g_i^*(\gamma_1),\overline{X}_{[i]}+h_i^*(\gamma_2)\right), \tag{21.28}$$

where $g_i^*(\gamma_1)=G_n^{-1}(\gamma_1^{1/i})-\mu_0$ and $h_i^*(\gamma_2)=-G_n^{-1}(1-\gamma_2^{1/(k-i+1)})+\mu_0$. As we saw earlier, for $0<\gamma_1,\gamma_2<1$ such that $\gamma_1+\gamma_2=\alpha+1$ (α fixed, $0<\alpha<1$), I_T as a confidence interval for $\theta_{[i]}$ has a coverage probability not less than α.

Let $L(\gamma_1)$ denote the length of the interval I_T in (21.28). Then $L(\gamma_1)=h_i^*(\gamma_2)+g_i^*(\gamma_1)$. Using $L(\gamma_1)$ one can numerically approximate that $\gamma_1^*(\alpha<\gamma_1^*<1)$ which minimizes $L(\gamma_1)$. This furnishes one reasonable way of selecting the best interval from the class in (21.28). Dudewicz (1972a) has made some comparisons of I_T and the interval used by Lal Saxena and Tong (1969) whose result is given as follows.

Theorem 21.2 Let $\alpha(0<\alpha<1)$ be fixed. Suppose that $g_n(y)=g_n(-y)$ and that $g_n(y-\theta)$ has a monotone likelihood ratio in y. Then the interval

$$I=\left(\overline{X}_{[k]}-d,\overline{X}_{[k]}+d\right) \tag{21.29}$$

as a confidence interval for $\mu_{[k]}$ has minimal coverage probability equal to $G_n^k(d)-G_n^k(-d)$.

If we set d so that the minimal coverage probability equals α, we can compare this with the interval I_T in (21.28). We can see that $g_k^*(\gamma_2)=h_k^*(\gamma_2)$ if and only if $\gamma_2=\gamma_1^k$. Taking this to be the case, the minimum coverage probability of I_T is $\gamma_1-[1-\gamma_1^{1/k}]^k>\alpha$. Consider the special case where π_i is $N(\mu_i,\sigma^2)$ where σ is known. In this case, we have

$$I_T=\left(\overline{X}_{[k]}-\frac{\sigma}{\sqrt{n}}\Phi^{-1}(\gamma_1^{1/k}),\overline{X}_{[k]}+\frac{\sigma}{\sqrt{n}}\Phi^{-1}(\gamma_2)\right) \tag{21.30}$$

with $0<\gamma_1,\gamma_2<1$, and $\gamma_1+\gamma_2=\alpha+1$, and

$$I=\left(\overline{X}_{[k]}-\frac{\sigma}{\sqrt{n}}d,\overline{X}_{[k]}+\frac{\sigma}{\sqrt{n}}d\right) \tag{21.31}$$

where $\Phi^k(d)-\Phi^k(-d)=\alpha$. Both the intervals have minimum coverage probability α. Let $L(\gamma_1)$ and L^* denote the lengths of the intervals I_T and

I, respectively. Dudewicz (1972a) has shown that

$$\lim_{k\to\infty}\frac{L(\gamma_1)}{L^*}=\frac{1}{2} \tag{21.32}$$

and for any $k \geqslant 2$ with $\gamma_1+\gamma_2=(1+\alpha)/2$,

$$\lim_{\alpha\uparrow 1}\frac{L(\gamma_1)}{L^*}=1. \tag{21.33}$$

We discussed in §21.3.2 some general results of Alam, Lal Saxena, and Tong (1973) for the interval I given in (21.22). When $G_n(t,\theta)=G_n(t-\theta)$, we set $g_1(\theta)=\theta-d_2$ and $g_2(\theta)=\theta+d_1$ where d_1 and d_2 are real numbers such that $d_1+d_2>0$, and consider the class of intervals

$$I(d_1,d_2)=(T_{[k]}-d_1, T_{[k]}+d_2) \tag{21.34}$$

with length $L=d_1+d_2$.

Dudewicz and Tong (1971) have investigated the problem of optimal choice of d_2 (or d_1) when L is fixed. The optimality criterion used is the infimum of the coverage probability. Thus we require the value of d_2 which maximizes the infimum of coverage probability. Dudewicz and Tong (1971) have shown that this optimal value d_0 is $L/2$ when $k=1$ or 2, and is less than $L/2$ for $k \geqslant 3$. Therefore for $k \geqslant 3$, the unsymmetric interval $I(d_1,d_2)$ performs better than the symmetric interval of Lal Saxena and Tong (1969).

Consider the special case where π_i is $N(\theta_i,\sigma_i^2)$. We assume that the σ_i^2 are known and that $\tau^2=\sigma_i^2/n_i, i=1,\ldots,k$, so that the sample means T_i (based on n_i observations from π_i) have equal known variance τ^2. The optimal interval is given by

$$I=\left(\overline{X}_{[k]}-(L-\tau x_0),\overline{X}_{[k]}+\tau x_0\right). \tag{21.35}$$

Dudewicz and Tong (1971) have tabulated the x_0-values for $k=3(1)6(2)14$ and $c=L/\tau=1.0(0.1)4.0$.

Now going back to the interval $I(c,d)$ defined by (21.34), Alam, Lal Saxena, and Tong (1973) have considered minimizing L without the assumption of the symmetry of G_n. They assume that $G_n^r(y)$ is unimodal for $r=1$ and k. Given $\gamma(0<\gamma<1)$, let

$$c_r^*=\inf\{c: G_n^r(c)\geqslant\gamma\}, \qquad r=1,k. \tag{21.36}$$

For a given value of $c \geqslant c_r^*$, let

$$d_r(c) = \inf\{d : G_n^r(c) - G_n^r(-d) = \gamma\}. \tag{21.37}$$

Let $d(c) = \max[d_1(c), d_k(c)]$.

Now, let c_r and d_r denote the values of c and d, respectively, minimizing $L = c + d$ subject to $G_n^r(c) - G_n^r(-d) \geqslant \gamma$. Let $L_r = c_r + d(c_r)$ and define $r' = 1$ if $L_1 \leqslant L_k$ and $r' = k$ if $L_1 > L_k$. Then the optimal confidence interval is given by

$$I = \left[\overline{X}_{[k]} - c_{r'}, \overline{X}_{[k]} + d(c_{r'})\right]. \tag{21.38}$$

21.4.2 Scale Parameters

In the notations of §21.3, we have $G_n(t, \theta) = G_n(t/\theta_i)$ for all $t \in (T_*, T^*)$ and $\theta \in (0, \infty)$. Then the lower, the upper and the two-sided confidence intervals for $\theta_{[i]}$ with coverage probabilities γ_1, γ_2, and $\gamma_1 + \gamma_2 - 1 = \alpha$ are given, respectively, by

$$I_L = \left(\frac{T_{[i]}}{G_n^{-1}(\gamma_1^{1/i})}, \infty\right),$$

$$I_U = \left(0, \frac{T_{[i]}}{G_n^{-1}(1 - \gamma_2^{1/(k-i+1)})}\right), \tag{21.39}$$

$$I_T = \left(\frac{T_{[i]}}{G_n^{-1}(\gamma_1^{1/i})}, \frac{T_{[i]}}{G_n^{-1}(1 - \gamma_2^{1/(k-i+1)})}\right).$$

Consider the particular example of normal variances. Let π_i be $N(\mu_i, \sigma_i^2), i = 1, \ldots, k$. The σ_i^2 are unknown and we require an interval estimate of $\sigma_{[i]}^2$. We assume that the μ_i are not known. Let s_i^2 be the sample variance based on n independent observations from π_i so that $(n-1)s_i^2/\sigma^2$ has a chi-square distribution with $\nu = (n-1)$ degrees of freedom. Here s_i^2 plays the role of T_i and $G_n(t, \sigma_i^2)$ is the cdf of s_i^2. The general results of Chen (1977a) give the following results for $\sigma_{[i]}^2$.

For any $i (1 \leqslant i \leqslant k)$, an upper confidence interval for $\sigma_{[i]}^2$ with coverage probability not less than γ_2 is given by

$$I_U = \left(0, \frac{s_{[i]}^2}{b_n}\right) \tag{21.40}$$

where b_n is the solution of

$$(1 - G_n(b_n))^{k-i+1} = \gamma_2 \tag{21.41}$$

and $G_n(y)$ is the cdf of χ^2_ν/ν. A lower confidence interval for $\sigma^2_{[i]}$ with coverage probability not less than γ_1 is given by

$$I_L = \left(\frac{s^2_{[i]}}{a_n}, \infty \right) \tag{21.42}$$

where a_n is the solution of

$$\left[G_n(a_n) \right]^i = \gamma_1. \tag{21.43}$$

Chen (1977a) has tabulated the values of a_n and b_n for $k=1(1)5, i=1(1)k$, $\nu=1(1)20$, 25, 30, 40, 60, 100, and γ_1 (or γ_2)=0.95, 0.99. Additional tables for selected values of k and γ_1 (or γ_2) are given by Chen and Dudewicz (1973a).

Returning to the general results of §21.3.2, we set $g_1(\theta)=d\theta$ and $g_2(\theta)=cd\theta$, $d>0$, $c>1$. The two-sided interval for $\theta_{[k]}$ is

$$I = \left((cd)^{-1} T_{[k]}, d^{-1} T_{[k]} \right). \tag{21.44}$$

For selecting the best interval from this class, Alam, Lal Saxena, and Tong (1973) used the criterion of choosing c as small as possible subject to the requirement that the coverage probability be not less than a preassigned number γ. They have given a method of constructing the optimal interval. Define

$$Q(c,d) = \min\left[H_1(c,d), H_k(c,d) \right] \tag{21.45}$$

where $H_r(c,d) = G_n^r(cd) - G_n^r(d), r=1,k$. For a given c, let $d_0(c) = \max_d Q(c,d)$. Now choose the smallest c, say $c_0(\gamma)$, such that $Q\{c_0(\gamma), d_0(c_0(\gamma))\} \geq \gamma$. Then the optimal interval is given for $c=c_0(\gamma)$ and $d=d_0(c_0(\gamma))$. A similar algorithm can be used for the optimum choice of c and d in the confidence interval for $\theta_{[1]}$ given by $J=((cd)^{-1}T_{[1]}, d^{-1}T_{[1]})$.

Suppose π_i is $N(\mu_i, \sigma_i^2)$. The role of T_i is played by $s_i^2 = n^{-1}\Sigma_j(X_{ij}-\mu_i)^2$ or $(n-1)^{-1}\Sigma_j(X_{ij}-\bar{X}_i)^2$ according as whether the μ_i are known or unknown. Alam, Lal Saxena, and Tong (1973) have tabulated the optimum values of c and d in the cases of both $\sigma^2_{[1]}$ and $\sigma^2_{[k]}$ for $k=2(1)8(2)12, \gamma=0.75$, 0.90, and $\nu=10(10)60$ where $\nu=n$ or $n-1$ depending on whether the μ_i are known or unknown.

Lal Saxena (1971) has considered intervals of the type

$$I_n = (a_n T_{[k]}, b_n T_{[k]}) \tag{21.46}$$

where $0 \leqslant a_n < 1 < b_n \leqslant \infty$ for all n. It is assumed that (a) $\{\frac{1}{\theta} g_n(\frac{t}{\theta})\}$ has a monotone likelihood ratio in t, and (b) a_n and b_n are so chosen that $G_n(a_n^{-1}) + G_n(b_n^{-1}) \leqslant 1$, $0 < a_n < 1 < b_n < \infty$. Then

$$\inf_{\Omega} \Pr\{\theta_{[k]} \in I_n\} = G_n^k(a_n^{-1}) - G_n^k(b_n^{-1}). \tag{21.47}$$

Case I. Suppose that n and γ are preassigned. Then we must choose a_n and b_n such that

$$G_n(a_n^{-1}) + G_n(b_n^{-1}) \leqslant 1 \tag{21.48}$$

and

$$G_n^k(a_n^{-1}) - G_n^k(b_n^{-1}) = \gamma. \tag{21.49}$$

We see that (21.48) and (21.49) do not give a unique solution. One can take a particular solution such that

$$G_n(a_n^{-1}) + G_n(b_n^{-1}) = 1. \tag{21.50}$$

Using (21.50) in (21.49), we get

$$G_n^k(a_n^{-1}) - [1 - G_n(a_n^{-1})]^k = \gamma. \tag{21.51}$$

For fixed k, n, and γ, (21.51) gives the unique choice of a_n which in turn gives the unique choice of b_n from (21.50).

Case II. Suppose that $a_n = a < 1$ and γ are preassigned. We consider the choice of the smallest n and b_n to be such that

$$G_n(a^{-1}) + G_n(b_n^{-1}) = 1 \tag{21.52}$$

and

$$G_n^k(a^{-1}) - [1 - G_n(a^{-1})]^k \geqslant \gamma. \tag{21.53}$$

Since the left-hand side of (21.53) is an increasing function of $G_n(a^{-1})$, there exists a number $c = c(k,\gamma)$ such that (21.53) is satisfied whenever $G_n(a^{-1}) \geqslant c$. Since $\lim_{n\to\infty} G_n(a^{-1}) = 1$, there exists a smallest value of n, say n^*, such that (21.52) holds. This n^* is the chosen value of n. The corresponding b_{n*} is obtained from (21.52). In some cases when γ is very small, b_{n*} could be less than unity. In that case we choose $b_{n*}=1$. The result in (21.47) is still valid.

Generally, the interval estimation of $\theta_{[1]}$ can be handled in a way analogous to that of $\theta_{[k]}$.

21.5 INTERVAL ESTIMATION OF ORDERED MEANS FROM k NORMAL POPULATIONS

Let π_i be $N(\mu_i, \sigma_i^2), i=1,\ldots,k$. When the σ_i^2 are known, we have discussed interval estimation of $\mu_{[i]}$ (in particular, $\mu_{[1]}$ and $\mu_{[k]}$) based on a single-stage sampling procedure as a particular case of estimation of ordered location parameters. These results are contained in Alam, Lal Saxena, and Tong (1973), Chen and Dudewicz (1973a), Dudewicz (1970a, 1972a), Dudewicz and Tong (1971), and Lal Saxena and Tong (1969). Further, the preceding results are for equal sample sizes. In this section, we discuss the cases of unequal sample sizes, multistage procedures, and interval estimation based on a hybrid point estimator.

21.5.1 Interval Estimation of the Ordered Means of Two Normal Populations Based on Hybrid Estimators: Known Variances

Let X and Y be means of samples of sizes n_1 and n_2 from π_1 and π_2, respectively. We assume that X and Y have a common known variance $\tau^2=\sigma_1^2/n_1=\sigma_2^2/n_2$. Let $\delta=|\mu_1-\mu_2|$. We want an interval estimate of $\mu_{[2]}$. Blumenthal and Cohen (1968a, b) have considered the class of hybrid estimators of $\mu_{[2]}$ [see (21.5)] defined by

$$T=\begin{cases} \frac{1}{2}(X+Y), & \text{if } |X-Y|<a, \\ M, & \text{otherwise,} \end{cases} \tag{21.54}$$

where $M=\max(X,Y)$ and $a\geqslant 0$ is a fixed number. Tong (1971) has considered interval estimation of $\mu_{[2]}$ based on T.

For fixed $a\geqslant 0$, and fixed d_1, d_2, d_3, and d_4 such that $d_1+d_2>0$ and $d_3+d_4>0$, let the random interval I have the form

$$I=\begin{cases} (T-d_1, T+d_2), & \text{if } T=(X+Y)/2, \\ (T-d_3, T+d_4), & \text{if } T=M. \end{cases} \tag{21.55}$$

Note that d_1+d_2 is not necessarily equal to d_3+d_4; hence the length of I is not assumed to be fixed. Also, the signs of the d_i are not assumed; so T is not necessarily contained in I. The one-sided intervals can, of course, be obtained by setting $d_1=d_3=\infty$ or $d_2=d_4=\infty$. Let γ be the coverage probability and define

$$c=\frac{a}{\tau},\quad \theta=\frac{\delta}{\tau},\quad \lambda_i=\frac{d_i}{\tau}\qquad (i=1,2,3,4). \tag{21.56}$$

It is seen from the expression for γ that it depends on a, δ, and d_i only through c, θ, and λ_i. Also γ depends on μ_1 and μ_2 only through δ. Tong (1971) has obtained the following results.

Theorem 21.3. For every fixed $\theta \geqslant 0$,

a. if $\lambda_1+\lambda_2=\lambda_3+\lambda_4<\infty$ and $\lambda_1 \geqslant \lambda_2 \geqslant 0$, then $\gamma=\gamma_\theta(c)$ is strictly decreasing for c satisfying $2(\lambda_3-\lambda_1)<c<2\{\theta+(\lambda_1-\lambda_4)\}$, and strictly increasing otherwise; the supremum of $\gamma_\theta(c)$ is achieved at either $c=\max\{0, 2(\lambda_3-\lambda_1)\}$ or $c=\infty$;

b. if $\lambda_1=\lambda_3=\infty$, then $\gamma_\theta(c)$ is strictly decreasing for $c>2(\lambda_2-\lambda_4)$ and strictly increasing otherwise; the supremum of $\gamma_\theta(c)$ is achieved at $c=0$ when $\lambda_2 \leqslant \lambda_4$ or at $c=2(\lambda_2-\lambda_4)$ when $\lambda_2>\lambda_4$;

c. if $\lambda_2=\lambda_4=\infty$, then $\gamma_\theta(c)$ is strictly decreasing for $c<2(\lambda_3-\lambda_1)$, and strictly increasing otherwise; the supremum of $\gamma_\theta(c)$ is achieved at $c=\infty$ when $\lambda_3 \leqslant \lambda_1$ or at either $c=0$ or $c=\infty$ when $\lambda_3>\lambda_1$.

As a consequence of Theorem 21.3, we see that for the construction of one-sided confidence interval, no matter what the true value of θ is, we should always use $T=M$ for the interval with $\lambda_1=\lambda_3=\infty$ and $\lambda_2=\lambda_4$, and use $T=(X+Y)/2$ for the interval with $\lambda_2=\lambda_4=\infty$ and $\lambda_1=\lambda_3$.

For the rest of the discussion we consider two-sided intervals and assume that $\lambda_1=\lambda_3 \geqslant \lambda_2=\lambda_4 \geqslant 0$. We have

$$\begin{aligned}\alpha(\theta)&=\gamma_\theta(0)-\gamma_\theta(\infty)\\ &=\{\Phi(\lambda_2)\Phi(\theta+\lambda_2)-\Phi(-\lambda_2)\Phi(\theta-\lambda_2)\}\\ &\quad-\left[\Phi\left\{\left(\frac{\theta}{2}+\lambda_1\right)\sqrt{2}\right\}-\Phi\left\{\left(\frac{\theta}{2}-\lambda_2\right)\sqrt{2}\right\}\right].\end{aligned} \tag{21.57}$$

As a consequence of Theorem 21.3 we should either always use the maximum of the sample means or always use the pooled sample mean, depending on whether $\alpha(\theta)$ is positive or negative. Tong (1971) has shown that, under the assumption that $\lambda_1=\lambda_3 \geqslant \lambda_2=\lambda_4 \geqslant 0$ and $0<\lambda_1+\lambda_2<\infty$, either $\alpha(\theta)>0$ for every $\theta \geqslant 0$ or there exists a $\theta_0=\theta_0(\lambda_1,\lambda_2)$ such that $\alpha(\theta)<, =,$ or >0 according as $\theta<, =, >\theta_0$. When the maximum of the sample means is used, Dudewicz and Tong (1971) have proved that for fixed $\lambda_1+\lambda_2$ the choice $\lambda_1=\lambda_2$ is optimal in the sense that the coverage probability under the least favorable configuration is maximized. When the pooled sample mean is used, the symmetric interval is approximately optimal for small θ. Therefore in most applications the symmetric interval with $\lambda_1=\lambda_2=\lambda$, say, should be suitable. Tong (1971) has tabulated the values of $\theta_0(\lambda)$ for $\lambda=0.1(0.1)4.0$.

Tong (1971) has also compared the performances of different procedures, including procedures based on the maximum of sample means, the hybrid estimators, the Pitman estimator, and the Bayes estimators, for the interval estimation of $\mu_{[k]}$. We consider symmetric intervals and first take up the situation in which we have partial information about θ, namely, whether $\theta<\theta_0$ or $\theta>\theta_0$ where $\theta_0=\theta_0(\lambda)$ is the point at which $\alpha(\theta)$ changes sign. From the earlier discussion, we know that the procedure R_1 (i.e., always use the pooled sample mean if $\theta<\theta_0$ and always use the maximum of the sample means if $\theta>\theta_0$) is optimal among the class of all procedures based on hybrid estimators. In terms of coverage probability the procedure R_1 can do almost as well as if θ were known. It also turns out that R_1 is better than the procedure R (i.e., always use the maximum of the sample means) when the partial information about θ is available and when θ is small.

Now consider the situation in which no information about θ is available. In this case, the procedure R_1 cannot be applied. One natural way is to use the procedure R_2 which uses the hybrid estimator with $a=\tau\theta_0$. Numerical results show that R_2 is better than R if and only if θ is small. We can compare R_2 with the procedures based on the Pitman estimator and Bayes estimators. Based on Monte Carlo studies it is found that (a) for small θ the Bayes estimators are better than R_2 which in turn is better than the Pitman estimator, (b) for moderate or large θ the Pitman estimator is better than R_2 which in turn is better than Bayes estimators, and (c) the procedure R_2 is the most stable one in the sense that the change of the coverage probabilities over different θ values is the smallest. There is none that is found to be uniformly the best.

21.5.2 Single-Stage Procedure with Unknown Equal Variances

Let $\bar{X}_i$ be the mean of the sample of size n from $\pi_i, i=1,\ldots,k$, and s_ν^2 be the usual pooled unbiased estimate of σ^2 on $\nu=k(n-1)$ degrees of freedom. Chen and Dudewicz (1973a) have discussed one-sided and two-sided intervals for $\mu_{[i]}$. The one-sided intervals are defined by

$$I_{1L}=\left(\bar{X}_{[i]}-\frac{a_i s_\nu}{\sqrt{n}},\infty\right),$$

$$I_{1U}=\left(-\infty,\bar{X}_{[i]}+\frac{b_i s_\nu}{\sqrt{n}}\right), \tag{21.58}$$

where a_i and b_i are to be determined so that the coverage probability is not less than $\gamma(0<\gamma<1)$. It is shown that a_i and b_i are given by

$$F_{i,\nu}(a_i,\ldots,a_i|I_i)=\gamma,$$

$$F_{k-i+1,\nu}(b_i,\ldots,b_i|I_{k-i+1})=\gamma, \tag{21.59}$$

where $F_{r,\nu}(y,\ldots,y|I_r)$ is the cdf of an r-variate t distribution with ν degrees of freedom and identity correlation matrix I_r of order r. Values of a_i and b_i are tabulated by Krishnaiah and Armitage [1966] for $k=1(1)10$, $\nu=5(1)35$, and $\gamma=.90, .95, .975, .99$.

The one-sided intervals in (21.58) can be combined to obtain a two-sided interval with minimum probability coverage α. Let $\gamma_1,\gamma_2(0<\gamma_1,\gamma_2<1)$ be such that $\gamma_1+\gamma_2=\alpha+1$ (α fixed and $0<\alpha<1$). Then

$$I=\left(\bar{X}_{[i]}-\frac{a_i(\gamma_1)s_\nu}{\sqrt{n}},\bar{X}_{[i]}+\frac{b_i(\gamma_2)s_\nu}{\sqrt{n}}\right) \tag{21.60}$$

is a two-sided interval for $\mu_{[i]}$ with minimum coverage probability α.

Let us consider a two-sided interval for $\mu_{[k]}$ which is of the form

$$I(d_1,d_2)=\left(\bar{X}_{[k]}-\frac{d_1s_\nu}{\sqrt{n}},\bar{X}_{[k]}+\frac{d_2s_\nu}{\sqrt{n}}\right). \tag{21.61}$$

In the case of the symmetric interval with $d_1=d_2=d$, Chen and Dudewicz (1973a) have shown that the minimum coverage probability is γ if d is chosen as the solution of

$$F_{k,\nu}(d,\ldots,d|I_k)-F_{k,\nu}(-d,\ldots,-d|I_k)=\gamma. \tag{21.62}$$

It has also been shown that

$$\inf_{\Omega(\mu_{[k]})}\Pr\{\mu_{[k]}\in I(d_1,d_2)\}\geqslant$$

$$\min\left[\{F_{k,\nu}(d_1,\ldots,d_1|I_k)-F_{k,\nu}(-d_2,\ldots,-d_2|I_k)\},\right.$$
$$\left.\{F_\nu(d_1)-F_\nu(-d_2)\}\right], \tag{21.63}$$

where $F_\nu(\cdot)$ is the cdf of a univariate student t-distribution with ν degrees of freedom.

In the case of variances equal and known, the length of the two-sided confidence interval for $\mu_{[k]}$ proposed by Dudewicz (1972a) is shorter than that of the symmetric one proposed by Lal Saxena and Tong (1969) for most values of k, and the asymmetric confidence interval given by Dudewicz and Tong (1971) is optimum. Similar results in the case of equal but unknown variances are only conjectured.

21.5.3 Two-Stage Procedure with Unknown Variances

Let us first assume that the variances are equal to σ^2. For $\nu=1,2,\ldots,$ let the sequence of real numbers $\{a_\nu\}$ satisfy

$$F_{k,\nu}(a_\nu,\ldots,a_\nu|I_k)-F_{k,\nu}(-a_\nu,\ldots,-a_\nu|I_k)=\gamma, \tag{21.64}$$

where $F_{k,\nu}$ is defined as in (21.59). Following the idea of Stein [1945], Tong (1970) has proposed the following two-stage procedure for constructing a symmetric two-sided interval of width $2d$ and minimum coverage probability γ.

Let $\mathbf{X}_j=(X_{1j},X_{2j},\ldots,X_{kj})$, $j=1,2,\ldots,$ be a sequence of independent vector observations, each vector being a set of observations from the k normal populations. Let $\overline{X}_i(n)$ be the mean of the first n observations from π_i and $s_\nu^2=\nu^{-1}\Sigma_{i=1}^k\Sigma_{j=1}^n[X_{ij}-\overline{X}_i(n)]^2$ with $\nu=k(n-1)$.

PROCEDURE R_t: (a) Let $n_0\geqslant 2$ be a preassigned positive integer. Observe the sequence $\{\mathbf{X}_j\}$ for $j=1,\ldots,n_0$. Compute s^2 with $\nu=k(n_0-1)$. (b) Observe the sequence $\{\mathbf{X}_j\}$ for $j=n_0+1,\ldots,N$ where N is the smallest integer n satisfying

$$n\geqslant\max\left\{n_0,\frac{a_\nu^2s_\nu^2}{d^2}\right\}. \tag{21.65}$$

(c) Take the interval to be

$$I=\left(\overline{X}_{[k]}-d,\overline{X}_{[k]}+d\right), \tag{21.66}$$

where $\overline{X}_{[k]}$ is based on N observations.

Lal Saxena and Tong (1969) have used the two-sided interval in (21.66) using a single-stage sample of size n. Let us refer to this procedure as R_0. The minimum sample size required is the smallest integer $n\geqslant N_0$, where N_0 is the smallest integer greater than or equal to σ^2z^2/d^2 and $\Phi^k(z)-\Phi^k(-z)=\gamma$. For every $\lambda=\sigma/d>0$, if we disregard that N_0 must be an integer, the efficiency of R_t with respect to R_0 can be defined as

$$\eta_k(\lambda)=\frac{E(N)}{N_0}. \tag{21.67}$$

Note that the efficiency of R_t is $\leqslant 1$ if $\eta_k(\lambda) \geqslant 1$. Now let $C_k(\lambda)$ denote the infimum of the coverage probability of R_t. Then the results relating to R_t and its efficiency are summarized in the following theorem of Tong (1970).

Theorem 21.4. Under R_t, for every k and n_0 we have

a. $C_k(\lambda) \geqslant \gamma$ for every $\lambda > 0$,
b. $\eta_k(\lambda) \geqslant (a_\nu / z)^2$ for every $\lambda > 0$,
c. $\eta_k(\lambda) \to (a_\nu / z)^2$ as $\lambda \to \infty$.

Tong (1970) has also given an approximation to a_ν based on the Student's t-distribution which could be helpful for practical purposes. It should be noted that at least for $k \leqslant 2$, the efficiency of R_t is uniformly less than one. For $k=1$, the relative inefficiency of the two-stage procedure can be explained as a consequence of the fact that the observations in the second stage are not utilized in estimating the unknown parameter σ^2.

A different two-stage procedure has been proposed by Chen and Dudewicz (1976). Their approach is to use one-sided intervals to obtain a two-sided interval. We now assume that the variances σ_i^2 are unknown and possibly unequal. Let us first describe the sampling procedure.

Take an initial random sample $X_{i1}, \ldots, X_{in_0}$ of size $n_0 (\geqslant 2)$ from $\pi_i (i = 1, \ldots, k)$ and define

$$n_i = \max\left\{ n_0 + 1, \left\langle \frac{s_i^2}{h^2} \right\rangle \right\} \tag{21.68}$$

where s_i^2 is the sample variance based on $(n_0 - 1)$ degrees of freedom, $h = h(k, \gamma, n_0, a, b)$, and $\langle y \rangle$ denotes the smallest integer $\geqslant y$. Take $n_i - n_0$ additional observations $X_{i, n_0+1}, \ldots, X_{i, n_i}$ from π_i and define

$$\tilde{\bar{X}}_i = \sum_{j=1}^{n_i} a_{ij} X_{ij}, \qquad i = 1, \ldots, k. \tag{21.69}$$

Here the a_{ij} are to be chosen such that

$$\sum_{j=1}^{n_i} a_{ij} = 1, a_{i1} = \cdots = a_{in_0}, \quad \text{and } s_i^2 \sum_{j=1}^{n_i} a_{ij}^2 = h^2. \tag{21.70}$$

The one-sided intervals for $\mu_{[k]}$ proposed by Chen and Dudewicz (1976) are

$$I_L = \left(\tilde{\bar{X}}_{[k]} - a, \infty \right),$$

$$I_U = \left(-\infty, \tilde{\bar{X}}_{[k]} + b \right), \tag{21.71}$$

where the values of a and b for which each of the intervals will have minimum coverage probability γ are given by

$$F_\nu^k\left(\frac{a}{h}\right)=F_\nu\left(\frac{b}{h}\right)=\gamma \tag{21.72}$$

where $F_\nu(\cdot)$ is a univariate t distribution with $\nu=n_0-1$ degrees of freedom. The h-values are tabulated by Chen and Dudewicz (1976) for $a=b=1,2$, $k=1(1)10$, $\nu=n_0-1=1(1)20$, 25, 30, 40, 60, 100, and $\gamma=0.90$, 0.95, 0.99.

For constructing a two-sided interval, given $\alpha\in(0,1)$ and $L>0$, let γ_1 minimize $f(\gamma)=F_\nu^{-1}(\gamma^{1/k})+F_\nu^{-1}(1+\alpha-\gamma)$ subject to $\gamma\in(\alpha,1)$. Let $h=L/f(\gamma_1)$, and $a=hF_\nu^{-1}(\gamma_1^{1/k})$, and let n_i be given by (21.68). Then

$$I_T=\left(\tilde{\bar{X}}_{[k]}-a,\tilde{\bar{X}}_{[k]}+L-a\right) \tag{21.73}$$

is a two-sided confidence interval for $\mu_{[k]}$ of length L with confidence probability not less than α.

Chen and Dudewicz (1976) have also discussed a symmetric confidence interval for $\mu_{[k]}$, namely, $I_{TS}=(\tilde{\bar{X}}_{[k]}-d,\tilde{\bar{X}}_{[k]}+d)$. They have shown that I_{TS} has minimum coverage probability γ if h is the solution of $F_\nu^k(d/h)-F_\nu^k(-d/h)=\gamma$, where $\nu=n_0-1$. The values of h are given for $d=1$, $k=1,2,3$, $\nu=1(1)20$, 25, 40, 60, 100, and $\gamma=0.90$, 0.95, 0.99.

Chen (1977b) has used the approach of Chen and Dudewicz (1976) to define a class of two-sided intervals for $\mu_{[i]}$. Given length $L>0$, $\alpha\in(0,1)$, and any $\gamma\in(\alpha,1)$, the interval $(\tilde{\bar{X}}_{[i]}-a,\tilde{\bar{X}}_{[i]}+L-a)$ has a minimum confidence probability α when $h=L/f(\gamma)$ and $a=hF_\nu^{-1}(\gamma^{1/i})$ where $f(\gamma)=F_\nu^{-1}(\gamma^{1/i})+F_\nu^{-1}(1+\alpha-\gamma)^{1/(k-i+1)}$ subject to $\gamma\in(\alpha,1)$. Chen (1977b) has shown that, subject to $\gamma\in(\alpha,1)$, $f(\gamma)$ has a unique minimum occurring at $\gamma=\gamma_0$ when $\alpha>0.8413$. The values of h and a are tabulated by Chen (1977b) for $L=1$; $k=2(1)10$, 12; $\nu=2(1)25$, 27, 30(5)60, 80, 120; and $\alpha=0.80$, 0.90, 0.95, 0.975, 0.99, where γ is chosen such that $f(\gamma)$ reaches its minimum.

21.5.4 Efficiency of Some Multistage Procedures

In §21.5.3, we discussed a two-stage procedure R_T of Tong (1970) and noted that at least for $k\leqslant 2$ it is inefficient relative to R_0 [the procedure of Lal Saxena and Tong (1969) when the common variance σ^2 is known]. Tong (1970), in fact, has considered a class of multistage procedures and their efficiencies relative to R_0. We continue the notations used earlier.

Let R be any multistage procedure having the following features.

1. Initially $n_0(\geqslant 2)$ observations are taken from each population.
2. At any stage with observed $(\mathbf{X}_1, \mathbf{X}_2, \ldots, \mathbf{X}_n)$, the stopping rule N depends on $(\mathbf{X}_1, \mathbf{X}_2, \ldots, \mathbf{X}_n)$ only through $\mathbf{V}_n = (s^2_{k(n_0-1)}, s^2_{kn_0}, \ldots, s^2_\nu)$ where

$$s_\nu^2 = \nu^{-1} \sum_{i=1}^{k} \sum_{j=1}^{n} \left(X_{ij} - \overline{X}_i\right)^2 \tag{21.74}$$

with $\nu = k(n-1)$ for every n.

3. $\Pr[N < \infty | \boldsymbol{\mu}, \sigma^2, R] = 1$ and $E(N) < \infty$ for every $\sigma^2 > 0$.
4. For every observed value n of N, the terminal decision rule is given by $I = (\overline{X}_{[k]} - d, \overline{X}_{[k]} + d)$.

Let $\beta_k(y) = \Phi^k(y) - \Phi^k(-y)$, $y > 0$. It is concave for $y > 0$ when $k \leqslant 2$. For $k > 2$, there exists a y_k such that $\beta_k(y)$ is a concave function for $y > y_k$. The following theorem of Tong (1970) gives an upper bound for the coverage probability $C_k(\lambda)$ of any multistage procedure R with the features described previously.

Theorem 21.5. When $k \leqslant 2$,

$$C_k(\lambda) < \beta_k\left(\sqrt{E(N)}\ /\lambda\right) \qquad \text{for every } \lambda > 0, \tag{21.75}$$

and when $k > 2$, (21.75) holds for all $\lambda \leqslant \sqrt{n_0}\ / y_k$.

The efficiency of R relative to R_0 is defined by $\eta_k(\lambda)$ in (21.67). As a corollary to Theorem 21.5, we get $C_k(\lambda) \geqslant \gamma$ for every $k \leqslant 2$, $\lambda > 0$ and for every $k > 2$, $\lambda \leqslant \sqrt{n_0}\ / y_k$, only if $\eta_k(\lambda) > 1$. In other words, for $k \leqslant 2$ and for $k > 2$ with σ^2 not very large, the multistage procedure cannot do better than the single-stage procedure (with the advantage of knowing σ^2) if it must meet the minimum coverage probability requirement.

21.5.5 A Sequential Procedure

We remarked in §21.5.3 that for $k = 1$ the relative inefficiency of the two-stage procedure R_t can be explained as a consequence of the fact that the observations obtained in the second stage are not utilized in estimating the unknown parameter σ^2. This suggests the idea of performing the experiment so that σ^2 can be estimated sequentially. A random stopping rule for this purpose was developed by Chow and Robbins [1965]; its performance has been considered by Starr [1966] for the interval estimation of one population mean. Tong (1970) considered its application to the interval estimation of $\mu_{[k]}$. It should be noted that this solves the problem only asymptotically. If λ is not very large, $C_k(\lambda)$ may be less than γ. The sequential procedure R_s is described as follows. We continue the notations of §§21.5.3 and 21.5.4.

PROCEDURE R_s: (a) Let $n_0 \geqslant 2$ be a preassigned positive integer. Observe the sequence $\{\mathbf{X}_j\}$ one vector at a time. Stop with $\mathbf{X}_N$ where N is the first integer $n \geqslant n_0$ such that

$$s_\nu^2 \leqslant nd^2/a_\nu^2, \tag{21.76}$$

where $\nu = k(n-1)$, a_ν satisfies (21.64), and s_ν^2 is given by (21.74).

(b) For the observed value n of N in (a), apply the decision rule $I = (\overline{X}_{[k]} - d, \overline{X}_{[k]} + d)$.

The asymptotic behavior of R_s established by Tong (1970) is contained in the following theorem.

Theorem 21.6. Under R_s, for every k and every n_0, we have

a. $\lim_{\lambda\to\infty} N/\lambda^2 z^2 = 1$ a.s.,
b. $\lim_{\lambda\to\infty} \eta_k(\lambda) = 1$,
c. $\lim_{\lambda\to\infty} C_k(\lambda) = \gamma$.

21.5.6 Interval Estimation of the Largest Mean: Unequal Sample Sizes

The single-stage procedures discussed so far were based on samples of equal sizes from the k populations. We now discuss procedures based on samples of unequal sizes.

We first consider the case of a common known variance σ^2. Chen and Dudewicz (1973b) have considered one-sided intervals for $\mu_{[k]}$ given by

$$I_{01} = \left(\max_i \left[\overline{X}_i - \frac{a\sigma}{\sqrt{n_i}} \right], \infty \right),$$

$$I_{02} = \left(-\infty, \max_i \left[\overline{X}_i + \frac{b\sigma}{\sqrt{n_i}} \right] \right), \tag{21.77}$$

where the constants a and b subject to the minimum coverage probability requirement are given by

$$\Phi^k(a) = \gamma \quad \text{and} \quad \Phi(b) = \gamma. \tag{21.78}$$

A two-sided interval is given by

$$I_{03} = \left(\max_i \left[\overline{X}_i - \frac{d\sigma}{\sqrt{n_i}} \right], \max_i \left[\overline{X}_i + \frac{d\sigma}{\sqrt{n_i}} \right] \right) \tag{21.79}$$

where the constant d is to be chosen as the solution of

$$\Phi^k(d)+\Phi^k(-d)=\gamma. \tag{21.80}$$

The constant a in (21.78) and the constant d in (21.80) can be computed using the algorithm of Milton and Hotchkiss [1969] for $\Phi(\cdot)$.

One can also combine the results for the one-sided intervals to obtain a two-sided interval

$$I_{04}=\left(\max_i\left[\bar{X}_i-\frac{a(\gamma_1)\sigma}{\sqrt{n_i}}\right], \max_i\left[\bar{X}_i+\frac{b(\gamma_2)\sigma}{\sqrt{n_i}}\right]\right) \tag{21.81}$$

which is the intersection of I_{01} and I_{02} with coverage probabilities γ_1 and γ_2, respectively. If γ_1 and γ_2 are such that $\gamma_1+\gamma_2=\alpha+1$ (α fixed, $0<\alpha<1$), then I_{04} is a two-sided interval for $\mu_{[k]}$ with a coverage probability not less than α.

Let $L>0$ be given and define

$$I_{05}=\left(\max_i\left[\bar{X}_i-\frac{(L-d)\sigma}{\sqrt{n_i}}\right], \max_i\left[\bar{X}_i+\frac{d\sigma}{\sqrt{n_i}}\right]\right), \tag{21.82}$$

where $d<L$. This can be taken as a two-sided interval for $\mu_{[k]}$. The optimal choice of d is the value that maximizes $\inf_\Omega \Pr\{\mu_{[k]}\in I_{05}\}$. Chen and Dudewicz (1976) have shown that the optimum d is equal to $L/2$ for $k=1,2$ and is less than $L/2$ for $k>2$.

In the case of $n_1=\cdots=n_k=n$, I_{01}, I_{02}, and I_{04} reduce to the intervals of Dudewicz (1970a, 1972a), I_{03} becomes the interval of Lal Saxena and Tong (1969), and I_{05} reduces to the interval of Dudewicz and Tong (1971).

Other possible intervals are

$$\begin{aligned}
I_{11}&=\left(\bar{X}_{[k]}-a(\gamma_1)\sigma, \infty\right),\\
I_{12}&=\left(-\infty, \bar{X}_{[k]}+b(\gamma_2)\sigma\right),\\
I_{13}&=\left(\bar{X}_{[k]}-d(\gamma)\sigma, \bar{X}_{[k]}+d(\gamma)\sigma\right),\\
I_{14}&=\left(\bar{X}_{[k]}-a(\gamma_1)\sigma, \bar{X}_{[k]}+b(\gamma_2)\sigma\right)
\end{aligned} \tag{21.83}$$

with coverage probabilities at the least equal to γ_1, γ_2, γ, and

$\alpha=\gamma_1+\gamma_2-1$, respectively. The constants are given by,

$$\prod_{i=1}^{k}\Phi\big(a(\gamma_1)\sqrt{n_i}\,\big)=\gamma_1,\ \Phi\big(b(\gamma_2)\sqrt{n_{[1]}}\,\big)=\gamma_2,$$

$$\prod_{i=1}^{k}\Phi\big(d(\gamma)\sqrt{n_i}\,\big)-\prod_{i=1}^{k}\Phi\big(-d(\gamma)\sqrt{n_i}\,\big)=\gamma, \tag{21.84}$$

$$\prod_{i=1}^{k}\Phi\big(a(\gamma_1)\sqrt{n_i}\,\big)+\Phi\big(b(\gamma_2)\sqrt{n_{[1]}}\,\big)-1=\alpha,$$

where $n_{[1]}=\min(n_1,\ldots,n_k)$. However, there is no available result comparing the intervals in (21.83) and the earlier set $\{I_{01},I_{02},I_{03},I_{04}\}$.

Chen and Dudewicz (1976) have similarly considered the case of unknown common variance σ^2.

21.6 INTERVAL ESTIMATION OF THE LARGEST α-QUANTILE

Let $\pi_i(i=1,\ldots,k)$ have the unknown continuous distribution F_i and let $\xi_{\alpha i}$ be its unique α-quantile for $0<\alpha<1$. Define $\theta=\max_i\xi_{\alpha i}$. Rizvi and Lal Saxena (1972) have discussed a distribution-free confidence interval I such that

$$\inf_{\Omega}\Pr\{\theta\in I\}\geqslant\gamma \tag{21.85}$$

where Ω is the set of k-tuples $\{F_1,\ldots,F_k\}$.

Let $Y_{ri}(1\leqslant r\leqslant n)$ denote the rth order statistic based on n independent observations from π_i. Define $Y_r=\max_{1\leqslant i\leqslant k}Y_{ri}, r=1,\ldots,n$. We set $Y_0=-\infty$ and $Y_{n+1}=\infty$. For $s<t$, let

$$I=(Y_s,Y_t). \tag{21.86}$$

Then

$$\Pr\{\theta\in I\}=\prod_{i=1}^{k}G_s(F_i(\theta))-\prod_{i=1}^{k}G_t(F_i(\theta)), \tag{21.87}$$

where

$$G_r(p)=r\binom{n}{r}\int_0^p u^{r-1}(1-u)^{n-r}\,du, \qquad r=1,\ldots,n,$$

$$G_0(\cdot)\equiv1, \qquad \text{and } G_{n+1}(\cdot)\equiv0. \tag{21.88}$$

We get upper and lower confidence intervals (denoted by I_U and I_L) by taking (a) $s=0$, $t<n+1$, and (b) $s>0$, $t=n+1$. Rizvi and Lal Saxena (1972) have shown that

$$\inf_{\Omega} \Pr\{\theta \in I\} = \begin{cases} G_s^k(\alpha), & \text{when } I = I_L, \\ 1 - G_t(\alpha), & \text{when } I = I_U. \end{cases} \tag{21.89}$$

Thus, to meet the coverage probability requirement, we have to take the largest s such that $G_s^k(\alpha) \geqslant \gamma$ with $0<\gamma<\{1-(1-\alpha)^n\}^k$ and the smallest integer t such that $1-G_t(\alpha) \geqslant \gamma$ with $0<\gamma<1-\alpha^n$. For fixed k and α, we can find I_L and I_U for any γ provided n is taken large enough.

For the two sided interval $I(0<s<t<n+1)$,

$$\inf_{\Omega} \Pr\{\theta \in I\} = min\left[G_s(\alpha) - G_t(\alpha), G_s^k(\alpha) - G_t^k(\alpha) \right]. \tag{21.90}$$

For specified k, α, γ, and n, the choice of integers s and t such that the two-sided interval satisfies the probability requirement may not be unique unless some optimality criterion is introduced. Rizvi and Lal Saxena (1972) have used the optimality requirement that the ranks s and t be as close as possible. The following algorithm is given by them.

Let c be a positive integer $<n$. Consider $I=(Y_s, Y_{s+c})$ and let $Q(s,c)= \inf_{\Omega} \Pr\{Y_s<\theta<Y_{s+c}\}$. For every fixed c, let $s_0(c)$ be the value of $s(1 \leqslant s \leqslant n-c)$ for which $Q(s,c)$ is maximum. Now choose the smallest c (call it c_0) such that $Q(s_0(c),c) \geqslant \gamma$. Then the optimal interval is given by I with $c=c_0$ and $s=s_0(c_0)$.

For large n, one can use the normal approximation. We can see that, for the one-sided intervals, s is the largest integer such that $s \leqslant n\alpha + [n\alpha(1-\alpha)]^{\frac{1}{2}} \Phi^{-1}(1-\gamma^{1/k})$ and t is the smallest integer such that $t \geqslant n\alpha + [n\alpha(1-\alpha)]^{\frac{1}{2}} \Phi^{-1}(\gamma)$. For the two-sided optimum interval we have

$$Q(s,c) = \min\{\Phi(d-x) - \Phi(-x), \Phi^k(d-x) - \Phi^k(-x)\}, \tag{21.91}$$

where

$$d = c\left[n\alpha(1-\alpha) \right]^{\frac{1}{2}} \quad \text{and} \quad x = (t - n\alpha)\left[n\alpha(1-\alpha) \right]^{-\frac{1}{2}}. \tag{21.92}$$

Dudewicz and Tong (1971) showed that for any d, the value of x, say x_0, that maximizes the right-hand side of (21.91) is either the value for which the two terms within the braces of (21.91) are equal or the value which maximizes the second term. For $k=1$ and 2, $x_0=d/2$. Table 1 of Dudewicz and Tong (1971) gives x_0 for $k=3(1)6(2)14$ and $d=1(0.1)4$.

21.7 COMBINED GOALS OF SELECTION AND ESTIMATION

In this section we discuss two problems. The first is a problem of selecting the best of k Poisson processes and simultaneously estimating its parameter. The second is a problem of confidence interval with simultaneous selection in the cases of location and scale parameter families.

21.7.1 A Selection and Estimation Problem for Poisson Processes

Let $\pi_i (i=1,\ldots,k)$ be a Poisson process with parameter (rate of occurrence) λ_i. The goal considered by Alam and Thompson (1973) is to select the best process (the one associated with $\lambda_{[k]}$) and to estimate its parameter. Let $\pi_{(k)}$ denote the best population. Given $\theta > 1$, $\alpha > 0$, $0 < \beta < 1$, we want to select the population associated with $\lambda_{[k]}$ and seek an estimator δ of $\lambda_{[k]}$ such that

$$P = \Pr\left\{ \pi_{(k)} \text{ is selected and } \left| \frac{\delta}{\lambda_{[k]}} - 1 \right| \leqslant \alpha \right\} \geqslant \beta \text{ whenever } \frac{\lambda_{[k]}}{\lambda_{[k-1]}} \geqslant \theta > 1. \tag{21.93}$$

Note that the accuracy of the estimate is not important if an incorrect selection has been made. We assume that $\alpha < 1$; otherwise, the condition on the error of estimation is trivally satisfied for $\delta \equiv 0$.

Alam and Thompson (1973) have considered two cases, namely, (i) the continuous case in which the processes are observed continuously in time, and (ii) the discrete case in which they are observed at successive intervals of time $t = 1, 2, \ldots$. Following are two procedures R and S applicable to the continuous and discrete cases, respectively.

PROCEDURE R: Let $t_{i,N}$ denote the waiting time for the Nth occurrence from π_i, where N is a fixed positive integer. Let $t_N = \min(t_{1,N}, \ldots, t_{k,N})$. Observe t_N. Let $t_{i,N} = t_N$. Select π_i as the best process, and set $\delta = c/t_N$, where $c > 0$ is a predetermined number.

PROCEDURE S: Let $U(m,n)$ denote the mth smallest observation in a random sample of n observations from a uniform distribution on $(0,1)$. Let $X_{i,t}$ denote the number of occurrences form π_i until time t, and let $T_{i,N}$ denote the smallest positive integer value of t for which $X_{i,t} \geqslant N$. Let $T_N = \min(T_{1,N}, \ldots, T_{k,N})$. Observe T_N. Let I denote the set of integers i for which $T_{i,N} = T_N$. For each $i \in I$, observe $U(m_i, n_i)$ where $m_i = N - X_{i,T_N - 1}$ and $n_i = X_{i,T_N} - X_{i,T_N - 1}$. Let $U_i = T_N - 1 + U(m_i, n_i)$ and $U = \min(U_j, j \in I)$. Let $U_i = U$. Select π_i as the best process and set $\delta = c/U$.

If we consider k Poisson populations with means $\lambda_1,\ldots,\lambda_k$, the procedure S leads to a sequential procedure for selecting the population with mean $\lambda_{[k]}$ and estimating $\lambda_{[k]}$.

In the continuous case, the waiting time for the Nth occurrence from a Poisson process with parameter λ has a gamma distribution with the density

$$g_\lambda(t)=\left[\Gamma(N)\right]^{-1}\lambda^N t^{N-1}e^{-\lambda t}, t>0. \tag{21.94}$$

Therefore, t_N (defined for R) represents the smallest observed value in a sample of k observations taken one from each of k gamma distributions with parameters $\lambda_1,\ldots,\lambda_k$. In the discrete case, it can be shown that U has the same distributions as t_N. Thus, for both R and S,

$$P=\int_{c/(1+\alpha)}^{c/(1-\alpha)}\prod_{i=1}^{k-1}\left[1-G\left(\frac{\lambda_{[i]}x}{\lambda_{[k]}}\right)\right]dG(x) \tag{21.95}$$

where $G(t)$ is the cdf corresponding to $g_\lambda(t)$ with $\lambda=1$.

It has been shown by Alam and Thompson (1973) that P is minimized when $\lambda_{[1]}=\cdots=\lambda_{[k-1]}=\theta^{-1}\lambda_{[k]}$.

The value of P at a least favorable configuration depends only on k, c, α, θ, and N. Let this value be denoted by Q. Given k, α, θ, and N, let $c(N)$ be the value of c that maximizes Q and let $P(N)$ denote this maximum value of P. Alam and Thompson (1973) have tabulated the values of $P(N)$ and $c(N)$ for $k=2(1)5$, $\theta=1.1$, 1.2, 1.4, $\alpha=0.1$, 0.2, and $N=2(2)20(5)40$. It is conjectured that $P(N)$ is monotone increasing in N. Given $\beta(0<\beta<1)$, the smallest value of N equal to n, say, can be determined such that $P(N)\geqslant\beta$ for $N\geqslant n$. The values of n are given for $k=2(1)5$, $\theta=1.1$, 1.2, 1.4, $\alpha=0.1$, 0.2, and $P(N)=0.80(0.05)0.95$, 0.99.

21.7.2 Interval Estimation and Simultaneous Selection: Location and Scale Parameters

Let $\{F(x;\theta)\}$ be a family of absolutely continuous distribution functions on the real line indexed by parameter θ and $f(x,\theta)$ denote the density corresponding to $F(x,\theta)$. Let $\pi_i, i=1,\ldots,k$, be k populations from the preceding family with parameters $\theta_1,\ldots,\theta_k$, respectively. For $1\leqslant t\leqslant k$, Rizvi and Lal Saxena (1974) have discussed the problem of selecting all the populations with $\theta_i\geqslant\theta_{[k-t+1]}$ and simultaneously give an interval I such that $\theta_{[k-t+1]}\in I$. Let CS denote the (correct) selection of all π_j with $\theta_{[j]}$, $j=k-t+1,\ldots,k$, and CD (correct dedision) denote the inclusion of $\theta_{[k-t+1]}$ in I. Let $P(\boldsymbol{\theta})$ denote $\Pr\{\text{CS}\cap\text{CD}|\text{R}\}$. Then the procedure R, for some preassigned $\gamma(0<\gamma<1)$, is required to satisfy

$$\inf_{\Omega(\psi)} P(\boldsymbol{\theta}) \geq \gamma, \tag{21.96}$$

where $\Omega(\psi)=\{\boldsymbol{\theta}:\theta_{[k-t]}\leq\psi(\theta_{[k-t+1]})\}$ and ψ is a given function on the real line such that $\psi(x)<x$.

Let $X_1,\ldots,X_k$ be a set of observations from $\pi_1,\ldots,\pi_k$, respectively. Consider two suitably chosen continuous increasing functions h_1 and h_2 (with inverses g_2 and g_1, respectively). The procedure R of Rizvi and Lal Saxena (1974) selects the populations corresponding to $X_{[j]}, j=k-t+1, \ldots,k$, and choose $I=(h_1(X_{[k-t+1]}),h_2(X_{[k-t+1]}))$.

The following theorem gives the result regarding the infimum of $P(\boldsymbol{\theta})$.

Theorem 21.7. Suppose $F(x,\theta)$ is differentiable for all θ in the parameter space. Also, suppose that (a) $(\partial/\partial\theta)F(x,\theta)\leq 0$ for all x, (b) $(\partial/\partial\theta)F(x,\theta)/(\partial/\partial\theta)F(x',\theta)$ is an increasing function of θ for every $x>x'$, and (c) $F(x,\theta^*)=0$ for all x, where θ^* is the largest possible value of the parameter ($+\infty$ included). Then

$$\inf_{\Omega^*(\psi)} P(\boldsymbol{\theta}) = \min_{r=0,1,\ldots,t-1} P(\boldsymbol{\theta}^{(r)}) \tag{21.97}$$

where $\Omega^*(\psi)=\{\boldsymbol{\theta}:\theta_{[k-t]}\leq\psi(\theta_{[k-t+1]}),\theta_{[k-t+1]}=\theta \text{ held fixed}\}$ and $\boldsymbol{\theta}^{(r)}$ is a vector $\boldsymbol{\theta}$ with the first $(k-t)$ ordered components equal to $\psi(\theta),(r+1)$ among the best t ordered components equal to θ and the rest equal to θ^*.

If the conditions of Theorem 21.7 hold and

$$F(g_1(\theta),\theta)+F(g_2(\theta),\theta)\geq 1, \tag{21.98}$$

then it has been proved by Rizvi and Lal Saxena (1974) that the minimum on the right-hand side of (21.97) occurs for $r=t-1$.

If the X_i are consistent estimators, it is possible to choose $g_1(\theta)$ and $g_2(\theta)$ so that the infimum of $P(\theta)$ over $\Omega^*(\psi)$ goes to unity for any fixed θ as the common sample size n tends to infinity. This requires choosing $g_1(\theta)$ and $g_2(\theta)$ such that $g_1(\theta)<\theta, g_2(\theta)>\theta$ and (21.98) holds.

The condition (a) is satisfied when θ is a location or a scale parameter. The condition (b) reduces in these cases to that of a monotone likelihood ratio. We briefly mention some explicit results in these two cases. In what follows we write $F_n(x,\theta)$ for a typical $F(x,\theta)$.

Location Parameter Case. Let $F_n(x,\theta)=F_n(x-\theta)$, $\psi(\theta)=\theta-\delta$, $g_1(\theta)=\theta-a$, $g_2(\theta)=\theta+b$, where $\delta\geq 0$ and a and b with $a+b>0$ are preassigned constants. If the density $f_n(x-\theta)$ has a monotone likelihood ratio in x, and

a and b are chosen such that

$$F_n(-a)+F_n(b)\geqslant 1, \tag{21.99}$$

then for $1\leqslant t\leqslant k$,

$$\inf_{\Omega(\psi)} P(\boldsymbol{\theta})=t\int_{-a}^{b} F_n^{k-t}(x+\delta)\left[1-F_n(x)\right]^{t-1}dF_n(x). \tag{21.100}$$

If the X_i are consistent and both a and b are chosen to be positive constants satisfying (21.99), then the right-hand side of (21.100) goes to unity as $n\to\infty$ for any fixed k and t. Thus for a preassigned γ, there exists a smallest value of n so that (21.96) is satisfied. If $F_n(\cdot)$ is symmetric about the origin, a and b can be any pair with $b\geqslant a>0$. In general, the choice of a and b may depend on n.

Scale Parameter Case. Let $F_n(x,\theta)=F_n(x/\theta), x>0, \theta\geqslant 0, F_n(0)=0$, $\psi(\theta)=c\theta$, $g_1(\theta)=\theta/a$ and $g_2(\theta)=\theta/b$, where a,b,c are preassigned constants such that $0\leqslant b<a$ and $0<c\leqslant 1$. If $(1/\theta)f_n(x/\theta)$ has a monotone likelihood ratio in x for θ, and constants a and b are chosen such that

$$F_n(a^{-1})+F_n(b^{-1})\geqslant 1, \tag{21.101}$$

then for $1\leqslant t\leqslant k$,

$$\inf_{\Omega(\psi)} P(\boldsymbol{\theta})=t\int_{a^{-1}}^{b^{-1}} F_n^{k-t}\left(\frac{x}{c}\right)\left[1-F_n(x)\right]^{t-1}dF_n(x). \tag{21.102}$$

If the X_i are consistent and constants a and b are chosen (possibly depending on n) such that $0\leqslant b<1<a$ in addition to (21.101), the right-hand side of (21.102) goes to unity as $n\to\infty$. Thus there exists a smallest value of n so that (21.96) is satisfied.

REMARK 21.2 The formulation of the problem includes as a special case the indifference zone formulation of the ranking problem by taking $a=b=\infty$ in the case of location parameters, and $a=\infty$ and $b=0$ in the case of scale parameters. It also includes as a special case the confidence interval formulation for the largest location or scale parameters by setting $t=k$. □

21.8 TOLERANCE INTERVALS FOR STOCHASTICALLY ORDERED DISTRIBUTIONS

Let $F_1,\ldots,F_k$ be unknown continuous distributions belonging to a family of stochastically ordered distributions. Let $F_{[1]}\leqslant F_{[2]}\leqslant\cdots\leqslant F_{[k]}$ denote the ordered cdf's so that $F_{[1]}$ is the stochastically largest distribution. Lal

Saxena (1976b) has discussed distribution-free tolerance intervals for $F_{[j]}$, $1 \leqslant j \leqslant k$. Let $X_{i1},\ldots,X_{in}$ be a random sample from $F_i, i=1,\ldots,k$. For fixed j, let $I_j = I_j(X_{i1},\ldots,X_{in})$ denote a tolerance interval for $F_{[j]}$. Since I_j is a random interval, the probability coverage of I_j by $F_{[j]}$, denoted by $P_{(j)}(I_j)$, is itself a random variable. Let $\Omega = \{\mathbf{F} : \mathbf{F} = (F_1,\ldots,F_k)\}$.

Definition 21.6. An interval I_j is a *β-expectation tolerance interval* for $F_{[j]}$ if

$$\inf_{\Omega} E_{\mathbf{F}}\big(P_{(j)}(I_j)\big) \geqslant \beta. \tag{21.103}$$

Definition 21.7. An interval I_j is a *β-content tolerance interval* for $F_{[j]}$ at confidence level γ if

$$\inf_{\Omega} P_{\mathbf{F}}\big(P_{(j)}(I_j) \geqslant \beta\big) \geqslant \gamma. \tag{21.104}$$

Let $Y_{i:r,n}$ denote the rth order statistic in the sample from F_i. For every i, we define $Y_{i:0,n} = -\infty$ and $Y_{i:n+1,n} = \infty$. For $i \leqslant i'$ and $r \leqslant r'$, (with at least one strict inequality), let us consider the following tolerance intervals for $F_{[j]}$.

$$\begin{aligned} I_{1j} &= (-\infty, Y_{[i']:r',n}) \text{ for } i' \geqslant k-j+1; \\ I_{2j} &= (Y_{[i]:r,n}, \infty) \text{ for } i \leqslant k-j+1; \\ I_{3j} &= (Y_{[i]:r,n}, Y_{[i']:r',n}) \text{ for } i \leqslant k-j-1 \leqslant i', r \leqslant r' \end{aligned} \tag{21.105}$$

with at least one strict inequality.

Then

$$\begin{aligned} P_{(j)}(I_{1j}) &= F_{[j]}(Y_{[i']:r',n}), \\ P_{(j)}(I_{2j}) &= 1 - F_{[j]}(Y_{[i]:r,n}), \\ P_{(j)}(I_{3j}) &= F_{[j]}(Y_{[i']:r',n}) - F_{[j]}(Y_{[i]:r,n}). \end{aligned} \tag{21.106}$$

Lal Saxena (1976b) has obtained the infima of $E_{jm} \equiv E_{\mathbf{F}}(P_{(j)}(I_{mj}))$ and $P_{jm} \equiv P_{\mathbf{F}}(P_{(j)}(I_{mj}) \geqslant \beta), m = 1,2,3$. To state his results, let $Z_{[i],j}(r,n)$ denote the ith order statistic in a random sample of size j from a beta distribution with density

$$g(z;r,n) = \frac{\Gamma(n+1)}{\Gamma(r)\Gamma(n-r+1)} z^{r-1}(1-z)^{n-r}, 0<z<1,$$

and with the cdf $G(z;r,n)$. The results of Lal Saxena (1976b) are given in the following theorem.

Theorem 21.8.

a. For $i' \geqslant k-j+1$,

$$\inf_{\Omega} E_{j1} = EZ_{[i'-k+j],j}(r',n) \text{ and}$$

$$\inf_{\Omega} P_{j1} = 1 - G(G(\beta;r',n);i-k+j,j).$$

b. For $i \leqslant k-j+1$,

$$\inf_{\Omega} E_{j2} = 1 - EZ_{[i],k-j+1}(r,n) \text{ and}$$

$$\inf_{\Omega} P_{j2} = G(G(1-\beta;r,n);i,k-j+1).$$

c. For $i \leqslant k-j+1 \leqslant i'$, $r \leqslant r'$ with at least one strict inequality,

$$\inf_{\Omega} E_{j3} \geqslant EZ_{[i'-k+j],j}(r',n) - EZ_{[i],k-j+1}(r,n) \text{ and}$$

$$\inf_{\Omega} P_{j3} > G\left(G\left(\frac{1-\beta}{2};r,n \right);i,k-j+1 \right) - G\left(G\left(\frac{1+\beta}{2};r',n \right);i'-k+j,j \right).$$

As regards the choices of i, i', r, and r', if the interest is in the criterion E_{jm}, then it is desirable to have i as large as possible and i' as small as possible to keep the interval as small as possible. So we take $i=k-j+1=i'$. Also, we would want r' as small as possible and r as large as possible. On the other hand, if one is interested in P_{jm}, it is still desirable to choose i as large as possible and i' as small as possible. So we take as before $i=k-j+1=i'$. For deciding on r and r' for the two-sided tolerance interval it is suggested that the largest r and the smallest r' be chosen so that

$$G\left(\frac{1-\beta}{2};r,n \right) \geqslant \left(\frac{1+\gamma}{2} \right)^{1/(k-j+1)}$$

and

$$1 - \left[1 - G\left(\frac{1+\beta}{2};r',n \right) \right]^{j} \leqslant \frac{1-\gamma}{2},$$

provided γ is between 0 and $\min_{t=j,k-j+1}[2\{1-((1+\beta)/2)^n\}^t - 1]$.

21.9 NOTES AND REMARKS

Alam and Lal Saxena (1974) have also discussed the special case of binomial parameters. Chen (1977a) has applied his results for the stochastically ordered distributions to the cases of binomial and Poisson parameters. The nonexistence of an exact β-confidence interval for a scale parameter is shown by Chen (1977c).

CHAPTER 22

General Theory of Some Multiple Decision Problems and Some Miscellaneous Topics

22.1 INTRODUCTION

In this chapter we discuss some general theory of certain multiple decision problems and other related topics such as slippage tests and selection of variables in linear regression. Our discussions of these topics are in no way meant to be exhaustive because of the size of the literature in these areas. However, we briefly discuss certain salient features of these problems since they are of related interest.

22.2 GENERAL THEORY OF SOME MULTIPLE DECISION PROBLEMS

Lehmann (1957a) has discussed a class of multiple decision procedures and has shown the members of this class to be possessing uniformly minimum risk among all procedures that are unbiased with respect to a certain loss function. In a companion paper (1957b), he has extended the theory to problems in which it is permitted not to come to a definite conclusion regarding one or more of the questions under consideration.

22.2.1 A Class of Multiple Decision Problems

Let X be a random observable, the distribution of which depends on the parameter θ. We are interested in a number of different hypotheses concerning θ, say H_γ: $\theta \in \omega_\gamma$, $\gamma \in \Gamma$. The class of alternatives K_γ to H_γ is that θ lies in the complement of ω_γ, which is denoted by ω_γ^{-1}. When considering these hypotheses simultaneously, we wish to determine which of these hypotheses hold and which do not. One is therefore faced with a multiple decision problem in which the different possible decisions correspond to the statements that a certain set of the hypotheses H_γ is correct while the remaining ones are false, or equivalently, that θ lies in a certain set, say Ω_i, $i \in I$, determined by these conditions. The sets Ω_i, which are the atoms of the field of sets generated by sets ω_γ, are formally given by

$$\Omega_i = \bigcap_{\gamma \in \Gamma} \omega_\gamma^{x_{i\gamma}} \tag{22.1}$$

where the x's indicate which of the hypotheses are true ($x_{i\gamma} = 1$) and which are false ($x_{i\gamma} = -1$) for the given Ω_i. Some of the intersections formally defined by (22.1) could be empty, in which case we restrict the Ω's to denote the nonempty ones and require that none of the possible decisions should correspond to the empty intersections.

To specify a loss function, we assume that the losses for the individual testing problems are a_γ and b_γ for falsely rejecting and accepting H_γ, respectively. We further assume that in the simultaneous consideration of these problems the losses are additive. If $\theta \in \Omega_i$ and the decision d_k is taken that $\theta \in \Omega_k$, the resulting loss is

$$w_{ik} = \sum_{\gamma \in \Gamma} (\epsilon_{ik\gamma} a_\gamma + \epsilon_{ki\gamma} b_\gamma), \tag{22.2}$$

where

$$\epsilon_{ik\gamma} = \begin{cases} 1 & \text{if } x_{i\gamma} = 1,\ x_{k\gamma} = -1, \\ 0 & \text{otherwise.} \end{cases} \tag{22.3}$$

In other words, w_{ik} is the sum of all those a_γ for which H_γ is true but rejected plus the sum of all the b_γ for which H_γ is falsely accepted. The risk is thus a weighted sum of the probabilities of error. One can define w_{ik} slightly more generally by setting

$$w_{ik} = \sum_{\gamma \in \Gamma} \nu_\gamma (\epsilon_{ik\gamma} a_\gamma + \epsilon_{ki\gamma} b_\gamma)$$

in the case of finite Γ, using any positive weights ν_i, and

$$w_{ik} = \int_\Gamma (\epsilon_{ik\gamma} a_\gamma + \epsilon_{ki\gamma} b_\gamma)\, d\mu(\gamma) \tag{22.4}$$

in the general case.

For problems of interest to us here, we can restrict ourselves to nonrandomized procedures with no essential loss of generality. Then a decision procedure for the given multiple decision problem is a partition of the sample space into sets D_i such that, when the observation falls into D_i, the decision d_i: $\theta \in \Omega_i$ is taken. We can relate these decision procedures to the tests of hypotheses H_γ. Let $\mathcal{C}$ be a family of such tests with acceptance regions A_γ for H_γ and the rejection regions A_γ^{-1} and consider the decision procedure induced by $\mathcal{C}$ through the relation

$$D_i = \bigcap_{\gamma \in \Gamma} A_\gamma^{x_{i\gamma}}. \tag{22.5}$$

It is possible for $P(\cup_{i \in I} D_i)$ to be less than unity. The partition (22.5) is then not a procedure for the given decision problem. The family $\mathcal{C}$ is said to be *compatible* when the induced procedure satisfies

$$P\Big(\bigcup_{i \in I} D_i \Big) = 1 \qquad \text{for all} \quad \theta. \tag{22.6}$$

Compatibility is equivalent to the condition that the simultaneous application of the tests A_γ not lead to any inconsistencies.

When Γ is uncountable, it is possible that a set of measurable A_γ's gives rise to a nonmeasurable D_i. In specific problems such difficulties can be eliminated and so we assume that such difficulties do not arise. Lehmann (1957a) has shown that the relation (22.5) defines a 1–1 correspondence between the decision procedures for the problem induced by the hypotheses H_γ, $\gamma \in \Gamma$, and the compatible families $\mathcal{C}$ of tests of these hypotheses.

Lehmann (1957a) has illustrated the preceding general concepts with examples of three-decision problems relating to a single parameter, classification problems for two independent parameters, and comparison of several populations. The simplest three-decision problem is to decide whether a real-valued parameter θ is less than, equal to, or greater than a specified value θ_0. This may be generated by the hypotheses H_1: $\theta \geqslant \theta_0$ and H_2: $\theta \leqslant \theta_0$, which according to (22.1) lead to the choice from the three parameter sets $\Omega_0 = \omega_1\omega_2$: $\theta = \theta_0$; $\Omega_1 = \omega_1\omega_2^{-1}$: $\theta > \theta_0$; $\Omega_2 = \omega_1^{-1}\omega_2$: $\theta < \theta_0$. Frequently one wishes to determine only whether or not θ is approximately equal to θ_0. The corresponding three-decision problem of deciding whether $\theta < \theta_1$, $\theta_1 \leqslant \theta \leqslant \theta_2$ or $\theta > \theta_2$ (where $\theta_1 < \theta_0 < \theta_2$), can be generated by the hypotheses H_1: $\theta \geqslant \theta_1$ and H_2: $\theta \leqslant \theta_2$.

Let ξ, η be two real-valued parameters. The problem of classifying ξ and η as being $\leqslant$ or $>$ than ξ_0 and $\leqslant$ or $>$ than η_0, respectively, leads to the choice from the four parameter sets

$$\Omega_1\colon \xi \leqslant \xi_0, \eta \leqslant \eta_0; \quad \Omega_2\colon \xi \leqslant \xi_0, \eta > \eta_0;$$

$$\Omega_3\colon \xi > \xi_0, \eta \leqslant \eta_0; \quad \Omega_4\colon \xi > \xi_0, \eta > \eta_0.$$

This can be generated by the hypotheses H_1: $\xi \leqslant \xi_0$, H_2: $\eta \leqslant \eta_0$.

For comparing several populations, let samples be given from k populations with distributions depending on the parameters $\theta_1, \ldots, \theta_k$ and possibly on certain nuisance parameters, which may or may not be common to the different populations. We may consider the hypotheses H_{ij}: $\theta_i \leqslant \theta_j$ or H'_{ij}: $\theta_i \leqslant \theta_j + \Delta$, $(\Delta > 0)$.

The problem of point estimation can also be formulated as a multiple decision problem. Let θ be a continuous real-valued parameter, and consider the decision problem generated by the set of hypotheses $H(\theta_0)$: $\theta \leqslant \theta_0$. If θ^* denotes the true value of the parameter, the hypothesis $H(\theta_0)$ with $\theta^* \leqslant \theta_0$ are true while those with $\theta^* > \theta_0$ are false. The associated intersection (22.1) is therefore

$$\bigcap_{\theta_0 \geqslant \theta^*} \omega(\theta_0) \cap \bigcap_{\theta_0 < \theta^*} \omega^{-1}(\theta_0). \tag{22.7}$$

The set (22.7) obviously consists of the single point θ^*, or in the case that nuisance parameters are present, of the totality of points for which $\theta = \theta^*$. The possible decisions in the induced multiple decision problem correspond to the possible true values of θ.

In some situations one may be interested in obtaining a point estimate of θ after testing and rejecting some hypothesis concerning it. If, for example, a new treatment is compared with a standard one, we may test the hypothesis that the new treatment does not represent an improvement, that is, that $\theta = \eta - \xi \leqslant 0$, where η and ξ denote the means of the new and old treatments. In case the hypothesis is rejected one requires an estimate of $\eta - \xi$. A procedure for testing H: $\theta \leqslant \theta_0$, and estimating θ in case of rejection, can be generated by the set of hypotheses $H(\theta_1)$: $\theta \leqslant \theta_1$, with $\theta_1 \geqslant \theta_0$.

Lehmann (1957a) has also discussed some nonparametric problems which can be treated by simultaneous tests of hypotheses. Let $X_1, \ldots, X_n$ be independently distributed with cdf F. The problem of goodness of fit is to decide whether $F = F_0$. In case this is rejected, suppose we want to determine the sets of points u for which $F(u)$ is $<$, $=$, and $> F_0(u)$. This problem may be generated by the set of hypotheses $H_+(u)$: $F(u) \geqslant F_0(u)$ and $H_-(u)$: $F(u) \leqslant F_0(u)$. One can similarly handle the two-sample problem and the problem of estimating a cumulative distribution function.

In all the examples explained previously, Lehmann (1957a) has given procedures for testing the corresponding hypotheses. For many distributions of interest (normal, binomial, Poisson, etc.) the indicated procedures possess the optimum property of uniformly minimizing the risk among all unbiased procedures, provided the levels α_γ of the test φ_γ of H_γ are related to the losses a_γ and b_γ through the equation

$$\alpha_\gamma = \frac{b_\gamma}{(a_\gamma + b_\gamma)}. \tag{22.8}$$

22.2.2 Restricted Products of Decision Problems

The method, described in §22.2.1, of generating a multiple decision problem from a set of hypotheses is a special case of the following process. Let $\mathcal{P} = \{P_\theta, \theta \in \Omega\}$ be a family of distributions, $\mathcal{D} = \{d\}$, a space of possible decisions, and $W(\theta, d)$, a loss function. Suppose that $\mathcal{P}$ is fixed but that two different decision spaces $\mathcal{D}'$, $\mathcal{D}''$ with the loss functions W', W'' are of interest. From the two associated decision problems we form a new problem by considering simultaneously the two given ones. This new problem is called the *product* of the two. Its decision space is the Cartesian product $\mathcal{D}' \times \mathcal{D}''$ and the loss resulting from the decision $d = (d', d'')$ is

$$W(\theta, d) = W'(\theta, d') + W''(\theta, d''), \tag{22.9}$$

or slightly more generally $W(\theta, d) = \rho W'(\theta, d') + (1-\rho) W''(\theta, d'')$. If in a product problem some of the decision pairs (d', d'') are omitted from $\mathcal{D}' \times \mathcal{D}''$, so that $\mathcal{D}$ is a subset of $\mathcal{D}' \times \mathcal{D}''$, we refer to this as a *restricted product*. Given any procedures δ', δ'' for the component problems with decision spaces $\mathcal{D}'$ and $\mathcal{D}''$, let $\delta = (\delta', \delta'')$ be the procedure that takes the decision (d',d'') when $\delta'=d'$ and $\delta''=d''$. The pair (δ',δ'') is said to be *compatible* with the given set of restrictions if $P_\theta\{(\delta'(X),\delta''(X))\in\mathfrak{D}\}=1$ for all θ.

22.2.3. Decision Procedures Allowing Partial Conclusions

Lehmann (1957b) extended the theory of his earlier paper (1957a) to problems in which it is allowed not to come to a definite conclusion regarding one or more of the questions under consideration.

In earlier discussion we were concerned with testing a set of hypotheses H_γ: $\theta\in\omega_\gamma$, so that in each component problem the choice was between the two decisions $\theta\in\omega_\gamma$ and $\theta\in\omega_\gamma^{-1}$. Instead of this one may want to know only whether the data reject the hypothesis. In case they do not, no alternative positive statement is required. The choice then lies between the statements $\theta\in\omega_\gamma^{-1}$ and $\theta\in\omega_\gamma^{0}=\Omega$.

If we consider simultaneously a number of such problems, we have a multiple decision problem in which the different possible decisions correspond to the statements that a certain number of hypotheses H_γ are false, while nothing is said regarding the remaining hypotheses. This is equivalent to

$$\Omega_i=\bigcap_\gamma \omega_\gamma^{y_{i\gamma}} \tag{22.10}$$

where the y's are -1 for each rejected H_γ and 0 for the others. When all the y's are zero, we have $\Omega_i=\Omega$, and thus make no statement whatever about the position of θ. As before, some of the formal intersections (22.10) may be empty and we shall then restrict the Ω's to denote the nonempty ones and require that none of the possible decisions should correspond to these empty intersections. If we assume that in the simultaneous consideration of a number of problems the losses are additive so that the total loss is the sum (or a weighted sum) of the losses of the component problems, and if we suppose the losses to be a_γ for rejecting H_γ when $\theta\in\omega_\gamma$, b_γ for not rejecting it when $\theta\in\omega_\gamma^{-1}$, and zero in the other two cases, the total loss is again given by (22.2) and (22.3) with

$$x_{i\gamma}=2y_{i\gamma}+1. \tag{22.11}$$

It now leads to the statement $\theta\in\omega_\gamma^{-1}$ or $\theta\in\omega_\gamma^{0}$ as $X\in A_\gamma^{-1}$, or $X\in A_\gamma$, where A_γ is the acceptance region of the best unbiased test of H_γ. Carrying out simultaneously a number of these tests leads to a procedure in which the decision d_i: $\theta \in \Omega_i$ is taken when X falls in the set

$$D_i = \bigcap_{\gamma} A^{x_{i\gamma}} \tag{22.12}$$

where Ω_i is now defined by (22.10) and (22.11).

The sets Ω_i, which define the possible decisions, no longer constitute a partition of the parameter space. It is possible that $\Omega_i = \Omega_j$, $i \neq j$. However, the losses resulting from decisions d_i and d_j when θ is in some Ω_k will usually not be the same. For this reason one must consider all formal intersections (22.10) as distinct. In applications of interest to us, this difficulty is overcome by a natural further restriction of the decision space. We eliminate from the list of permissible decisions not only the formal intersections (22.10) that are empty, but also those for which the intersection Ω_i equals some other Ω_j satisfying $y_{j\gamma} \leqslant y_{i\gamma}$ for all γ.

22.2.4 Some Further Results on Multiple Decision Problems

Schaafsma (1969) has also discussed minimax risk and unbiasedness for multiple decision problems. His paper starts from a point of view similar to Lehmann (1957a). However, unlike Lehmann, he does not define the loss for all $\theta \in \Omega$. The basic setup is as follows: On the basis of an observation x in the sample space $\mathcal{X}$ of a random observable X, we choose a decision d out of the finite space $\mathcal{D} = \{d_0, \ldots, d_a\}$ of all interesting decisions. It is known that X has the density p_θ out of the class $\mathcal{P} = \{p_\theta\colon \theta \in \Omega\}$ of admissible densities with respect to some σ-finite measure μ over $\mathcal{X}$. The parameter space Ω is partitioned into $m+1$ mutually exclusive subsets so that $\Omega = \Omega_0 \cup \ldots \cup \Omega_m$ where Ω_0 is an indifference zone ($\Omega_0 = \varnothing$ is permitted). Let $\Omega' = \Omega - \Omega_0$ and the loss function L: $\Omega' \times \mathcal{D} \to [0, \infty)$ be defined by $L(\theta, d) = w_{ij}$ for $d = d_j$ and $\theta \in \Omega_i$ ($i = 1, \ldots, m$; $j = 0, \ldots, a$). For fixed d, the loss is constant over each Ω_i whereas no loss is defined for $\theta \in \Omega_0$. We are concerned with multiple decision problems of Type I; that is, the parameter space Ω is a subset of the Euclidean space such that (i) $E_\theta\{\varphi(X)\}$ is a continuous function of $\theta (\theta \in \Omega)$ for each test function φ and (ii) $\Omega_0' \neq \varnothing$ where Ω_0' is the intersection of the closures of $\Omega_1, \ldots, \Omega_m$.

Schaafsma (1969) has considered several problems. In particular, he has considered the problem of selecting a nonempty subset of k normal populations, $N(\mu_i, 1)$, $i = 1, \ldots, k$, so that the selected subset contains the population with the largest mean. Here $\Omega = \Omega_0 \cup \ldots \cup \Omega_k$ consists of all possible $\theta = (\mu_1, \ldots, \mu_k)$; Ω_i is defined by $\mu_i > \max_{j \neq i} \mu_j$ ($i = 1, \ldots, k$). The decision space $\mathcal{D}$ consists of all $2^k - 1$ decisions $d(h_1, \ldots, h_k)$. Here $h_i = 0$ or 1 ($i = 1, \ldots, k$), $\Sigma h_i > 0$ and $d(h_1, \ldots, h_k)$ corresponds to selecting all the populations for which $h_i = 1$. If $\theta \in \Omega_i$ and $h_i = 0$, an error of the first kind is committed resulting in a loss of b units. If $\theta \in \Omega_i$ and $h_j = 1$ in $d(h_1, \ldots, h_k)$ for $j \neq i$, it results in a loss of a units. Accordingly, the loss function becomes

$$L\{\theta, d(h_1, \ldots, h_k)\} = b(1 - h_i) + a\left(\sum_{j=1}^{k} h_j - h_i\right), \theta \in \Omega_i, \qquad i = 1, \ldots, k. \tag{22.13}$$

The general results of Schaafsma (1969) yield the following particular results in this case.

(1) If $b > (k-1)a$, there exists a unique minimax risk procedure which assigns $d(1, 1, \ldots, 1)$ with probability 1 to (almost) all $x \in \mathcal{X}$. This trivial procedure is not attractive because the risk is everywhere equal to $(k-1)a$.

(2) If $b < (k-1)a$, a minimax risk procedure never selects a subset consisting of more than one population.

22.3 SELECTION OF VARIABLES IN LINEAR REGRESSION

In applications of regression analysis for prediction purposes, we often have a large number of independent variables. In such situations, for "adequate" prediction, it may be sufficient to use only a subset of these variables. Thus arises the problem of determining the "best" subset of variables. This problem has long been of interest to applied statisticians, one of the earliest authors being Hotelling (1940). He considered the example of predicting the ultimate weight of a mature ox by using one of two available variables, namely, the length at an early age and the calf's girth at his heart. The problem of selecting the best set of independent variables has received considerable attention in recent years thanks to the availability of high-speed computers. Several authors have dealt with various aspects of this problem including the application of the subset selection formulation discussed in earlier chapters.

The typical regression user has not apparently been benefited much from several techniques in the literature for reducing the number of predictor variables in the model. One reason for the lack of a definitive solution is that there is not a single problem but rather several of them. Recently Hocking (1976) has provided a nice summary of the state of the art with a long list of relevant references. Thompson (1978) has made a brief review and evaluation of significant methods. Our aim here is to sketch only briefly the essential features of the development.

The problem of selecting a subset of independent or predictor variables is usually described under the assumption that the set of variables under consideration contains all important variables and variable functions, and that the analyst has available to him "good data" on which to base the conclusions. The assumption of good data includes the usual assumptions such as homogeneity of variance. A serious problem arises out of multicollinearity among the independent variables.

Suppose there are $n \geqslant t+1$ observations on a t-vector of input variables, $\mathbf{x}=(x_1,\ldots,x_t)$, and a scalar response, y, such that the jth response ($j=1,\ldots,n$) is determined by

$$y_j=\beta_0+\sum_{i=1}^{t}\beta_i x_{ij}+e_j. \tag{22.14}$$

The residuals e_j are assumed to be independent and identically distributed $N(0,\sigma^2)$. Let RSS_p denote the residual sum of squares for the least squares fit for a p-term subset model. The p-term model with minimum RSS_p among all possible p-term models is called the best model of size p.

22.3.1 Computational Techniques

Several computational techniques have been proposed to identify the best model. If t is not too large one can consider fitting all possible models. The availability of high-speed computers has led to efforts in this direction and a number of efficient algorithms are available for evaluating all possible regression. Obviously, the computational task increases enormously as t increases. Various methods have been proposed for evaluating only a small number of subsets by either adding or deleting variables one at a time according to a specific criterion. These procedures, generally referred to as stepwise methods, consist of variations on two basic ideas called forward selection (FS) and backward elimination (BE). The FS technique starts with no variables in the equation and adds one variable at a time until either all variables are in or a stopping criterion is satisfied. The BE technique, on the other hand, starts with the equation in which all the variables are included and eliminates the variables one at a time until either all variables are eliminated or a stopping criterion is satisfied. One popular combination of the two techniques is basically FS with the possibility at each step of deleting a variable as in BE.

The stepwise procedures are not satisfactory for several reasons. They do not generally assure that the best subset of a given size will be obtained. Also they imply an order of importance to the variables. This can be misleading. For example, the first variable included by the FS technique may be quite unnecessary in the presence of other variables. Also, the first variable deleted by the BE technique may be the first variable selected by the FS technique. Further, the subsets revealed by the stepwise methods do not satisfy any reasonable optimality criterion.

Several authors have developed algorithms designed to ensure that the best subset of each size for $1 \leqslant p \leqslant t+1$ will be identified with only a small fraction of the 2^t subsets evaluated. Among the contributors in this direction are Beale (1970), Beale, Kendall, and Mann (1967), Furnival and Wilson (1974), Hocking and Leslie (1967), Kirton (1967), and LaMotte and Hocking (1970). As a compromise between the stepwise procedures which have limited output and the optimal procedures, some suboptimal methods have been proposed by Gorman and Toman (1966) and Barr and Goodnight (1971).

Hoerl and Kennard (1970) have suggested the use of the biased ridge estimator for problems involving nonorthogonal predictors. Ridge regression is not designed for the purpose of variable selection; however, there is an inherent deletion of variables. From the computational point of view ridge regression is found to be quite efficient.

22.3.2 Selection Criteria

The criterion to be employed in the decision regarding the choice of the subset or subsets should naturally be governed by the intended use. For example, if the objective is to obtain a good description of the response variable, and a least-squares fit is used, then the residual sums of squares is a good criterion. The other possible uses of the regression equations are prediction and estimation, extrapolation, estimation of parameters, model building, and control of output level by varying the input levels. To suit these different purposes, several criteria have been suggested. The more common ones are [see Hocking (1976) for references]:

1. The residual mean square, $\text{RMS}_p = \text{RSS}_p/(n-p)$,
2. The squared multiple correlation coefficient,

$$R_p^2 = 1 - (\text{RSS}_p/\text{TSS}), \text{ where TSS is the total sum of squares,}$$

3. The adjusted R^2,

$$\bar{R}_p^2 = 1 - (n-1)(1 - R_p^2)(n-p)^{-1},$$

4. The average prediction variance,

$$J_p = \frac{(n+p)\text{RMS}_p}{n},$$

5. The total squared error, $C_p = (\text{RSS}_p/\hat{\sigma}^2) + 2p - n$,
6. The average prediction mean squared error,

$$S_p = \frac{\text{RMS}_p}{(n-p-1)},$$

7. The standardized residual sum of squares,

$$\text{RSS}_p^* = \mathbf{e}_p' D_p^{-1} \mathbf{e}_p, \text{ where } \mathbf{e}_p = \mathbf{Y} - \hat{\mathbf{Y}} \text{ and}$$

$$D_p = \text{diag}\left(I - X_p (X_p' X_p)^{-1} X_p'\right),$$

8. The prediction sum of squares,

$$\text{PRESS} = \mathbf{e}_p' D_p^{-2} \mathbf{e}_p.$$

Hocking (1976) has also discussed biased estimation, relation of ridge to principal component estimators, and analysis of subsets with biased estimators.

Recently, Hawkins (1976) has studied two methods for reducing the dimensionality by discarding irrelevant variables in multivariate analysis of variance. One method is exact and laborious and hence is suitable only in the cases of data of low dimensionality. The other is a stepwise procedure whose motivation and algorithmic aspects are closely related to stepwise regression. As we have seen earlier, this cannot guarantee that a globally optimal subset will be found.

22.3.3 Subset Selection Problems for Variances with Applications to Regression Analysis

In §14.2.2, we discussed selection of the best component of a multivariate normal population in terms of the variances. We now continue the discussion and consider some applications to regression analysis.

Let $\mathbf{W}' = ((\mathbf{W}_1'), (\mathbf{W}_2'), \ldots, (\mathbf{W}_k'))$ be a normally distributed random vector of length kn with mean zero and covariance matrix Σ. Each $\mathbf{W}_i$ is a random vector of length n with

$$E(\mathbf{W}_i \mathbf{W}_i') = \sigma_i^2 \Sigma_{ii} = \sigma_i^2 I,$$

and

$$E(\mathbf{W}_i \mathbf{W}_j') = \sigma_i \sigma_j \Sigma_{ij},$$

with Σ_{ij} positive definite. Thus $\Sigma=(\sigma_i\sigma_j\Sigma_{ij})$ and it may be singular. It is assumed that Σ_{ij}'s are known. Here π_i refers to the distribution of $\mathbf{W}_i$. For selecting a subset containing the population associated with $\sigma^2_{[1]}$, Arvesen and McCabe (1975) employed a procedure similar to that of Gupta and Sobel (1962a) in the case of independent populations discussed in §12.4.1. The rule R selects π_i if and only if

$$\mathrm{SS}_i \leqslant \frac{\mathrm{SS}_{[1]}}{c}, \qquad i=1,\ldots,k, \tag{22.15}$$

where $\mathrm{SS}_i=\mathbf{W}_i'W_i$. The problem of calculating a lower bound for the PCS involves the joint distribution of a set of dependent chi-square random variables.

For $k=2$, we can simplify the problem by transforming to canonical variates. Thus, we assume that $\Sigma_{11}=\Sigma_{22}=I$, and $\Sigma_{12}=\Sigma_{21}=\mathrm{diag}(\rho_1,\ldots,\rho_n)$. In this case, the solution can be obtained by iterative methods on a computer; however, the convergence will be slow if some of the $\{\rho_i\}$ are large.

In the special case of $\rho_1=\cdots=\rho_n=\rho$, further simplification is possible. For arbitrary k, Arvesen and McCabe (1975) have discussed the asymptotic case using an Edgeworth-type approximation.

Now, to discuss the application of these results to the problem of selecting the best regression equation, we assume the following standard linear model,

$$\mathbf{Y}=X\boldsymbol{\beta}+\boldsymbol{\epsilon}, \tag{22.16}$$

where X is an $N\times p$ known matrix of rank $p\leqslant N$, $\boldsymbol{\beta}$ is a $p\times 1$ parameter vector, and $\boldsymbol{\epsilon}\sim N(0,\sigma^2 I_N)$. The model (22.16) which includes p independent variables is considered as the "true" model. Now, consider the models

$$\mathbf{Y}=X_i\boldsymbol{\beta}_i+\boldsymbol{\epsilon}_i, \tag{22.17}$$

where X_i is an $N\times t$ matrix (of rank t), $\boldsymbol{\beta}_i$ is a $t\times 1$ parameter vector, and $\boldsymbol{\epsilon}_i\sim N(0,\sigma_i^2 I_N)$, where $i=1,\ldots,k=\binom{p}{t}$ over all possible partitions. The goal is to select that design X_i (or set of independent variables) associated with $\sigma^2_{[1]}$.

Let

$$\begin{aligned}\mathrm{SS}_i&=\mathbf{Y}'\{I-X_i(X_i'X_i)^{-1}X_i'\}\mathbf{Y}\\&=\mathbf{Y}'Q_i\mathbf{Y}, \text{ say.}\end{aligned} \tag{22.18}$$

Then SS_i/σ^2 has a chi-square distribution with $N-t$ degrees of freedom and a noncentrality parameter $(X\boldsymbol{\beta})'Q_i(X\boldsymbol{\beta})/2\sigma^2$.

The procedure R of Arvesen and McCabe (1975) selects the design X_i if and only if

$$SS_i \leqslant \frac{SS_{[1]}}{c}, \tag{22.19}$$

where $0<c=c(p,t,N,P^*)<1$ is to be determined so that $P(CS|R)\geqslant P^*$. It has been shown by Arvesen and McCabe that the PCS is asymptotically minimized when $\beta_1=\cdots=\beta_p=0$. This reduces the problem to that discussed first.

McCabe and Arvesen (1974) have provided an algorithm for determining c for given P^* and X using Monte Carlo methods. A write-up for a FORTRAN program implementing this algorithm is given by McCabe, Arvesen, and Pohl (1973). This procedure has been applied to certain data sets by Arvesen and McCabe (1973).

Gupta and Huang (1977d) considered a formulation in which the goal is to exclude all "inferior" independent variables. The "true" model is given by

$$\mathbf{Y}=X\boldsymbol{\beta}+\boldsymbol{\epsilon}, \tag{22.20}$$

where X is an $N\times p$ known matrix of rank $p\leqslant N$, $\boldsymbol{\beta}$ is a $p\times 1$ parameter vector, and $\boldsymbol{\epsilon}\sim N(0,\sigma_0^2 I_N)$. All vectors are column vectors and we have $X=(\mathbf{1},\mathbf{X}_1,\ldots,\mathbf{X}_{p-1})$, $\boldsymbol{\beta}'=(\beta_0,\ldots,\beta_{p-1})$ and $\mathbf{1}'=(1,\ldots,1)$.

Consider the models

$$\mathbf{Y}=X_{(i)}\beta_{(i)}+\boldsymbol{\epsilon}_i, \qquad i=1,\ldots,k, \tag{22.21}$$

where $X_{(i)}=(\mathbf{1},\mathbf{X}_1,\ldots,\mathbf{X}_{i-1},\mathbf{X}_{i+1},\ldots,\mathbf{X}_k)$,$\boldsymbol{\beta}'_{(i)}=(\beta_0,\beta_1,\ldots,\beta_{i-1},\beta_{i+1},\ldots,\beta_k)$, $\boldsymbol{\epsilon}_i\sim N(0,\sigma_i^2 I_k)$, and $k=p-1$. $X_{(i)}$ associated with model (22.21) is referred to as population π_i. Any population π_i is said to be inferior if $\sigma_0^2<\delta_2^*\sigma_i^2$ where $0<\delta_2^*<1$ is a known constant.

Define $SS_i=\mathbf{Y}'\{I-X_{(i)}(X'_{(i)}X_{(i)})^{-1}X'_{(i)}\}\mathbf{Y}$, $i=1,\ldots,k$. When σ_0^2 is known, Gupta and Huang (1977d) have proposed a rule R which rejects π_i if and only if

$$SS_i \geqslant \frac{(N-k)c}{\delta_2^*} \tag{22.22}$$

where $\delta_2^*<c<1$. The constant c is to be determined so that the probability that all inferior populations are rejected is at least P^*. It has been shown that asymptotically the worst configuration is given by $\boldsymbol{\beta}=\mathbf{0}$. Asymptotic case discussion is similar to that of Arvesen and McCabe (1975).

22.3.4. An Indifference Zone Approach for Selecting the Best Predictor Variate

Here we are concerned with selecting one variate from a set of predictor variates. Let $\mathbf{X}=(X_0,X_1,\ldots,X_k)$ have a multivariate normal distribution with unknown mean vector $\boldsymbol{\mu}$ and unknown covariance matrix $\Sigma=((\sigma_{ij}))$, where X_0 is the variate to be predicted and the other X_i are predictor variates. Let $\sigma_{0.i}^2$ denote the conditional variance of X_0 given X_i. Ramberg (1977) considered the problem of selecting the best variate which is defined to be the variate associated with the smallest $\sigma_{0.i}^2$. Since $\sigma_{0.i}^2=\sigma_{00}(1-\rho_{0i}^2)$, the problem is equivalent to selecting the variate associated with the largest ρ_{0i}^2. Based on a sample of N independent vector-observations, Ramberg (1977) proposed the natural rule R which selects the variate that yields the smallest sample conditional variance $s_{0.i}^2$. The sample size N must be determined so that $\text{PCS}\geqslant P^*$ whenever $\sigma_{0.[1]}^2\leqslant\sigma_{0.[2]}^2/\theta^*$, $\theta^*>1$.

Ramberg (1977) has given an asymptotic solution for $k=2$, namely,

$$N_a=2+\left[\frac{2\Phi^{-1}(P^*)}{\log\theta^*}\right]^2. \tag{22.23}$$

He has also considered some approximations to N_a for $k>3$.

22.4 SLIPPAGE TESTS

In Chapter 1 we discussed very briefly the slippage tests as the first step in the direction of more realistic formulations overcoming certain deficiencies in the classical tests of homogeneity. We now discuss the slippage problems in a little more detailed manner.

The list of early authors who studied these problems includes among others Mosteller [1948], Paulson [1952], Truax [1953], Doornbos and Prins [1956], and Kudô [1956b]. Although these problems are related to selection and ranking problems, one has to bear in mind that these problems are formulated as simultaneous tests of hypotheses.

The problems of Mosteller [1948] and Paulson [1952] were stated in §1.2. We now give a decision-theoretic formulation of the slippage problem.

22.4.1 A Decision-Theoretic Formulation

We observe a k-dimensional random variable $X=(X_1,X_2,\ldots,X_k)$ distributed according to a density function $f(x_1,\ldots,x_k;\ \theta_1,\theta_2,\ldots,\theta_k)$ which is known except for the parameter point $\theta=(\theta_1,\ldots,\theta_k)$ where the θ_i are real numbers. We assume the following symmetry of the density: $f(x_{\pi 1},x_{\pi 2},\ldots,x_{\pi k};\ \theta_{\pi 1},\theta_{\pi 2},\ldots,\theta_{\pi k})=f(x_1,x_2,\ldots,x_k;\ \theta_1,\theta_2,\ldots,\theta_k)$ for all permutations $(1,2,\ldots,k)\to(\pi 1,\pi 2,\ldots,\pi k)$. As we know, there are $(k+1)$ available actions, denoted by $a_0,a_1,\ldots,a_k$, and the loss in taking action a_j when θ is the true parameter point is $L_j(\theta)$. The loss functions are assumed to have the following properties.

(1) $L_{\pi j}(\theta_{\pi 1},\theta_{\pi 2},\ldots,\theta_{\pi k})=L_j(\theta_1,\theta_2,\ldots,\theta_k)$ for all permutations $(1,2,\ldots,k)\to(\pi 1,\pi 2,\ldots,\pi k)$ and for all j. (We may include the case $j=0$ by defining $\pi 0=0$.)

(2) $L_j(\theta)<L_i(\theta)$ for all $i\neq j$ if $\theta_j=\omega+\Delta$ for some real ω and $\Delta>0$, and $\theta_i=\omega$ for all $i\neq j$. $L_0(\theta)<\min_{1\leqslant i\leqslant k}L_i(\theta)$ if $\theta_i=\omega$ for some real ω, $i=1,2,\ldots,k$. Otherwise, $L_0(\theta)=L_1(\theta)=\cdots=L_k(\theta)$.

A *symmetric decision function* $\varphi=(\varphi_0,\varphi_1,\ldots,\varphi_k)$ satisfies the conditions: (i) $0\leqslant\varphi_i(x)\leqslant 1$, (ii) $\Sigma_{i=0}^{k}\varphi_i(x)=1$, and (iii) $\varphi_{\pi j}(x_{\pi 1},\ldots,x_{\pi k})=\varphi_j(x_1,\ldots,x_k)$ for all permutations π of the set $(1,\ldots,k)$ and all j (including $j=0$ with $\pi 0=0$).

Karlin and Truax [1960] have shown that any symmetric Bayes solution is Bayes against a symmetric distribution. Let $f_i(x_1,\ldots,x_k;\ \omega,\Delta)$ denote the density of X when $\theta_i=\omega+\Delta$, and $\theta_j=\omega$ for $j\neq i$. Let us make the following assumption (A) on the densities.

A. For $j>0$ and $k>1$,

$$f_j(x_1,x_2,\ldots,x_k;\ \omega,\Delta)\geqslant f_k(x_1,x_2,\ldots,x_k;\ \omega,\Delta)$$

if and only if $x_j\geqslant x_k$, or strict inequality in both instances.

The following theorem of Karlin and Truax [1960] gives the general form of the Bayes procedures, but does not make explicit the nature of the set R_0 (the set where H_0 will be accepted) besides the fact that R_0 is symmetric.

Theorem 22.1. If $f(\mathbf{x};\boldsymbol{\theta})$ satisfies the assumption (A), then any symmetric Bayes procedure for the slippage problem has the form $\varphi_0(\mathbf{x})=1$ if $\mathbf{x}\in R_0$, a symmetric set, and $\varphi_i(x)=1$ if $\mathbf{x}\notin R_0$, and $x_i>x_j$ for all $j\neq i$, $i=1,2,\ldots,k$. If $\mathbf{x}\notin R_0$ and $\max_{1\leqslant i\leqslant k}x_j=x_{i_1}=x_{i_2}=\cdots=x_{i_r}$, then $\varphi_{i_1}(\mathbf{x})+\varphi_{i_2}(\mathbf{x})+\cdots+\varphi_{i_r}(\mathbf{x})=1$.

Karlin and Truax [1960] have studied the form of the Bayes solution in the particular cases of normal means (variance known or unknown) and gamma distributions. In most of the situations they have considered both the "uncontrolled" and the "controlled" cases. The controlled case refers to the situation where in addition to an observation from each of the k densities, we have an observation from a density whose parameter is known not to have slipped.

Karlin and Truax [1960] have also discussed the general form of the invariant Bayes solutions for the translation and the scale parameters problems. Further they have proved, subject to slight smoothness restrictions, that each symmetric invariant Bayes procedure involving a single critical parameter is uniformly most powerful among the class of all symmetric invariant procedures, having a prescribed error associated with hypothesis H_0. The loss functions are assumed to be zero or one, according as a correct or incorrect decision was made.

22.4.2 Some Additional Results

Doornbos [1966] has surveyed some of the important results on slippage tests. The list of topics covered by him includes slippage tests for one outlier as well as more than one outliers. The procedures discussed are both parametric and nonparametric. Specific distributions considered include normal, gamma, binomial, Poisson, and negative binomial. He has also discussed optimum properties and efficiency of the procedures. These results relate to the papers of several authors among whom are Doornbos, Kesten and Prins [1956], Doornbos and Prins [1956, 1958], Kudô [1956b], Paulson [1952], and Truax [1953].

Conover [1968] has discussed two nonparametric tests for the k-sample slippage problem. One of them is intended for samples of equal size and the other for samples of unequal sizes. Both can be considered as generalizations of the test proposed by Mosteller [1948].

Srivastava (1966) studied some asymptotically efficient sequential procedures for a univariate and a multivariate slippage problem using the results of Chow and Robbins [1965].

Hall and Kudô [1968a] have discussed slippage problems without assuming the condition of invariance on the procedures. They have given a generalization of the Neyman–Pearson lemma to slippage problems and obtained conditions under which a uniformly most powerful test exists. In a companion paper [1968b], they have applied the concept of similarity in hypotheses testing to slippage tests.

CHAPTER 23

Guide to Tables

23.1 INTRODUCTION

While discussing several procedures in the earlier chapters, we have described relevant tables available in the literature. Some of these tables are of interest in other problems too. In this chapter, we describe again many of the important tables and provide references to procedures of the earlier chapters which use these. There are several other computational results available in the literature that are not described here. Many of these are important in specific contexts and have been described in relevant places in this monograph.

In each case, the notations are explained; however, consistent with our earlier practice, $\Phi(\cdot)$ and $\varphi(\cdot)$ denote the cdf and the density of a standard normal distribution unless otherwise specified.

23.2 TABLES RELATING TO MULTIVARIATE NORMAL INTEGRALS

I. For given integers k and t such that $k \geqslant 2$ and $1 \leqslant t \leqslant k-1$, let

$$t\int_{-\infty}^{\infty}[\Phi(y+d)]^{k-t}[1-\Phi(y)]^{t-1}\varphi(y)dy = P^*. \tag{23.1}$$

Suppose $Y_1, Y_2, \ldots, Y_k$ are independent random variables where $Y_i \sim N(\mu, \sigma^2), i=1,\ldots,k-t$, and $Y_i \sim N(\mu+\delta, \sigma^2), i=k-t+1,\ldots,k, \delta>0$. Then the left-hand side of (23.1) is

$$\Pr\{\max(Y_1,\ldots,Y_{k-t}) < \min(Y_{k-t+1},\ldots,Y_k)\}$$

where $d=\delta/\sigma$. The values of d satisfying (23.1) have been tabulated by Bechhofer (1954) for the following combinations of k and t:

k	t
2(1)10	$1(1)\left[\frac{k}{2}\right]$
11	2(1)5
12	3(1)5
13, 14	5

The selected values of P^* are: 0.05(.05)0.80(.02)0.90(.01)0.99, 0.995, 0.999, 0.9995. For any given combination of k and t, the smallest value of P^* chosen is such that $P^* \geqslant [t!(k-t)!]/k!$

The value of d determines the smallest common sample size n for the procedure R_1 of Bechhofer (1954) described in §2.1 in the special case of selecting the populations associated with the t largest means from a set of k normal populations with unknown means and common known variance. For the rule R_1, $d=\sqrt{n}\,\delta/\sigma$. For the special case of selecting the best $(t=1)$ population, (23.1) reduces to

$$\int_{-\infty}^{\infty}\Phi^{k-1}(y+d)\varphi(y)dy = P^* \tag{23.2}$$

II. Let $Z_1, Z_2, \ldots, Z_M$ be correlated standard normal variates with correlations $\rho_{ij} = E(Z_iZ_j), i \neq j$. Let $f(z_1, z_2, \ldots, z_M; \{\rho_{ij}\})$ denote the joint density of $Z_1, Z_2, \ldots, Z_M$ and

$$F_M(H_1, H_2, \ldots, H_M; \{\rho_{ij}\})$$
$$= \Pr\{Z_1 \leqslant H_1, \ldots, Z_M \leqslant H_M\}$$
$$= \int_{-\infty}^{H_1} \int_{-\infty}^{H_2} \cdots \int_{-\infty}^{H_M} f(z_1, z_2, \ldots, z_M; \{\rho_{ij}\}) dz_1 dz_2 \ldots dz_M. \tag{23.3}$$

If the correlation matrix $\{\rho_{ij}\}$ of the Z_i has the structure $\rho_{ij} = \alpha_i \alpha_j (i \neq j)$ where $-1 \leqslant \alpha_i \leqslant 1$, then the variates Z_i can be generated from $M+1$ independent standard normal variates $W_0, W_1, \ldots, W_M$ by the transformation

$$Z_i = (1 - \alpha_i^2)^{\frac{1}{2}} W_i + \alpha_i W_0 \tag{23.4}$$

and it follows that (see Dunnett and Sobel [1954])

$$F_M(H_1, \ldots, H_M; \{\rho_{ij}\}) = \int_{-\infty}^{\infty} \prod_{i=1}^{M} \Phi\left(\frac{H_i - \alpha_i z}{(1 - \alpha_i^2)^{\frac{1}{2}}} \right) \varphi(z) dz. \tag{23.5}$$

In particular, when $\rho_{ij} = \rho > 0$ and $H_1 = \cdots = H_M = H$, (23.5) reduces to

$$F_M(H, \ldots, H; \{\rho\}) = \int_{-\infty}^{\infty} \Phi^M \left[\frac{z\rho^{\frac{1}{2}} + H}{(1-\rho)^{\frac{1}{2}}} \right] \varphi(z) dz. \tag{23.6}$$

Evaluation of multivariate normal probability integrals has been discussed in detail by Gupta [1963a] who has given an extensive bibliography on this and related topics in a companion paper [1963b]. We denote $F_M(H, \ldots, H; \{\rho\})$ by $F_M(H; \rho)$. The values of $H = H(M, P^*, \rho)$ satisfying

$$F_M(H; \rho) = P^* \tag{23.7}$$

have been tabulated correct to three decimal places by Gupta [1963a] for $M = 1(1)50$, and $P^* = 0.75, 0.90, 0.95, 0.975, 0.99$ in the case of $\rho = 0.5$. It should be noted that, when $\rho = 0.5$, $H(k-1, P^*, \frac{1}{2}) = d/\sqrt{2}$ where d is given by (23.2). Thus the tables of Bechhofer (1954) for the case of $t = 1$ are related to the tables of Gupta [1963a] for the case of $\rho = 0.5$. Gupta, Nagel, and Panchapakesan [1973] have tabulated the values of $H(M, P^*, \rho)$ correct to four decimal places for $M = 1(1)50$, $P^* = 0.75, 0.90, 0.95, 0.975, 0.99$, and $\rho = 0.100$, 0.125, 0.200, 0.250, 0.300, 1/3, 0.375, 0.400, 0.500, 0.600, 0.625, 2/3, 0.700, 0.750, 0.800, 0.875, 0.900.

The values of $F_M(H;\rho)$ have been tabulated by Gupta [1963a] correct to five decimal places for $M = 1(1)12$, $H = -3.5(0.1)3.5$, and $\rho = 0.100$, 0.125, 0.200, 0.250, 0.300, 1/3, 0.375, 0.400, 0.500, 0.600, 0.625, 2/3, 0.700, 0.750, 0.800, 0.875, 0.900. In the special case of $\rho = 0.5$, Milton [1963] has given the values of $F_M(H;0.5)$ correct to eight decimal places for $M = 2(1)9(5)24$ and $H = 0.00(.05)5.15$. The probability $F_M(H;0.5)$ is a particular case of rank order probabilities (see Milton [1970], p. 5).

The tables of H-values (especially in the special case of $\rho = 0.5$) and the tables of d-values satisfying (23.1) are applicable to many procedures that have been discussed in earlier chapters. In the cases of some of these procedures, the tables are relevant to asymptotic solutions. The following list provides the contexts in which the tables are useful. The notation R_1(§13.2.1), for example, refers to the procedure R_1 of §13.2.1.

1. Procedure R_1 (§2.1) of Bechhofer (1954) for selecting the populations associated with t largest means from a set of k normal populations with unknown means and common known variance σ^2.
2. Procedure R_1 (§12.2.1) of Gupta (1956) for selecting a subset containing the population with the largest (smallest) mean from a set of k normal populations with unknown means and common known variance.
3. Procedure R_1 (§14.2.1) of Gnanadesikan (1966) for selecting a subset of the components of multivariate random vector to include the component with the largest expectation when the covariance matrix Σ is known and the correlations are equal and positive.
4. Procedure R_{12} (§12.3) of Rizvi (1971) for selecting a subset containing the population with the largest absolute value of the mean from a set of k normal populations with unknown means and a common known variance.
5. Procedure R_1 (§20.2) of Paulson (1952) for comparing k normal populations with a control normal population in terms of their means in the case of common known variance.
6. Procedure R_1 (§20.5.1) of Gupta and Sobel (1958a) for selecting a subset of normal populations to include all those populations that are better than a control normal population assuming that they all have a common known variance.
7. Approximate solution for the common number of observations per process required for the procedure R_1 (§4.1) of Sobel and Huyett (1957) for

selecting the binomial process associated with the largest probability of a success.

8. Approximate solution for the procedure R_1 (§4.9.1) of Bechhofer, Elmaghraby, and Morse (1959) for selecting the most probable multinomial cell.

9. Approximate solution for the two-stage subsampling procedure R (§5.1.1) of Bechhofer (1967) for selecting the finite population with the largest population average.

10. Solution of the constant involved in the stopping rule of the sequential procedure R_3 (§6.2) of Robbins, Sobel, and Starr (1968) for selecting the population associated with the largest mean from a set of k normal populations with unknown means and a common unknown variance.

11. Approximate solution for the constant of the procedure R_1 (§13.2.1) of Gupta and Sobel (1960) for selecting a subset containing the binomial population with the largest success probability.

Suppose that $H_r = rd$, $r = 1,\dots,M$, and $\rho_{ij} = 0.5$ in (23.3). In this case, $F_M(H_1,\dots,H_M;\{0.5\})$ has been tabulated by Olkin, Sobel, and Tong (1976) for $d = 0.00(0.02)0.20(0.10)4.00$ and $M = 1(1)9$, 11, 14 and 19. This integral appears in the discussion of the estimation of PCS in §5.3.1.

III. Suppose that $Y_1, Y_2, \dots, Y_k$ are correlated standard normal variates with the correlation matrix $\{\rho_{ij}\}$, where for $m =$ the integral part of $(k+1)/2$,

$$\rho_{ij} = \begin{cases} \dfrac{1}{2}, & i,j = 1,\dots,m \text{ or} \\ & i,j = m+1,\dots,k, (i \neq j) \\ -\dfrac{1}{2}, & i = 1,\dots,m; j = m+1,\dots,k, \text{ or} \\ & i = m+1,\dots,k; j = 1,\dots,m. \end{cases} \tag{23.8}$$

Tong (1969) has tabulated the values of $b = b(k, P^*)$ satisfying the equation

$$\Pr\{Y_1 \leqslant b, \dots, Y_k \leqslant b\} = P^* \tag{23.9}$$

for $k = 1(1)10(2)20$ and $P^* = 0.50, 0.75, 0.90, 0.95, 0.975, 0.99$. These values are used to determine the smallest sample size required for the procedure R_1 (§20.6.1) of Tong (1969) for partitioning a set of normal populations with respect to a control in terms of their means when they are assumed to have a common known variance.

IV. Let $Z_1,\ldots,Z_k$ be independent random variables which are normally distributed with means $\mu_1,\ldots,\mu_k$ and common variance σ^2. Let

$$B_i = \Pr\left\{ \max_{j\neq i} (Y_j - Y_i) \leqslant H \right\}, \qquad i=1,\ldots,k. \tag{23.10}$$

Deely and Gupta (1968) have computed these probabilities when $\sigma^2 = n^{-1}$, the means have a slippage configuration in which $\mu_k - \mu_i = \delta > 0$ for $i = 1,\ldots,k-1$, and $H = d/\sqrt{n}$, where $d > 0$ is the solution of (23.2). In this case, we have

$$B_i = \int_{-\infty}^{\infty} \Phi^{k-2}(z+d)\Phi(z+d-\delta\sqrt{n})\varphi(z)\,dz, \qquad i=1,\ldots,k-1,$$

$$B_k = \int_{-\infty}^{\infty} \Phi^{k-1}(z+d+\delta\sqrt{n})\varphi(z)\,dz. \tag{23.11}$$

Let B denote the common value of $B_1,\ldots,B_{k-1}$, and define

$$T = k^{-1}[B_k + (k-1)B]. \tag{23.12}$$

Deely and Gupta (1968) have tabulated the values of B_k, B, and T for $k = 2(1)10, 25, 50$; $P^* = 0.75, 0.90, 0.95$, and $\delta\sqrt{n} = 0.00, 0.50, 1.00(0.25)3.00, 4.00, 5.00, \infty$.

As we have pointed out earlier, the constant d satisfying (23.2) is used in the procedure R_1 (§12.2.1) of Gupta (1956) for selecting a subset containing the population with the largest mean from a set of normal populations with unknown means and a common known variance σ^2. When $\sigma^2 = 1$ and $\mu_1 = \cdots = \mu_{k-1} = \mu_k - \delta$, the probability B_k denotes the PCS using R_1. The probability of including any particular nonbest population is given by B. The value of T is the expected proportion of the given set of populations that is included in the selected subset.

V. Let $X_1, X_2, \ldots, X_k$ be independent random variables that are normally distributed with means $\mu_1, \mu_2, \ldots, \mu_k$, respectively, and a common variance σ^2/n, where n is a positive integer. Let $W_i = |X_i|$ and $\theta_i = |\mu_i|, i = 1, 2, \ldots, k$. Then the cdf of W_i is given by

$$G(y,\theta_i) = \begin{cases} \Phi\left\{ \frac{\sqrt{n}}{\sigma}(y-\theta_i) \right\} - \Phi\left\{ -\frac{\sqrt{n}}{\sigma}(y+\theta_i) \right\}, & y > 0, \\ 0 & \text{otherwise}. \end{cases} \tag{23.13}$$

The distribution of W_i is folded normal with folding at the origin. It is also a noncentral chi-distribution with one degree of freedom and noncentrality parameter $\sqrt{n}\ \theta_i/\sigma$.

Now, we assume that $\sigma=1$, and that the θ_i are such that

$$\theta_1=\cdots\theta_{k-t}=0;\theta_{k-t+1}=\cdots=\theta_k=\delta^*,\ \delta^*>0. \tag{23.14}$$

We are interested in

$$P(k,t,n,\delta^*)=\Pr\{\max(W_1,\ldots,W_{k-t})<\min(W_{k-t+1},\ldots,W_k)\}. \tag{23.15}$$

This probability

$$\begin{aligned}&=t\int_0^\infty\left[\Phi(\sqrt{n}\,y)-\Phi(-\sqrt{n}\,y)\right]^{k-t}\\&\quad\times\left[\Phi(-\sqrt{n}\,y-\sqrt{n}\,\delta^*)+\Phi(-\sqrt{n}\,y+\sqrt{n}\,\delta^*)\right]^{t-1}\\&\quad\times\sqrt{n}\left[\varphi(\sqrt{n}\,y-\sqrt{n}\,\delta^*)+\varphi(\sqrt{n}\,y+\sqrt{n}\,\delta^*)\right]dy.\end{aligned}$$

Let $\lambda=\sqrt{n}\ \delta^*$, change the variable of the integration by $z=\sqrt{n}\ y$ and then integrate by parts. Now the probability in (23.15) can be written as

$$P(k,t,\lambda)=2(k-t)\int_0^\infty\left[2\Phi(z)-1\right]^{k-t-1}\left[\Phi(-z+\lambda)+\Phi(-z-\lambda)\right]^t\varphi(z)dz.$$

Rizvi (1971) has tabulated the values of $\lambda=\lambda(k,P^*)$ satisfying the equation

$$P(k,t,\lambda)=P^* \tag{23.16}$$

for $k=2(1)10$, and $P^*=0.5000$, 0.7500, 0.9000, 0.9500, 0.9750, 0.9900, 0.9950, 0.9990, 0.9995, 0.9999 in the two special cases: $t=1$ and $t=k-1$.

The λ-value staisfying (23.16) determines the minimum sample size required to guarantee that PCS $\geqslant P^*$ for the procedure R_8 (§2.3) of Rizvi (1971) for selecting the population associated with the t largest absolute values of the means from a set of k normal populations with a common known variance. For this procedure the least favorable configuration is given by (23.14). For the two special values of t, Rizvi (1971) has also tabulated the PCS for $k=2(1)10$ and $\lambda=0.0(0.5)2.0(0.2)7.0$.

VI. Let $X_1, X_2, \ldots, X_k$ be independent random variables that are normally distributed with means $\theta_1, \theta_2, \ldots, \theta_k$, respectively, and common known variance σ^2. Carroll and Gupta (1977) have developed algorithms for computing probabilities of rankings of k populations with the associated distribution functions differing in a location or a scale parameter. The numerical values are approximate because of approximating a continuous distribution by a discrete distribution. They have computed the approximate values of

a. $\Pr\{X_1 < X_2 < \cdots < X_k\}$ when $\theta_{i+1} - \theta_i = \delta, i = 1, 2, \ldots, k-1$,
b. $\Pr\{X_1 < X_2 < \cdots < X_k\}$ when $\theta_{i+1} - \theta_1 = \delta i(i+1)/2, i = 1, 2, \ldots, k-1$,

and

c. $\Pr\{(X_1, X_2) < (X_3, X_4) < \cdots < (X_{k-1}, X_k)\}$, when k is even, $\theta_1 = \theta_2, \theta_3 = \theta_4, \ldots, \theta_{k-1} = \theta_k$, and $\theta_3 - \theta_2 = \theta_5 - \theta_4 = \cdots = \theta_{k-1} - \theta_{k-2} = \delta$. Here $(X_1, X_2) < (X_3, X_4)$ means that $\max(X_1, X_2) < \min(X_3, X_4)$.

The ranges of k and δ in the preceding computations are as follows.

Case (a): $k = 2(1)10$ and $\delta = 0.00(0.10)4.40$,

Case (b): $k = 2(1)6$ and $\delta = 0.25(0.25)3.75$,

Case (c): $k = 4(2)10$ and $\delta = 1.60(0.20)4.80$.

As Carroll and Gupta (1977) have indicated, these table values provide lower bounds on the probability of a correct ranking when, for example, $\theta_{[i+1]} - \theta_{[i]} \geqslant \Delta\sigma$ in the case of (a) and $\theta_{[i+1]} - \theta_{[i]} \geqslant i\Delta\sigma$ in the case of (b).

VII. Let $X_1, X_2, \ldots, X_k$ be independent and normally distributed with means $\theta_1, \theta_2, \ldots, \theta_k$, respectively, and common variance σ^2/n. Let t and s be integers such that $1 \leqslant t \leqslant s \leqslant k$. Suppose the θ_i have the configuration

$$\theta_1 = \cdots = \theta_{k-t} = \theta_{k-t+1} - \delta^* = \cdots = \theta_k - \delta^*. \tag{23.17}$$

Let $X_{[1]} \leqslant X_{[2]} \leqslant \cdots \leqslant X_{[k]}$ and define

$$P(k, s, t, \delta^*) = \Pr\{X_i \geqslant X_{[k-s+1]}, i = k-t+1, \ldots, k\}. \tag{23.18}$$

Letting $\lambda = \sqrt{n}\,\delta^*/\sigma$, this probability can be written as

$$P(k, s, t, \lambda) = \binom{k-t}{k-s} \int_{-\infty}^{\infty} [1 - \Phi(x - \lambda)]^t [1 - \Phi(x)]^{s-t} d\Phi^{k-s}(x). \tag{23.19}$$

Desu and Sobel (1968) have tabulated the values of $\lambda=\lambda(k,s,t,P^*)$ satisfying the equation

$$P(k,s,t,\lambda)=P^* \tag{23.20}$$

for $P^*=0.900, 0.950, 0.975, 0.990, 0.999$, and for the following sets of values of k, t, and s:

k	t	s	k	t	s
2	1	1	8	1	1(1)7
3	1	1, 2		2	2(1)7
4	1	1(1)3		3	3(1)7
	2	2, 3		4	4(1)7
5	1	1(1)4		5	6, 7
	2	2(1)4		6	7
	3	4			
6	1	1(1)5	9	1	1(1)8
	2	2(1)5		2	2(1)8
	3	3(1)5		3	3(1)8
	4	5		4	4(1)8
7	1	1(1)6		5	6(1)8
	2	2(1)6		6	7, 8
	3	3(1)6		7	8
	4	5, 6			
	5	6			

The value of λ determines the minimum sample size required for the fixed subset size procedure R_1 (§9.2.1) of Mahamunulu (1967) for selecting a subset of s populations from k normal populations with common known variance σ^2 so that the selected subset contains the populations associated with the t largest means.

23.3 TABLES RELATING TO MULTIVARIATE t

Let $\mathbf{X}=(X_1,X_2,\ldots,X_p)$ have a p-variate normal distribution with the mean vector $\boldsymbol{\mu}=(\mu_1,\mu_2,\ldots,\mu_p)$ and correlation matrix $\Omega=(\rho_{ij})$. It is assumed that $V(X_i)=\sigma^2, i=1,\ldots,p$. Also, let $\nu s^2/\sigma^2$ be a chi-square variate with ν degrees of freedom distributed independently of $X_1,X_2,\ldots,X_p$. Then the joint distribution of $t_1,t_2,\ldots,t_p$ where $t_i=X_i/s$ is called a central or noncentral p-variate t distribution with n degrees of freedom according as the mean vector $\boldsymbol{\mu}$ is a zero vector or not. Dunnett and Sobel [1955] have obtained the density function of the central p-variate distribution in the form

$$g(t_1,\ldots,t_p)=\frac{|A|^{\frac{1}{2}}\Gamma[(\nu+p)/2]}{(\nu\pi)^{p/2}\Gamma(\nu/2)}\left[1+\frac{1}{\nu}\sum_i\sum_j a_{ij}t_it_j\right]^{-\frac{(\nu+p)}{2}} \tag{23.21}$$

where $A=((a_{ij}))$ is the inverse of the correlation matrix Ω. The preceding expression was derived independently by Cornish [1954].

I. Tables in the Bivariate Central Case

Let

$$G_\nu(h;\rho)=\Pr\{t_1\leqslant h,\ t_2\leqslant h\}. \tag{23.22}$$

The values of $G_\nu(h;\rho)$ have been tabulated correct to six decimal places by Krishnaiah, Armitage, and Breiter [1969] for $h=1.0(0.1)5.5$, $\rho=-0.9(0.1)0.9$, and $\nu=5(1)35$. These values have also been tabulated by Dunnett and Sobel [1954] in the case of $\rho=\pm 0.5$ for $h=0.00(0.25)2.50(0.50)10$ and $\nu=1(1)30(3)60(15)120, 150, 300, 600, \infty$. For these special values of ρ, Dunnett and Sobel [1954] have also tabulated the values of $h=h(\nu,P^*)$ such that

$$G_\nu(h;\rho)=P^* \tag{23.23}$$

for $\nu=1(1)30(3)60(15)120, 150, 300, 600, \infty$, and $P^*=0.50, 0.75, 0.90, 0.95, 0.99$.

II. Tables in the Multivariate Central Case.

Let $\rho_{ij}=\rho$ for $i\neq j$. Now

$$\begin{aligned}G_\nu(h;\rho)&=\Pr\{t_1\leqslant h,\ldots,t_p\leqslant h\}\\&=\int_{-\infty}^{h}\cdots\int_{-\infty}^{h} g(t_1,\ldots,t_p)\,dt_1,\ldots,dt_p,\end{aligned} \tag{23.24}$$

where $g(t_1,\ldots,t_p)$ is given by (23.21).

Krishnaiah and Armitage [1966] have given tables of upper 1 and 5 percent points for $p=1(1)10$, $\rho=0.0(0.1)0.9$, and $\nu=5(1)35$. For the special case of $\rho=\frac{1}{2}$, tables have been constructed earlier by Dunnett (1955) and Gupta and Sobel (1957). Dunnett has given the upper 1 and 5 percent points for $p=2(1)9$, and $\nu=5(1)20, 24, 30, 40, 60, 120, \infty$.

Gupta and Sobel (1957) considered the distribution of

$$Z=\frac{\max\limits_{1\leqslant i\leqslant p-1}(X_i-X_p)}{s}. \tag{23.25}$$

If we let $t_i=(X_i-X_p)/\sqrt{2}\,s$, $i=1,\dots,p-1$, then

$$\Pr\{Z \leqslant d\}=\Pr\left\{t_1 \leqslant \frac{d}{\sqrt{2}},\dots,t_{p-1} \leqslant \frac{d}{\sqrt{2}}\right\}, \tag{23.26}$$

where $\mathbf{t}=(t_1,\dots,t_{p-1})$ has the multivariate t distribution with $\rho_{ij}=\frac{1}{2}$. The probability in (23.26) is equal to

$$\int_0^{\infty}\int_{-\infty}^{\infty}\Phi^{p-1}(u+dy)\varphi(u)q_\nu(y)\,du\,dy, \tag{23.27}$$

where $q_\nu(y)$ is the density of $Y=\sqrt{\chi_\nu^2/\nu}$. Gupta and Sobel (1957) have tabulated the values of $d=d(p,\nu,P^*)$ satisfying the equation

$$\Pr\{Z \leqslant d\}=G_\nu\left(d/\sqrt{2}\,;\tfrac{1}{2}\right)=P^* \tag{23.28}$$

for $p=2,5,10(1)16(2)20(5)40,50$; $\nu=15(1)20,24,30,36,40,48,60(20)120,360,\infty$; and $P^*=0.75,0.90,0.95,0.99$.

Tong (1969) has tabulated the values of $h=h(p,\nu,P^*)$ satisfying

$$G_\nu\left(h;\{\rho_{ij}\}\right)=P^* \tag{23.29}$$

for $p=2(1)6(2)12(4)20$, $P^*=0.50,0.75,0.90,0.95,0.975,0.99$, and $\nu=5(1)10(2)20(4)60(30)120,\infty$ when the ρ_{ij} are given by (23.8).

Following are the procedures described in earlier chapters to which these tables are relevant.

1. Two-sample procedure R_2 (§2.1) of Bechhofer, Dunnett and Sobel (1954) for selecting the population with the largest mean from a set of k normal populations with variances $\sigma_i^2=a_i\sigma^2, i=1,\dots,k$, where the a_i are known positive integers and σ^2 is unknown.
2. Procedure R_2 (§12.2.2) of Gupta (1956) for selecting a subset containing the population with the largest mean from a set of k normal populations with a common unknown variance based on equal sample sizes.
3. Procedure R_{12}' (§12.3) of Rizvi (1971) for selecting a subset containing the population with the largest absolute value of the mean from a set of k normal populations with a common unknown variance.
4. Procedure R (§20.4) of Dunnett (1955) for choosing all the experimental populations that are better than a control when all the populations are normal with unknown means and common unknown variance, and the comparisons are in terms of the means.

5. Procedure R_1 (§20.5.1) of Gupta and Sobel (1958a) for selecting a subset containing all populations better than a control when all populations are normal with a common unknown variance and the comparisons are in terms of the means (equal sample sizes).

6. Two-stage procedure R_2 (§20.6.1) of Tong (1969) for partitioning a set of populations with respect to a control when all the populations involved are normal with a common unknown variance and the comparisons are in terms of the means.

7. Procedure (§20.10.1) of Dunnett (1955) for constructing one-sided simultaneous confidence intervals for $\mu_i - \mu_0, i = 1,\ldots,k$, where the μ_i are the means of k normal populations and μ_0 is the mean of a control normal population, and the variances are assumed to be equal and unknown.

8. Procedure (§21.5.2) of Chen and Dudewicz (1973a) for constructing one-sided confidence interval for the ith smallest among the means of k normal populations which have a common unknown variance.

III. Tables of Multivariate $|t|$ Distribution

Let $|t| = (|t_1|,\ldots,|t_p|)$, where $t_1,\ldots,t_p$ have the joint density $g(t_1,\ldots,t_p)$ given by (23.21). Let

$$H_\nu(h;\{\rho_{ij}\}) = \Pr\{|t_1| \leqslant h,\ldots,|t_p| \leqslant h\}. \tag{23.30}$$

The values of h satisfying the equation $H_\nu(h;\{\rho_{ij}\}) = P^*$ have been tabulated by Dunnett (1955) for $p = 2(1)9$, $P^* = 0.95, 0.99$; and $\nu = 5(1)20, 24, 30, 40, 60, 120, \infty$ in the special case where $\rho_{ij} = 0.5$ for all $i \neq j$. These values give approximate solutions except when $p = 2$, since they were computed by equating to P^* the right-hand side of the inequality

$$\begin{aligned} &\Pr\{|t_1| \leqslant h,\ldots,|t_p| \leqslant h | \rho_{ij} = \rho\} \\ &\qquad\qquad \geqslant \left[\Pr\{|t_1| \leqslant h, |t_2| \leqslant h | \rho_{12} = \rho\}\right]^{p/2}. \end{aligned} \tag{23.31}$$

In a later paper, Dunnett (1964) has tabulated the exact values of $h = h(\nu, p, P^*)$ for $\rho = 0.5$, $p = 2(1)12, 15, 20$; $P^* = 0.95, 0.99$; and $\nu = 5(1)20, 24, 30, 40, 60, 120, \infty$. In these tables, the value shown as a superscript multiplied by $(1\text{-}2\rho)/(1-\rho)$ gives in each case the percentage increase required in the critical value of t when $0.125 < \rho < 0.5$.

The h-values are used in the procedure (§20.10.1) of Dunnett (1955) for constructing simultaneous two-sided confidence intervals for $\mu_i - \mu_0, i = 1,\ldots,k$, where the μ_i are the means of experimental populations, μ_0 is the mean of the control population, and the populations are all normal with common unknown variance.

IV. Tables Related to Univariate t

Let $G_\nu(x)$ be the cdf of a Student's t random variable with ν degrees of freedom. Also, let

$$G_\nu^k\left(\frac{a}{h}\right)=\gamma,$$

$$G_\nu\left(\frac{b}{h}\right)=\gamma,$$

$$G_\nu^k\left(\frac{d}{h}\right)-G_\nu^k\left(\frac{-d}{h}\right)=\gamma. \tag{23.32}$$

Chen and Dudewicz (1976) have tabulated the values of h satisfying each of the preceding equations for $\gamma=0.90, 0.95, 0.99$, and $\nu=1(1)20, 25, 40, 60, 100$ and the following choices of a, b, or d depending on the case: $a=1,2$; $b=1,2$; $d=1$. The range for k is: $k=1(1)3$ in the case of the last equation and $k=1(1)10$ in the other cases. These values are used in procedures (§21.5.3) of Chen and Dudewicz (1976) for constructing one-sided and two-sided confidence intervals for $\mu_{[k]}$, the largest among the means of k normal populations all having a common unknown variance.

V. Other References

Dunnett and Sobel [1955] have given some approximations for the evaluation of the probability integral of the multivariate t distribution when the correlation matrix has the structure $\rho_{ij}=c_ic_j$. John [1961] suggested some recurrence relations for the evaluation of the integral. In later papers, he [1964, 1966] has discussed alternative methods. Pillai and Ramachandran [1954] have tabulated upper 5 percent points of the multivariate t with an identity correlation matrix for $p=1(1)8$, and $\nu=3(1)10(2)14, 15, 16, 18, 20, 24, 30, 40, 60, 120, \infty$.

23.4 TABLES FOR THE STUDENTIZED LARGEST AND SMALLEST CHI-SQUARE DISTRIBUTIONS

Let $X_1, X_2, \ldots, X_p$ be p independently distributed chi-square variates with ν degrees of freedom. Let X_0 be another chi-square variate with ν' degrees of freedom distributed independently of the other X_i's. Define

$$U=\frac{\nu' X_{\max}}{\nu X_0}, \qquad V=\frac{\nu' X_{\min}}{\nu X_0}, \tag{23.33}$$

where $X_{\max} = \max(X_1,\ldots,X_p)$ and $X_{\min} = \min(X_1,\ldots,X_p)$. The distributions of U and V are known as the studentized largest and smallest chi-square distributions.

Let $c = c(p,\nu,\nu',P^*)$ be the upper $100(1-P^*)$ percent point of the distribution of U. Then

$$\Pr\{U \leqslant c\} = P^*. \tag{23.34}$$

It is easy to see that

$$\Pr\{U \leqslant c\} = \int_0^{\infty} G^p\left(\frac{x\nu c}{\nu'};\nu\right) g(x;\nu')\,dx$$

where $G(x;\nu)$ and $g(x;\nu)$ are the cdf and the density of a chi-square distribution with ν degrees of freedom.

Armitage and Krishnaiah [1964] have tabulated the values of $c = c(p,\nu,\nu',P^*)$ for (i) $P^* = 0.90, 0.95, 0.975$, $\nu = 1(1)19$, $\nu' = 5(1)45$, $p = 1(1)12$, and (ii) $P^* = 0.99$, $\nu = 1(1)19$, $\nu' = 6(1)45$, $p = 1(1)12$. Gupta (1963) has given tables of the values of $b = c^{-1}$ in the case of $\nu = \nu'$ for $p = 1(1)10$, $\nu = 2(2)50$, and $P^* = 0.75, 0.90, 0.95, 0.99$. He has also given approximate values of b in selected cases based on normal approximation for large ν. The values of b are given also by Guttman and Milton (1969) for $p = 1(1)3$, $\nu = 2(2)60(5)100$, and $P^* = 0.75, 0.90, 0.95, 0.99$.

Some very special cases of the tables of the studentized largest chi-square distribution have been given by Finney [1941], Nair [1948], Pearson and Hartley [1954], David [1956], and Ramachandran [1956]. The square roots of the upper 5 percent points are given by Pillai and Ramachandran [1954] for $\nu = 1$, $p = 1(1)8$, and $\nu' = 5(5)20, 24, 30, 40, 60, 120, \infty$.

Let us now consider the distribution of V, the studentized smallest chi-square. In this case,

$$\Pr(V > d) = \int_0^{\infty}\left[1 - G\left(\frac{d\nu x}{\nu'};\nu\right)\right]^p g(x;\nu')\,dx. \tag{23.35}$$

The upper $100P^*$ percent point of the distribution of V is given by $d = d(p,\nu,\nu',P^*)$ satisfying the equation

$$\Pr(V > d) = P^*. \tag{23.36}$$

Krishnaiah and Armitage [1964] have tabulated the values of d satisfying (22.36) for $p = 1(1)12$, $\nu = 1(1)20$, $\nu' = 5(1)45$, and $P^* = 0.90, 0.95, 0.975, 0.99$. In the case of $\nu = \nu'$, Gupta and Sobel (1962b) have given these values of d for $p = 1(1)10$, $\nu = 2(2)50$, and $P^* = 0.75, 0.90, 0.95, 0.99$. On the other hand,

Guttman and Milton (1969) have tabulated the reciprocals of d for $p=1(1)3$, $\nu=2(2)60(5)100$, and $P^*=0.75, 0.90, 0.95, 0.99$. The lower 5 percent points of the distribution of V have been constructed by Ramachandran [1958].

The tables of Gupta (1963), Armitage and Krishnaiah [1964], and Guttman and Milton (1969) are applicable in the following cases.

1. Analogs of the procedures R_1 (§4.8) and R_3 (§4.8) of Alam (1971b) for selecting the Poisson process associated with the largest mean arrival time.
2. Procedure R_{16} (§12.6) of Gupta (1963) for selecting a subset containing the gamma population having the largest scale parameter from a set of gamma populations having the same shape parameter.
3. Procedure R_1 (§14.3.1) of Gupta (1966b) and Gupta and Studden (1970) for selecting from multivariate normal populations a subset containing the population associated with the largest Mahalanobis distance.

Procedures employing the tables of Gupta and Sobel (1962b), Krishnaiah and Armitage [1964], and Guttman and Milton (1969) are:

1. Procedures R_1 (§4.8) and R_3 (§4.8) of Alam (1971b) for selecting the Poisson process having the smallest mean arrival time,
2. Procedure R_{13} (§12.4.1) of Gupta and Sobel (1962a) for selecting a subset containing the population having the smallest variance from a set of k normal populations,
3. Analog of the procedure R_1 (§14.3.1) of Gupta (1966b) for selecting a subset from multivariate normal populations to contain the population having the smallest Mahalanobis distance,
4. A special case of the procedure R_3 (§14.3.2) of Gnanadesikan and Gupta (1970) for selecting a subset from bivariate normal populations to include the population having the smallest generalized variance.

For some other applications of these tables, see Krishnaiah and Armitage [1964] and Armitage and Krishnaiah [1964].

23.5 ADDITIONAL TABLES FOR CHI-SQUARE DISTRIBUTIONS

I. Let $X_1, X_2, \ldots, X_k$ be independent random variables such that $Z_i = X_i/\sigma_i^2 (i=1,\ldots,k)$ has a chi-square distribution with ν degrees of freedom. We are interested in the evaluation of the probabilities of two types of events, namely,

A. $\Pr\{\max(X_1,\ldots,X_t)<\min(X_{t+1},\ldots,X_k)\}$
when $\sigma_1^2=\cdots=\sigma_t^2=\sigma^2$ and $\sigma_{t+1}^2=\cdots=\sigma_k^2=\theta^*\sigma^2, \theta^*\geqslant 1$, and
B. $\Pr\{X_1<X_2<\cdots<X_t<\min(X_{t+1},\ldots,X_k)\}$
when $\sigma_{[i+1]}^2=\theta_{i+1,i}^*\sigma_{[i]}^2$, θ_{i+1}^*, $i\geqslant 1, i=1,\ldots,t$, and $\sigma_{[t+1]}^2=\cdots=\sigma_{[k]}^2$.

We note that the two types A and B are the same for $t=1$. Bechhofer and Sobel (1954a) have evaluated these probabilities in the following cases.

k	t	Type	Range for ν and θ^*
2	1	A(=B)	$\nu=1(1)20; \theta^*=1.0(0.1)2.2$
3	1	A(=B)	same as above
4	1	A(=B)	same as above
3	2	A	same as above
3	3	B	same as above with $\theta^*_{3,2}=\theta^*_{2,1}=\theta^*$

The contexts of these evaluations are the procedures R_6 and R_7 (§2.2) of Bechhofer and Sobel (1954a) for dividing a set of k normal populations with variances $\sigma_1^2,\ldots,\sigma_k^2$, into two groups, namely, the t best (i.e., those associated with $\sigma_{[1]}^2,\ldots,\sigma_{[t]}^2$) and the k-t worst. This is called Goal I. Goal II requires in addition that the t best populations be completely ranked. The configurations of the σ_i^2 specified under A and B are the least favorable configurations for the procedures R_6 and R_7 corresponding to Goals I and II, respectively. The probabilities evaluated become the probabilities of a correct decision.

II. Ratio of Two Noncentral Chi-Square Variates

Let X_1 and X_2 be independent and identically distributed each having a noncentral chisquare distribution with p degrees of freedom and noncentrality parameter λ. Let

$$P=\Pr\left(\frac{X_1}{X_2}\geqslant c\right). \tag{23.37}$$

Gupta (1966b) has tabulated the values of P for $\lambda=0(0.2)1,2,5$, $p=2(2)6$, and $c=0.05(0.05)1.00$.

III. Distribution of χ_ν^2/ν

Let $G_\nu(y)$ be the cdf of X/ν, where X has a chi-square distribution with ν degrees of freedom. Given integers k and i such that $k\geqslant 1$ and $1\leqslant i\leqslant k$, let

$$\left[G_\nu(a)\right]^i=\gamma_1,$$

$$\left[1-G_\nu(b)\right]^{k-i+1}=\gamma_2. \tag{23.38}$$

Chen (1977a) has tabulated the values of a and b for $k=1(1)5$, $i=1(1)k$, $\nu=1(1)20,25,30,40,60,100$, and $\gamma_1=\gamma_2=0.95,0.99$. These values are used in the procedures (§21.4.2) of Chen (1977a) for constructing lower and upper confidence intervals for $\sigma^2_{[i]}$, the ith smallest among the variances of k normal populations. Additional tables for selected values of k and γ_1 (or γ_2) are given by Chen and Dudewicz (1973a).

23.6 DISTRIBUTIONS OF MAXIMUM AND MINIMUM OF F RATIOS

Let $Y_0, Y_1, \dots, Y_p$ be independent and identically distributed each having a central F distribution with r and s degrees of freedom. Define

$$U=\max\left\{\frac{Y_1}{Y_0},\dots,\frac{Y_p}{Y_0}\right\},$$

$$V=\min\left\{\frac{Y_1}{Y_0},\dots,\frac{Y_p}{Y_0}\right\}.$$

Let $c=c(p,r,s,P^*)>0$ be defined by the relation

$$\Pr\left\{U\leqslant\frac{1}{c}\right\}=P^* \tag{23.39}$$

which can be rewritten as

$$\int_0^\infty\left[F_{r,s}\left(\frac{x}{c}\right)\right]^p f_{r,s}(x)\,dx=P^*, \tag{23.40}$$

where $F_{r,s}(x)$ and $f_{r,s}(x)$ denote the cdf and the density of an F variate with r and s degrees of freedom.

Let us now denote by $d=d(p,r,s,P^*)$ the upper $100P^*$ percent point of the distribution of V. Then

$$\int_0^\infty[1-F_{r,s}(xd)]^p f_{r,s}(x)\,dx=P^*. \tag{23.41}$$

By using the fact that $F_{r,s}(y)=1-F_{s,r}(1/y)$, we can rewrite (23.41) in the form

$$\int_0^\infty F^p_{s,r}\left(\frac{y}{d}\right)f_{s,r}(y)\,dy=P^*. \tag{23.42}$$

Thus $c=c(p,r,s,P^*)$ which satisfies (23.39) and $d=d(p,r,s,P^*)$ which satisfies (23.41) are related by the equation

$$d(p,r,s,P^*) = c(p,s,r,P^*). \tag{23.43}$$

The values of $c(p,s,r,P^*)$ satisfying (23.39) have been tabulated by Gupta and Panchapakesan (1969a) for $p=1(1)4$, $P^*=0.75, 0.90, 0.95, 0.99$, and the following combinations of $q=r/2$ and $m=s/2$: $m=1(1)8$ for $q=1(1)4$, and $m=1(1)4$ for $q=5(1)8$.

These tables are applicable in determining the constants defining the following procedures.

1. Procedure R_2 (§14.3.1) of Gupta and Studden (1970) for selecting a subset from a set of multivariate normal populations to include the population having the largest (smallest) Mahalanobis distance, assuming that the covariance matrices Σ_i are unknown.
2. Procedure D (§14.3.3) of Gupta and Panchapakesan (1969a) for selecting a subset from a set of multivariate normal populations to include the population having the largest (smallest) multiple correlation coefficient.

23.7 DISTRIBUTIONS OF MAXIMUM AND MINIMUM OF RATIOS AND DIFFERENCES OF ORDER STATISTICS

Let $Y_1, Y_2, \ldots, Y_k$ be independent and identically distributed as the jth order statistic based on n independent observations from a continuous distribution G. The distribution function of $Y_i (i=1,\ldots,k)$ is given by

$$G_j(y) = \sum_{t=j}^{n} \binom{n}{t} G^t(y)[1-G(y)]^{n-t}, \qquad 1 \leq j \leq n. \tag{23.44}$$

If the support of G is the positive real line, we are interested in the distributions of

$$U = \max\left(\frac{Y_1}{Y_k}, \frac{Y_2}{Y_k}, \ldots, \frac{Y_{k-1}}{Y_k}\right) \tag{23.45}$$

and

$$V = \min\left(\frac{Y_1}{Y_k}, \frac{Y_2}{Y_k}, \ldots, \frac{Y_{k-1}}{Y_k}\right). \tag{23.46}$$

If the support of G extends over the real line, we are interested in the distributions of

$$W = \max(Y_1 - Y_k, Y_2 - Y_k, \dots, Y_{k-1} - Y_k) \tag{23.47}$$

and

$$Z = \min(Y_1 - Y_k, Y_2 - Y_k, \dots, Y_{k-1} - Y_k). \tag{23.48}$$

Tables of percentage points or reciprocals of percentage points are available for the distributions of U and V when G is exponential. Reciprocals of the percentage points of the distribution of U are available when G is folded normal. When G is logistic, percentage points have been tabulated in the case of W. We consider each of these in the following paragraphs.

Let $c = c(k,n,j,P^*)$ be the reciprocal of the upper $100(1-P^*)$ percent point of the distribution of U. Then

$$\int_0^\infty \left[G_j\left(\frac{x}{c}\right)\right]^{k-1} dG_j(x) = P^*. \tag{23.49}$$

When G is exponential ($G(x) = 1 - e^{-x}, x \geqslant 0$), the values of c (which, in this case, will be called c_1) have been tabulated by Barlow, Gupta, and Panchapakesan (1969) for $P^* = 0.75, 0.90, 0.95, 0.99$ and the following values of k, n, and j:

(i) $j = 1$, $k = 2(1)11$ (in this case c_1 is independent of n),

(ii) $k = 2(1)4$, $n = 5(1)15, j = 2(1)n$,

(iii) $k = 5$, $n = 5(1)12, j = 2(1)n$,

(iv) $k = 6$, $n = 5(1)10, j = 2(1)n$.

In the case of G being folded normal ($G(x) = \Phi(x) - \Phi(-x), x \geqslant 0$), Gupta and Panchapakesan (1975) have computed the values of c (now denoted by c_2) for $k = 2(1)10$, $n = 5(1)10$, $j = 1(1)n$, and $P^* = 0.75$, $0.90, 0.95, 0.99$.

The values of c_1 and c_2 are needed for the procedure R_1 (§16.2) of Barlow and Gupta (1969), and Gupta and Panchapakesan (1975) for selecting a subset containing the population having the largest α-quantiles from a set of populations that are star-shaped with respect to G when G is (i) exponential and (ii) folded normal, respectively. When G is exponential, the selection is from a set of IFRA distributions. When G is folded normal, the set belongs to a subfamily of IFRA distributions.

Let $d=d(k,n,j,P^*)$ be the upper $100P^*$ percent point of the distribution of V. Then

$$\int_0^{\infty}\left[1-G_j(xd)\right]^{k-1}dG_j(x)=P^*. \tag{23.50}$$

In the case of $G(x)=1-e^{-x}, x\geqslant 0$, Barlow, Gupta, and Panchapakesan (1969) have tabulated the values of d for $P^*=0.75, 0.90, 0.95, 0.99$ and the following combinations of the values of k, n, and j:

(i) $j=1$, $k=2(1)11$ (d is independent of n),

(ii) $k=2(1)5$, $n=5(1)15, j=2(1)n$,

(iii) $k=6$, $n=5(1)12, j=2(1)n$.

The values of d are used in the procedure R_2 (§16.2) of Barlow and Gupta (1969) for selecting a subset containing the population having the smallest α-quantiles from a set of IFRA distributions.

Let $D=D(k,n,j,P^*)$ be the upper $100(1-P^*)$ percent point of the distribution of W. Then

$$\int_{-\infty}^{\infty}G_j^{k-1}(t+D)dG_j(t)=P^*. \tag{23.51}$$

The values of D are given by Gupta and Panchapakesan (1974) for $k=2(1)10$, $n=5(2)15$, $j=(n+1)/2$, and $P^*=0.75, 0.90, 0.95, 0.99$ when G is logistic distribution, $G(x)=(1+e^{-x})^{-1}$. These values are applicable in the case of the procedure R_3 (§16.3) of Gupta and Panchapakesan (1974) for selecting a subset containing the population having the largest median from a set of distributions which are tail-ordered with respect to G.

23.8 TABLES RELATING TO DISCRETE DISTRIBUTIONS

In this section we consider tables concerning binomial, hypergeometric, and multinomial distributions.

23.8.1 Binomial Distribution

A. Let $X_1, X_2, \ldots, X_k$ be independent and let X_i have the binomial distribution $B(n,p_i)$, where $p_1=\cdots=p_{k-1}=p_k-\delta^*, \delta^*>0$. Define

$$P(n,p_1,\ldots,p_k)=\Pr\{X_k>\max(X_1,\ldots,X_{k-1})\}. \tag{23.52}$$

Sobel and Huyett (1957) have tabulated the exact and approximate values of n required so that

$$P(n,p_1,\ldots,p_k) \geqslant P^* \tag{23.53}$$

for $k=2,3,4,10$, $\delta^*=0.05(0.05)0.50$, and $P^*=0.50,0.60,0.75(0.05)0.95,0.99$. The value of n gives the number of observations required per process in the procedure R_1 (§4.1) of Sobel and Huyett (1957) for selecting the binomial process having the largest success probability.

B. Let d be the smallest positive integer for which

$$\inf_{0<\theta<1} \sum_{\alpha=0}^{n} b(\alpha;n,\theta)\left[B(\alpha+d;n,\theta)\right]^{k-1} \geqslant P^*, \tag{23.54}$$

where $b(\alpha;n,\theta)=\binom{n}{\alpha}\theta^{\alpha}(1-\theta)^{n-\alpha}$ and $B(y;n,\theta)=\sum_{\alpha=0}^{y} b(\alpha;n,\theta)$. Gupta and Sobel (1960) have tabulated the values of d for $k=2(1)20(5)50$, $n=1(1)20(5)50(10)100(25)200(50)500$, and $P^*=0.75,0.90,0.95,0.99$. The values are exact only for $k=2$.

The values of d are used in the procedure R_1 (§13.2.1) of Gupta and Sobel (1960) for selecting a subset of k binomial populations to include the population with the largest success probability. As pointed out previously, the values are approximate in the case of $k>2$. Better (approximate) values of d using conditional distribution have been given by Gupta, Huang, and Huang (1976) for $k=3(1)15$, $n=1(1)10$, and $P^*=0.75,0.90,0.95,0.99$.

C. Let d be the smallest integer for which (23.54) is satisfied. The left-hand side of (23.54) is the infimum of the PCS for the procedure R_1 (§13.2.1). Now, define

$$E(S)=\sum_{\alpha=0}^{n} b(\alpha;n,p+\delta)\left[B(\alpha+d;n,p)\right]^{k-1}$$
$$+(k-1)\sum_{\alpha=0}^{n} b(\alpha;n,p)\left[B(\alpha+d;n,p)\right]^{k-2}B(\alpha+d;n,p+\delta). \tag{23.55}$$

Then $E(S)$ denotes the expected number of populations included in the selected subset when $p_1=\cdots p_{k-1}=p$ and $p_k=p+\delta$. Gupta and Sobel (1960) have tabulated the values of $E(S)$ for $\delta=0.00,0.10,0.25,0.50$; $p+\delta=0.50,0.75,0.95,1.00$; $k=2,3,5,10$; $n=5(5)25$; and $P^*=0.75,0.90,0.95$.

D. Consider a random variable X_T which has the distribution specified by

$$\Pr\{X_T=x\}=\binom{T}{x}\left(\frac{1}{k}\right)^{x}\left(1-\frac{1}{k}\right)^{T-x}, \qquad x=0,1,\ldots,T. \tag{23.56}$$

For given P^*, k, and T, let c_T be the smallest integer such that

$$\Pr(X_T > c_T) \leqslant P^* \text{ and } \Pr(X_T \geqslant c_T) > P^*. \tag{23.57}$$

Define

$$\rho = \frac{P^* - \Pr(X_T > c_T)}{\Pr(X_T = c_T)}. \tag{23.58}$$

Gupta and Nagel (1971) have tabulated the values of c_T and ρ for $k=2,3,5$, $T=0(1)50$, and $P^*=0.75, 0.90, 0.95, 0.99$. The values of c_T and ρ are used in the randomized conditional procedure R_5 (§13.3.4) of Gupta and Nagel (1971) for selecting a subset containing the best Poisson population. If $Y_1, \ldots, Y_k$ are the observations from the k Poisson populations with common parameter, the distribution in (23.56) arises when conditioning on ΣY_i.

23.8.2. Hypergeometric Distribution

A. Let Z_T have the hypergeometric distribution for which

$$\Pr(Z_T = i) = \frac{\binom{n}{i}\binom{(k-1)n}{T-i}}{\binom{kn}{T}}, \qquad i = 1, \ldots, n; T \leqslant kn. \tag{23.59}$$

For given P^*, k, n, and T, let c_T be the smallest integer such that

$$\Pr(Z_T > c_T) \leqslant P^* \quad \text{and} \quad \Pr(Z_T \geqslant c_T) > P^*. \tag{23.60}$$

Define

$$\rho = \frac{P^* - \Pr(Z_T > c_T)}{\Pr(Z_T = c_T)}. \tag{23.61}$$

The values of c_T and ρ have been tabulated by Gupta and Nagel (1971) for $k=2,3,5$, $n=5,10$, and $P^*=0.75, 0.90, 0.95, 0.99$, with T in each case going from 0 to nk. These values are used in the randomized (conditional) procedure R_5 (§13.2.2) of Gupta and Nagel (1971) for selecting a subset containing the best binomial population.

B. Let $X_1, \ldots, X_k$ be independent and identically distributed random variables where X_i is the number of nondefectives in a random sample of size n drawn without replacement from a finite population consisting of N

units of which $N\theta(0<\theta<1)$ are nondefectives. The distribution of X_i is hypergeometric. Let $c=c(k,N,n,P^*)$ be the smallest integer such that

$$\Pr\{X_k \geqslant X_j - c \quad \text{for} \quad j=1,\ldots,k-1\} \geqslant P^*. \tag{23.62}$$

Bartlett and Govindarajulu (1970) have tabulated the values of c for $k=2,3$; $N=5(5)50,\infty$; $n=1(1)\min(15,[N/2])$; and $P^*=0.75, 0.90, 0.95, 0.99$, where $[s]$ denotes the largest integer less than or equal to s. These tables are needed for the procedure R (§13.5) of Bartlett and Govindarajulu (1970) for selecting a subset containing the hypergeometric distribution associated with the largest θ_i, from k populations with unknown proportions $\theta_1,\ldots,\theta_k$ of nondefectives.

23.8.3 Multinomial Distribution

Let p_i denote the probability associated with the ith cell of a multinomial distribution. For a sample of N observations, let X_i denote the number of observations that fall into the ith cell. We consider two types of configurations of the cell probabilities.

A. Case I: Let $p_i=p$ for $i=1,\ldots,k-1$, and $p_k=Ap$, $A\geqslant 1$. For a given integer $D(0\leqslant D\leqslant N)$, let

$$\begin{aligned} P_k &= \Pr\{X_k \geqslant X_j - D, j=1,\ldots,k-1\}, \\ P_1 &= \Pr\{X_1 \geqslant X_j - D, j=2,\ldots,k\}, \\ B &= \sum_{i=1}^{k} \Pr\{X_i \geqslant X_j - D, j=1,\ldots,k; j\neq i\} \\ &= P_k + (k-1)P_1. \end{aligned} \tag{23.63}$$

Gupta and Nagel (1967) have tabulated the values of P_k, P_1 and B/k for $A=1,3,5$; $D=0,1,2$; $k=2(1)10$; $N=2(1)15$. For the procedure R_1 (§13.6.1) of Gupta and Nagel (1967) for selecting a subset containing the most probable cell when the cell probabilities have the configuration $(p,\ldots,p,Ap)$, the probabilities P_k and P_1 denote the probabilities of selecting the best cell and a fixed nonbest cell, respectively. The expected proportion of cells included in the selected subset is given by B/k. For fixed N, k, and P^*, the minimum value of D required to guarantee the P^*-condition is given by Gupta and Nagel (1967) for $k=2(1)10$; $N=2(1)15$; and $P^*=0.75, 0.90$.

B. Case II: Let $p_1=p/A$, $A\geqslant 1$, and $p_i=p$ for $i=2,\ldots,k$. For a given integer $C(0\leqslant C\leqslant N)$, let

$$Q_1=\Pr\{X_1\leqslant X_j+C, j=2,\ldots,k\},$$

$$Q_k=\Pr\{X_k\leqslant X_j+C, j=1,\ldots,k-1\}, \tag{23.64}$$

$$B'=\sum_{i=1}^{k}\Pr\{X_i\leqslant X_j+C, j=1,\ldots,k; j\neq i\}$$

$$=Q_1+(k-1)Q_k.$$

The values of Q_1, Q_k, and B'/k have been tabulated by Gupta and Nagel (1967) for $A=1,3,5$; $C=0,1,2$; $k=2(1)10$; and $N=2(1)15$. For the procedure R' (§13.6.1) of Gupta and Nagel (1967) for selecting a subset containing the least probable cell when the cell probabilities have the configuration $(p/A,p,\ldots,p)$, the probabilities Q_1 and Q_k denote the probabilities of selecting the best cell and a fixed nonbest cell, respectively. The expected proportion of cells included in the selected subset is given by B'/k. For fixed N, k, and P^*, the minimum value of C that guarantees the P^*-condition is given by Gupta and Nagel (1967) for $k=2(1)10$, $N=2(1)15$, and $P^*=0.75, 0.90$.

C. Let $X_{\max}=\max(X_1,\ldots,X_k)$ and $X_{\min}=\min(X_1,\ldots,X_k)$. Gupta and Nagel (1967) have also tabulated the expected value and the variance of $X_{\max}$ and $X_{\min}$ corresponding to the two configurations $(p,\ldots,p,Ap)$ and $(p/A,p,\ldots,p)$ of the cell probabilities in each case for $k=2(1)10$, $N=2(1)15$, and $A=1,2,3,5$.

Neuts and Carson (1975) have discussed several numerical algorithms developed for evaluating probabilities related to multinomial trials.

23.9 SOME MISCELLANEOUS TABLES

In this section we discuss tables relating to several distributions that have been constructed in special contexts of procedures discussed in the earlier chapters.

I. Let

$$g_N(t)=\frac{1}{\Gamma(N)}t^{N-1}e^{-t}, t>0. \tag{23.65}$$

and

$$Q=\int_{c/(1+\alpha)}^{c/(1-\alpha)}\left[1-G_N(x)\right]\left[1-G_N\left(\frac{x}{\theta}\right)\right]^{k-2}dG_N(x), \qquad (23.66)$$

where $c>0$, $\theta>1$, $0<\alpha<1$, and $G_N(x)$ is the cdf corresponding to the density g_N. For given α, θ, k, and N, we consider maximizing Q as a function of c. Let $c=c_0$ be the value for which Q is maximum. Alam and Thompson (1973) have tabulated the values of c_0 and the maximum value $Q(c_0)$ for $k=2(1)5$; $\theta=1.1, 1.2, 1.4$; $\alpha=0.1, 0.2$; $N=2(2)20(5)40$. These values are used in the procedures R and S (§21.7.1) of Alam and Thompson (1973) for a combined goal of selection and estimation for Poisson processes.

II. Given positive integers k and n, define

$$P_1=\binom{n}{r}\sum_{i=0}^{k-1}\frac{(-1)^i\binom{k-1}{i}}{\binom{n(i+1)}{r}} \qquad (23.67)$$

for $1\leqslant r\leqslant n$. Rizvi and Sobel (1967) have provided a short table of P_1-values in the case of r equal to the greatest integer $\leqslant(n+1)/2$ for $k=2(1)10$. The n-value starts from 1 and increases in steps of 2 until a value (depending on k) for which P_1 is very close to unity.

This table is relevant to the procedure R_1 (§15.2) of Rizvi and Sobel (1967) for selecting from a set of k continuous distributions a subset containing the distribution with the largest α-quantile. The procedure can be constructed subject to the P^*-condition only if P^* does not exceed P_1 in (23.67). Generally, r is chosen such that $r\leqslant(n+1)\alpha<r+1$. The table values are for the case of $\alpha=\frac{1}{2}$.

III. Let

$$G_r(u)=\frac{\Gamma(n+1)}{\Gamma(r)\Gamma(n-r+1)}\int_0^u t^{r-1}(1-t)^{n-r}dt \qquad (23.68)$$

where r and n are positive integers such that $1\leqslant r\leqslant n$. $G_r(u)$ is the standard incomplete beta function denoted by $I_u(r,n-r+1)$. Let c be the smallest integer such that $1\leqslant c\leqslant r-1$ and

$$\int_0^1 G_{r-c}^{k-1}(u)\,dG_r(u)\geqslant P^*. \qquad (23.69)$$

Rizvi and Sobel (1967) have tabulated the largest value of $r-c$ for $r=(n+1)/2$, $k=2(1)10$, $n=5(10)95(50)495$, and $P^*=0.75, 0.90, 0.95, 0.975, 0.99$. These values are applicable to their procedure R_1 (§15.2) for selecting a subset of continuous distributions in terms of their α-quantiles when $\alpha=\frac{1}{2}$.

IV. Let X_{ij}, $j=1,\ldots,n$, be independent observations from a population $\pi_i (i=1,\ldots,k)$ which has the associated continuous distribution F_i. Let R_{ij} denote the rank of X_{ij} (rank 1 denotes the smallest) among the ordered observations of the pooled sample of size kn. Define $T_i=\Sigma_{j=1}^{n} R_{ij}$ which is the rank-sum statistic associated with π_i. In the case of $k=2$, this is the well-known Wilcoxon–Mann–Whitney statistic.

McDonald (1973) has discussed the distribution $U=\max_{1\leqslant j\leqslant k} T_j - T_1$, when the distributions F_i are identical. The values of $\Pr\{U \leqslant b\}$ have been tabulated by him for $k=2, n=2(1)20$; $k=3, n=2(1)8$; $k=4, n=2(1)5$; and $k=5, n=2,3$. McDonald (1973) has also provided a table of asymptotic values of b satisfying $\Pr\{U\leqslant b\}=P^*$ for $P^*=0.75, 0.90, 0.95, 0.975, 0.99$ and the following combinations of k and n: $k=2, n=10(5)20$; $k=3, n=6(1)8$; $k=4, n=3(1)5$; and $k=5, n=3$.

V. For positive integers $k \geqslant s \geqslant t$, let

$$A(k,s,t,c^*,n)$$

$$=(k-s)\binom{k-t}{s-t}\int_0^1 \left[1-G(c^*G^{-1}(x,n),n)\right]^t x^{k-s-1}(1-x)^{s-t}dx \tag{23.70}$$

where

$$G(y,n)=\int_0^y \frac{e^{-x}x^{n-1}}{(n-1)!}dx. \tag{23.71}$$

Hooper and Santner (1979) have tabulated the smallest odd integer n for which

$$A(k,s,t,c^*,n) \geqslant P^* \tag{23.72}$$

for $c^*=0.65, 0.70$, $P^*=0.75, 0.90, 0.95$, $k=2(1)9$, $t=1(1)[k/2]$ and $s=t(1)[k/2]$, where $[\cdot]$ denotes the largest integer function. The solution is the sample size for the means procedure for selecting from k exponential populations $[F(x,\theta_i)=1-e^{-x/\theta_i}, x\geqslant 0, \theta_i>0]$ a subset of size s containing at least t preferred populations (i.e., those for which $\theta_i \geqslant c^*\theta_{[k-t+1]}$).

CHAPTER 24

Illustrative Examples

24.1 INTRODUCTION

In this chapter we discuss some examples to illustrate some of the important procedures we have discussed so far. We also describe some specific applications of these procedures. In their book, Gibbons, Olkin, and Sobel (1977) have discussed several examples under the indifference zone formulation. Our main emphasis is on the subset selection approach. In the course of these examples, we also include some additional discussions. For referring to a procedure in an earlier chapter, we use the notation R_2 (§13.2.2) to refer to the procedure R_2 of §13.2.2 of this monograph.

24.2 NORMAL MEANS

The problem of comparing normal means is one that arises most often in several contexts. We consider here the data of Duncan [1955] from a randomized block experiment with six blocks and seven treatments (varieties of barley). The conventional analysis of variance test for homogeneity of the populations at level $\alpha = 0.05$ yields in this case a significant F-ratio of 4.61. The ordered sample means (yield in bushels per acre) for the seven varieties A, B, C, D, E, F, and G (relabeled) are $\bar{x}_{[1]} = 49.6$, $\bar{x}_{[2]} = 58.1$, $\bar{x}_{[3]} = 61.0$, $\bar{x}_{[4]} = 61.5$, $\bar{x}_{[5]} = 67.6$, $\bar{x}_{[6]} = 71.2$, and $\bar{x}_{[7]} = 71.3$. The error mean square is $s^2 = 79.64$ with $\nu = 30$ degrees of freedom. Since the F-ratio is significant, one might use, following some classical analysis, the least significant difference (LSD) and group all the varieties which are not significantly different. For the data under consideration LSD $= s_\nu \sqrt{2/n}\, t_{\alpha/2,\nu} = 10.51$ and the grouping obtained is: (A, B), (B, C, D, E), and (C, D, E, F, G). This technique is not very satisfactory for different reasons. One drawback is that the groups are not necessarily nonoverlapping. Several other methods such as Newman–Keuls' multiple range test, Tukey's honestly significant difference method and multiple-range test, Scheffé's method, Duncan's multiple-range test, and Bayesian K-ratio LSD rule, are available in literature. All these are multiple comparison procedures and are summarized and are applied to Duncan's data by Chew (1977b). Chew (1977b) has also applied the subset selection procedure of Gupta (1956) to the same data. Using Gupta's procedure, the selected subset consists of varieties C, D, E, F, and G with $P^* = .95$. Thus we can assert with (confidence) probability $P^* = .95$ that varieties corresponding to (C, D, E, F, G) contain the best variety in terms of the highest true mean yield.

We now give another example for the problem of selecting large (true) unknown means subset containing the best.

EXAMPLE 1. The study of Carrier, Orton, and Malpass [1962] is concerned with comparison of the anxiety levels of four groups (B = Bright, N = Normal, NEMH = Noninstitutionalized Educable Mentally Handicapped, and IEMH = Institutionalized Educable Mentally Handicapped) of children aged 10–14 years in five Southern Illinois communities. The grouping was based on their scores on the Wechsler Intelligence Scale for Children. A random sample of children was then taken from each group and each child was scored on the Children's Manifest Anxiety (CMA) scale. Large scores represent greater anxiety. We use the simulated data of Gibbons, Olkin, and Sobel (GOS), p. 33.

Group	Sample size	Sample mean
B	9	12.9
N	9	21.2
NEMH	9	22.4
IEMH	9	30.1

Here we assume that the CMA scores for the four groups are distributed normally with a common (unknown) variance. The "best" group is the one with the largest true (population) average. We are interested in selecting a subset containing the best with $P^*=0.95$. The pooled sample variance is $s^2=27.7$. Using the procedure R_2 (§12.2.2) of Gupta (1956), we include the group yielding sample mean $\bar{X}_i$ if and only if

$$\bar{X}_i \geqslant \max_{1 \leqslant j \leqslant 4} \bar{X}_j - d_2 \frac{s}{\sqrt{n}}.$$

To obtain the d_2-value corresponding to $k=4$ and $\nu=k(n-1)=32$ from the table of Gupta and Sobel (1957), we need to interpolate in terms of both k and ν; however, we get $d_2/\sqrt{2}=2.14$ from the table of Krishnaiah and Armitage [1966]. Hence the procedure R_2 selects a group if the corresponding sample mean is at least as large as 24.79 ($=30.10-5.31$). Thus we select only one group, namely, IEMH.

Suppose we have available to us a hypothetical value of $\sigma^2=30$ based on past experience. In this case, we use the procedure R_1 (§12.2.1) of Gupta (1956). This procedure selects all groups whose means are at least equal to $30.1-d_1\sigma/\sqrt{n}$, where $d_1=\sqrt{2}\times 2.0621$ from the table of Gupta, Nagel, and Panchapakesan (1973). In other words, groups whose means are at least as large as 24.78 are selected. The decision will then be to select only the group IEMH.

The performance of the procedure R_1 can be evaluated in terms of the true PCS and $E(S)$ for various true configurations of the means. A single measure of efficiency taking both the PCS and $E(S)$ into account is $\text{eff}(R_1)=\text{PCS}/E(S)$. We consider two types of configurations, namely, (A) slippage configuration: $\mu_{[1]}=\cdots=\mu_{[k-1]}=\mu_{[k]}-\delta$, and (B) equally spaced configurations: $\mu_{[i+1]}-\mu_{[i]}=\delta$.

Under the slippage configuration, the efficiency of R_1 can be obtained from Table 3 of Deely and Gupta (1968). Table 24.1 considers selected values of k and $\delta\sqrt{n}/\sigma$ for $P^*=0\cdot 95$. The efficiency is directly available

Table 24.1. $\text{Eff}(R_1)$ ***for slippage configuration*** $\mu_{[1]}=\cdots=\mu_{[k-1]}=\mu_{[k]}-\delta$
$P^*=0.95$

k \ $\frac{\delta\sqrt{n}}{\sigma}$	0.5	1.0	1.5	2.0	2.5	3.0	4.0	5.0
2	.52	.55	.58	.63	.69	.76	.89	.97
3	.35	.36	.39	.42	.47	.55	.73	.91
4	.26	.27	.29	.31	.35	.41	.60	.83
5	.21	.22	.23	.25	.28	.33	.50	.75
10	.10	.11	.11	.12	.13	.15	.25	.46

for these selected values from Table 6 of Deverman (1969) in the case of $P^*=0.90$.

For a fixed configuration, as one can see, the efficiency decreases as the number of populations increases, as is to be expected. For a fixed number of populations, the efficiency increases as the slippage distance increases. In our example, $k=4$, $n=9$, and $\sigma^2=30$. Suppose $\mu_{[1]}=\mu_{[2]}=\mu_{[3]}=\mu_{[4]}-7$. Then $\delta\sqrt{n}/\sigma=3.83$ and the efficiency is approximately .60.

Now, let us consider the equally spaced means configuration $(\mu,\mu+\delta,\ldots,\mu+(k-1)\delta)$. Table 24.2 gives the efficiency of R_1 for $P^*=.95$ and selected values of k and $\delta\sqrt{n}/\sigma$. The entries of this table are obtained using Table 1B of Gupta (1965).

It should be noted that for $k=2$, both slippage configuration and equally spaced means configuration coincide, and hence the first rows of Tables 24.1 and 24.2 are same. Table 24.2 shows that the procedure R_1 performs well when the means are far apart from one another.

The procedure R_1 is defined for every sample size n. The constant d_1 depends upon k and P^*. One could impose some additional requirement which can be met only by samples of certain minimum size. For example, one may require that the expected proportion of the populations selected should not be more than a specified fraction when $\mu_{[1]}=\cdots=\mu_{[k-1]}\leqslant\mu_{[k]}-\delta$. This expected proportion is maximum when $\mu_{[1]}=\cdots=\mu_{[k-1]}=\mu_{[k]}-\delta$ and so we must control the expected proportion corresponding to this configuration. In our example, let it be required that the expected proportion of populations selected should not be more than 0.65 when $\mu_{[1]}=\mu_{[2]}=\mu_{[3]}=\mu_{[4]}-3$. To determine the minimum n required we use Table 3 of Deely and Gupta (1968) where expected proportion of populations selected is tabulated for $P^*=.95$, and selected values of k and $\delta\sqrt{n}/\sigma$. For $k=4$, we see that $\delta\sqrt{n}/\sigma\geqslant 3$ will satisfy the requirement. In other words, $n\geqslant 9\sigma^2/\delta^2=30$. Thus we can meet the requirements on PCS and the expected proportion of populations selected by taking 30 observations from each population. Although we considered so far equal sample

Table 24.2. Eff(R_1) ***for equally spaced configuration*** $\mu_{[i+1]}-\mu_{[i]}=\delta$, $i=1,2,\ldots,k-1$; $P^*=0.95$

k \ $\delta\sqrt{n}/\sigma$	0.5	1.0	1.5	2.0	3.0	4.0	5.0
2	.52	.55	.58	.63	.76	.89	.97
3	.36	.39	.46	.54	.70	.85	.95
4	.27	.33	.42	.51	.67	.82	.93
5	.23	.30	.40	.49	.65	.80	.92
10	.16	.27	.37	.46	.61	.75	.88

Table 24.3. Simulated scores on the children's manifest anxiety scale for four groups of girls

B	N	NEMH	IEMH
19.0	20.2	24.2	31.7
8.0	18.2	21.7	27.6
17.2	22.4	20.2	31.7
2.0	37.8	21.9	29.9
9.8	22.7	24.0	24.7
10.0	19.4	16.7	27.5
15.2			31.4
14.8			31.0

sizes, often we do not have equal sample sizes as was really the case with the study of Carrier, Orton, and Malpass [1962]. Data of Table 24.3 are adopted from the data of GOS (p. 33) by leaving out some of the simulated observations.

Our goal is still to select a subset including the best population with $P^*=0.95$. We use the procedure R_4 (§12.2.3) of Gupta and D. Y. Huang (1976a). We have $k=4$, $n_{[1]}=n_{[2]}=6$, and $n_{[3]}=n_{[4]}=8$. From the table of Gupta and D. Y. Huang (1976a), we have $d_4=2.074$. The procedure R_4 also selects only one group, namely, IEMH.

As we have noted earlier, the subset selection approach is designed to include as many populations as necessary for the guaranteed PCS. The example illustrates the fact that we will not be selecting too many if the data show enough evidence that the population means are far apart.

EXAMPLE 2. This example illustrates the use of the selection procedure using the data of an experiment on the yield of 35 varieties of oats by Dr. G. K. Middleton, and given in Chapter 3 of Bose, Clatworthy, and Shrikhande [1954]. The intra-block estimates of the 35 treatment effects are as follows:

Treatment no.	Intra-block estimate	Treatment no.	Intra-block estimate
1	428.933	19	385.222
2	432.800	20	362.089
3	446.045	21	311.667
4	508.822	22	355.556
5	497.534	23	288.955
6	509.622	24	291.489
7	571.622	25	263.000
8	473.445	26	425.178
9	479.600	27	339.578
10	339.045	28	264.067

Treatment no.	Intra-block estimate	Treatment no.	Intra-block estimate
11	494.690	29	401.934
12	358.222	30	365.445
13	396.111	31	384.089
14	385.312	32	437.511
15	275.200	33	359.711
16	515.667	34	261.045
17	423.400	35	433.645
18	454.756		

Since the experiment was a partially balanced one, the estimates, though unbiased, are not truly equicorrelated but as the variance of the estimated difference between two treatments remains approximately the same whether the treatments are first or second associates, we may consider the treatments approximately equicorrelated and take the variance of the difference to be the weighted average of the variances of the difference between first associates and the variances of the difference between second associates (the weights being the number of pairs which are first and second associates, respectively). This is given by

$$\operatorname{Var}(\hat{t}_i - \hat{t}_j) = 2s^2(1-\rho') = 1957.255,$$

so that

$$s\sqrt{1-\rho'} = 31.28.$$

Since the number of degrees of freedom on which the estimate s^2 of σ^2 is based is 56, which is large, we can reasonably consider σ^2 known. Then consulting Table A1 of Gupta (1956) or Table I of Gupta [1963a] or the Table for $\rho = \frac{1}{2}$ in Gupta, Nagel, and Panchapakesan [1973], we find that corresponding to $k=35$ $(N=34)$, the d-values for $P^* = .75$, .90, .95, .975 and .99, are 2.839, 3.524, 3.939, 4.301, and 4.726, respectively. Hence the selection procedure that selects a population if $\bar{x}_i \geqslant \bar{x}_{\max} - ds\sqrt{1-\rho'}$, selects subsets of desirable populations as follows:

P^*	Numbers of the varieties selected	Size of the selected set
.99	1, 2, 3, 4, 5, 6, 7, 8, 9, 11, 16, 18, 26, 32, 35	15
.975	3, 4, 5, 6, 7, 8, 9, 11, 16, 18, 32	11
.95	4, 5, 6, 7, 8, 9, 11, 16, 18	9
.90	4, 5, 6, 7, 8, 9, 11, 16	8
.75	4, 5, 6, 7, 11, 16	6

EXAMPLE 3. To compare five different types of phosphorescent coatings of airplane instrument dials, each type was assigned randomly to 9 dials.

The coated dials were then excited with an ultraviolet light. The following table shows the number of minutes each dial glowed after the light source was turned off.

Type of Coating

I	II	III	IV	V
45.7	51.7	45.9	54.8	65.9
48.4	46.4	54.8	55.6	65.4
51.9	49.8	62.9	63.5	60.0
57.0	52.7	64.7	61.6	70.1
41.0	48.1	54.3	55.7	69.5
61.4	54.8	57.9	59.2	64.0
47.0	54.0	53.9	53.2	56.0
51.1	49.1	51.7	56.9	68.1
56.9	44.7	47.9	66.7	61.4

We assume that the duration of glow for each type is normally distributed and for the purpose of this example, we assume that the standard deviation for each type is 5 minutes. Our goal is to select a subset containing the best type (i.e., the type with the largest true mean duration of glow) but with the restriction that we do not select more than 3 types. The PCS is required to be at least .90 when $\mu_{[k]}-\mu_{[k-1]}\geqslant 4$. We use the procedure R_{11} (§12.2.7) of Gupta and Santner (1973). Using their Table II, we can see that the PCS requirement is guaranteed by selecting the ith type if and only if $\bar{X}_i \geqslant \max(\bar{X}_{[3]}, \bar{X}_{[5]} - 0.4\sigma/\sqrt{n}\,)$. The sample means are 51.2, 50.1, 54.9, 58.6, and 64.5, respectively. The procedure selects the ith type if and only if $\bar{X}_i \geqslant \max(54.9, 63.8)$. Our decision, therefore, will be to select the type V only.

The preceding example illustrates an interesting application of subset selection where the size of the selected subset is also controlled. It combines the subset selection with the indifference zone approach.

24.3 NORMAL VARIANCES

The problem of comparing normal variances is a special case of comparison of gamma populations; however, the emphasis here is that the basic observations arise out of normal populations.

EXAMPLE 4. An experiment was conducted to compare the variability in the same grade of gasoline sold by four dealers of different brands. Nine tank trucks are sampled for each dealer and the octane rating is determined for each sample. The ratings are given as follows.

Brand			
I	II	III	IV
31.2	35.4	39.1	38.2
33.3	33.6	40.2	28.2
25.4	25.8	33.4	35.1
29.8	27.7	42.2	31.4
37.1	36.7	43.1	30.2
42.3	26.2	37.2	27.4
41.6	36.3	39.5	34.2
32.4	34.2	27.4	31.4
42.2	37.6	35.7	29.7

The goal is to select a subset containing the best (the dealer with the smallest variation) with $P^*=.90$. The procedure to be applied here is R_{13} (§12.4.1) of Gupta and Sobel (1962a) which selects the ith brand if and only if $s_i^2 \leqslant c^{-1}s_{[1]}^2$ where the value of $c=.2610$ from Table 3 of Gupta and Sobel (1962b). For the data, $s_1^2=37.05$, $s_2^2=22.26$, $s_3^2=23.68$, and $s_4^2=12.15$. Thus the rule selects Brand i if $s_i^2 \leqslant 46.55$; in other words, the selected subset includes all the brands. Our decision indicates that the brands are not very much different from one another as far as this measure of variability (variance) is concerned.

24.4 BINOMIAL PROPORTIONS

EXAMPLE 5. An important problem in marketing management is to pretest comparative effectiveness of advertising copies. The most commonly used method for this is called *monadic design* [Wolfe et al. [1963], pp. 24–26] in which each individual in the sample is shown only one of the k different test ads typically along with advertisements of one or more competitors. Thus each ad is exposed to a randomly chosen group of n individuals. A measure of effectiveness of an ad is the (population) proportion of people who will recall it. This provides a binomial model, with k populations, $B(n,p_i)$, $i=1,\dots,k$, where p_i is the proportion for the ith population.

Let us consider the following data for five different ads.

Ad no.	1	2	3	4	5
Sample size (n)	10	10	10	10	10
Number who recalled (x_i)	6	3	4	8	2

Our goal is to select a subset to include the most effective ad with a P^*-value equal to .90. The procedure R_1 (§13.2.1) of Gupta and Sobel (1960) selects the ith ad if and only if $x_i \geqslant 8-d$, where $d=4$ from Table 2 of Gupta and Sobel (1960). Our decision will, therefore, be to select ad Nos. 1, 3, and 4.

Let us look at the performance of R_1 (§13.2.1) in terms of the expected proportion of populations retained in the selected subset when the true configuration of the p_i is given by $p_{[1]}=\cdots=p_{[k-1]}=p$ and $p_{[k]}=p+\delta$, $\delta>0$. Table 24.4 is excerpted from Table 3 of Gupta and Sobel (1960).

Suppose that four of the ads are equally effective with $p=.45$ and the fifth was the most effective with its true proportion equal to 0.95. Then we see from the Table 24.4 that the procedure R_1 in the long run will select on the average .502 or half the number of ads.

If we use the conditional procedure R_4 (§13.2.2) of Gupta, Huang, and Huang (1976), we will select the ith ad if and only if $x_i \geqslant 8-D(t)$, where observed value of $\Sigma X_i = t = 23$. The value of $D(t)$ here is needed for $P_1^*=.975$. From values in Table IIC of Gupta, Huang, and Huang (1976) for $P_1^*=.95$ and .99, we get a conservative value of $D(t)=7$. Our decision based on this conditional rule will then be to include all ads.

Another conditional rule R_5 (§13.2.2) of Gupta and Nagel (1971) has the selection probabilities

$$p_i(\mathbf{x}) = \begin{cases} 1 & \text{if } x_i > 3, \\ .78 & \text{if } x_i = 3, \\ 0 & \text{if } x_i < 3. \end{cases}$$

Thus according to this rule the ad Nos. 1, 3, and 4 will be included with probability 1, and ad No. 2 will be included with probability .78 on the basis of an auxiliary experiment.

Table 24.4. Expected proportion of populations retained in the selected subset by R_1 (§13.2.1) for the true configuration $p_{[1]}=\cdots=p_{[k-1]}=p$; $p_{[k]}=p+\delta$ for $k=5$, $n=10$, $P^*=0.90$, and selected values of δ and $p+\delta$.

δ \ $p+\delta$	.50	.75	.95	1.00
.00	.934	.974	1.000	1.000
.10	.923	.946	.994	.999
.25	.853	.849	.912	.937
.50	.502	.505	.502	.502

24.5 POISSON PROCESSES

EXAMPLE 6. To compare the performance of five different machines that produce sheet of metal in continuous rolls, data may be collected on the number of defects, X, that are spotted in several time segments of, say, 5 minute duration. It is reasonable to assume that X has a Poisson distribution with parameters λ_i, $i=1,\ldots,5$, for the four machines. Our goal is to choose (eliminate as undesirable) the machines that have high rate of defects. In other words, we wish to select a subset that will contain the population associated with $\lambda_{[k]}$ with a minimum guaranteed probability $P^*=0.90$. The output of each machine was inspected for 10 randomly chosen time segments. The data are summarized below:

Machine	I	II	III	IV	V
Total no. of defects(x_i)	12	8	4	9	11

We apply the conditional procedure R_5 (§13.3.3) of Gupta and Nagel (1971) which assigns selection probabilities $(p_1(\mathbf{x}),\ldots,p_5(\mathbf{x}))$ where

$$p_i(\mathbf{x})=\begin{cases}1 & \text{if } x_i>c_T\\ \rho & \text{if } x_i=c_T\\ 0 & \text{if } x_i<c_T\end{cases}$$

where T is the total number of defectives for all the machines. For our data, $T=44$. From Table 2 of Gupta and Nagel (1971), we get $c_T=5$ and $\rho_T=.90$. Thus we see that all machines except III will be selected ("eliminated" as highly undesirable).

It should be pointed out that in the preceding example we are choosing the subset of worst populations if large values of λ'_is are undesirable. This is *not the same* as claiming that those machines that are not selected provide a subset containing the best (i.e., the one with the smallest value of λ_i) with a guaranteed probability of P^*.

EXAMPLE 7. There are five different counters with unknown mean arrival rates. It is assumed that the arrivals occur according to the Poisson distribution. Our goal is to select a subset containing the process with the largest mean rate. The five processes are continuously observed until there are 10 arrivals at any one of the counters. The following table gives the data at termination:

Counter	1	2	3	4	5
Number of arrivals	2	10	6	3	4

The procedure R_7 (§13.3.5) of Gupta and Wong (1977) can be applied. This procedure selects a counter if the number of arrivals is at least equal to $10-c_1$, where the constant c_1 can be obtained from Table I of Gupta and Wong (1977). For $P^*=0.90$, the value of c_1 is 7, which means that all counters except the first are selected. If we choose $P^*=0.75$, then $c=5$. In this case, we select only two counters. The tables also show that in the preceding cases, the actual minimum PCS values are 0.9429 and 0.7737, respectively. These higher probabilities are due to the fact that we are dealing with discrete distributions.

24.6 GAMMA POPULATIONS

EXAMPLE 8. The data for this example were obtained by simulation consisting of five observations from each of three gamma populations having the densities $f(x;\lambda_i)=\lambda_i^2 xe^{-\lambda_i x}/\Gamma(2)$, $x \geqslant 0$, $i=1,2,3$. If the gamma distributions are used as a model for life distributions, then the mean life of the ith population is $2/\lambda_i$. Our goal is to select a subset that contains the population associated with $\lambda_{[1]}$; in other words, we want to select a subset that contains the population associated with the largest mean life. The procedure R_{16} (§12.6) of Gupta (1963) is applicable here. This rule selects the ith population if and only if $\overline{X}_i \geqslant b\overline{X}_{[k]}$, where $\overline{X}_i$ is the ith sample mean and $0<b<1$ is determined to guarantee a specified minimum PCS (P^*).

Data

Population	Observations (arranged in increasing order)				
1	3.0925,	5.2966,	5.6316,	17.8707,	24.8762
2	0.0618,	0.3192,	1.0409,	1.2856,	2.2585
3	2.1589,	3.1773,	6.0540,	6.4198,	7.9153

Let us take $P^*=0.90$. From Table 1B of Gupta (1963), we obtain $b=.492$. The sample means are 11.35352, 0.9932, 5.14506. The procedure selects a population if the corresponding sample mean is at least 5.5859. Our conclusion will, therefore, be to select only the first population.

24.7 MULTINOMIAL CELLS

EXAMPLE 9. A grocer wants to compare consumer preferences for five different brands of cereal he is interested in selling. He would like to select a subset of highly preferred brands. He surveys a random sample of 15 customers regarding their most preferred brand. The results are:

Brand	A	B	C	D	E
Number of customers	3	1	7	1	3

For $P^*=.90$, the procedure R_1 (§13.6.1) of Gupta and Nagel (1967) selects a brand if the number x_i of customers in favor of it is at least equal to $x_{max}-D=7-5=2$. Thus the brands A, C, and E are selected.

The value of D for implementing R_1 (§13.6.1) is obtained from Table 1B of Gupta and Nagel (1967). Suppose we use $P^*=0.75$ in the preceding problem. Then $D=3$ and the decision will be to select only brand C.

To illustrate the performance of the previous procedure R_1 with the help of the existing tables, suppose $k=3$, $N=7$, and $P^*=0.75$. Then the value of D is 2. Table 24.5 provides the true PCS and the efficiency $(=\text{PCS}/E(S))$ of R_1 when $p_{[1]}=\cdots=p_{[k-1]}=p_{[k]}/A$, for $A=1,3,5$.

Table 24.5. Performance of R_1 (§13.6.1) $k=3$, $N=7$, $P^*=0.75$, and $p_{[1]}=\cdots=p_{[k-1]}=p_{[k]}/A$.

A	True PCS	$E(S)$	Efficiency $=P(CS)/E(S)$
1	.781	2.344	0.333
3	.982	1.794	0.547
5	.997	1.407	0.709

It is clear from Table 24.5 that if $A=3$, that is, if the true unknown cell-probabilities are .2, .2, and .6, then the actual probability of selecting the cell associated with .6 is .982 and the procedure will select on the average a proportion .60 of the cells in the long run.

24.8 NONPARAMETRIC PROCEDURES

In this section we discuss some nonparametric procedures applied to traffic fatality data and also a related parametric procedure.

EXAMPLE 10. McDonald (1977b) has given data relating to motor-vehicle traffic fatalities per year per 100 million vehicle miles. His data are for the years 1960 through 1976 and cover the 48 contiguous states and the District of Columbia in the United States. The states are our populations and we have 17 observations per population. The goal is to select a subset of the 49 states (counting D.C. as a state) to include the state with the largest motor vehicle fatality rate (MFR) with a guaranteed minimum probability $P^* = 0.90$. Let $T_i = \Sigma_{j=1}^{17} R_{ij}$, where R_{ij} is the rank of the observation for state i among the 49 observations for the jth year. The procedure R_1 (a special case of $R'_1(h,G)$ of §15.3) selects State i if and only if $T_i \geqslant \max_{1 \leqslant j \leqslant 49} T_j - b$. Since k and n are large, we can use normal approximation to obtain $b = 215.09$. In obtaining the scores T_i, ties are resolved by midranks (averaged rank). One can alternatively use the procedure R_2 (a special case of $R_2'(g,G)$ of §15.3) which selects State i if and only if $T_i \geqslant c$. This is equivalent to (using normal approximation in evaluating c) selecting State i if and only if $T_i \geqslant 350.27$. In applying these procedures to the data R_1 selects 16 states whereas R_2 chooses 30 states.

McDonald (1977b) has applied the counterparts of R_1 and R_2 (R_1' and R_2', say) to select the states with small fatality rates. In this case, R_1' chooses 12 states but R_2' selects 31 states.

Sobel (1977) has applied his procedure R_1 (§15.2) to the motor vehicle fatality data to compare the states in terms of medians of the distributions of their MFRs. His procedure chooses 17 states. For the dual problem of selecting a subset containing the state with the smallest median rate, an analog of R_1 chooses 2 states.

We now discuss a parametric approach to handle the MFR data.

Example 11. Gupta and Hsu (1977b) have considered a parametric approach to the motor vehicle traffic fatality problem. Let X_{ij} denote the MFR for the ith state and the jth year, $i = 1, \ldots, 49$; $j = 1, \ldots, 17$. The index i denotes the state in alphabetic order and j denotes the year in increasing order. Our goal is to select the states having the lowest (highest) "average" MFR. An appropriate model is the two-way layout:

$$X_{ij} = m + a_i + b_j + (ab)_{ij} + \epsilon_{ij}, \qquad i = 1, \ldots, 49; \quad j = 1, \ldots, 17$$

where $\Sigma_{i=1}^{49} a_i = 0$, $\Sigma_{j=1}^{17} b_j = 0$, $\Sigma_{i=1}^{49} (ab)_{ij} = 0$, $\Sigma_{j=1}^{17} (ab)_{ij} = 0$, and the ϵ_{ij} are independently distributed with mean zero.

To achieve our goal, besides good estimates of the a_i, we need to have estimates of the variances of these estimates. So it is generally necessary to have $(ab)_{ij} = 0$ for all $i.j$. The analysis of McDonald (1977b) shows that the transformation $Y_{ij} = \log_e(X_{ij} - 1)$ is such that for the transformed data the interaction can be assumed to be zero. Thus for the transformed data, a reasonable model is:

$$Y_{ij} = \mu + \alpha_i + \beta_j + \epsilon_{ij}, \qquad i = 1, \ldots, 49; \quad j = 1, \ldots, 17$$

where $\Sigma_{i=1}^{49}\alpha_i=0$, $\Sigma_{j=1}^{17}\beta_j=0$, and the ϵ_{ij} are independently distributed with mean zero. The goal is to select the states having the smallest (largest) $\mu+\alpha_i$. To apply Gupta's normal means procedure R_2 (§12.2) we estimate each $\mu+\alpha_i$ by $y_{i.}=\Sigma_{j=1}^{17}y_{ij}/17$ and σ^2 by $s^2=\Sigma\Sigma(y_{ij}-y_{i.}-y_{.j}+y_{..})^2/(48\times 16)$. For $P^*=0.90$, the means procedure selects only one state as having the lowest MFR, and chooses a subset of 6 states as having the highest MFR. It is important though not surprising to note that the corresponding figures for McDonald (1977b) using rank-sum statistic are 12 and 16.

24.9 RESTRICTED FAMILIES OF DISTRIBUTIONS

In this section, we consider two examples of selection from IFRA distributions in terms of medians and an example of selection in terms of medians from distributions that are tail-ordered with respect to logistic distribution when centered at their respective medians. One of the IFRA examples is concerned with selecting a subset containing the population with the largest median and the other relates to selecting a subset whose size is restricted with a goal of selecting good populations.

EXAMPLE 12. For this example, we use the data of Example 8. The gamma distributions of this example are distributions with increasing failure rate (IFR) and therefore are also distributions with increasing failure rate on the average (IFRA). Hence we are justified in using the data as observations from three IFRA distributions. Our goal is to select the distribution with the largest α-quantile; in particular, we take $\alpha=0.5$. We apply the procedure R_1 (§16.2) of Barlow and Gupta (1969) to select a subset of the IFRA populations to contain the population with the largest median life with a guaranteed minimum probability $P^*=0.90$. Our procedure is based on the sample medians T_i and it selects the ith population if and only if $T_i \geqslant c_1 T_{[k]}$, where c_1 is found to be .25464 from Table 2B of Barlow, Gupta, and Panchapakesan (1969). The sample medians are 5.6316, 1.0409, and 6.0540. We select a population if the corresponding sample median is at least 1.5416; this results in the selection of populations 1 and 3. Compare this with the selection in terms of large mean life discussed in Example 8 of Section 24.6 in which only the first population was selected.

EXAMPLE 13. Suppose we have three IFRA distributions F_1, F_2, and F_3. Let ξ_i denote the median of F_i. Any population F_i is a good or preferred population if $\xi_i \geqslant c^*\xi_{[3]}$, where we take $c^*=0.7$. The goal is to select a subset consisting of not more than two populations to contain at least one good population with a guaranteed probability $P^*=0.75$. For this problem,

Hooper and Santner (1979) have proposed the procedure R which selects π_i if and only if $T_i \geqslant \max[T_{[2]}, \psi_n^{-1}(T_{[3]})]$ where the T_i are sample medians based on n observations from each population and $\psi_n(y) = 4^{1/\sqrt{n}}y$. Hooper and Santner (1979) have used this choice of $\psi_n(y)$ in constructing their tables, but theoretically $\psi_n(y) = y\rho^{e(n)}$, where $\rho > 1$ and $\{e(n)\}$ is a sequence of positive numbers converging to zero as $n \to \infty$. The problem is to determine the minimum sample size needed. For $P^* = 0.75$, and the preceding special choice of $\psi_n(y)$, the minimum odd sample size needed is 17 [Table 4.2 of Hooper and Santner (1979)]. The application of the procedure is now straightforward.

EXAMPLE 14. Let F_i be a continuous distribution with median Δ_i, $i = 1, \ldots, 5$, such that F_i centered at Δ_i is tail-ordered with respect to the logistic distribution $G(x) = [1 + e^{-x}]^{-1}$. The goal is to select a subset of these five distributions so that the population with the largest Δ_i is included in the selected subset with a guaranteed minimum probability $P^* = 0.90$. The following data were generated from logistic distributions $F_i(x) = [1 + \exp\{-(x - \Delta_i)/\theta_i\}]^{-1}$, $i = 1, \ldots, 5$, where we chose $(\Delta_i, \theta_i) = (2, .7)$, $(5, .4)$, $(-2, .3)$, $(3, .1)$, and $(1, .8)$, respectively.

Samples

I	II	III	IV	V
2.3171	4.8000	−2.1149	3.0341	1.5696
2.7735	5.0868	−2.1503	2.9811	2.5208
1.6297	4.9064	−1.6616	3.0499	−0.7520
1.6570	5.5220	−1.5542	2.6596	2.1544
3.5113	4.3148	−3.1440	3.2830	3.1816
2.9660	4.9872	−1.6814	3.0579	1.0712
1.4400	4.5524	−1.6769	3.1498	0.9888
2.2807	3.9756	−1.3172	2.8442	0.1968
2.4655	4.2784	−3.1688	3.0205	0.9960

The procedure R_3 (§16.3) of Gupta and Panchapakesan (1974) is applicable here. This procedure is based on the sample medians T_i, $i = 1, \ldots, 5$. For our data, $T_1 = 2.3171$, $T_2 = 4.8000$, $T_3 = -1.6814$, $T_4 = 3.0341$, and $T_5 = 1.0712$. The procedure R_3 selects the distribution F_i if and only if $T_i \geqslant 4.8000 - D$, where the D-value from Table 1 of Gupta and Panchapakesan (1974) is 1.7296. Thus the procedure selects F_i if and only if the corresponding sample median is at least as large as 3.0704. Our selected subset thus consists only of one population, namely, the one that gave rise to the sample median 4.8000, that is, the population with $(\Delta, \theta) = (5, .4)$.

24.10 MULTIVARIATE NORMAL POPULATIONS

EXAMPLE 15. Becker et al. [1965] have studied the problem of talker identification. Their speech spectrographic data consisted of seven replicate utterances of ten words by ten speakers. The within speaker covariance matrix for each word and each speaker provided a measure of variation among the repeated utterances of the same word by a speaker. It is of interest to find for each speaker the word with the least variation. A reasonable measure of variation is the generalized variance. Our goal is to select a subset of the words which would contain the word with the smallest generalized variance with a minimum guaranteed probability $P^* = 0.95$.

For one speaker the generalized variances for the ten words are 3.4323, 1.5411, 748.5346, 31300.2675, 306138.2997, 197.3952, 3624.5861, 720616.4465, 7.89846, 3.50126. The vector of log (generalized variance) is [1.2330, .4324, 6.6181, 10.3513, 12.6318, 5.2851, 9.0623, 13.6879, 2.0666, 1.2531]. The procedure R_3 (§14.3.2) of Gnanadesikan and Gupta (1970) selects the ith word if and only if

$$\log Z_i \leqslant \log Z_{[1]} - \log c_3,$$

where $Z_i = |S_i|$ is the sample generalized variance for the ith word. With the help of the approximation in (14.29), $\log c_3$ is found to be -6.9266. Thus the rule includes the ith word if $\log Z_i$ is in the interval [.4324, 7.359]. This results in selection of six words which have small generalized variances and can thus be used for the talker identification problem.

EXAMPLE 16. Fisher [1936] studied the data consisting of four measurements on samples of flowers from three species of iris: *Virginia*, *versicolor*, and *setosa*. The four measurements are sepal width, sepal length, petal length, and petal width. Thus we have $k = 3$ populations each with $p = 4$ variates. We assume that for each species these variables follow the multivariate normal distribution. The goal is to select a subset of the species of iris which includes the species having the largest (weighted) Mahalanobis distance $\lambda_i = \boldsymbol{\mu}_i' \Sigma_i^{-1} \boldsymbol{\mu}_i$ with a minimum guaranteed probability of P^* ($=0.90$), where $\boldsymbol{\mu}_i' = (\mu_{i1}, \mu_{i2}, \mu_{i3}, \mu_{i4})$, $i = 1, 2, 3$, denotes the true unknown vector of the means of the four variables for the ith species and Σ_i is the covariance matrix. The Σ_i are assumed to be unknown. Suppose we have $n = 20$ observations from each population. The procedure R_2 (§14.3.1) of Gupta and Studden (1970) is applicable here. This procedure R_2 selects π_i if and only if $c_2 T_i \geqslant T_{[k]}$, where $T_i = \mathbf{X}_i' S_i^{-1} \mathbf{X}_i$, $\mathbf{X}_i$ is the sample mean vector, and S_i is the sample covariance matrix for the sample from π_i. The values of c_2 are tabulated in Gupta and Panchapakesan (1969a) (see Table I with $p=2q$ and $n-p=2m$). It should be pointed out that this problem is equivalent to the subset selection problem for the largest of the noncentrality parameters λ_i of k independent noncentral F distributions with p and $n-p$ degrees of freedom.

EXAMPLE 17. Suppose we want to compare three groups (populations) of persons in terms of the association between verbal ability and the set of two factors, namely, chronological age and years of formal education. In other words, we are ranking $k=3$ populations in terms of the multiple correlation coefficient between the variable X_1 (=verbal ability score) and the set (X_2,X_3) where X_2=chronological age, and X_3=years of formal education. In the subset selection formulation, we are interested in selecting a subset containing the population with the largest true unknown multiple correlation coefficient $\rho_{1.23}$ with a minimum probability $P^*=0.75$. We take a sample of $n=19$ individuals from each population and compute the sample multiple correlation coefficients R_1,R_2,R_3. Let $R_i^{*2}=R_i^2/(1-R_i^2)$, $i=1,2,3$. The procedure D (§14.3.3) of Gupta and Panchapakesan (1969a) selects the ith population π_i if and only if $R_i^{*2} \geqslant c_4R_{[3]}^{*2}$. Suppose for the sample observations, $R_1^2=.447$, $R_2^2=.581$, and $R_3^2=.817$. Then $R_1^{*2}=.808$, $R_2^{*2}=1.387$, and $R_3^{*2}=4.464$. The value of $c_4=.18794$ from Table 1 of Gupta and Panchapakesan (1969a). Thus the procedure D selects any population for which R^{*2} is at least as large as .839. Our decision, therefore, is to select π_2 and π_3.

24.11 SELECTION OF THE LARGEST LOCATION PARAMETER: PROCEDURES BASED ON SAMPLE MEDIANS

In this section, we consider the problem of selecting a subset containing the best from k populations which have unknown location parameters θ_i, $i=1,\ldots,k$. The best population is the one associated with the largest unknown θ_i. Our purpose here is to illustrate the use of procedures based on sample medians in the cases of normal and double exponential populations and point out the efficiency of the median procedure relative to the means procedure in these cases. One can expect the median procedure to be "robust" when dealing with logistic and double exponential distributions which have heavier tails compared to normal distributions. Extreme observations from these populations are more likely than those from the normal. Since the mean is very sensitive to extreme observations, we might expect that in the cases of logistic and double exponential populations, the medians procedure performs better than in the normal case.

EXAMPLE 18. Random normal deviates were generated from the four normal populations $N(\mu_i, 1)$, $i=1,\ldots,4$, with means $\mu_i=-3.0$, 1.0, 2.0, and 4.0, respectively. In each case a sample of size $n=11$ was considered. The observed sample medians came out as follows: -3.294, 2.121, 1.921, and 4.060. The procedure of Gupta and Singh (1977b), denoted here by R_M, selects π_i if and only if its sample median is at least as large as $4.060-d_1$ where the value of $d_1=.982$ [Table I of Gupta and Singh (1977b)] for $P^*=0.90$. Thus the selected subset will consist of only one population, namely, the population with mean 4.0 which yielded a sample with median 4.060. From symmetry considerations, the subset selected to contain the smallest unknown true mean consists of populations whose medians lie in the interval $[-3.294, -3.294+.982]$. This also results in the selection of only one population, namely, the one with $\mu=-3.0$.

For the preceding problem we can also use the procedure R_1 (§12.2.1) of Gupta (1956). The asymptotic efficiency of R_M relative to R_1 has been discussed by Gupta and Singh (1977b). Let us recall the definition of asymptotic relative efficiency (ARE) of R_M relative to R_1. For a given ϵ $(0<\epsilon<1)$, let $N_{R_i}(\epsilon)$ be the number of observations needed so that $E(S'|R_i)=\epsilon$ for the procedure R_i, $i=1,M$, where S' is the number of nonbest populations selected. Then ARE of R_M relative to R_1 is $\text{ARE}(R_M,R_1)=\lim_{\epsilon\downarrow 0}(N_{R_1}(\epsilon)/N_{R_M}(\epsilon))$. Let $\text{ARE}(R_M,R_1;\Delta)$ denote the asymptotic relative efficiency corresponding to the slippage configuration of the means given by $\mu_{[1]}=\cdots=\mu_{[k-1]}=\mu_{[k]}-\Delta$. Gupta and Singh (1977b) have shown that $\text{ARE}(R_M,R_1;\Delta)=2/\pi=0.64$. Suppose that π_i is not $N(\mu_i,1)$ but a mixture (contamination model) of $N(\mu_i,1)$ and $N(\mu_i,b)$. In other words, if $f_1(x,\mu_i)$ is the density of $N(\mu_i,1)$ and $f_2(x,\mu_i)$ is the density of $N(\mu_i,b)$, then the density associated with π_i is

$$f(x,\mu_i)=\alpha f_1(x,\mu_i)+(1-\alpha)f_2(x,\mu_i), \qquad 0<\alpha<1.$$

In this case, Gupta and Singh (1977b) have shown that $\text{ARE}(R_M,R_1;\Delta)=(2/\pi b)[\alpha+(1-\alpha)b][1-\alpha+\alpha\sqrt{b}\,]^2$ which increases indefinitely as b increases. This shows that depending on the values of α and b, the medians procedure can be expected to perform much better than the means procedure in the case of contamination model. Note that the ARE's are independent of Δ.

EXAMPLE 19. Suppose we have $k=5$ Laplace or double exponential populations with densities

$$f(x,\theta_i)=\tfrac{1}{2}e^{-|x-\theta_i|}, \quad -\infty<x<\infty, \quad \theta_i>0, \qquad i=1,\ldots,5.$$

We are interested in selecting a subset containing the population associated with the largest θ_i. The following data were simulated by Gupta and Leong (1977) with $\theta_1=0$, $\theta_2=2.5$, $\theta_3=3.4$, $\theta_4=-2.0$, and $\theta_5=-0.65$ for the five double exponential distributions.

π_1	π_2	π_3	π_4	π_5
−3.4839	−0.9839	−0.0839	−5.4839	−4.1339
−2.6762	−0.1762	0.7238	−4.6762	−3.3262
−0.3129	2.1871	3.0871	−2.3129	−0.9629
−0.2264	2.2736	3.1736	−2.2264	−0.8764
−0.1761	2.3239	3.2239	−2.1761	−0.8261
0.1462	2.6462	3.5462	−1.8538	−0.5038
0.3033	2.8033	3.7033	−1.6967	−0.3467
1.6160	4.1160	5.0160	−0.3840	0.9660
5.6924	8.1924	9.0924	3.6924	5.0424

Let Y_i denote the sample median based on nine observations from π_i, $i=1,\ldots,5$. The procedure of Gupta and Leong (1977), denoted here by R_M^*, selects π_i if and only if $Y_i \geqslant \max_{1\leqslant j\leqslant 5} Y_j - d$. For $P^*=0.95$, the value of $d=1.3210$ [Table of Gupta and Leong (1977)]. The sample medians are $Y_1=-0.1761$, $Y_2=2.3239$, $Y_3=3.2239$, $Y_4=-2.1761$, and $Y_5=-0.8261$. Thus the procedure R_1 selects a population if its sample median is at least as large as $3.2239-1.3210=1.9029$. The decision will, therefore, be to select the populations π_2 and π_3.

Gupta and Singh (1977b) have compared the performances of the procedure R_M^* and the means procedure R_1 for the Laplace distributions. They have shown that under the slippage configuration $\theta_{[1]}=\cdots=\theta_{[k-1]}=\theta_{[k]}-\Delta$, $\Delta>0$, $\text{ARE}(R_M^*, R_1; \Delta)=2$.

For similar comparisons of medians and means procedures for logistic distributions, see Lorenzen and McDonald (1978).

Bibliography

Alam, K. (1967). A two-sample estimate of the largest mean. *Ann. Inst. Statist. Math.* **19**, 271–283.

Alam, K. (1970a). Monotonicity properties of the multinomial distribution. *Ann. Math. Statist.* **41**, 315–317.

Alam, K. (1970b). A two-sample procedure for selecting the population with the largest mean from *k* normal populations. *Ann. Inst. Statist. Math.* **22**, 127–136.

Alam, K. (1971a). On selecting the most probable category. *Technometrics* **13**, 843–850.

Alam, K. (1971b). Selection from Poisson processes. *Ann. Inst. Statist. Math.* **23**, 411–418.

Alam, K. (1973). On a multiple decision rule. *Ann. Statist.* **1**, 750–755.

Alam, K. (1977). A Bayes rule for selecting the most probable multinomial event. Tech. Report No. 241, Dept. of Math. Sciences, Clemson University, Clemson, South Carolina.

Alam, K., and Rizvi, M. H. (1966). Selection from multivariate populations. *Ann. Inst. Statist. Math.* **18**, 307–318.

Alam, K., Rizvi, M. H., and Solomon, H. (1975). Selection of largest multiple correlation coefficients: exact sample size case. *Ann. Statist.* **4**, 614–620.

Alam, K., and Saxena, K. M. Lal (1974). On interval estimation of a ranked parameter. *J. Roy. Statist. Soc. Ser. B* **36**, 277–283.

Alam, K., Saxena, K. M. Lal, and Tong, Y. L. (1973). Optimal confidence interval for a ranked parameter. *J. Amer. Statist. Assoc.* **68**, 720–725.

Alam, K., Seo, K., and Thompson, J. R. (1971). A sequential sampling rule for selecting the most probable multinomial event. *Ann. Inst. Statist. Math.* **23**, 365–374.

Alam, K., and Thompson, J. R. (1971). A selection procedure based on ranks. *Ann. Inst. Statist. Math.* **23**, 253–262.

Alam, K., and Thompson, J. R. (1972). On selecting the least probable multinomial event. *Ann. Math. Statist.* **43**, 1981–1990.

Alam, K., and Thompson, J. R. (1973). A problem of ranking and estimation with Poisson process. *Technometrics* **15**, 801–808.

Alam, K., and Wallenius, K. T. (1977). On selecting the best population. *Commun. Statist.-Theor. Meth.* **A6**, 1091–1104.

Anderson, P. O., Bishop, T. A., and Dudewicz, E. J. (1977). Indifference-zone ranking and selection: confidence intervals for true achieved $P(CD)$. *Commun. Statist.-Theor. Meth.* **A6**, 1121–1132.

Arvesen, J. N. (1971). A subset selection procedure for selecting the largest multiple correlation coefficient. Mimeo. Ser. No. 269, Dept. of Statist., Purdue University, West Lafayette, Indiana.

Arvesen, J. N., and McCabe, G. P., Jr. (1973). Variable selection in regression analysis, *Proceedings of the University of Kentucky Conference on Regression with a Large Number of Predictors* (Ed. W. O. Thompson and F. B. Cady), Dept. of Statist., Univ. of Kentucky, Lexington.

Arvesen, J. N., and McCabe, G. P., Jr. (1975). Subset selection problems of variances with applications to regression analysis. *J. Amer. Statist. Assoc.* **70**, 166–170.

Bahadur, R. R. (1949). On A Class of Decision Problems in the Theory of k Populations. Ph.D. Thesis. Univ. of North Carolina, Chapel Hill.

Bahadur, R. R. (1950). On the problem in the theory of k populations. *Ann. Math. Statist.* **21**, 362–375.

Bahadur, R. R., and Goodman, L. (1952). Impartial decision rules and sufficient statistics. *Ann. Math. Statist.* **23**, 553–562.

Bahadur, R. R., and Robbins, H. (1950). The problem of the greater mean. *Ann. Math. Statist.* **21**, 469–487. Correction, **22** (1951), 310.

Bangdiwala, I. S. (1958). Some Sequential Procedures for Ordering Populations According to Means, Variances and Regression Coefficients. Ph.D. Thesis. North Carolina State College.

Barlow, R. E., and Gupta, S. S. (1969). Selection procedures for restricted families of distributions. *Ann. Math. Statist.* **40**, 905–917.

Barlow, R. E., Gupta, S. S., and Panchapakesan, S. (1969). On the distribution of the maximum and minimum of ratios of order statistics. *Ann. Math. Statist.* **40**, 918–934.

Barr, A. J., and Goodnight, J. H. (1971). *Statistical Analysis System.* Raleigh Student Supply Store, North Carolina State Univ., Raleigh.

Barr, D. R., and Rizvi, M. H. (1966a). An introduction to selection and ranking procedures. *J. Amer. Statist. Assoc.* **61**, 640–646.

Barr, D. R. (Capt.), and Rizvi, M. H. (1966b). Ranking and selection problems of uniform distributions. *Trabajos Estadist.* **17**, 15–31.

Barron, A. M. (1968). A Class of Sequential Multiple Decision Procedures. Ph.D. Thesis (Mimeo. Ser. No. 169). Dept. of Statist., Purdue Univ., West Lafayette, Indiana.

Barron, A. M., and Gupta, S. S. (1972). A class of non-eliminating sequential multiple decision procedures. *Operations Research-Verfahren* (Ed. Henn, Künzi, and Schubert), Verlag Anton Hain, Meisenheim am Glan, Germany, pp. 11–37.

Barron, A. M., and Mignogna, E. (1977). Sequential selection procedures for choosing the better of two binomial populations. *Commun. Statist.-Theor. Meth.* **A6**, 525–552.

Bartlett, N. S. (1965). Ranking and Selection Procedures for Continuous and Certain Discrete Populations. Ph.D. Thesis. Case Inst. of Technology, Cleveland, Ohio.

Bartlett, N. S., and Govindarajulu, Z. (1968). Some distribution-free statistics and their application to the selection problem. *Ann. Inst. Statist. Math.* **20**, 79–97.

Bartlett, N. S., and Govindarajulu, Z. (1970). Selecting a subset containing the best hypergeometric population. *Sankhyā Ser. B* **32**, 341–352.

Bawa, V. S. (1970a). Asymptotically Optimal Ranking and Selection Procedures. Ph.D. Thesis (Tech. Report No. 102). Dept. of Operations Res., Cornell Univ., Ithaca, New York.

Bawa, V. S. (1970b). Asymptotic efficiency of one multifactor experiment relative to several one-factor experiments for selecting the normal population with the largest mean. Tech. Report No. 121, Dept. of Operations Res., Cornell Univ., Ithaca, New York.

Bawa, V. S. (1972). Asymptotic efficiency of one R-factor experiment relative to R one-factor experiments for selecting the best normal population. *J. Amer. Statist. Assoc.* **67**, 660–661.

Beale, E. M. L. (1970). Selecting an optimal subset. *Integer and Nonlinear Programming* (Ed. J. Abadie), North-Holland, Amsterdam, Holland, pp. 451–562.

Beale, E. M. L., Kendall, M. G., and Mann, D. W. (1967). The discarding of variables in multivariate analysis. *Biometrika* **54**, 357–366.

Bechhofer, R. E. (1954). A single-sample multiple decision procedure for ranking means of normal populations with known variances. *Ann. Math. Statist.* **25**, 16–39.

Bechhofer, R. E. (1955). Multiple decision procedures for ranking means. *Proc. Ninth Annual Convention of Amer. Soc. Qual. Contr.*, pp. 513–519.

Bechhofer, R. E. (1958). A sequential multiple decision procedure for selecting the best one of several normal populations with a common unknown variance and its use with various experimental designs. *Biometrics* **14**, 408–429.

Bechhofer, R. E. (1967). A two-stage subsampling procedure for ranking means of finite populations with an application to bulk-sampling problems. *Technometrics* **9**, 355–364.

Bechhofer, R. E. (1968a). Single-stage procedures for ranking multiply-classified variances of normal populations. *Technometrics* **10**, 693–714.

Bechhofer, R. E. (1968b). Multiple comparisons with a control for multiply-classified variances of normal populations. *Technometrics* **10**, 715–718.

Bechhofer, R. E. (1968c). Designing factorial experiments to rank variances. *Transactions of the Twenty-Second Annual Technical Conference of the American Society for Quality Control*, pp. 69–73.

Bechhofer, R. E. (1969). Optimal allocation of observations when comparing several treatments with a control. *Multivariate Analysis–II (Ed. P. R. Krishnaiah), Academic, New York, pp.* 463–473.

Bechhofer, R. E. (1970a). An undesirable feature of a sequential multiple-decision procedure for selecting the best one of several normal populations with a common known variance. Correction note. *Biometrics* **26**, 347–349.

Bechhofer, R. E. (1970b). On ranking the players in a 3-player tournament. *Nonparametric Techniques in Statistical Inference* (Ed. M. L. Puri), Cambridge University Press, London, pp. 545–549.

Bechhofer, R. E. (1974). A two-sample procedure for selecting the population with the largest mean from several normal populations with unknown variances:

some comments on Ofosu's paper. Tech. Report. No. 233. Dept. of Operations Res., Cornell Univ., Ithaca, New York.

Bechhofer, R. E. (1975). Ranking and selection procedures. *Proceedings of the Twentieth Conference on the Design of Experiments in Army Research Development and Testing*. ARO Report 75-2, Part 2, Dept. of the Army, pp. 929–949.

Bechhofer, R. E. (1977). Selection in factorial experiments. *Proceedings of the* 1977 *Winter Simulation Conference* (Ed. H. J. Highland, R. G. Sargent, and J. W. Schmidt) held at the National Bureau of Standards, Gaithersburg, Maryland, pp. 65–70.

Bechhofer, R. E., and Blumenthal, S. (1962). A sequential multiple decision procedure for selecting the best one of several normal populations with a common unknown variance, II: Monte Carlo sampling results and new computing formulae. *Biometrics* **18**, 52–67.

Bechhofer, R. E., Dunnett, C. W., and Sobel, M. (1953). A two-sample multiple decision procedure for ranking means of normal populations with unknown variances (preliminary report). Abstract. *Ann. Math. Statist.* **24**, 136.

Bechhofer, R. E., Dunnett, C. W., and Sobel, M. (1954). A two-sample multiple-decision procedure for ranking means of normal populations with a common unknown variance. *Biometrika* **41**, 170–176.

Bechhofer, R. E., Elmaghraby, S., and Morse, N. (1959). A single-sample multiple decision procedure for selecting the multinomial event which has the highest probability. *Ann. Math. Statist.* **30**, 102–119.

Bechhofer, R. E., Kiefer, J., and Sobel, M. (1968). *Sequential Identification and Ranking Procedures*. The University of Chicago Press, Chicago.

Bechhofer, R. E., and Nocturne, D. J. (1972). Optimal allocation of observations when comparing several treatments with a control, II: 2-sided comparisons. *Technometrics* **14**, 423–436.

Bechhofer, R. E., Santner, T. J., and Turnbull, B. W. (1977). Selecting the largest interaction in a two-factor experiment. *Statistical Decision Theory and Related Topics–II* (Eds. S. S. Gupta and D. S. Moore), Academic, New York, pp. 1–18.

Bechhofer, R. E., and Sobel, M. (1953). A sequential multiple-decision procedure for ranking means of normal populations with known variances (prelim. report), Abstract. *Ann. Math. Statist.* **24**, 136–137.

Bechhofer, R. E., and Sobel, M. (1954a). A single-sample multiple-decision procedure for ranking variances of normal populations. *Ann. Math. Statist.* **25**, 273–289.

Bechhofer, R. E., and Sobel, M. (1954b). On a sequential ranking procedure (prelim. report). Abstract, *Bull. Amer. Math. Soc.* **60**, 34–35.

Bechhofer, R. E., and Sobel, M. (1956a). A sequential multiple-decision procedure for selecting the population with the largest mean from k normal populations with a common unknown variance (prelim. report). Abstract, *Ann. Math. Statist.* **27**, 218–219.

Bechhofer, R. E., and Sobel, M. (1956b). A scale invariant sequential multiple-decision procedure for selecting the population with the smallest variance (prelim. report). Abstract, *Ann. Math. Statist.* **27**, 219.

Bechhofer, R. E., and Sobel, M. (1956c). A sequential multiple-decision procedure for selecting the multinomial event with the largest probability (preliminary report). Abstract. *Ann. Math. Statist.* **27**, 861.

Bechhofer, R. E., and Sobel, M. (1958a). *Sequential Multiple-Decision Procedures for Ranking Parameters of Koopman–Darmois Populations with Special Reference to Ranking Means of Normal Populations*. Cornell Univ., Ithaca, New York [earlier version of Bechhofer, R. E., Kiefer, J., and Sobel, M. (1958)].

Bechhofer, R. E., and Sobel, M. (1958b). Non-parametric multiple-decision procedures for selecting that one of k populations which has the highest probability of yielding the largest observation (prelim. report). Abstract. *Ann. Math. Statist.* **29**, 325.

Bechhofer, R. E., and Turnbull, B. W. (1971). Optimal allocation when comparing several treatments with a control, III: Globally best one-sided intervals for unequal variances. *Statistical Decision Theory and Related Topics* (Eds. S. S. Gupta and J. Yackel), Academic, New York, pp. 41–78.

Bechhofer, R. E., and Turnbull, B. W. (1977). On selecting the process with the largest fraction of conforming product. *Proceedings of the* 31*st Technical Conference of the American Society for Quality Control*, pp. 568–573.

Bechhofer, R. E., and Turnbull, B. W. (1978). Two $(k+1)$-decision selection procedures for comparing k normal means with a specified standard. *J. Amer. Statist. Assoc.* **73**, 385–392.

Becker, W. A. (1961). Comparing entries in random sample tests. *Poultry Sci.* **40**, 1507–1514.

Becker, W. A. (1962). Ranking all-or-none traits in random sample tests. *Poultry Sci.* **41**, 1437–1438.

Becker, W. A. (1964). Changes in performance of entries in random sample tests. *Poultry Sci.* **43**, 716–722.

Berger, R. L. (1977a). Minimax, Admissible, and Gamma-minimax Multiple Decision Rules. Ph.D. Thesis (Mimeo. Ser. No. 489). Dept. of Statist., Purdue Univ., West Lafayette, Indiana.

Berger, R. L. (1977b). Minimax subset selection for loss measured by subset size. Mimeo. Ser. No. 507, Dept. of Statist., Purdue Univ., West Lafayette, Indiana.

Berger, R. L., and Gupta, S. S. (1977). Minimax subset selection with applications to unequal variance problems. Mimeo. Ser. No. 495, Dept. of Statist., Purdue Univ., West Lafayette, Indiana. To appear in *Scand. J. Statist.*

Berry, D. A., and Sobel, M. (1973). An improved procedure for selecting the better of two binomial populations. *J. Amer. Statist. Assoc.* **68**, 979–984.

Berry, D. A., and Young, D. H. (1977). A note on inverse sampling procedures for selecting the best binomial populations. *Ann. Statist.* **5**, 235–236.

Bhalla, P. (1973). A Bayes Selection Procedure with Particular Results for Binomial Probability Parameters. Ph.D. Thesis. Dept. of Statist., Univ. of Southwestern Louisiana, Lafayette, Louisiana.

Bhapkar, V. P., and Gore, A. P. (1971). Some selection procedures based on *U*-statistics for the location and scale problems. *Ann. Inst. Statist. Math.* **23**, 375–386.

Bhattacharyya, P. K. (1956). Comparison of the means of k normal populations. *Calcutta Statist. Assoc. Bull.* **7**, 1–20.

Bhattacharyya, P. K. (1958). On the large sample behaviour of a multiple-decision procedure. *Calcutta Statist. Assoc. Bull.* **8**, 43–47.

Bickel, P. J., and Yahav, J. A. (1977). On selecting a subset of good populations. *Statistical Decision Theory and Related Topics–II* (Eds. S. S. Gupta and D. S. Moore), Adademic, New York, pp. 37–55.

Bishop, T. A. (1976). Selecting the better of two negative exponential populations: Bayes and minimax procedures. Tech. Report No. 127, Dept. of Statist., Ohio State Univ., Columbus, Ohio.

Bishop, T. A., and Dudewicz, E. J. (1977). Complete ranking of reliability-related distributions. *IEEE Transactions on Reliability*, **26**, 362–365.

Bjørnstad, J. F. (1978). On Optimal Subset Selection Procedures. Ph.D. Thesis, Dept. of Statistics, Univ. of California, Berkeley.

Bland, R. P. (1961). A minimum average risk solution for the problem of choosing the largest mean. Mimeo. Ser. No. 280. Inst. of Statist., Univ. of North Carolina, Chapel Hill.

Bland, R. P., and Bratcher, T. L. (1968). A Bayesian approach to the problem of ranking binomial populations. *SIAM J. Appl. Math.* **16**, 843–850.

Blumenthal, S. (1975). Sequential estimation of the largest normal mean when the variance is unknown. *Comm. Statist.* **4**, 655–669.

Blumenthal, S. (1976). Sequential estimation of the largest normal mean when the variance is known. *Ann. Statist.* **4**, 1077–1087.

Blumenthal, S., Christie, T., and Sobel, M. (1971). Best sequential tests for an identification problem with partially overlapping distributions. *Sankhyā Ser. A* **33**, 185–192.

Blumenthal, S., and Cohen, A. (1968a). Estimation of the larger translation parameter. *Ann. Math. Statist.* **39**, 502–516.

Blumenthal, S., and Cohen, A. (1968b). Estimation of the larger of two normal means. *J. Amer. Statist. Assoc.* **63**, 861–876.

Blumenthal, S., and Patterson, D. W. (1969). Rank order procedures for selecting a subset containing the population with the smallest scale parameter. *Sankhyā Ser. A* **31**, 37–42.

Bofinger, E. (1976). Least favorable configuration when ties are broken. *J. Amer. Statist. Assoc.* **71**, 423–424.

Boyce, D. E., Farhi, A., and Weischedel, R. (1974). Optimal Subset Selection-Multiple Regression, Interdependence and Optimal Network Algorithms. *Lecture Notes in Economics and Mathematical Systems. No.* 103. Springer-Verlag, Berlin-Heidelberg-New York.

Bratcher, T. L. (1969). A Bayesian Treatment of a Multiple Comparison Problem for Binomial Probabilities. Ph.D. Thesis. Southern Methodist Univ., Texas.

Bratcher, T. L., and Bhalla, P. (1974). On the properties of an optimal selection procedure. *Comm. Statist.* **3**, 191–196.

Bratcher, T. L., and Bland, R. P. (1975). On comparing binomial probabilities from a Bayesian viewpoint. *Comm. Statist.* **4**, 975–985.

Bross, I. (1949). Maximum Utility Statistics. Ph.D. Thesis. North Carolina State College.

Bross, I. (1950). Two-choice selection. *J. Amer. Statist. Assoc.* **45**, 530–540.

Broström, G. (1976). Admissibility of subset selection procedures. Statist. Res. Report No. 1976-8, Inst. of Math. and Statist., Univ. of Umeå, Umeå, Sweden.

Broström, G. (1977a). Subset selection with a reliability application. Statist. Res. Report No. 1977-2. Inst. of Math. and Statist., Univ. of Umeå, S-90187 Umeå, Sweden.

Broström, G. (1977b). An improved procedure for selecting all populations better than a standard. Tech. Report 1977-3, Inst. of Math. and Statist., Univ. of Umeå, Sweden.

Broström, G. (1978). Sequentially rejective subset selection procedures. Statistical Research Report No. 1978-1, Inst. of Math. Statistics, University of Umeå, Umeå, Sweden.

Bühlmann, H., and Huber, P. J. (1963). Pairwise comparison and ranking in tournaments. *Ann. Math. Statist.* **34**, 501–510.

Burlacu, V. (1972). Metode ale Selectiei Bazate pe Ordonare. *Stud. Cerc. Mat.* **24**, 1193–1207 (English summary).

Cacoullos, T. (1965a). Comparing Mahalanobis distances I: comparing distances between k known normal populations and another unknown. *Sankhyā Ser. A* **27**, 1–22.

Cacoullos, T. (1965b). Comparing Mahalanobis distances II: Bayes procedures when the mean vectors are unknown. *Sankhyā Ser. A* **27**, 23–32.

Cacoullos, T., and Sobel, M. (1966). An inverse-sampling procedure for selecting the most probable event in a multinomial distribution. *Multivariate Analysis* (Ed. P. R. Krishnaiah), Academic, New York, pp. 423–455.

Cane, G. J. (1979). Note on a result of Seeger in partitioning normal populations. *Ann. Statist.* **7**, 917–919.

Canner, P. L. (1970). Selecting one of two treatments when the responses are dichotomous. *J. Amer. Statist. Assoc.* **65**, 293–306.

Carroll, R. J. (1974a). Some Contributions to the Theory of Parametric and Nonparametric Sequential Ranking and Selection. Ph.D. Thesis (Mimeo. Ser. No. 359). Dept. of Statist., Purdue Univ., West Lafayette, Indiana.

Carroll, R. J. (1974b). Asymptotically nonparametric sequential selection procedures. Mimeo. Ser. No. 944, Inst. of Statist., Univ. of North Carolina, Chapel Hill.

Carroll, R. J. (1974c). Some approximations in selection theory. Mimeo. Ser. No. 957, Inst. of Statist., Univ. of North Carolina, Chapel Hill.

Carroll, R. J., and Gupta, S. S. (1977). On the probabilities of rankings of k populations with applications. *J. Statist. Comput. Simul.* **5**, 145–157.

Carroll, R. J., Gupta, S. S., and Huang, D. Y. (1975). On selection procedures for the t best populations and some related problems. *Comm. Statist.* **4**, 987–1008.

Carroll, R. J., and Santner, T. J. (1975). A note on a minimization result for sequential ranking procedures. Tech. Report No. 258. Dept. of Operations Res., Cornell Univ., Ithaca, New York.

Chambers, M. L., and Jarratt, P. (1964). Use of double sampling for selecting best population. *Biometrika* **51**, 49–64.

Chambers, M. L., and Mack, C. (1966). On confidence limits for the minimum of two normal means; a new inference principle. *J. Statist. Operational. Res.* **2**, 14–27.

Chen, H. J. (1973). Estimation of Ordered Parameters and Subset Selection. Ph.D. Thesis. Univ. of Rochester, New York.

Chen, H. J. (1975). Strong consistency and asymptotic unbiasedness of a natural estimator for a ranked parameter. *Sankhyā Ser. B* **30**, 92–94.

Chen, H. J. (1977a). Estimation of ordered parameters from k stochastically increasing distributions. *Naval. Res. Logist. Quart.* **24**, 269–280.

Chen, H. J. (1977b). A class of fixed-width confidence intervals for a ranked normal mean. *Commun. Statist.-Simula. Computa.* **B6** (2), 137–156.

Chen, H. J. (1977c). Non-existence of an exact β-confidence interval for a scale parameter. *Austral. J. Statist.* **19**, 43–55.

Chen, H. J. (1977d). Estimation of the best alternative. *Proceedings of the* 1977 *Winter Simulation Conference* (Eds. H. J. Highland, R. G. Sargent, and J. W. Schmidt) held at the National Bureau of Standards, Gaithersburg, Maryland, pp. 73–79.

Chen, H. J. (1977e). Tables of percentage points for interval estimation of a ranked normal mean with known variances. *Biom. Z.* **19**, 627–635.

Chen, H. J., and Dudewicz, E. J. (1973a). Estimation of Ordered Parameters and Subset Selection. Tech. Report. Division of Statistics, Ohio State Univ., Columbus, Ohio.

Chen, H. J., and Dudewicz, E. J. (1973b). Interval estimation of a ranked mean of k normal populations with unequal sample sizes. Tech. Report No. 73-21, Dept. of Math. Sciences, Memphis State Univ., Memphis, Tennessee.

Chen, H. J., and Dudewicz, E. J. (1976). Procedures for fixed-width interval estimation of the largest normal mean. *J. Amer. Statist. Assoc.* **71**, 752–756.

Chen, H. J., Dudewicz, E. J., and Lee, Y. J. (1976). Subset selection procedures for normal means under unequal sample sizes. *Sankhyā Ser. B* **38**, 249–255.

Chen, H. J., Tsai, P. J., and Wang, M. W. (1978). Confidence intervals for the largest mean from k correlated normal populations. *Commun. Statist.-Simula. Computa.* **B7**(3), 205–222.

Chernoff, H. and Yahav, J. (1977). A subset selection problem employing a new criterion. *Statistical Decision Theory and Related Topics–II* (Eds. S. S. Gupta and D. S. Moore), Academic, New York, pp. 93–119.

Chew, V. (1977a). Statistical hypothesis testing: an academic exercise in futility. *Proc. Fla. State Hort. Soc.* **90**, 214–215.

Chew, V. (1977b). Comparisons Among Treatment Means in an Analysis of Variance. Report ARS/H/6, Research Service, U. S. Dept. of Agriculture, Washington, D.C.

Chiu, W. K. (1974a). A generalized selection goal for a ranking problem. *S. Afr. Statist. J.* **8**, 45–48.

Chiu, W. K. (1974b). A generalized selection goal for a sequential ranking procedure. *Nanta. Math.* **7**, 42–46.

Chiu, W. K. (1974c). The ranking of means of normal populations for a generalized selection goal. *Biometrika* **61**, 579–584.

Chiu, W. K. (1974d). Selecting the m populations with largest means from k normal populations with unknown variances. *Austral. J. Statist.* **16**, 144–147. Corrigendum: *Austral. J. Statist.* **17** (1975), 114.

Chiu, W. K. (1977). On correct selection for a ranking problem. *Ann. Inst. Statist. Math.* **29**, 59–66.

Chow, Y. S., Moriguti, S., Robbins, H., and Samuels, S. M. (1964). Optimal selection based on relative rank (the "secretary" problem). *Israel J. Math.* **2**, 81–90.

Christiansen, H. D. (1974). Distribution-free reliability analysis of two-component systems by means of Rasch's item analysis model. *Scand. J. Statist.* **1**, 33–35.

Cohen, D. S. (1959). A Two-Sample Decision Procedure for Ranking Means of Normal Populations with a Common Known Variance. M.S. Thesis, Dept. of Operations Research, Cornell Univ., Ithaca, New York.

Colton, T. (1963). A model for selecting one of two medical treatments. *J. Amer. Statist. Assoc.* **58**, 388–400.

Dalal, S. R., and Hall, W. J. (1977). Most-economical robust selection procedures for location parameters. Tech. Report, Dept. of Statist., Univ. of Rochester, Rochester, New York.

Dalal, S. R., and Srinivasan, V. (1977). Determining sample size for pretesting comparative effectiveness of advertising copies. *Management Sci.* **23**, 1284–1294.

Dariyal, L. D. and Dudewicz, E. J. (1977). Sobel's nonparametric selection procedure for quantiles: extensions and tables. Tech. Report. Dept. of Statist., Ohio State Univ., Columbus, Ohio.

David, H. A. (1956). The ranking of variances in normal populations. *J. Amer. Statist. Assoc.* **51**, 621–626.

Deely, J. J. (1965). Multiple Decision Procedures from an Empirical Bayes Approach. Ph.D. Thesis (Mimeo. Ser. No. 45). Dept. of Statist., Purdue Univ., West Lafayette, Indiana.

Deely, J. J. (1966). Sequential selection of the best of k populations. Mimeo. Ser. No. 91, Dept. of Statist., Purdue Univ., West Lafayette, Indiana.

Deely, J. J. and Gupta, S. S. (1968). On the properties of subset selection procedures. *Sankhyā Ser. A* **30**, 37–50.

Desu, M. M. (1970). A selection problem. *Ann. Math. Statist.* **41**, 1596–1603.

Desu, M. M., Narulla, S. C., and Villarreal, B. (1977). A two-stage procedure for selecting the best of k exponential distributions. *Commun. Statist.-Theor. Method* **A6**(12), 1223–1230.

Desu, M. M., and Sobel, M. (1968). A fixed subset-size approach to a selection problem. *Biometrika* **55**, 401–410. Corrections and amendments: **63** (1976), 685.

Desu, M. M., and Sobel, M. (1971). Nonparametric procedures for selecting fixed-size subsets. *Statistical Decision Theory and Related Topics* (Eds. S. S. Gupta and J. Yackel), Academic, New York, pp. 255–273.

Deverman, J. N. (1969). A General Selection Procedure Relative to t Best Populations. Ph.D. Thesis. Dept. of Statist., Purdue Univ., West Lafayette, Indiana.

Deverman, J. N., and Gupta, S. S. (1969). On a selection procedure concerning the t best populations. Tech. Report. Sandia Laboratories, Livermore, California. Also Abstract: *Ann. Math. Statist.* **40**, 1870.

Dixon, D. O., and Bland, R. P. (1971). A Bayes solution for the problem of ranking Poisson parameters. *Ann. Inst. Statist. Math.* **23**, 119–124.

Dudewicz, E. J. (1966). The Efficiency of a Non-parametric Selection Procedure:

Largest Location Parameter Case. M.S. Thesis (Tech. Report No. 14). Dept. of Operations Res., Cornell Univ., Ithaca, New York.

Dudewicz, E. J. (1968). A categorized bibliography on multiple decision (ranking and selection) procedures. Tech. Report. Dept. of Statist., Univ. of Rochester, Rochester, New York.

Dudewicz, E. J. (1969a). An approximation to the sample size in selection problems. *Ann. Math. Statist.* **40**, 492–497.

Dudewicz, E. J. (1969b). Selection procedures: an introduction for practitioners. *Estadist.* **27**, 377–382.

Dudewicz, E. J. (1969c). Estimation of Ordered Parameters. Ph.D. Thesis. (Tech. Report No. 60). Dept. of Operations Res., Cornell Univ., Ithaca, New York.

Dudewicz, E. J. (1970a). Confidence intervals for ranked means. *Naval. Res. Logist. Quart.* **17**, 69–78.

Dudewicz, E. J. (1970b). Review of *Sequential Identification and Ranking Procedures*, by R. E. Bechhofer, J. Kiefer, and M. Sobel. *Math. Reviews* **39** #6445.

Dudewicz, E. J. (1971a). Non-existence of a single-sample selection procedure whose $P(CS)$ is independent of the variances. *S. Afr. Statist. J.* **5**, 37–39.

Dudewicz, E. J. (1971b). A nonparametric selection procedure's efficiency: largest location parameter case. *J. Amer. Statist. Assoc.* **66**, 152–161.

Dudewicz, E. J. (1971c). Maximum likelihood estimators for ranked means. *Z. Wahrscheinlichkeitstheorie Verw. Geb.* **19**, 29–42.

Dudewicz, E. J. (1972a). Two-sided confidence intervals for ranked means. *J. Amer. Statist. Assoc.* **67**, 462–464.

Dudewicz, E. J. (1972b). Point estimation of ordered parameters: the general location parameter case. *Tamkang J. Math.* **3**, 101–114.

Dudewicz, E. J. (1973). Maximum probability estimators for ranked means. *Ann. Inst. Statist. Math.* **25**, 467–477.

Dudewicz, E. J. (1974a). A note on selection procedures with unequal observation numbers. *Zastos. Mat.* **14**, 31–35.

Dudewicz, E. J. (1974b). Estimation of ordered parameters: the range of k normal means. *Comm. Statist.* **3**, 279–285.

Dudewicz, E. J. (1976a). Generalized maximum likelihood estimators for ranked means. *Z. Wahrscheinlichkeitstheorie Verw. Geb.* **35**, 283–297.

Dudewicz, E. J. (1976b). Statistics in simulation: how to design for selecting the best alternative. *Proceedings of the* 1976 *Bicentennial Winter Simulation Conference* (Eds. H. J. Highland, T. J. Schriber, and R. G. Sargent), pp. 67–71.

Dudewicz, E. J. (1977). New procedures for selection among (simulated) alternatives. *Proceedings of the* 1977 *Winter Simulation Conference* (Eds. H. J. Highland, R. G. Sargent, and J. W. Schmidt) held at the National Bureau of Standards, Gaithersburg, Maryland, pp. 59–62.

Dudewicz, E. J., and Dalal, S. R. (1975). Allocation of observations in ranking and selection with unequal variances. *Sankhyā Ser. B* **37**, 28–78.

Dudewicz, E. J., and Fan, C. (1973). Further light on non-parametric selection efficiency. *Nav. Res. Logist. Quart.* **20**, 737–743.

Dudewicz, E. J., Ramberg, J. S., and Chen, H. C. (1975). New tables for multiple comparisons with a control (unknown variances). *Biom. Z.* **17**, 13–26.

Dudewicz, E. J., and Taneja, V. S. (1978). A multivariate solution of the multivariate ranking and selection problem. Tech. Report No. 167, Dept. of Statistics, Ohio State University, Columbus, Ohio.

Dudewicz, E. J., and Tong, Y. L. (1971). Optimum confidence intervals for the largest location parameter. *Statistical Decision Theory and Related Topics* (Eds. S. S. Gupta and J. Yackel), Academic, New York, pp. 363–376.

Dudewicz, E. J., and Zaino, N. A., Jr. (1971). Sample size for selection. *Statistical Decision Theory and Related Topics* (Eds. S. S. Gupta and J. Yackel), Academic, New York, pp. 347–362.

Dudewicz, E. J., and Zaino, N. A., Jr. (1973). Ranking and selection procedures with correlated observations and simulation. *Proceedings of the Computer Science and Statistics Seventh Annual Symposium on the Interface* (Ed. W. J. Kennedy), Statistical Laboratory, Iowa State University, Ames, Iowa, pp. 150–155.

Dudewicz, E. J., and Zaino, N. A., Jr. (1977). Allowance for correlation in setting simulation run-length via ranking-and-selection procedures. *TIMS Studies in the Management Sciences* **7**, 51–61.

Duncan, D. B. (1965). A Bayesian approach to multiple comparisons. *Technometrics* **7**, 171–222.

Dunnett, C. W. (1955). A multiple comparison procedure for comparing several treatments with a control. *J. Amer. Statist. Assoc.* **50**, 1096–1121.

Dunnett, C. W. (1960). On selecting the largest of k normal population means (with discussion). *J. Roy. Statist. Soc., Ser. B* **22**, 1–40.

Dunnett, C. W. (1964). New tables for multiple comparisons with a control. *Biometrics* **20**, 482–491.

Eaton, M. L. (1966). Some Optimal Properties of Ranking Procedures with Applications in Multivariate Analysis. Ph.D. Thesis (Tech. Report No. 22). Dept. of Statist., Stanford Univ., Stanford, California.

Eaton, M. L. (1967a). Some optimum properties of ranking procedures. *Ann. Math. Statist.* **38**, 124–137.

Eaton, M. L. (1967b). The generalized variance: testing and ranking problem. *Ann. Math. Statist.* **38**, 941–943.

Eckardt, W. L., Jr. (1973). A Sequential Selection Procedure: Binomial Populations. Ph.D. Thesis. Dept. of Applied Math. and Computer Sciences, Washington Univ., St. Louis, Missouri.

Eckardt, W. L., Jr. (1976). A sequential method for selecting the best of three binomial populations. *J. Amer. Statist. Assoc.* **71**, 473–474.

Fabian, V. (1962). On multiple decision methods for ranking population means. *Ann. Math. Statist.* **33**, 248–254.

Fabian, V. (1970a). A note on Paulson's sequential ranking procedure. Tech. Report RM-260, VF-9, Statistical Laboratory, Michigan State Univ., East Lansing.

Fabian, V. (1970b). Review of *Sequential Identification and Ranking Procedures* by R. E. Bechhofer, J. Kiefer, and M. Sobel, 1968. *J. Amer. Statist. Assoc.* **65**, 459–461.

Fabian, V. (1974a). Note on Anderson's sequential procedures with triangular boundary. *Ann. Statist.* **2**, 170–176.

Fabian, V. (1974b). Acknowledgement of priority to 'Note on Anderson's sequential procedures with triangular boundary.' *Ann. Statist.* **2**, 1063.

Fairweather, W. R. (1968). Some extensions of Somerville's procedure for ranking means of normal populations. *Biometrika* **55**, 411–418.

Farrell, E. J. (1961). References on ranking, selection and multiple comparisons. Tech. Report EF2459-19, Remington Rand Univac, St. Paul, Minnesota.

Federer, W. T. (1963). Procedures and designs useful for screening material in selection and allocation, with a bibliography. *Biometrics* **19**, 553–587.

Fisher, D. M. (1972). Classification, Selection and Testing Procedures for Asymmetric Distributions. Ph.D. Thesis, Dept. of Statist., Univ. of Iowa, Iowa City.

Flora, J. D. Jr. (1971). Bayes Nonparametric Selection Procedures. Ph.D. Thesis, Dept. of Statist., Florida State University, Tallahassee.

Freeman, H., Kuzmack, A., and Maurice, R. (1967). Multivariate t and the ranking problem. *Biometrika* **54**, 305–308.

Frischtak, R. M. (1973). Statistical Multiple-Decision Procedures for Some Multivariate Selection Problems. Ph.D. Thesis (Tech. Report No. 187), Dept. of Operations Res., Cornell Univ., Ithaca, New York.

Furnival, G. M., and Wilson, R. W., Jr. (1974). Regression by leaps and bounds. *Technometrics* **16**, 499–512.

Fushimi, M. (1973a). An improved version of a Sobel–Weiss Play-the-Winner procedure for selecting the better of two binomial populations. *Biometrika* **60**, 517–523.

Fushimi, M. (1973b). A nonsymmetric sequential procedure for selecting the better of two binomial populations. Tech. Report No. 189, Dept. of Operations Res., Cornell Univ., Ithaca, New York.

Garside, M. J. (1965). The best subset in multiple regression analysis. *J. Roy. Statist. Soc., Ser. C, Appl. Statist.* **14**, 196–200.

Gaynor, J. A. (1976). Asymptotic Efficiencies of Selection Procedures. Ph.D. Thesis. Dept. of Math., Univ. of Illinois, Urbana.

Gaynor, J. A. (1977). A large deviation approach to asymptotic efficiency of selection procedures. Mimeo. Ser. No. 474, Dept. of Statist., Purdue Univ., West Lafayette, Indiana.

Geertsema, J. C. (1972). Nonparametric sequential procedure for selecting the best of k populations. *J. Amer. Statist. Assoc.* **67**, 614–616.

Ghosh, M. (1973). Nonparametric selection procedure for symmetric location parameter populations. *Ann. Statist.* **1**, 773–779.

Gibbons, J. D., Olkin, I., and Sobel, M. (1977). *Selecting and Ordering Populations*. Wiley, New York.

Gnanadesikan, M. (1966). Some Selection and Ranking Procedures for Multivariate Normal Populations. Ph.D. Thesis. Dept. of Statist., Purdue Univ., West Lafayette, Indiana.

Gnanadesikan, M., and Gupta, S. S. (1970). Selection procedures for multivariate normal distributions in terms of measures of dispersion. *Technometrics* **12**, 103–117.

Goel, P. K. (1972). A note on the non-existence of subset selection procedure for Poisson populations. Mimeo. Ser. No. 303. Dept. of Statist., Purdue Univ., West Lafayette, Indiana.

Goel, P. K., and Rubin, H. (1977). On selecting a subset containing the best population—A Bayesian approach. *Ann. Statist.* **5**, 969–983.

Gore, A. P. (1975). Some nonparametric tests and selection procedures for main effects in two-way layouts. *Ann. Inst. Statist. Math.* **27**, 487–500.

Gorman, J. W., and Toman, R. J. (1966). Selection of variables for fitting equations to data. *Technometrics* **8**, 27–51.

Govindarajulu, Z., and Gore, A. P. (1971). Selection procedures with respect to measures of association. *Statistical Decision Theory and Related Topics* (Eds. S. S. Gupta and J. Yackel), Academic, New York, pp. 313–345.

Govindarajulu, Z., and Harvey, C. (1974). Bayesian procedures for ranking and selection problems. *Ann. Inst. Statist. Math.* **26**, 35–53.

Groves, J. E. (1972). On Partitioning a Set of Populations with Respect to a Control. Ph.D. Thesis. Kansas State Univ., Manhattan.

Grundy, P. M., Healy, M. J. R., and Rees, D. H. (1956). Economic choice of the amount of experimentation. *J. Roy. Statist. Soc. Ser. B* **18**, 32–49.

Gulati, C. M. (1971). Optimal Allocation of Observations for Selecting One among Several Populations. Ph.D. Thesis, Dept. of Statist., Carnegie-Mellon Univ., Pittsburgh, Pennsylvania.

Gupta, S. S. (1956). On a decision rule for a problem in ranking means. Ph.D. Thesis (Mimeo. Ser. No. 150). Inst. of Statist., Univ. of North Carolina, Chapel Hill.

Gupta, S. S. (1963). On a selection and ranking procedure for gamma populations. *Ann. Inst. Statist. Math.* **14**, 199–216.

Gupta, S. S. (1965). On some multiple decision (selection and ranking) rules. *Technometrics* **7**, 225–245.

Gupta, S. S. (1966a). Selection and ranking procedures and order statistics for the binomial distribution. *Classical and Contagious Discrete Distributions* (Ed. G. P. Patil), Pergamon, New York, pp. 219–230.

Gupta, S. S. (1966b). On some selection and ranking procedures for multivariate normal populations using distance functions. *Multivariate Analysis* (Ed. P. R. Krishnaiah), Academic, New York, pp. 457–475.

Gupta, S. S. (1967). On selection and ranking procedures. *Trans. Amer. Soc. Qual. Contrl.*, pp. 151–155.

Gupta, S. S. (1970). Multiple decision (selection and ranking) procedures and multivariate distribution theory. Report No. ARL 70-0012, Aerospace Research Laboratories, U. S. Air Force Base, Dayton, Ohio.

Gupta, S. S. (1972). Review of *Sequential Identification and Ranking Procedures* by R. E. Bechhofer, J. Kiefer, and M. Sobel, 1968. *Ann. Math. Statist.* **43**, 697–700.

Gupta, S. S. (1974). Selection and ranking procedures for multivariate normal populations. *Proceedings of the Nineteenth Conference on the Design of Experiments in Army Research Development and Testing*. ARO Report 74-1, Dept. of the Army, pp. 13–22.

Gupta, S. S. (1977). Selection and ranking procedures: a brief introduction. *Commun. Statist.-Theor. Meth.* **A6**, 993–1001.

Gupta, S. S. and Hsu, J. C. (1977a). On the monotonicity of Bayes subset selection procedures. *Proceedings of the 41st Session of the International Statistical Institute,* Vol. 47, Book 4, pp. 208–211.

Gupta, S. S. and Hsu, J. C. (1977b). Subset selection procedures with special reference to the analysis of two-way layout: Application to motor-vehicle fatality data. *Proceedings of the 1977 Winter Simulation Conference* (Eds. H. J. Highland, R. G. Sargent, and J. W. Schmidt) held at the National Bureau of Standards, Gaithersburg, Maryland, pp. 81–85.

Gupta, S. S. and Hsu, J. C. (1978). On the performance of some subset selection procedures. *Commun. Statist.-Simula. Computa.* **B7**(6), 561–591.

Gupta, S. S., and Huang, D. Y. (1974). Nonparametric subset selection procedures for the t best populations. *Bull. Inst. Math. Acad. Sinica* **2**, 377–386.

Gupta, S. S., and Huang, D. Y. (1975a). On subset selection procedures for Poisson populations and some applications to the multinomial selection problems. *Applied Statistics* (Ed. R. P. Gupta), North-Holland, Amsterdam, pp. 97–109.

Gupta, S. S., and Huang, D. Y. (1975b). On some parametric and nonparametric sequential subset selection procedures. *Statistical Inference and Related Topics,* Vol. 2 (Ed. M. L. Puri), Academic, New York, pp. 101–128.

Gupta, S. S., and Huang, D. Y. (1975c). On Γ-minimax classification procedures. *Proceedings of the 40th Session of the International Statistical Institute,* Vol. 46, Book 3, pp. 330–335.

Gupta, S. S., and Huang, D. Y. (1976a). Selection procedures for the means and variances of normal populations: unequal sample sizes case. *Sankhyā Ser. B* **38**, 112–128.

Gupta, S. S., and Huang, D. Y. (1976b). Subset selection procedures for the entropy function associated with the binomial populations. *Sankhyā Ser. A* **38**, 153–173.

Gupta, S. S., and Huang, D. Y. (1976c). On some methods for constructing optimal subset selection procedures. Mimeo. Ser. No. 470, Dept. of Statist., Purdue Univ., West Lafayette, Indiana.

Gupta, S. S., and Huang, D. Y. (1977a). On some Γ-minimax subset selection and multiple comparison procedures. *Statistical Decision Theory and Related Topics–II* (Eds. S. S. Gupta and D. S. Moore), Academic, New York, pp. 139–155.

Gupta, S. S., and Huang, D. Y. (1977b). On some optimal sampling procedures for selection problems. *The Theory and Applications of Reliability* (Eds. C. P. Tsokos and I. N. Shimi), Academic, New York, pp. 495–505.

Gupta, S. S., and Huang, D. Y. (1977c). Some multiple decision problems in analysis of variance. *Commun. Statist.-Theor. Meth.* **A6**, 1035–1054.

Gupta, S. S., and Huang, D. Y. (1977d). Nonparametric selection procedures for selecting the regression equations with larger slope. *J. Chinese Statist. Assoc.* **15**, 5717–5724.

Gupta, S. S., and Huang, D. Y. (1977e). On selecting an optimal subset of regression variables. Mimeo. Ser. No. 501, Dept. of Statist., Purdue Univ., West Lafayette, Indiana.

Gupta, S. S., and Huang, D. Y. (1978a). A note on optimal subset selection procedures. Mimeo. Series No. 470 (Revised). Dept. of Statist., Purdue Univ., West Lafayette, Indiana.

Gupta, S. S., and Huang, D. Y. (1978b). An essentially complete class of multiple decision procedures. Mimeo. Ser. No. 78-12, Dept. of Statist., Purdue Univ., West Lafayette, Indiana.

Gupta, S. S., Huang, D. Y., and Huang, W. T. (1976). On ranking and selection procedures and tests of homogeneity for binomial populations. *Essays in Probability and Statistics* (Eds. S. Ikeda et al.), Shinko Tsusho Co. Ltd., Tokyo, Japan, pp. 501–533.

Gupta, S. S., Huang, D. Y., and Nagel, K. (1979). Locally optimal subset selection procedures based on ranks. *Optimizing Methods in Statistics* (Ed. J. S. Rustagi), Academic, New York. pp. 251–260.

Gupta, S. S., and Huang, W. T. (1974a). A note on selecting a subset of normal populations with unequal sample sizes. *Sankhyā Ser. A* **36**, 389–396.

Gupta, S. S., and Huang, W. T. (1974b). On a maximin strategy for sampling based on selection procedures from several populations. *Comm. Statist.* **3**, 325–342.

Gupta, S. S., and Huang, W. T. (1975). Some selection procedures based on the Robbins-Monro stochastic approximation. *Tamkang J. Math.* **6**, 301–314.

Gupta, S. S., and Huang, W. T. (1977). On selection rules for finite mixtures of distributions. Mimeo. Ser. No. 504, Dept. of Statist., Purdue Univ., West Lafayette, Indiana.

Gupta, S. S., Huyett, M. J., and Sobel, M. (1957). Selection and ranking problems with binomial populations. *Trans. Amer. Soc. Qual. Contr.*, 635–644.

Gupta, S. S., and Leong, Y. K. (1977). Some results on subset selection procedures for double exponential populations. *Proceedings of the Workshop on Decision Information for Tactical Command and Control*† (Eds. R. M. Thrall, C. P. Tsokos, and J. C. Turner). Robert M. Thrall and Associates, 12003 Pebble Hill Dr., Houston, Texas 77024, pp. 139–166.

Gupta, S. S., Leong, Y. K., and Wong, W. Y. (1978). On subset selection procedures for Poisson populations. Mimeo. Ser. No. 78-3, Dept. of Statist., Purdue Univ., West Lafayette, Indiana.

Gupta, S. S., and Lu, M. W. (1977). Subset selection procedures for restricted families of probability distributions. Mimeo. Ser. No. 518, Dept. of Statist., Purdue Univ., West Lafayette, Indiana. (Revised 1978.) To appear in *Ann. Inst. Statist. Math.* **31**, (1979), 111–128.

Gupta, S. S., and McDonald, G. C. (1970). On some classes of selection procedures based on ranks. *Nonparametric Techniques in Statistical Inference* (Ed. M. L. Puri), Cambridge University Press, London, pp. 491–514.

Gupta, S. S., and McDonald, G. C. (1972). Some selection procedures with applications to reliability problems. *Operations Research and Reliability* (Proceedings of the NATO Conference, Italy, 1969), Gordon and Breach, New York, pp. 421–439.

Gupta, S. S., and Miescke, K. J. (1978). On subset selection procedures for ranking means of three normal populations. Mimeo. Ser. No. 78-19. Dept. of Statist., Purdue Univ., W. Lafayette, Indiana. To appear in *Sankhyā Ser. A*.

Gupta, S. S., and Nagel, K. (1967). On selection and ranking procedures and order statistics from the multinomial distribution. *Sankhyā Ser. B* **29**, 1–34.

Gupta, S. S., and Nagel, K. (1971). On some contributions to multiple decision theory. *Statistical Decision Theory and Related Topics* (Eds. S. S. Gupta and J. Yackel), Academic, New York, pp. 79–102.

†To be published by Academic Press under the title *Decision Information.*

Gupta, S. S., and Panchapakesan, S. (1969a). Some selection and ranking procedures for multivariate normal populations. *Multivariate Analysis–II* (Ed. P. R. Krishnaiah), Academic, New York, pp. 475–505.

Gupta, S. S., and Panchapakesan, S. (1969b). Selection and ranking procedures. *The Design of Computer Simulation Experiments* (Ed. T. H. Naylor), Duke University Press, Durham, North Carolina, pp. 132–160.

Gupta, S. S., and Panchapakesan, S. (1971). Contributions to multiple decision (subset selection) rules, multivariate distribution theory and order statistics. Report No. ARL 71-0218, Aerospace Research Laboratories, U. S. Air Force Base, Dayton, Ohio.

Gupta, S. S., and Panchapakesan, S. (1972a). On a class of subset selection procedures. *Ann. Math. Statist.* **43**, 814–822.

Gupta, S. S., and Panchapakesan, S. (1972b). On multiple decision (subset selection) procedures. *J. Mathematical and Physical Sci.* **6**, 1–72.

Gupta, S. S., and Panchapakesan, S. (1974). Inference for restricted families: (a) multiple decision procedures; (b) order statistics inequalities. *Reliability and Biometry* (Eds. F. Proschan and R. J. Serfling), SIAM, Philadelphia, pp. 503–596.

Gupta, S. S., and Panchapakesan, S. (1975). On a quantile selection procedure and associated distribution of ratios of order statistics from a restricted family of probability distributions. *Reliability and Fault Tree Analysis* (Eds. R. E. Barlow, J. B. Fussell, and N. D. Singpurwalla), SIAM, Philadelphia, pp. 557–576.

Gupta, S. S., and Santner, T. J. (1973). On selection and ranking procedures—a restricted subset selection rule. *Proceedings of the 39th Session of the International Statistical Institute*, Vol. 45, Book 1, pp. 409–417.

Gupta, S. S., and Singh, A. K. (1977a). On selection of populations close to a control or standard. Mimeo. Ser. No. 508, Dept. of Statist., Purdue Univ., West Lafayette, Indiana.

Gupta, S. S., and Singh, A. K. (1977b). On rules based on sample medians for selection of the largest location parameter. Mimeo. Ser. No. 514, Dept. of Statist., Purdue Univ., West Lafayette, Indiana.

Gupta, S. S. and Singh, A. K. (1978). On selection procedures for normal populations with a common known coefficient of variation with an application to multivariate normal populations. Mimeo. Ser. No. 78-14, Dept. of Statist., Purdue Univ., West Lafayette, Indiana.

Gupta, S. S., and Sobel, M. (1957). On a statistic which arises in selection and ranking problems. *Ann. Math. Statist.* **28**, 957–967.

Gupta, S. S., and Sobel, M. (1958a). On selecting a subset which contains all populations better than a standard. *Ann. Math. Statist.* **29**, 235–244.

Gupta, S. S., and Sobel, M. (1958b). On the distribution of a certain statistic based on ordered uniform chance variables. *Ann. Math. Statist.* **29**, 274–281.

Gupta, S. S., and Sobel, M. (1960). Selecting a subset containing the best of several binomial populations. *Contributions to Probability and Statistics* (Eds. I. Olkin et al.), Stanford University Press, Stanford, California, pp. 224–248.

Gupta, S. S., and Sobel, M. (1962a). On selecting a subset containing the population with the smallest variance. *Biometrika* **49**, 495–507.

Gupta, S. S., and Sobel, M. (1962b). On the smallest of several correlated *F*-statistics. *Biometrika* **49**, 509–523.

Gupta, S. S., and Studden, W. J. (1966). On some aspects of selection and ranking procedures with applications. Mimeo. Ser. No. 81, Dept. of Statist., Purdue Univ., West Lafayette, Indiana.

Gupta, S. S., and Studden, W. J. (1970). On some selection and ranking procedures with applications to multivariate populations. *Essays in Probability and Statistics* (Eds. R. C. Bose et al.), University of North Carolina Press, Chapel Hill, pp. 327–338.

Gupta, S. S., and Wong, W. Y. (1976a). Subset selection procedures for finite schemes in information theory. *Colloquia Mathematica Societatis János Bolyai*, Topics in Information Theory. Conference held at Keszthely, August 1975, pp. 279–291.

Gupta, S. S., and Wong, W. Y. (1976b). Subset selection procedures for the means of normal populations with unequal variances: unequal sample sizes case. Mimeo. Ser. No. 473, Dept. of Statist., Purdue Univ., West Lafayette, Indiana.

Gupta, S. S., and Wong, W. Y. (1977). On subset selection procedures for Poisson processes and some applications to the binomial and multinomial problems. *Operations Research Verfahren* (Eds. Henn et al.), Verlog Anton Hain, Meisenheim am Glan, Germany, pp. 49–70.

Guttman, I. (1961). Best population and tolerance regions. *Ann. Inst. Statist. Math.* **13**, 9–26.

Guttman, I. (1963). A sequential procedure for the best population. *Sankhyā Ser. A* **25**, 25–28.

Guttman, I. (1970). *Statistical Tolerance Regions: Classical and Bayesian*. Griffin's Statistical Monographs & Courses, No. 26, Charles Griffin and Company, London.

Guttman, I., and Milton, R. C. (1969). Procedures for a best population problem when the criterion of bestness involves a fixed tolerance region. *Ann. Inst. Statist. Math.* **21**, 149–161.

Guttman, I., and Tiao, G. C. (1964). A Bayesian approach to some best population problems. *Ann. Math. Statist.* **35**, 825–835.

Hadian, A., and Sobel, M. (1970). Selecting the *t*th largest using binary comparisons. *Combinatorial Theory and its Applications, II* (Proc. Colloq. Balatonfüred, 1969), North-Holland, Amsterdam, pp. 585–599.

Hall, W. J. (1954). An optimum property of Bechhofer's single-sample multiple-decision procedure for ranking means and some extensions. Mimeo. Ser. No. 118, Inst. of Statist., Univ. of North Carolina, Chapel Hill.

Hall, W. J. (1958). Most economical multiple-decision rules. *Ann. Math. Statist.* **29**, 1079–1094.

Hall, W. J. (1959). The most economical character of Bechhofer and Sobel decision rules. *Ann. Math. Statist.* **30**, 964–969.

Halmgren, E. G. (1972). Contributions to the Estimation of Ordered Parameters. Ph. D. Thesis. Dept. of Statist., Univ. of Iowa, Iowa City.

Hawkins, D. M. (1976). The subset problem in multivariate analysis of variance. *J. Roy. Statist. Soc. Ser. B* **38**, 132–139.

Hirunraks, A. (1969). An Information Theory Approach to Selection Problems Involving the Multinomial Distribution. Ph.D. Thesis. Dept. of Statist., Oregon State Univ., Cornwallis.

Hocking, R. R. (1976). The analysis and selection of variables in linear regression. *Biometrics* **32**, 1–49.

Hocking, R. R., and Leslie, R. N. (1967). Selection of the best subset in regression analysis. *Technometrics* **9**, 531–540.

Hoel, D. G. (1971a). A method for construction of sequential selection procedures. *Ann. Math. Statist.* **42**, 630–642.

Hoel, D. G. (1971b). A selection procedure based upon Wald's double dichotomy test. Tech. Report No. TM-3237. Oak Ridge National Laboratory, Oak Ridge, Tennessee.

Hoel, D. G. (1972). An inverse stopping rule for play-the-winner sampling. *J. Amer. Statist. Assoc.* **67**, 148–151.

Hoel, D. G., and Mazumdar, M. (1968). An extension of Paulson's selection procedure. *Ann. Math. Statist.* **39**, 2067–2074.

Hoel, D. G., and Sobel, M. (1972). Comparisons of sequential procedures for selecting the best binomial population. *Proceedings of the Sixth Berkeley Symposium on Mathematical Statistics and Probability*, Vol. IV (Eds. L. Le Cam, J. Neyman, and E. L. Scott), University of California Press, Berkeley and Los Angeles, pp. 53–69.

Hoel, D. G., Sobel, M., and Weiss, G. H. (1972). A two-stage procedure for choosing the better of two binomial populations. *Biometrika* **59**, 317–322.

Hoel, D. G., Sobel, M., and Weiss, G. H. (1975a). Comparison of sampling methods for choosing the best binomial populations with delayed observations. *J. Statist. Comput. Simul.* **3**, 299–313.

Hoel, D. G., Sobel, M., and Weiss, G. H. (1975b). A survey of adaptive sampling for clinical trials. *Perspectives in Biometry* (Ed. R. M. Elashoff), Academic, New York, pp. 29–61.

Hoel, D. G., and Weiss, G. H. (1974). Comparison of methods for choosing the better of two negative exponential lifetime distributions. *Reliability and Biometry* (Eds. F. Proschan, and R. J. Serfling), SIAM, Philadelphia, pp. 563–583.

Hoerl, A. E., and Kennard, R. W. (1970). Ridge regression: biased estimation in non-orthogonal problems. *Technometrics* **12**, 55–67.

Hooper, J. H. (1977). Selection procedures for ordered families of distributions. Ph.D. Thesis (Tech. Report No. 339). Dept. of Operations Res., Cornell Univ., Ithaca, New York.

Hooper, J. H., and Santner, T. J. (1979). Design of Experiments for selection from ordered families of distributions. *Ann. Statist.* **7**, 615–643.

Hotelling, H. (1940). The selection of variates for use in prediction with some comments on the general problem of nuisance parameters. *Ann. Math. Statist.* **11**, 271–283.

Hsu, J. C. (1977). On Some Decision-Theoretic Contributions to the Problem of Subset Selection. Ph.D. Thesis (Mimeo. Ser. No. 491). Dept. of Statist., Purdue Univ., West Lafayette, Indiana.

Hsu, J. C. (1978a). Robust and nonparametric subset selection procedures. Tech. Report No. 156, Ohio State Univ., Columbus.

Hsu, J. C. (1978b). A class of nonparametric subset selection procedures. Technical Report No. 166, Dept. of Statist., Ohio State Univ., Columbus.

Huang, D. Y. (1974). On some optimal subset selection procedures for Model I and Model II in treatments versus control problems. Mimeo. Ser. No. 373, Dept. of Statist., Purdue Univ., West Lafayette, Indiana.

Huang, D. Y. (1975a). Some Contributions to Fixed Sample and Sequential Multiple Decision (Selection and Ranking) Theory. Ph.D. Thesis (Mimeo. Ser. No. 363). Dept. of Statist., Purdue Univ., West Lafayette, Indiana.

Huang, D. Y. (1975b). A note on a sequential selection procedure with applications to regression analysis. Mimeo. Ser. No. 439, Dept. of Statist., Purdue Univ., West Lafayette, Indiana.

Huang, D. Y. (1976). A note on an efficiency of the Bechhofer–Kiefer–Sobel sequential selection procedure. *Bull. Inst. Math. Acad. Sinica* **4**, 249–252.

Huang, D. Y., and Panchapakesan, S. (1976). A modified subset selection formulation with special reference to one-way and two-way layout experiments. *Commun. Statist.-Theor. Meth.* **A5**, 621–633.

Huang, D. Y., and Panchapakesan, S. (1978). A subset selection formulation of the complete ranking problem. *J. Chinese Statist. Assoc.* **16**, 5801–5810.

Huang, W. T. (1972). Some Contributions to Sequential Selection and Ranking Procedures. Ph.D. Thesis (Mimeo. Ser. No. 295). Dept. of Statist., Purdue Univ., West Lafayette, Indiana.

Huang, W. T. (1973a). On partitioning k multivariate normal populations with respect to a control. *Bull. Inst. Math. Acad. Sinica* **1**, 191–206.

Huang, W. T. (1973b). Some selection procedures for Poisson populations conditioned on the sum of observations. *Tamkang J. Math.* **4**, 195–202.

Huang, W. T. (1974). Selection procedures of Pareto populations. *Sankhyā Ser. B* **36**, 417–426.

Huang, W. T. (1975a). Bayes approach to a problem of partitioning k normal populations. *Bull. Inst. Math. Acad. Sinica* **3**, 87–97.

Huang, W. T. (1975b). On some optimal properties of selection procedures for finite populations. Mimeo. Ser. No. 431. Dept. of Statist., Purdue Univ., West Lafayette, Indiana.

Huber, P. J. (1963a). A remark on a paper of Trawinski and David entitled: "Selection of the best treatment in a paired-comparison experiment." *Ann. Math. Statist.* **34**, 92–94.

Huber, P. J. (1963b). Pairwise comparison and ranking: optimum properties of the sum procedure. *Ann. Math. Statist.* **34**, 511–520.

Jensen, D. R. (1976). The comparison of several response functions with a standard. *Biometrics* **32**, 51–59.

Johnson, N. L., and Maurice, R. J. (1963). Minimax-regret procedure for choosing between two populations using sequential sampling. *J. Roy. Statist. Soc. Ser. B* **25**, 297–304.

Kappenman, R. F. (1972). A note on selection of the greatest excedence probability. *Technometrics* **14**, 219–222.

Kapur, M. N. (1958). On a class of decision procedures for ranking normal populations according to their variances. *J. Indian Soc. Agric. Statist.* **10**, 99–106.

Kesten, H., and Morse, N. (1959). A property of the multinomial distribution. *Ann. Math. Statist.* **30**, 120–127.

Kiefer, J. (1977). Conditional confidence and estimated confidence in multidecision problems (with applications to selection and ranking). *Multivariate Analysis—IV* (Ed. P. R. Krishnaiah), North-Holland, Amsterdam, New York, and Oxford, pp. 143–158.

Kiefer, J. E., and Weiss, G. H. (1971). A truncated test for choosing the better of two binomial populations. *J. Amer. Statist. Assoc.* **66**, 867–871.

Kiefer, J. E., and Weiss, G. H. (1974). Truncated version of a play-the-winner rule for choosing the better of two binomial populations. *J. Amer. Statist. Assoc.* **69**, 807–809.

Kim, W. C. (1978). On a selection problem in reliability theory. Mimeo. Ser. No. 78-16, Dept. of Statist., Purdue Univ., West Lafayette, Indiana.

Kirton, H. C. (1967). "Best" models in multiple regression analysis. N. S. W. Dept. of Agriculture, Sydney, Australia.

Kiviat, P. (1960). An I.B.M. 704 E.D.P.M. program for the Monte Carlo simulation of a sequential multiple-decision procedure for selecting the best one of several normal populations with a common known variance. Tech. Report. Dept. of Industrial Engineering, Cornell Univ., Ithaca, New York.

Kleijnen, J. P. C. (1975). *Statistical Techniques in Simulation*, Vol. 2, Marcel Dekker, New York.

Kleijnen, J. P. C. (1976). Robustness of multiple ranking procedures: a Monte Carlo experiment illustrating design and analysis techniques. Tech. Report No. 76.015, Katholieke Hogeschool, Tilburg, The Netherlands.

Kleijnen, J. P. C. (1977). Robustness of a multiple ranking procedure: a Monte Carlo experiment illustrating design and analysis techniques. *Commun. Statist.-Simula. Computa.* **B6**(3), 235–262.

Kleijnen, J. P. C., and Naylor, T. H. (1969). The use of multiple ranking procedures to analyse simulations of business and economic systems. *Proceeding of Business and Economics Section of the Amer. Statist. Assoc.*, Washington, D.C., pp. 605–615.

Kleijnen, J. P. C., Naylor, T. H., and Seaks, T. G. (1972). The use of multiple ranking procedures to analyze simulations of management systems. *A Tutorial Management Series, Application Series 18*, **B**, 245–257.

Krishnaiah, P. R. (1967). Selection procedures based on covariance matrices of multivariate normal populations. *Blanch Anniversary Volume*, Aerospace Research Laboratories, U.S. Air Force, Dayton, Ohio, pp. 147–160.

Krishnaiah, P. R., and Rizvi, M. H. (1966). Some procedures for selection of multivariate normal populations better than a control. *Multivariate Analysis* (Ed. P. R. Krishnaiah), Academic, New York, pp. 477–490.

Kulldorf, G. (1977). *Bibliography on Selection Procedures*. Institute of Mathematics and Statistics, Univ. of Umeå, Sweden.

Lam, K., and Chiu, W. K. (1976). On the probability of correctly selecting the best of several normal populations. *Biometrika* **63**, 410–411.

LaMotte, L. R., and Hocking, R. R. (1970). Computational efficiency in the selection of regression variables. *Technometrics* **12**, 83–93.

Lawing, W. D., Jr. (1965). Multiple Decision Sequential Procedures. Ph.D. Thesis, Dept. of Statist., Iowa State Univ., Ames.

Lee, Y. J. (1977a). On selecting the best contestant. Tech. Report No. TR-77-24. Dept. of Math., Univ. of Maryland, College Park.

Lee, Y. J. (1977b). Winner selection. *Proceedings of the 1977 Winter Simulation Conference* (Eds. H. J. Highland, R. G. Sargant, and J. W. Schmidt) held at the National Bureau of Standards, Gaithersburg, Maryland, pp. 87–91.

Lee, Y. J., and Dudewicz, E. J. (1974). Nonparametric ranking and selection procedures. Tech. Report No. 105, Dept. of Statist., Ohio State Univ., Columbus.

Lehmann, E. L. (1957a). A theory of some multiple decision problems, I. *Ann. Math. Statist.* **28**, 1–25.

Lehmann, E. L. (1957b). A theory of some multiple decision problems, II. *Ann. Math. Statist.* **28** 547–572.

Lehmann, E. L. (1961). Some Model I problems of selection. *Ann. Math. Statist.* **32**, 990–1012.

Lehmann, E. L. (1963). A class of selection procedures based on ranks. *Math. Ann.* **150**, 268–275.

Lehmann, E. L. (1966). On a theorem of Bahadur and Goodman. *Ann. Math. Statist.* **37**, 1–6.

Leong, Y. K. (1976). Some Results on Subset Selection Problems. Ph.D. Thesis. (Mimeo, Ser. No. 475). Dept. of Statist., Purdue Univ., West Lafayette, Indiana.

Leong, Y. K., and Wong, W. Y. (1977a). A note on selection and ranking procedures from multinomial distribution. Research Report No. 16/77, Dept. of Math., Univ. of Malaya, Kuala Lumpur, Malaysia.

Leong, Y. K., and Wong, W. Y., (1977b). On Poisson selection problems. Research Report No. 23/77, Dept. of Math., Univ. of Malaya, Kuala Lumpur, Malaysia.

Leong, Y. K., and Wong, W. Y. (1978a). Subset selection procedures for selecting binomial populations better than a control. Research Report No. 11/78. Dept. of Math., Univ. of Malaya, Kuala Lumpur, Malaysia.

Leong, Y. K., and Wong, W. Y. (1978b). A class of subset selection procedures for binomial populations. Research Report No. 14/78, Dept. of Math., Univ. of Malaya, Kuala Lumpur, Malaysia.

Levy, K. J. (1975). Selecting the best population from among k binomial populations or the population with the largest correlation coefficient among k bivariate normal populations. *Psychometrika* **40**, 121–122.

Lindley, D. V. (1961). Dynamic programming and decision theory. *Appl. Statist.* **10**, 39–51.

Lorenzen, T. J., and McDonald, G. C. (1978). Selecting logistic populations using the sample medians. Research Report, General Motors Research Laboratories, Warren, Michigan.

Lu, M. W. (1976). On Some Multiple Decision Problems. Ph.D. Thesis (Mimeo. Ser. No. 464). Dept. of Statist., Purdue Univ., West Lafayette, Indiana.

Mahamunulu, D. M. (1966). On a Generalized Goal in Fixed Sample Ranking and Selection Problems. Ph.D. Thesis (Tech. Report No. 72). Dept. of Statist., Univ. of Minnesota, Minneapolis, Minnesota.

Mahamunulu, D. M. (1967). Some fixed-sample ranking and selection problems. *Ann. Math. Statist.* **38**, 1079–1091.

Male, L. M. (1971). A Multiple Comparison Procedure for Binomial Random Variables. Ph.D. Thesis. Cornell Univ., Ithaca, New York.

Matsui, T. (1972). On selecting the best one of k normal populations based on ranks. *J. Japan Statist. Soc.* **2**, 71–81.

Maurice, R. J. (1957). A minimax procedure for choosing between two populations using sequential sampling. *J. Roy. Statist. Soc. Ser. B* **19**, 255–261.

Maurice, R. J. (1958a). Ranking means of two normal populations with unknown variances. *Biometrika* **45**, 250–252.

Maurice, R. J. (1958b). Selection of the population with the largest mean when comparisons can be made only in pairs. *Biometrika* **45**, 581–586.

Maurice, R. J. (1959). A different loss function for the choice between two populations. *J. Roy. Statist. Soc. Ser. B* **21**, 203–213.

McCabe, G. P., Jr., and Arvesen, J. N. (1974). A subset selection procedure for regression variables. *J. Statist. Comput. Simul.* **3**, 137–146.

McCabe, G. P., Jr., Arvesen, J. N., and Pohl, R. J. (1973). A computer program for subset selection in regression analysis. Mimeo. Ser. No. 317. Dept. of Statist., Purdue Univ., West Lafayette, Indiana.

McCabe, G. P., Jr. and Ross, M. A. (1975). A stepwise algorithm for selecting regression variables using cost criteria. *Proceedings of Computer Science and Statistics: 8th Annual Symposium on the Interface*, Health Sciences Computing Facility, Univ. of California, Los Angeles.

McDonald, G. C. (1969a). On Some Distribution-free Ranking and Selection Procedures. Ph.D. Thesis (Mimeo. Ser. No. 174). Dept. of Statist., Purdue Univ., West Lafayette, Indiana.

McDonald, G. C. (1969b). A note on monotonicity in the gamma distributions. Res. Publ. GMR-930. General Motors Research Laboratories, Warren, Michigan.

McDonald, G. C. (1971). On approximating constants required to implement a selection procedure based on ranks. Statistical Decision Theory and Related Topics (Eds. S. S. Gupta and J. Yackel), Academic, New York, pp. 299–312.

McDonald, G. C. (1972). Some multiple comparison selection procedures based on ranks. *Sankhyā Ser. A* **34**, 53–64.

McDonald, G. C. (1973). The distribution of some rank statistics with applications in block design selection problems. *Sankhyā Ser. A* **35**, 187–203.

McDonald, G. C. (1975). Characteristics of three selection rules based on ranks in the 3×2 exponential case. *Sankhyā Ser. B* **36**, 261–266.

McDonald, G. C. (1976). The distribution of a variate based on independent ranges from a uniform population. *Technometrics* **18**, 343–349.

McDonald, G. C. (1977a). Subset selection procedures based on sample ranges. *Commun. Statist.-Theor. Meth.* **A6**, 1055–1079.

McDonald, G. C. (1977b). Nonparametric selection procedures applied to state traffic fataility rates. *Proceedings of the 1977 Winter Simulation Conference* (Eds. H. J. Highland, R. G. Sargent, and J. W. Schmidt) held at the National Bureau of Standards, Gaithersburg, Maryland, pp. 93–100.

McDonald, G. C. (1978). Subset selection rules based on quasi-ranges for uniform populations. Research Report GMR-1775 (Revised). General Motors Research Lab., Warren, Michigan.

Meilijson, I. (1969). A note on sequential multiple decision procedures. *Ann. Math. Statist.* **40**, 653–657.

Meyer, W. W. (1978). On a desirable property of range in relation to subset selection. Research Report GMR-2802. General Motors Research Lab., Warren, Michigan.

Miescke, K. J. (1978). Bayesian subset selection for additive and linear loss functions. Mimeo. Series No. 78-23, Dept. of Statist., Purdue Univ., West Lafayette, Indiana.

Miescke, K. J. (1979a). Identification and selection procedures based on tests. *Ann. Statist.* **7**, 207–219.

Miescke, K. J. (1979b). Simultaneous testing problems. Mimeo. Ser. No. 79-1, Dept. of Statist., Purdue Univ., West Lafayette, Indiana.

Mignogna, E. (1975). Procedures for Choosing the Best of k Bernoulli Populations. Ph.D. Thesis. Dept. of Math., Statist., and Comput. Sci., American Univ., Washington, D.C.

Miller, R. G., Jr. (1966). *Simultaneous Statistical Inference*. McGraw Hill, New York.

Nagel, K. (1966). On selecting a subset containing the best of several discrete distributions. Mimeo. Ser. No. 66. Dept. of Statist., Purdue Univ., West Lafayette, Indiana.

Nagel, K. (1970). On Subset Selection Rules with Certain Optimality Properties. Ph.D. Thesis (Mimeo. Ser. No. 222). Dept. of Statist., Purdue Univ., West Lafayette, Indiana.

Naik, U. D. (1967). Bayes rules for comparisons of scale parameters of certain distributions. *Ann. Inst. Statist. Math.* **19**, 19–37.

Naik, U. D. (1970). Bayes rules for selection of the best normal population. *J. Karnatak, Univ. Sci.* **15**, 70–87.

Naik, U. D. (1975). Some selection rules for comparing p processes with a standard. *Comm. Statist.* **4**, 519–535.

Naik, U. D. (1976). On selection procedures based on the excedence probability *Calcutta Statist. Assoc. Bull.* **25**, 29–40.

Naik, U. D. (1977). Some subset selection problems. *Commun. Statist.-Theor. Meth.* **A6**(10), 955–966.

Nebenzahl, E. (1970). Play-the-Winner Sampling in Selecting the Better of Two Binomial Populations. Ph.D. Thesis (Tech. Report No. 147). School of Statist., Univ. of Minnesota, Minneapolis.

Nebenzahl, E., and Sobel, M. (1972). Play-the-winner sampling for a fixed sample size binomial selection problem. *Biometrika* **59**, 1–8.

Neuts, M. F., and Carson, C. C. (1975). Some computational problems related to multinomial trials. *Canadian J. Statist.* **3**, 235–248.

Nomachi, Y. (1967). A closed sequential procedure selecting the best population in a family of populations with one parameter exponential distributions. *Bull. Math. Statist.* **12**, 21–34.

Nomachi, Y. (1969). A closed sequential procedure selecting the population with minimum variance from several normal populations. *Kumamoto J. Sci. Ser. A* **8**, 129–134.

Nordbrock, E. (1976). An improved play-the-winner sampling procedure for selecting the better of two binomial populations. *J. Amer. Statist. Assoc.* **71**, 137–139.

O'Brien, P. C. (1967). Selection Procedures Relating to Gamma, Weibull, Poisson and Normal Distributions. M.S. Thesis, Dept. of Statist., Iowa State Univ., Ames.

O'Brien, P. C. (1970). Procedures for selecting the Best of Several Populations. Ph.D. Thesis, Iowa State Univ., Ames.

Ofosu, J. B. (1972). A minimax procedure for selecting the population with the largest (smallest) scale parameter. *Calcutta Statist. Assoc. Bull.* **21**, 143–154. Errata **22** (1973), 129.

Ofosu, J. B. (1973). A two-sample procedure for selecting the population with the largest mean for several normal populations with unknown variances. *Biometrika* **60**, 117–124. Correction *Biometrika* **62** (1975), 221.

Ofosu, J. B. (1974a). Optimum sample size for a problem in choosing the population with the largest mean: some comments on Somerville's paper. *J. Amer. Statist. Assoc.* **69**, 270.

Ofosu, J. B. (1974b). On selection of procedures for exponential distributions. *Bull. Math. Statist.* **16**, 1–9.

Ofosu, J. B. (1975a). A two-stage minimax procedure for selecting the normal population with the smallest variance. *J. Amer. Statist. Assoc.* **70**, 171–174.

Ofosu, J. B. (1975b). On selecting a subset which includes the t best of k populations: scale parameter case. *Bull. Math. Statist.* **16**, 7–18.

Ofosu, J. B. (1975c). A Bayesian procedure for selecting the largest of k normal means. *Calcutta Statist. Assoc. Bull.* **24**, 43–51.

Olkin, I., Sobel, M., and Tong, Y. L. (1976). Estimating the true probability of correct selection for location and scale parameter families. Tech. Report No. 174. Dept. of Operations Research and Dept. of Statist., Stanford University, Stanford, California.

O'Neill, R., and Wethrill, G. B. (1971). The present state of multiple comparison methods. *J. Roy. Statist. Soc. Ser. B* **33**, 218–241, with discussion 241–250.

Panchapakesan, S. (1969). Some Contributions to Multiple Decision (Selection and Ranking) Procedures. Ph.D. Thesis (Mimeo. Ser. No. 192). Dept. of Statist., Purdue Univ., West Lafayette, Indiana.

Panchapakesan, S. (1971). On a subset selection procedure for the most probable event in a multinomial distribution. *Statistical Decision Theory and Related Topics* (Eds. S. S. Gupta, and J. Yackel), Academic, New York, pp. 275–298.

Panchapakesan, S. (1973). On a subset selection procedure for the best multinomial cell and related problems. Abstract, *Bull. Inst. Math. Statist.* **2**, 112–113.

Panchapakesan, S. (1978). On a monotonicity property relating to the gamma distribution. *J. Chinese Statist. Assoc.* **18**, 6003–6005.

Panchapakesan, S., and Santner, T. J. (1977). Subset selection procedures for Δ_p-superior populations. *Commun. Statist.-Theor. Meth.* **A6**, 1081–1090.

Patel, J. K. (1968). Selecting a subset containing the best one of several IFRA populations. Ph.D. Thesis, Univ. of Minnesota, Minneapolis.

Patel, J. K. (1974). On selecting a subset of IFRA populations which are better than a control. *Trabajos Estadist.* **25**, 89–101.

Patel, J. K. (1976). Ranking and selection of IFR populations based on means. *J. Amer. Statist. Assoc.* **71**, 143–146.

Patterson, D. W. (1966). Contributions to the Theory of Ranking and Selection Problems. Ph.D. Thesis. State University of Rutgers, New Brunswick, New Jersey.

Paulson, E. (1949). A multiple decision procedure for certain problems in analysis of variance. *Ann. Math. Statist.* **20**, 95–98.

Paulson, E. (1952). On the comparison of several experimental categories with a control. *Ann. Math. Statist.* **23**, 239–246.

Paulson, E. (1962). A sequential procedure for comparing several experimental categories with a standard or control. *Ann. Math. Statist.* **33**, 438–443.

Paulson, E. (1964a). A sequential procedure for selecting the population with the largest mean from k normal populations. *Ann. Math. Statist.* **35**, 174–180.

Paulson, E. (1964b). Sequential estimation and closed sequential decision procedures. *Ann. Math. Statist.* **35**, 1048–1058.

Paulson, E. (1967). Sequential procedures for selecting the best one of several binomial populations. *Ann. Math. Statist.* **38**, 117–123.

Paulson, E. (1969). A new sequential procedure for selecting the best one of k binomial populations. (Abstract). *Ann. Math. Statist.* **40**, 1865–1866.

Paulson, E. (1970). A note on a sequential procedure for comparing two Koopman–Darmois populations. *Ann. Math. Statist.* **41**, 1756–1759.

Perng, S. K. (1969). A comparison of the asymptotic expected sample sizes of two sequential procedures for ranking problem. *Ann. Math. Statist.* **40**, 2198–2202.

Perng, S. K., and Groves, B. D. (1977). Two sequential procedures for partitioning a set of one-parameter exponential populations with respect to a control. *Austral. J. Statist.* **19**, 56–60.

Pradhan, M., and Sathe, Y. S. (1973). Play-the-winner sampling for a fixed sample size with curtailment. *Biometrika* **60**, 424–427.

Pradhan, M., and Sathe, Y. S. (1974). Equivalence between fixed sample rule and Hoel's inverse sampling rule for play-the-winner. *J. Amer. Statist. Assoc.* **69**, 475–476.

Pradhan, M., and Sathe, Y. S. (1976). Analytical remarks on Canner's minimax method for finding the better of two binomial populations. *J. Amer. Statist. Assoc.* **71**, 239–241.

Puri, M. L., and Puri, P. S. (1969). Multiple decision procedures based on ranks for certain problems in analysis of variance. *Ann. Math. Statist.* **40**, 619–632.

Puri, P. S., and Puri, M. L. (1968). Selection procedures based on ranks: scale parameter case. *Sankhyā Ser. A* **30**, 291–302.

Raghavachari, M., and Starr, N. (1970). Selection problems for some terminal distributions. *Metron* **28**, 185–197.

Raifa, H., and Schlaifer, R. (1961). *Applied Statistical Decision Theory*. Graduate School of Business Administration, Harvard University, Boston, Massachusetts.

Ramberg, J. S. (1966). A Comparison of the Performance Characteristics of Two Sequential Procedures for Ranking Means of Normal Populations. M.S. Thesis (Tech. Report No. 4), Dept. of Operations Res., Cornell Univ., Ithaca, New York.

Ramberg, J. S. (1969a). A Multiple-Decision Approach to the Selection of the Best Set of Predictor Variates. Ph.D. Thesis (Tech. Report No. 27). Dept. of Operations Res., Cornell Univ., Ithaca, New York.

Ramberg, J. S. (1969b). Selection and ranking procedures: a comment. *The Design of Computer Simulation Experiments* (Ed. T. R. H. Naylor), Duke University Press, Durham, North Carolina, pp. 161–164.

Ramberg, J. S. (1972). Selection sample size approximations. *Ann. Math. Statist.* **43**, 1977–1980.

Ramberg, J. S. (1977). Selecting the best predictor variate. *Commun. Statist.-Theor. Meth.* **A6**, 1133–1147.

Ramberg, J. S. and Bechhofer, R. E. (1973). A comparison of two multiple decision procedures for ranking means of normal populations with common known variance. *Abstract. Inst. Math. Statist. Bull.* **2**, 108.

Ramey, J. T., Jr. (1974). Selecting the Most Probable Event. Ph.D. Thesis. Dept. of Math. Sciences, Clemson Univ., Clemson, South Carolina.

Ramey, J. T., Jr., and Alam, K. (1978). A Bayes sequential procedure for selecting the most probable multinomial event. Tech. Report No. 293, Dept. of Mathematical Sciences, Clemson University, Clemson, South Carolina.

Ramey, J. T., Jr., and Alam, K. (1979). A sequential procedure for selecting the most probable multinomial event. *Biometrika*, **66**, 171–173.

Randles, R. H. (1969). Gamma-Minimax Procedures for the Use of Incomplete Prior Information in Selection and Classification Problems. Ph.D. Thesis. Dept. of Statist., Florida State Univ., Tallahassee.

Randles, R. H. (1970). Some robust selection procedures. *Ann. Math. Statist.* **41**, 1640–1645.

Randles, R. H., and Hollander, M. (1971). Γ-minimax selection procedures in treatment versus control problems. *Ann. Math. Statist.* **42**, 330–341.

Randles, R. H., Ramberg, J. S., and Hogg, R. V. (1973). An adaptive procedure for selecting the population with the largest location parameter. *Technometrics* **15**, 769–789.

Rasmussen, W. T., and Robbins, H. (1975). The candidate problem with unknown population size. *J. Appl. Probability* **12**, 692–701.

Regier, M. H. (1975). Choosing the most homogeneous multinomial. *Contributed Papers* (*40th Session of the International Statistical Institute*), pp. 681–685.

Regier, M. H. (1976). Simplified selection procedures for multivariate normal populations. *Technometrics* **18**, 483–489.

Rinott, Y. (1978). On two-stage procedures and related probability-inequalities. *Commun. Statist.-Theor. Meth.* **A8**, 799–811.

Rinott, Y., and Santner, T. J. (1977). An inequality for multivariate normal probabilities with application to a design problem. *Ann. Statist.* **6**, 1228–1234.

Rizvi, M. H. (1963). Ranking and Selection Problems of Normal Populations Using the Absolute Values of Their Means: Fixed Sample Case, Ph.D. Thesis (Tech. Report No. 31), Dept. of Statist., Univ. of Minnesota, Minneapolis.

Rizvi, M. H. (1971). Some selection problems involving folded normal distribution. *Technometrics* **13**, 355–369.

Rizvi, M. H., and Saxena, K. M. Lal (1972). Distribution-free interval estimation of the largest α-quantile. *J. Amer. Statist. Assoc.* **67**, 196–198.

Rizvi, M. H., and Saxena, K. M. Lal (1974). On interval estimation and simultaneous selection of ordered location or scale parameters. *Ann. Statist.* **2**, 1340–1345.

Rizvi, M. H., and Sobel, M. (1967). Nonparametric procedures for selecting a subset containing the population with the largest α-quantile. *Ann. Math. Statist.* **38**, 1788–1803.

Rizvi, M. H., Sobel, M., and Woodworth, G. G. (1968). Nonparametric ranking procedures for comparison with a control. *Ann. Math Statist.* **39**, 2075–2093.

Rizvi, M. H., and Solomon, H. (1973). Selection of largest multiple correlation coefficients: asymptotic case. *J. Amer. Statist. Assoc.* **68**, 184–188. Corrigendum, **69**, 288.

Rizvi, M. H., and Woodworth, G. G. (1970). On selection procedures based on ranks: counterexamples concerning the least favorable configurations. *Ann. Math. Statist.* **41**, 1942–1951.

Robbins, H., and Siegmund, D. (1968). Iterated logarithm inequalities and related statistical procedures. *Mathematics of the Decision Sciences, Part* 2 (Eds. G. B. Dantzig and A. F. Veinott, Jr), American Mathematical Society, Providence, Rhode Island, pp. 267–279.

Robbins, H., Sobel, M., and Starr, N. (1968). A sequential procedure for selecting the best of *k* populations. *Ann. Math. Statist.* **39**, 88–92.

Roberts, C. D. (1962). An Asymptotically Optimal Sequential Design for Comparing Several Experimental Categories with a Standard or Control. Ph.D. Thesis

(Mimeo. Series No. 344), *Inst. of Statist.*, Univ. of North Carolina, Chapel Hill.

Roberts, C. D. (1963). An asymptotically optimal sequential design for comparing several experimental categories with a control. *Ann. Math. Statist.* **34**, 1486–1493.

Roberts, C. D. (1964). An asymptotically optimal fixed sample size procedure for comparing several experimental categories with a control. *Ann. Math. Statist.* **35**, 1571–1575.

Roberts, C. D. (1965). An asymptotically optimal fixed sample size procedure for comparing several experimental categories with a control. (In Spanish. English summary) *Estadistica* **23**, 288–294.

Roth, A. J. (1978). A new procedure for selecting a subset containing the best normal population. *J. Amer. Statist. Assoc.* **73**, 613–617.

Ryan, T. A., Jr., and Antle, C. E. (1976). A note on Gupta's selection procedure. *J. Amer. Statist. Assoc.* **71**, 140–142.

Santner, T. J. (1973). A Restricted Subset Selection Approach to Ranking and Selection Problems. Ph.D. Thesis (Mimeo. Ser. No. 318). Dept. of Statist., Purdue Univ., West Lafayette, Indiana.

Santner, T. J. (1974). Selection criteria for partially specified worth parameters. Tech. Report No. 244, Dept. of Operations Res., Cornell Univ., Ithaca, New York.

Santner, T. J. (1975). A restricted subset selection approach to ranking and selection problems. *Ann. Statist.*, **3**, 334–349.

Santner, T. J. (1976a). A generalized goal in restricted subset selection theory. *Sankhyā Ser. B* **38**, 129–143.

Santner, T. J. (1976b). A two-stage procedure for selection of δ^*-optimal means in the normal case. *Commun. Statist.-Theor. Meth.* **A5**, 283–292.

Santner, T. J. (1978). Designing two factor experiments for selecting interactions. Tech. Report No. 376. Dept. of Operations Research, Cornell Univ., Ithaca, New York.

Saxena, K. M. Lal (1971). Interval estimation of the largest variance of k normal populations. *J. Amer. Statist. Assoc.* **66**, 408–410.

Saxena, K. M. Lal (1976a). A single-sample procedure for the estimation of the largest mean. *J. Amer. Statist. Assoc.* **71**, 147–148.

Saxena, K. M. Lal (1976b). Distribution-free tolerance intervals for stochastically ordered distributions. *Ann. Statist.* **4**, 1210–1218.

Saxena, K. M. Lal and Tong, Y. L. (1969). Interval estimation of the largest mean of k normal populations with known variances. *J. Amer. Statist. Assoc.* **64**, 296–299.

Saxena, K. M. Lal, and Tong, Y. L. (1971). Optimum interval estimation for the largest scale parameter. (abstract). *Optimizing Methods in Statistics* (Ed. J. S. Rustagi), Academic, New York, p. 477.

Schaafsma, W. (1969). Minimax risk and unbiasedness for multiple decision problems of Type I. *Ann. Math. Statist.* **40**, 1684–1720.

Schafer, R. E. (1974). A single-sample complete ordering procedure for certain populations. *Reliability and Biometry* (Eds. F. Proschan and R. J. Serfling), SIAM, Philadelphia, pp. 597–617.

Schafer, R. E. (1977). On selecting which of k populations exceed a standard. *The Theory and Applications of Reliability* (Eds. C. P. Tsokos and I. N. Shimi), Academic, New York, pp. 449–473.

Schafer, R. E., and Rutemiller, H. C. (1975). Some characteristics of a ranking procedure for population parameters based on chi-square statistics. *Technometrics* **17**, 324–331.

Seal, K. C. (1954). On a Class of Decision Procedures for Ranking Means. Ph.D. Thesis (Mimeo. Ser. No. 109), Inst. of Statist., Univ. of North Carolina, Chapel Hill.

Seal, K. C. (1955). On a class of decision procedures for ranking means of normal populations. *Ann. Math. Statist.* **26**, 387–398.

Seal, K. C. (1957). An optimum decision rule for ranking means of normal populations. *Calcutta Statist. Assoc. Bull.* **7**, 131–150.

Seal, K. C. (1958a). On ranking parameters of scale in Type III populations. *J. Amer. Statist. Assoc.* **53**, 164–175.

Seal, K. C. (1958b). Selection of a subset superior, inferior or equivalent to a control population. *Calcutta Statist. Assoc. Bull.* **8**, 20–30.

Seeger, P. (1972). A note on partitioning a set of normal populations by their locations with respect to two controls. *Ann. Math. Statist.* **43**, 2019–2023.

Sen, A. K. (1971). Tests for change in Mean and Sequential Ranking Procedure. Dept. of Math., Univ. of Toronto, Toronto, Canada.

Sen, A. K., and Srivastava, M. S. (1972). On a sequential procedure based on Wilcoxon statistic. *J. Statist. Res.* **6**, 43–51.

Sen, P. K., and Puri, M. L. (1972). On some selection procedures in two-way layouts. *Z. Wahrscheinlichkeitstheorie Verw. Geb.* **22**, 242–250.

Shorock, R. A. (1969). The Slippage Selection and Detection Problems. Ph.D. Thesis. Univ. of Washington, Seattle.

Sievers, S. R. (1972). Topics in Ranking Procedures. Ph.D. Thesis. Dept. of Math., Cornell Univ., Ithaca, New York.

Simon, R., and Weiss, G. H. (1975). A class of adaptive sampling schemes for selecting the better of two binomial populations. *J. Statist. Comput. Simul.* **4**, 37–47.

Simon, R., Weiss, G. H., and Hoel, D. G. (1975). Sequential analysis of binomial clinical trials. *Biometrika* **62**, 195–200.

Singh, A. K. (1977). On Slippage Tests and Multiple Decision (Selection and Ranking) Procedures. Ph.D. Thesis (Mimeo. Ser. No. 494), Dept. of Statist., Purdue Univ., W. Lafayette, Indiana.

Singh, J. (1963). Statistical Theory of Selection (Selection Problems). Ph.D. Thesis. Univ. of Toronto, Toronto, Canada.

Sitaramaiah, A. (1966). Sampling Procedures for Ranking (or Selecting Best Population) of Correlated Populations. Ph.D. Thesis, Univ. of Iowa, Iowa City.

Sitek, M. (1972). Application of the selection procedure R to unequal observation numbers. *Zastoswania Matematyki* **12**, 355–371.

Smith, A. V., Jr. (1967). Comparison of two Bernoulli processes by multiple stage sampling using Bayesian decision theory. *Ann. Inst. Statist. Math.* **19**, 505–518.

Smith, A. V., Jr. (1970). Comparison of the means of two normal processes by multiple stage sampling using Bayesian decision theory. *Ann. Inst. Statist. Math.* **22**, 137–143.

Smith, M. H. (1975). A secretary problem with uncertain employment. *J. Appl. Probability* **12**, 620–624.

Sobel, M. (1956). Sequential procedures for selecting the best exponential population. *Proceedings of the Third Berkeley Symposium on Mathematical Statistics and Probability* (Ed. J. Neyman), Vol. V, Univ. of California Press, Berkeley, pp. 99–110.

Sobel, M. (1963). Single sample ranking problems with Poisson populations. Tech. Report No. 19. Dept. of Statist., Univ. of Minnesota, Minneapolis.

Sobel, M. (1967). Nonparametric procedures for selecting the t populations with the largest α-quantiles. *Ann. Math. Statist.* **38**, 1804–1816.

Sobel, M. (1969). Selecting a subset containing at least one of the t best populations. *Multivariate Analysis–II* (Ed. P. R. Krishnaiah), Academic, New York, pp. 515–540.

Sobel, M. (1977). Selecting the population with the smallest dispersion in a nonparametric setting. *Proceedings of the* 1977 *Winter simulation Conference* (Eds. H. J. Highland, R. G. Sargent, and J. W. Schmidt) held at the National Bureau of Standards, Gaithersburg, Maryland, pp. 103–114.

Sobel, M., and Huyett, M. J. (1957). Selecting the best one of several binomial populations. *Bell System Tech. J.* **36**, 537–576.

Sobel, M., and Starr, N. (1966). The performance of the sequential procedure for choosing the largest of k means. Tech. Report No. 88, Dept. of Statist., Univ. of Minnesota, Minneapolis.

Sobel, M., and Starr, N. (1968). Selecting the t fairest coins. Tech. Report. No. 99, Dept. of Statist., Univ. of Minnesota, Minneapolis.

Sobel, M., and Starr, N. (1975). Selecting the coin with the greatest bias. *Sankhyā Ser. A* **37**, 197–210.

Sobel, M., and Tong, Y. L. (1971). Optimal allocation of observations for partitioning a set of normal populations in comparison with a control. *Biometrika* **58**, 177–181.

Sobel, M., and Tong, Y. L. (1975). Estimating the true PCS in ranking and selection. Tech. Report. Dept. of Math., Univ. of Nebraska, Lincoln.

Sobel, M., and Weiss, G. H. (1970a). Play-the-winner sampling for selecting the better of two binomial populations. *Biometrika* **57**, 357–365.

Sobel, M., and Weiss, G. H. (1970b). Inverse sampling and other selection procedures for tournaments with 2 or 3 players. *Nonparametric Techniques in Statistical Inference* (Ed. M. L. Puri), Cambridge University Press, London, pp. 515–543.

Sobel, M., and Weiss, G. H. (1971a). Play-the-winner rule and inverse sampling in selecting the better of two binomial populations. *J. Amer. Statist. Assoc.* **66**, 545–551.

Sobel, M., and Weiss, G. H. (1971b). A comparison of play-the-winner and vector-at-a-time sampling for selecting the better of two binomial populations with restricted parameter values. *Trabajos Estadist* **22**, 195–206.

Sobel, M., and Weiss, G. H. (1972a). Play-the-winner rule and inverse sampling for selecting the best of $k \geqslant 3$ binomial populations. *Ann. Math. Statist.* **43**, 1808–1826.

Sobel, M., and Weiss, G. H. (1972b). Recent results on using the play-the-winner sampling rule with binomial selection problems. *Proceedings of the Sixth Berkeley Symposium on Mathematical Statistics and Probability* (Eds. L. Le Cam, J. Neyman, and E. L. Scott), Vol. I, Univ. of California Press, Berkeley and Los Angeles, pp. 717–736.

Sobel, M., and Yen, S. P. (1972). An asymptotic comparison of subset selection procedures. *Ann. Inst. Statist. Math.*, **24**, 465–468.

Soller, M., and Putter, J. (1964). On the probability that the best chicken stock will come out best in a single random sample test. *Poultry Sci.* **43**, 1425–1427.

Soller, M., and Putter, J. (1965). Probability of correct selection of sires having highest transmitting ability. *J. Dairy Sci.* **48**, 747–748.

Somerville, P. N. (1953). Some Problems of Optimum Sampling. Ph.D. Thesis, Univ. of North Carolina, Chapel Hill.

Somerville, P. N. (1954). Some problems of optimum sampling. *Biometrika* **41**, 420–429.

Somerville, P. N. (1957). Optimum sampling in binomial populations. *J. Amer. Statist. Assoc.* **52**, 494–502.

Somerville, P. N. (1970). Optimum sample size for a problem in choosing the population with the largest mean. *J. Amer. Statist. Assoc.* **65**, 763–775.

Somerville, P. N. (1971a). A generalization of a fundamental theorem of ranking and selection. *Biometrika* **58**, 227–228. Correction, **63** (1976), 412.

Somerville, P. N. (1971b). A technique for obtaining probabilities of correct selection in a two-stage selection problem. *Biometrika* **58**, 615–623.

Somerville, P. N. (1974a). Rejoinder to Ofosu's Comments. *J. Amer. Statist. Assoc.* **69**, 270–271.

Somerville, P. N. (1974b). On allocation of resources in a two-stage selection procedure. *Sankhyā Ser. B* **36**, 194–203.

Somerville, P. N. (1975). Optimum sample size for choosing the population having the smallest variance (large sample results). *J. Amer. Statist. Assoc.* **70**, 852–858.

Srivastava, M. S. (1966). Some asymptotically efficient sequential procedures for ranking and slippage problems. *J. Roy. Statist. Soc. Ser. B.* **28**, 370–380.

Srivastava, M. S. (1967). Comparing distances between multivariate populations: the problem of minimum distance. *Ann. Math. Statist.* **38**, 550–556.

Srivastava, M. S., and Ogilvie, J. (1968). The performance of some sequential procedures for a ranking problem. *Ann. Math. Statist.* **39**, 1040–1047.

Srivastava, M. S., and Taneja, V. S. (1972). Some sequential procedures for ranking multivariate normal populations. *Ann. Inst. Statist. Math.* **24**, 455–464.

Starr, N., and Woodroofe, M. B. (1974). Approximations in large sample ranking problems. *Sankhyā Ser. B* **36**, 400–405.

Stein, C. (1948). The selection of the largest of a number of means. Abstract, *Ann. Math. Statist.* **19**, 429.

St-Pierre, J. (1954). Distribution of linear contrasts of order statistics. Ph.D Thesis. Inst. Statist., Univ. of North Carolina, Chapel Hill.

Studden, W. J. (1967). On selecting a subset of k populations containing the best. *Ann. Math. Statist.* **38**, 1072–1078.

Sverdrup, E. (1969). *Multiple Decision Theory*. Lecture Notes Series No. 15, Matematisk Institut, Aarhus University, Denmark.

Swanepoel, J. W. H. (1977). Nonparametric elimination selection procedures based on robust estimators. *S. Afr. Statist. J.* **11**, 27–41.

Swanepoel, J. W. H., and Geertsema, J. C. (1973a). Selection sequences for selecting the best of k normal populations. *S. Afr. Statist. J.* **7**, 11–21.

Swanepoel, J. W. H., and Geertsema, J. C. (1973b). Nonparametric sequential procedures for selecting the best of k populations. *S. Afr. Statist. J.* **7**, 85–94.

Swanepoel, J. W. H., and Venter, J. H. (1975a). On the construction of sequential selection procedures. *S. Afr. Statist. J.* **9**, 103–118.

Swanepoel, J. W. H., and Venter, J. H. (1975b). A class of elimination selection procedures based on ranks. *S. Afr. Statist. J.* **9**, 119 – 128.

Swanepoel, J. W. H., and Geertsema, J. C. (1976). Sequential procedures with elimination for selecting the best of k normal populations. *S. Afr. Statist. J.* **10**, 9–36.

Taheri, H., and Young, D. H. (1974). A comparison of sequential sampling procedures for selecting the better of two binomial populations. *Biometrika* **61**, 585–592.

Taheri, H., and Young, D. H. (1977). An inverse sampling procedure for selecting a subset containing the best binomial population. *Sankhyā Ser. B* **29**, 130–144.

Tamhane, A. C. (1975a). A Minimax Multistage Elimination Type Rules for Selecting the Largest Normal Mean. Ph.D. Dissertation. Also Tech. Rep. No. 259. Dept. of Operations Research, Cornell Univ., Ithaca, New York.

Tamhane, A. C. (1975b). A minimax two-stage permanent elimination type procedure for selecting the smallest normal variance. Tech. Rep. No. 260. Dept. of Operations Research, Cornell Univ., Ithaca, New York.

Tamhane, A. C. (1976). A three-stage elimination type procedure for selecting the largest normal mean (common unknown variance) *Sankhyā Ser. B* **38**, 339–349.

Tamhane, A. C. (1977). On a class of multistage selection procedures with screening for the normal means problem. Discussion Paper No. 27, Center for Statistics and Probability, Northwestern University, Evanston, Illinois.

Tamhane, A. C. (1978). Selecting the better Bernoulli treatment using a matched samples design. Discussion Paper No. 31, Center for Statistics and Probability, Northwestern University, Evanston, Illinois.

Tamhane, A. C., and Bechhofer, R. E. (1977). A two-stage minimax procedure with screening for selecting the largest normal mean. *Commun. Statist.-Theor. Meth.* **A6**, 1003–1033.

Tamhane, A. C., and Bechhofer, R. E. (1979). A two-stage minimax procedure with screening for selecting the largest normal mean (II): an improved PCS lower bound and associated tables. *Commun. Statist.–Theor. Meth.* **A8**, 337–358.

Tamura, R. (1968). Some distribution-free multiple comparison procedures. *Mem. Fac. Lit. Sci. Shimane Univ. Natur. Sci.* **1**, 1–7.

Tamura, R. (1969). Some multivariate comparison procedures based on ranks. *Ann. Math. Statist.* **40**, 1486–1491.

Tasaka, M. (1973). Successive multiple decision procedures for ordered parameters. *Bull. Math. Statist.* **15**, 1–20.

Taylor, R. J., and David, H. A. (1962). A multistage procedure for the selection of the best of several populations. *J. Amer. Statist. Assoc.* **57**, 785–795.

Thompson, M. L. (1978). Selection of variables in multiple regression: Part I. A review and evaluation. *Int. Statist. Rev.* **46**, 1–19.

Thornby, J. I. (1964). A Study of the Ranking Criteria for Selecting a Subset Containing the Best of k Bivariate Normal Populations. Ph.D. Thesis (Tech. Report No. 36). Dept. of Statist., Univ. of Minnesota, Minneapolis.

Tiao, G. C., and Afonja, B. (1976). Some Bayesian considerations of the choice of design for ranking, selection and estimation. *Ann. Inst. Statist. Math.* **28**, 167–185.

Tong, Y. L. (1967). On Partitioning a Set of Normal Populations by Their Locations with Respect to a Control Using a Single-Stage, a Two-Stage and a Sequential Procedure. Ph.D. Thesis (Tech. Report No. 90), Dept. of Statist., Univ. of Minnesota, Minneapolis.

Tong, Y. L. (1969). On partitioning a set of normal populations by their locations with respect to a control. *Ann. Math. Statist.* **40**, 1300–1324.

Tong, Y. L. (1970). Multi-stage interval estimation of the largest mean of k normal populations. *J. Roy. Statist. Soc. Ser. B* **32**, 272–277.

Tong, Y. L. (1971). Interval estimation of the ordered means of two normal populations based on hybrid estimators. *Biometrika* **58**, 605–613.

Tong, Y. L. (1972). On the consistency of single-stage ranking procedures. *Ann. Inst. Statist. Math.* **24**, 271–284.

Tong, Y. L. (1973). An asymptotically optimal sequential procedure for the estimation of largest mean. *Ann. Statist.* **1**, 175–179.

Tong, Y. L. (1977). Application of a probability inequality to ranking and selection and other related problems. *Commun. Statist.-Theor. Meth.* **A6**, 1105–1120.

Tong. Y. L. (1978). An adaptive solution to ranking and selection problems. *Ann. Statist.* **6**, 658–672.

Tong, Y. L., and Wetzell, D. E. (1979). On the behaviour of the probability function for selecting the best normal population. *Biometrika* **66**, 174–176.

Trawinski, B. J. (1961). Selection of the Best Treatment in a Paired-Comparison Experiment. Ph.D. Thesis, Dept. of Statist., Virginia Polytechnic Institute, Blacksburg.

Trawinski, B. J. (1969). Asymptotic approximation to the expected size of the selected subset. *Biometrika* **56**, 207–213.

Trawinski, B. J., and David, H. A., (1963). Selection of the best treatment in a paired-comparison experiment. *Ann. Math. Statist.* **34**, 75–91.

Turnbull, B. W. (1976). Multiple decision rules for comparing several populations with a fixed known standard. *Commun. Statist.-Theor. Meth.* **A5**, 1225–1244.

Uvell, S. (1977). A problem of selection and estimation for variances of normal populations. Statist. Res. Report No. 1977-9, Dept. of Math. Statist., Univ. of Umeå, S-90187 Umeå, Sweden.

Wackerly, D. D. and Rao, P. V. (1976). Sequential procedures based on sample medians for choosing the best of k populations. Tech. Report No. 107, Dept. of Statist., Univ. of Florida, Gainesville, Florida.

Walker, W. (1964). The revised control data 1604 program for the Monte Carlo simulation of a sequential multiple-decision procedure for selecting the best one of several normal populations with a common known variance. Unpublished report. Dept. of Industrial Engineering, Cornell Univ., Ithaca, New York.

Wethrill, G. B., and Ofosu, J. B. (1974). Selection of the best of k normal populations. *Appl. Statist.* **23**, 253–277.

Williams, G. W. (1974). On selecting a subset containing the best treatment in the case of matched samples. Tech. Report No. 1, Dept. of Biostatist., Univ. of Michigan, Ann Arbor.

Wong, W. Y. (1976). Some Contributions to Subset Selection Procedures. Ph.D. Thesis (Mimeo. Ser. No. 454). Dept. of Statist., Purdue Univ., West Lafayette, Indiana.

Wong, W. Y. (1977a). Selecting the t-largest means from several normal populations with unknown variances. Research Report No. 8/77, Dept. of Math., Univ. of Malaya, Kuala Lumpur, Malaysia.

Wong, W. Y. (1977b). On some nonparametric subset selection procedures for the location and scale parameters. Research Report No. 12/77, Dept. of Math., Univ. of Malaya, Kuala Lumpur, Malaysia.

Woodworth, G. G. (1965). An extension of a result of Lehmann on the asymptotic efficiency of selection procedures based on ranks. Tech. Report No. 66, Dept. of Statist., Univ. of Minnesota, Minneapolis.

Yao, Jing-Shing (1976). Game-theoretic aspects of selecting a subset of k populations containing the best. *Chinese J. Math.* **4**, 87–92.

Yen, S. P. (1969). Efficiency comparison for two subset selection procedures. Tech. Report No. 119. Dept. of Statist., Univ. of Minnesota, Minneapolis.

Young, D. H. (1973). A note on some asymptotic properties of the precedence test and applications to a selection problem concerning quantiles. *Sankhyā Ser. B.* **35**, 35–44.

Zinger, A. (1961). Detection of best and outlying normal populations with known variances. *Biometrika* **48**, 457–461.

Zinger, A., and St-Pierre, J. (1958). On the choice of the best among three normal populations with known variances. *Biometrika* **45**, 436–446.

Related References

Abdel-Aty, S. H. [1954]. Approximate formulae for the percentage points and the probability integral of the non-central χ^2 distribution. *Biometrika* **41**, 538–540.

Adichie, J. N. [1967a]. Asymptotic efficiency of a class of nonparametric tests for regression parameters. *Ann. Math. Statist.* **38**, 884–893.

Adichie, J. N. [1967b]. Estimates of regression parameters based on rank tests. *Ann. Math. Statist.* **38**, 894–904.

Anderson, T. W. [1958]. *An Introduction to Multivariate Statistical Analysis*. John Wiley, New York.

Anderson, T. W. [1960]. A modification of the sequential probability ratio test to reduce the sample size. *Ann. Math. Statist.* **31**, 165–197.

Anscombe, F. J. [1952]. Large sample theory of sequential estimation. *Proc. Camb. Phil. Soc.* **48**, 600–617.

Arimoto, S. [1971]. Information-theoretical conservations on estimation problems. *Inform. and Control* **19**, 181–194.

Armitage, P. [1957]. Restricted sequential procedures. *Biometrika* **44**, 9–26.

Armitage, J. V., and Krishnaiah, P. R. [1964]. Tables for the studentized largest chi-square distribution and their applications. ARL 64-188, Aerospace Research Laboratories, Wright-Patterson Air Force Base, Dayton, Ohio.

Barlow, R. E., Bartholomew, D. J., Bremner, J. M., and Brunk, H. D. [1972]. *Statistical Inference under Order Restrictions*. John Wiley, New York.

Barlow, R. E., Esary, J. D., and Marshall, A. W. [1966]. A stochastic characterization of wear-out for components and systems. *Ann. Math. Statist.* **37**, 816–825.

Barlow, R. E., Fussell, J. B., and Singpurwalla, N. D. [1975]. *Reliability and Fault Tree Analysis*. Society for Industrial and Applied Mathematics, Philadelphia.

Barlow, R. E., Marshall, A. W., and Proschan, F. [1963]. Properties of probability distributions with monotone hazard rate. *Ann. Math. Statist.* **34**, 375–389.

Barlow, R. E., and Proschan, F. [1965]. *Mathematical Theory of Reliability*. John Wiley, New York.

Barlow, R. E., and Proschan, F. [1966]. Inequalities for linear combinations of order statistics from restricted families. *Ann. Math. Statist.* **37**, 1574–1592.

Barlow, R. E., and Proschan, F. [1975]. *Statistical Theory of Reliability and Life Testing: Probability Models*. Holt, Rinehart and Winston, Inc., New York.

Bartlett, M. S., and Kendall, D. G. [1946]. The statistical analysis of variance heterogeneity and the logarithmic transformation. *J. Roy. Statist. Soc. Suppl.* **8**, 128–138.

Bechhofer, R. E., and Tamhane, A. C. [1974]. An iterated integral representation for a multivariate normal integral having block covariance structure. *Biometrika* **61**, 615–619.

Becker, M. H., Gnanadesikan, R., Mathews, M. V., Pinkham, R. S., Pruzansky, S., and Wilk, M. B. [1965]. Comparisons of some statistical distance measures for talker identification. Unpublished Bell Telephone Laboratories Memorandum.

Berk, B. H. [1967]. Review of 'Invariance of maximum likelihood estimators' by P. W. Zehana. *Math. Rev.* **33**, 342–343.

Berry, D. A. [1972]. A Bernoulli two-armed bandit. *Ann. Math. Statist.* **43**, 871–897.

Berry, D. A. [1978]. Modified two-armed bandit strategies for certain clinical trials. *J. Amer. Statist. Assoc.* **73**, 339–345.

Bose, R. C., Clatworthy, W. H., and Shrikhande, S. S. [1954]. Tables of Partially Balanced Designs with Two Associate Classes. Tech. Bulletin No. 107, North Carolina Agricultural Experiment Station, Raleigh.

Box, G. E. P., and Tiao, G. C. [1962]. A further look at robustness via Bayes' theorem. *Biometrika* **49**, 419–432.

Bradt, R. N., Johnson, S. M., and Karlin, S. [1956]. On sequential designs for maximizing the sum of n observations. *Ann. Math. Statist.* **27**, 1060–1074.

Cacoullos, T., and Olkin, I. [1965]. On the bias of functions of characteristic roots of a random matrix. *Biometrika* **52**, 87–94.

Carrier, N. A., Orton, K. D., and Malpass, L. F. [1962]. Responses of Bright, Normal, and EMH children to an orally-administered Children's Manifest Anxiety Scale. *J. Educational Psychology* **53**, 271–274.

Chapman, D. G. [1952]. On tests and estimates for the ratio of Poisson means. *Ann. Inst. Statist. Math.* **4**, 45–49.

Chow, Y. S., and Robbins, H. [1965]. On the asymptotic theory of fixed-width sequential confidence intervals for the mean. *Ann. Math. Statist.* **36**, 457–462.

Cochran, W. G. [1963]. *Sampling Techniques*. John Wiley, New York.

Conover, W. J. [1968]. Two k-sample slippage tests. *J. Amer. Statist. Assoc.* **63**, 614–626.

Cornish, E. A. [1954]. The multivariate small t-distribution associated with a set of normal sample deviates. *Austral. J. Phys.* **7**, 531–542.

Curnow, R. N. [1958]. The consequences of errors of measurement for selection from certain non-normal distributions. *Proceedings of the 31st Session of the International Statistical Institute*.

Davenport, A. S. [1971]. Classification of Distributions Based on Measures of Spread. Ph.D. Thesis. University of Iowa, Iowa City.

David, H. A. [1956]. On the application to statistics of an elementary theorem in probability. *Biometrika* **43**, 85–91.

David, H. A. [1963]. *The Method of Paired Comparisons*. Hafner Publishing Co., New York.

David, H. A. [1970]. *Order Statistics*. John Wiley, New York.

Deely, J. J., and Smith, M. H. [1975]. A secretary problem with finite memory. *J. Amer. Statist. Assoc.* **70**, 357–361.

Degroot, M. H. [1970]. Optimal Statistical Decisions. McGraw-Hill, New York.

Doksum, K. [1969]. Starshaped transformations and the power of rank tests. *Ann. Math. Statist.* **40**, 1167–1176.

Donelly, T. G. [1957]. A Family of Truncated Sequential Tests. Ph.D. Thesis. Univ. of North Carolina, Chapel Hill.

Doornbos, R. [1966]. *Slippage Tests.* Mathematical Research Center, Amsterdam.

Doornbos, R., Kesten, H., and Prins, H. J. [1956]. A class of slippage tests. Report S 206 (VP8) Mathematics Centre, Amsterdam.

Doornbos, R., and Prins, H. J. [1956]. Slippage tests for a set of gamma variates. *Indag. Math.* **18**, 329–337.

Doornbos, R., and Prins, H. J. [1958]. On slippage tests, I, II, and III. *Indag. Math.* **20**, 38–55, 438–447.

Duncan, D. B. [1947]. Significance Tests for Differences between Ranked Varieties Drawn from Normal Populations. Ph.D. Thesis. Iowa State College, Ames.

Duncan, D. B. [1951]. A significance test for differences between ranked treatments in an analysis of variance. *Virginia J. Sci.* **2**, 171–189.

Duncan, D. B. [1952]. On the properties of the multiple comparisons test. *Virginia J. Sci.* **3**, 49–67.

Duncan, D. B. [1955]. Multiple range and multiple F tests. *Biometrics* **11**, 1–42.

Dunnett, C. W., and Sobel, M. [1954]. A bivariate generalization of Student's t distribution, with tables for certain cases. *Biometrika* **41**, 153–169.

Dunnett, C. W., and Sobel, M. [1955]. Approximations to the probability integral and certain percentage points of a multivariate analogue of Student's t-distribution. *Biometrika* **42**, 258–260.

Ehrenberg, A. S. C. [1959]. The pattern of consumer purchases. *Appl. Statist.* **8**, 26–41.

Fabius, J., and Van Zwet, W. R. [1970]. Some remarks on the two-armed bandit. *Ann. Math. Statist.* **41**, 1906–1916.

Feldman, D. [1962]. Contributions to the "two-armed bandit" problem. *Ann. Math. Statist.* **33**, 847–856.

Fieller, E. C. [1944]. A fundamental formula in the statistics of biological assay and some applications. *Quart. J. Pharmacy Pharmacol.* **17**, 117–123.

Finney, D. J. [1941]. The joint distribution of variance ratios based on a common error mean square. *Ann. of Eugenics* **11**, 136–140.

Finney, D. J. [1952]. *Statistical Method in Biological Assay.* Hafner Publishing Co., New York.

Fisher, R. A. [1936]. The statistical utilization of multiple measurements. *Ann. Eugen. Lond.* **8**, 376.

Flehinger, B. J., and Louis, T. A. [1971]. Sequential treatment allocation in clinical trials. *Biometrika* **58**, 419–426.

Flehinger, B. J., and Louis, T. A. [1972]. Sequential Medical trials with data dependent treatment allocation. *Proceedings of the Sixth Berkeley Symposium on Mathematical Statistics and Probability.* Vol. IV (Eds. L. Le Cam, J.

Neyman, and E. L. Scott).University of California Press, Berkeley and Los Angeles, pp. 43–52.

Gilbert, J. P., and Mosteller, F. [1966]. Recognizing the maximum of a sequence. *J. Amer. Statist. Assoc.* **61**, 35–73.

Girschick, M. A. [1946]. Contributions to the theory of sequential analysis I. *Ann. Math. Statist.* **17**, 123–146.

Gordon, R. D. [1941]. Values of Mills' ratio of area to bounding ordinate and of the normal probability integral for large values of the argument. *Ann. Math. Statist.* **12**, 364–366.

Gupta, S. S. [1963a]. Probability integrals of the multivariate normal and multivariate *t*. *Ann. Math. Statist.* **34**, 792–828.

Gupta, S. S. [1963b]. Bibliography on the multivariate normal integrals and related topics. *Ann. Math. Statist.* **34**, 829–838.

Gupta, S. S., Nagel, K., and Panchapakesan, S. [1973]. On the order statistics from equally correlated normal random variables. *Biometrika* **60**, 403–413.

Hájek, J. [1962]. Asymptotically most powerful rank order tests. *Ann. Math. Statist.* **33**, 1124–1147.

Hall, I. J., and Kudô, A. [1968a]. On slippage tests, I: a generalization of Neyman–Pearson lemma. *Ann. Math. Statist.* **39**, 1693–1699.

Hall, I. J., and Kudô, A. [1968b]. On slippage tests, II: similar slippage tests. *Ann. Math. Statist.* **39**, 2029–2037.

Hall, W. J., Wijsman, R. A., and Ghosh, J. K. [1965]. The relationship between sufficiency and invariance with applications in sequential analysis. *Ann. Math. Statist.* **36**, 575–614.

Hewitt, E., and Savage, L. J. [1955]. Symmetric measures on cartesian products. *Trans. Amer. Math. Soc.* **80**, 470–501.

Hill, W. G. [1976]. Order statistics of correlated variables and implications in genetic selection programmes. *Biometrics* **32**, 889–902.

Hodges, J. L., Jr., and Lehmann, E. L. [1956]. The efficiency of some nonparametric competitors of the *t*-test. *Ann. Math. Statist.* **27**, 324–335.

Hodges, J. L., Jr., and Lehmann, E. L. [1963]. Estimates of location based on rank tests. *Ann. Math. Statist.* **34**, 598–611.

Hoel, P. G. [1937]. A significance test for component analysis. *Ann. Math. Statist.* **8**, 149–158.

Hogg, R. V. [1967]. Some observations on robust estimation. *J. Amer. Statist. Assoc.* **62**, 1179–1186.

Hogg, R. V. [1972]. More light on the kurtosis and related statistics. *J. Amer. Statist. Assoc.* **67**, 422–424.

Huber, P. J. [1964]. Robust estimation of a location parameter. *Ann. Math. Statist.* **35**, 73–101.

John, S. [1961]. On the evaluation of the probability integral of the multivariate *t*-distribution. *Biometrika* **48**, 409–417.

John, S. [1964]. Methods for evaluation of probabilities of polygonal and angular regions when the distribution is bivariate *t*. *Sankhyā Ser. A* **26**, 47–54.

John, S. [1966]. On the evaluation of convex polyhedra under multivariate normal and *t* distributions. *J. Roy. Statist. Soc. Ser. B* **28**, 366–369.

Karlin, S., and Truax, D. [1960]. Slippage problems. *Ann. Math. Statist.* **31**, 296–324.

Khamis, S. M. [1965]. *Tables of Incomplete Gamma Function Ratio*. Justus Von Liebig, Germany.

Kimball, A. W. [1951]. On dependent tests of significance in the analysis of variance. *Ann. Math. Statist.* **22**, 600–602.

Krishnaiah, P. R., and Armitage, J. V. [1964]. Distribution of the studentized smallest chi-square, with tables and applications. ARL 64-218. Aerospace Research Laboratories, Wright-Patterson Air Force Base, Dayton, Ohio.

Krishnaiah, P. R., and Armitage, J. V. [1966]. Tables for multivariate t-distribution. *Sankhyā Ser. B* **28**, 31–56.

Krishnaiah, P. R., Armitage, J. V., and Breiter, M. C. [1969]. Tables for the probability of integrals of the bivariate t distributions. Tech. Report ARL 69-0060. Office of Aerospace Research, U. S. Air Force Base, Dayton, Ohio.

Kudô, A. [1956a]. On the invariant multiple decision procedures. *Bull. Math. Statist., Res. Assoc. Statist. Sci.* **6**, 57–68.

Kudô, A. [1956b]. On the testing of outlying observations. *Sankhyā* **17**, 67–76.

Kullback, S. [1959]. *Information Theory and Statistics*. John Wiley, New York.

Lawrence, M. J. [1975]. Inequalities for s-ordered distributions. *Ann. Statist.* **3**, 413–428.

Lehmann, E. L. [1955]. Ordered families of distributions. *Ann. Math. Statist.* **26**, 399–419.

Lehmann, E. L. [1959]. *Testing Statistical Hypotheses*. John Wiley, New York.

Marshall, A. W., and Olkin, I. [1974]. Majorization in multivariate distribution. *Ann. Statist.* **2**, 1189–1200.

Marshall, A. W., and Olkin, I. [1979]. *Majorization and Schur Functions*. Academic, New York.

Marshall, A. W., Olkin, I., and Proschan, F. [1967]. Monotonicity of ratios of means and other applications of majorization. *Inequalities* (Ed. O. Shisha), Academic, New York, pp. 177–190.

Marshall, A. W., and Proschan, F. [1965]. An inequality for convex functions involving majorization. *J. Math. Anal. Appl.* **12**, 87–90.

McCabe, G. P., Jr. [1978]. Evaluation of regression coefficient estimates using α-acceptability. *Technometrics* **20**, 131–139.

Miller, R. G., Jr. [1977]. Developments in multiple comparisons. *J. Amer. Statist. Assoc.* **72**, 779–788.

Milton, R. C. [1963]. Tables of equally correlated multivariate normal probability integral. Tech. Report No. 27. Dept. of Statist., Univ. of Minnesota, Minneapolis.

Milton, R. C. [1970]. *Rank Order Probabilities: Two-Sample Normal Shift Alternatives*. John Wiley, New York.

Milton, R. C. [1972]. Computer evaluation of the multivariate normal integral. *Technometrics* **14**, 881–889.

Milton, R. C., and Hotchkiss, R. [1969]. Computer evaluation of the normal and inverse normal distribution functions. *Technometrics* **11**, 817–822.

Mosteller, F. [1948]. A k-sample slippage test for an extreme population. *Ann. Math. Statist.* **19**, 58–65.

Mucci, A. G. [1973]. On a class of secretary problems. *Ann. Prob.* **1**, 417–427.

Nair, K. R. [1948]. The studentized form of the extreme mean square tests in the analysis of variance. *Biometrika* **35**, 16–31.

Nevius, S. E., Proschan, F., and Sethuraman, J. [1977a]. Schur functions in statistics II: Stochastic majorization. *Ann. Statist.* **5**, 263–273.

Nevius, S. E., Proschan, F., and Sethuraman, J. [1977b]. A stochastic version of weak majorization, with applications. *Statistical Decision Theory and Related Topics*- II (Eds. S. S. Gupta and D. S. Moore), Academic, New York, pp. 281–296.

Olkin, I. [1972]. Monotonicity properties of Dirichlet integrals with applications to the multinomial distribution and the analysis of variance. *Biometrika* **59**, 303–307.

Olkin, I., and Sobel, M. [1965]. Integral expressions for tail probabilities of the multinomial and negative multinomial distributions. *Biometrika* **52**, 167–179.

Ostrowski, A. [1952]. Sur quelques applications des functions convexes et concaves au sens de I. Schur. *J. Math. Pure Appl.* **31**, 253–292.

Parzen, E. [1954]. On uniform convergence of families of sequences of random variables. *Univ. Calif. Publ. Statist.* **2**, 23–54.

Paulson, E. [1952]. An optimum solution to the k-sample slippage problem for the normal distribution. *Ann. Math. Statist.* **23**, 610–616.

Pearson, E. S., and Hartley, H. O. [1954]. *Biometrika Tables for Statisticians*, vol. I. Cambridge University Press, England.

Pfanzagl, J. [1959]. Ein Kombiniertes test und Klassifikationes-Problem. *Metrika* **2**, 11–45.

Pillai, K. C. S., and Ramachandran, K. V. [1954]. On the distribution of the ratio of the ith observation in an ordered sample from a normal population to an independent estimate of the standard deviation. *Ann. Math. Statist.* **25**, 565–572.

Proschan, F., and Serfling, R. J. [1974]. *Reliability and Biometry: Statistical Analysis of Lifelength.* Society for Industrial and Applied Mathematics, Philadelphia.

Proschan, F., and Sethuraman, J. [1977]. Schur functions in statistics. I: The preservation theorem. *Ann. Math. Statist.* **5**, 256–262.

Puri, M. L., and Sen, P. K. [1967]. On some optimum nonparametric procedures in two-way layouts. *J. Amer. Statist. Assoc.* **62**, 1214–1229.

Puri, M. L., and Sen, P. K. [1971]. *Nonparametric Methods in Multivariate Analysis.* John Wiley, New York.

Ramachandran, K. V. [1956]. On the simultaneous analysis of variance test. *Ann. Math. Statist.* **27**, 521–528.

Ramachandran, K. V. [1958]. On the studentized smallest chi-square. *J. Amer. Statist. Assoc.* **53**, 868–872.

Rawlings, S. O. [1976]. Order statistics for a special class of unequally correlated multinormal variates. *Biometrics* **32**, 875–887.

Robbins, H. [1952]. Some aspects of the sequential design of the experiments. *Bull. Amer. Math. Soc.* **55**, 527–535.

Robbins, H. [1955]. An empirical Bayes approach to statistics. *Proceedings of the Third Berkeley Symposium on Mathematical Statistics and Probability* (Ed. J. Neyman), Vol. I, Univ. of California Press, Berkeley, pp. 157–164.

Robbins, H. [1956]. A sequential decision problem with a finite memory. *Proc. Nat. Acad. Sci. U.S.A.* **42**, 920–923.

Robbins, H. [1963]. The empirical Bayes approach to testing statistical hypotheses. *Rev. Inst. Internat. Statist.* **31**, 195–208.

Robbins, H. [1964]. The empirical Bayes approach to statistical decision problems. *Ann. Math. Statist.* **35**, 1–20.

Robbins, H. [1970]. Statistical methods related to the law of the iterated logarithm. *Ann. Math. Statist.* **41**, 1397–1409.

Samuels, S. M. [1968]. Randomized rules for the two-armed bandit with finite memory. *Ann. Math. Statist.* **39**, 2103–2107.

Samuels, S. M., and Gianini, J. [1976]. The infinite secretary problem. *Ann. Prob.* **4**, 418–432.

Scheffé, H. [1953]. A method for judging all contrasts in the analysis of variance. *Biometrika* **40**, 87–104.

Shannon, C. E. [1948]. A mathematical theory of communication. *Bell System Tech. J.* **27**, 373–423, 623–656.

Slepian, D. [1962]. On the one-sided barrier problem for gaussian noise. *Bell System Tech. J.* **41**, 463–501.

Smith, C. V., and Pyke, R. [1965]. The Robbins-Isbell two-armed-bandit problem with finite memory. *Ann. Math. Statist.* **36**, 1375–1386.

Starr, N. [1966]. The performance of a sequential procedure for the fixed-width interval estimation of the mean. *Ann. Math. Statist.* **37**, 36–50.

Stein, C. [1945]. A two-sample test for a linear hypothesis whose power is independent of the variance. *Ann. Math. Statist.* **16**, 243–258.

Truax, D. R. [1953]. An optimum slippage test for the variances of normal distributions. *Ann. Math. Statist.* **24**, 669–673.

Tukey, J. W. [1949]. Comparing individual means in the analysis of variance. *Biometrics* **5**, 99–114.

Van Zwet, W. R. [1964]. *Convex Transformations of Random Variables*. Mathematical Center, Amsterdam.

Vogel, W. [1960]. A sequential design for the two-armed bandit. *Ann. Math. Statist.* **31**, 430–443.

Wald, A. [1947]. *Sequential Analysis*. John Wiley, New York.

Wald, A. [1950]. *Statistical Decision Functions*. John Wiley, New York.

Weiss, L., and Wolfowitz, J. [1966]. Generalized maximum likelihood estimators. *Teor. Veroyat. Primen.* **11**, 68–93.

Weiss, L., and Wolfowitz, J. [1967]. Maximum probability estimators. *Ann. Inst. Statist. Math.* **19**, 193–206.

Weiss, L., and Wolfowitz, J. [1968]. Generalized maximum likelihood estimators in a particular case. *Teor. Veroyat. Primen.* **13**, 657–662.

Weiss, L., and Wolfowitz, J. [1969]. Asymptotically minimax tests of composite hypothesis. *Z. Wahrscheinlichkeitstheorie Verw. Geb.* **14**, 161–168.

Weiss, L., and Wolfowitz, J. [1974]. *Maximum Probability Estimators and Related Topics*. Lecture Notes in Mathematics, No. 424. Springer-Verlag, Berlin.

Wolfe, H. D. et al. [1963]. *Pretesting Advertising*. Studies in Business Policy, No. 103, National Industrial Conference Board, New York.

Zehna, P. W. [1966]. Invariance of maximum likelihood estimators. *Ann. Math. Statist*. **37**, 744.

Monographs, Books and Special Issues of Journals Devoted Fully or Partly to Ranking and Selection Problems

Bechhofer, R. E., Kiefer, J., and Sobel, M. (1968). *Sequential Identification and Ranking Procedures*. The University of Chicago Press, Chicago.

Chow, Y. S., Robbins, H., and Sigmund, D. O. (1971). *Great Expectations: The Theory of Optimal Stopping*. Houghton Mifflin, Boston.

Communications in Statistics—Theory and Methods. (1977). Special Issue on Selection and Ranking Problems. Vol. A6, No. 11, S. S. Gupta (ed.).

Dudewicz, E. J. (1976). *Introduction to Statistics and Probability*. Holt, Rinehart and Winston, New York (Chapters 11 and 12).

Gibbons, J. D., Olkin, I., and Sobel, M. (1977). *Selecting and Ordering Populations*. John Wiley, New York.

Gupta, S. S. and Huang, D. Y (1980). *Multiple Statistical Decision Theory*. to be published.

Gupta, S. S. and Moore, D. S. (eds.). *Statistical Decision Theory and Related Topics—II*. Academic, New York, 1977.

Gupta, S. S. and Yackel, J. (eds.). *Statistical Decision Theory and Related Topics*. Academic, New York, 1971.

Kleijnen, J. P. C. (1975). *Statistical Techniques in Simulation*, Vol. 2. Marcel Dekker, New York.

Kulldorf, G. (1977). *Bibliography of Selection Procedures*. Institute of Mathematics and Statistics, University of Umeå, Sweden.

Miller, R. G., Jr. (1966). *Simultaneous Statistical Inference*. McGraw-Hill, New York.

Technometrics, vol. 7, Number 2, 1965, Fred C. Leone (Editor); has articles on multiple comparisons and multiple decision theory.

Author Index

Subject Index

Applied Probability and Statistics (Continued)

DUNN · Basic Statistics: A Primer for the Biomedical Sciences, *Second Edition*

DUNN and CLARK · Applied Statistics: Analysis of Variance and Regression

ELANDT-JOHNSON · Probability Models and Statistical Methods in Genetics

ELANDT-JOHNSON and JOHNSON · Survival Models and Data Analysis

FLEISS · Statistical Methods for Rates and Proportions

GALAMBOS · The Asymptotic Theory of Extreme Order Statistics

GIBBONS, OLKIN, and SOBEL · Selecting and Ordering Populations: A New Statistical Methodology

GNANADESIKAN · Methods for Statistical Data Analysis of Multivariate Observations

GOLDBERGER · Econometric Theory

GOLDSTEIN and DILLON · Discrete Discriminant Analysis

GROSS and CLARK · Survival Distributions: Reliability Applications in the Biomedical Sciences

GROSS and HARRIS · Fundamentals of Queueing Theory

GUPTA and PANCHAPAKESAN · Multiple Decision Procedures: Theory and Methodology of Selecting and Ranking Populations

GUTTMAN, WILKS, and HUNTER · Introductory Engineering Statistics, *Second Edition*

HAHN and SHAPIRO · Statistical Models in Engineering

HALD · Statistical Tables and Formulas

HALD · Statistical Theory with Engineering Applications

HARTIGAN · Clustering Algorithms

HILDEBRAND, LAING, and ROSENTHAL · Prediction Analysis of Cross Classifications

HOEL · Elementary Statistics, *Fourth Edition*

HOLLANDER and WOLFE · Nonparametric Statistical Methods

HUANG · Regression and Econometric Methods

JAGERS · Branching Processes with Biological Applications

JESSEN · Statistical Survey Techniques

JOHNSON and KOTZ · Distributions in Statistics
Discrete Distributions
Continuous Univariate Distributions–1
Continuous Univariate Distributions–2
Continuous Multivariate Distributions

JOHNSON and KOTZ · Urn Models and Their Application: An Approach to Modern Discrete Probability Theory

JOHNSON and LEONE · Statistics and Experimental Design in Engineering and the Physical Sciences, Volumes I and II, *Second Edition*

KALBFLEISCH and PRENTICE · The Statistical Analysis of Failure Time Data

KEENEY and RAIFFA · Decisions with Multiple Objectives

LANCASTER · An Introduction to Medical Statistics

LEAMER · Specification Searches: Ad Hoc Inference with Nonexperimental Data

McNEIL · Interactive Data Analysis

MANN, SCHAFER, and SINGPURWALLA · Methods for Statistical Analysis of Reliability and Life Data

MEYER · Data Analysis for Scientists and Engineers